ALL■IN■ONE

CompTIA
CySA+™
Cybersecurity Analyst
Certification
EXAM GUIDE
(Exam CS0-002)

ABOUT THE AUTHORS

Brent Chapman, GCIH, GCFA, GCTI, CISSP, CySA+, is an information security engineer with more than 15 years of experience in information technology and cybersecurity. He is a former cyber operations officer in the United States Army and has held a number of unique assignments, including researcher at the Army Cyber Institute, instructor in the Department of Electrical Engineering and Computer Science at the US Military Academy at West Point, and project manager at the Defense Innovation Unit in Silicon Valley. He is a professional member of the Association of Computing Machinery, FCC Amateur Radio license holder, and contributor to several technical and maker-themed publications.

Fernando J. Maymí, PhD, CISSP, is a consultant, educator, and author with more than 25 years of experience in information security. He currently leads teams of cybersecurity consultants, analysts, and red teamers in providing services around the world. Fernando was the founding deputy director of the Army Cyber Institute, a government think tank he helped create for the Secretary of the Army to solve future cyberspace operations problems affecting the whole country. He has served as advisor to congressional leaders, corporate executives, and foreign governments on cyberspace issues. Fernando taught computer science and cybersecurity at the US Military Academy at West Point for 12 years. Fernando has written extensively and is the co-author of the eighth edition of the bestselling *CISSP All-in-One Exam Guide*.

About the Technical Editor and Contributor

Bobby E. Rogers is a cybersecurity professional with over 30 years in the information technology and cybersecurity fields. He currently works for a major engineering company in Huntsville, Alabama, as a contractor for Department of Defense agencies, helping to secure, certify, and accredit their information systems. Bobby's specialties are cybersecurity engineering, security compliance, and cyber risk management, but he has worked in almost every area of cybersecurity, including network defense, computer forensics, incident response, and penetration testing. He is a retired Master Sergeant from the US Air Force, having served for over 21 years. Bobby has built and secured networks in the United States, Chad, Uganda, South Africa, Germany, Saudi Arabia, Pakistan, Afghanistan, and several other countries all over the world. He holds a Master of Science degree in Information Assurance and is currently writing his dissertation for a doctoral degree in cybersecurity. His many certifications include CISSP-ISSEP, CRISC, and CySA+. He has narrated and produced over 30 computer security training videos for several training companies, and is also the author of *CompTIA Mobility+ Certification All-In-One Exam Guide (Exam MB0-001)*, *CRISC Certified in Risk and Information Systems Control All-In-One Exam Guide*, *Mike Meyers' CompTIA Security+ Certification Guide (Exam SY0-401)*, and contributing author/technical editor for the popular *CISSP All-In-One Exam Guide, Eighth Edition*, all from McGraw Hill.

ALL · IN · ONE

CompTIA
CySA+™
Cybersecurity Analyst
Certification

EXAM GUIDE

Second Edition

(Exam CS0-002)

Brent Chapman
Fernando J. Maymí

New York   Chicago   San Francisco
Athens   London   Madrid   Mexico City
Milan   New Delhi   Singapore   Sydney   Toronto

McGraw Hill books are available at special quantity discounts to use as premiums and sales promotions, or for use in corporate training programs. To contact a representative, please visit the Contact Us pages at www.mhprofessional.com.

CompTIA CySA+™ Cybersecurity Analyst Certification All-in-One Exam Guide, Second Edition (Exam CS0-002)

2 3 4 5 6 7 8 9   LCR   24 23 22 21

Library of Congress Control Number: 2020945425

ISBN   978-1-260-46430-6
MHID   1-260-46430-X

Sponsoring Editor	**Technical Editor**	**Production Supervisor**
Lisa McClain	Bobby Rogers	Lynn M. Messina
Editorial Supervisor	**Copy Editor**	**Composition**
Janet Walden	Lisa Theobald	KnowledgeWorks Global Ltd.
Project Manager	**Proofreader**	**Illustration**
Sarika Gupta,	Rick Camp	KnowledgeWorks Global Ltd.
KnowledgeWorks Global Ltd.	**Indexer**	**Art Director, Cover**
Acquisitions Coordinator	Claire Splan	Jeff Weeks
Emily Walters		

To Gina and Carol, for being patient, supportive, and loving,
and for reminding us of what really matters in life.

CONTENTS AT A GLANCE

CONTENTS

ACKNOWLEDGMENTS

None of us accomplishes anything strictly on our own merits; we all but extend the work of those who precede us. In a way, we are all running a relay race. It would be disingenuous, if not inconsiderate, if we failed to acknowledge those whose efforts contributed to the book you are about to read. First and foremost, we want to acknowledge Deborah and Emma, who brought us into this world and, through years of loving sacrifice, embarked us upon what has turned out to be the adventures of our lives.

We also want to call out the staff and faculty of the US Military Academy, especially those in the Department of Electrical Engineering and Computer Science. They educated, trained, and inspired us to become who we are today. Their teaching and mentorship fueled the flames of our intellectual curiosity and gave us the space to experiment. We also acknowledge the cadets, who, when we returned as faculty members years later, kept us on our toes through their insightful questions in class and their need for ever more challenging exercises in the labs. Our country is in good hands with them.

Finally, we would like to recognize those who put their lives on the line every day in the service of our country. It is these heroes who are ultimately responsible for preserving the environment that makes this all possible: from going to school to leading companies. Thank you, for selflessly defending our way of life.

INTRODUCTION

This second edition of the *CompTIA CySA+ Certification All-in-One Exam Guide* represents a major overhaul of the material covered in the previous edition. Fully updated to reflect the new objectives for exam number CS0-002, this new book will not only prepare you for your certification but also serve as a desktop reference in your daily job as a cybersecurity analyst. Our ultimate goal, as before, is to ensure that you will be equipped to know and be able to take the right steps to improve the security posture of your organization immediately upon arrival. But how do you convey these skills to a prospective employer within the confines of a one- or two-page resume? Using the title CySA+, like a picture, can be worth a thousand words.

Why Become a CySA+?

To answer that question simply, having CySA+ at the end of your signature will elevate employers' expectations. Hiring officials oftentimes screen resumes by looking for certain key terms, such as *CySA+*, before referring them to technical experts for further review. Attaining this certification improves your odds of making it past the first filters and also sets a baseline for what the experts can expect from you during an interview. It lets them know they can get right to important parts of the conversation without first having to figure out how much you know about the role of a cybersecurity analyst. The certification sets you up for success.

It also sets you up for lifelong self-learning and development. Preparing for and passing this exam will not only elevate your knowledge, but it will also reveal to you how much you still have to learn. Cybersecurity analysts never reach a point where they know enough. Instead, this is a role that requires continuous learning, because both the defenders and attackers are constantly evolving their tools and techniques. The CySA+ domains and objectives provide you a framework of knowledge and skills on which you can plan your own professional development.

The CySA+ Exam

CompTIA indicates the relative importance of each domain with these weightings on the exam:

Domain	Percent of Examination
1.0 Threat and Vulnerability Management	22%
2.0 Software and Systems Security	18%
3.0 Security Operations and Monitoring	25%
4.0 Incident Response	22%
5.0 Compliance and Assessment	13%

The CySA+ exam is administered at authorized testing centers or via remote online proctoring and presently will cost you $359. It consists of a minimum of 85 questions, which must be answered in no more than 165 minutes. To pass, you must score 750 points out of a maximum possible 900 points. The test is computer-based and adaptive, which means different questions will earn you different numbers of points. The bulk of the exam consists of short, multiple-choice questions with four or five possible responses. In some cases, you will have to select multiple answers to receive full credit. Most questions are fairly straightforward, so you should not expect a lot of "trick" questions or ambiguity. Still, you should not be surprised to find yourself debating between two responses that both seem correct at some point.

A unique aspect of the exam is its use of scenario questions. You may see only a few of these, but they will require a lot of time to complete. In these questions, you will be given a short scenario and a network map. There will be hotspots in the map that you can click to obtain detailed information about a specific node. For example, you might click a host and see log entries or the output of a command-line tool. You will have to come up with multiple actions that explain an observation, mitigate threats, or handle incidents. Deciding which actions are appropriate will require that you look at the whole picture, so be sure to click every hotspot before attempting to answer any of these questions.

Your exam will be scored on the spot, so you will know whether you passed before you leave the test center. You will be given your total score, but not a breakdown by domain. If you fail the exam, you will have to pay the exam fee again, but you may retake the test as soon as you'd like. Unlike other exams, there is no waiting period for your second attempt, though you will have to wait 14 calendar days between your second and third attempts if you fail twice.

What Does This Book Cover?

This book covers everything you need to know to become a CompTIA-certified Cybersecurity Analyst (CySA+). It teaches you how successful organizations manage cyber threats to their systems. These threats will attempt to exploit weaknesses in the systems, so the book also covers the myriad of issues that go into effective vulnerability management.

As we all know, no matter how well we manage both threats and vulnerabilities, we will eventually have to deal with a security incident. The book next delves into cyber incident response, including forensic analysis. Finally, it covers security architectures and tools with which every cybersecurity analyst should be familiar.

Though the book gives you all the information you need to pass the test and be a successful CySA+, you will have to supplement this knowledge with hands-on experience on at least some of the more popular tools. It is one thing to read about Wireshark and Snort, but you will need practical experience with these tools to know how best to apply them in the real world. The book guides you in this direction, but you will have to get the tools as well as practice the material covered in these pages.

Tips for Taking the CySA+ Exam

Though the CySA+ exam has some unique aspects, it is not entirely unlike any other computer-based test you may have taken. The following is a list of tips in increasing order of specificity. Some may seem like common sense to you, but we still think they're important enough to highlight.

- Get lots of rest the night before.
- Arrive early at the exam site.
- Read all possible responses before making your selection, even if you are "certain" that you've already read the correct option.
- If the question seems like a trick one, you may be overthinking it.
- Don't second-guess yourself after choosing your responses.
- Take notes on the dry-erase sheet (which will be provided by the proctor) whenever you have to track multiple data points.
- If you are unsure about an answer, give it your best shot, mark it for review, and then go on to the next question; you may find a hint in a later question.
- When dealing with a scenario question, read all available information at least once before you attempt to provide any responses.
- Don't stress if you seem to be taking too long on the scenario questions; you will get only a handful of those.
- Don't expect the exhibits (for example, log files) to look like real ones; they will be missing elements you'd normally expect but contain all the information you need to respond.

How to Use This Book

Much effort has gone into putting all the necessary information into this book. Now it's up to you to study and understand the material and its various concepts. To benefit the most from this book, you may want to use the following study method:

- Study each chapter carefully and make sure you understand each concept presented. Many concepts must be fully understood, and glossing over a couple here and there could be detrimental to you.

- Make sure to study and answer all the questions. If any questions confuse you, go back and study those sections again.

- If you are not familiar with specific topics, such as firewalls, reverse engineering, and protocol functionality, use other sources of information (books, articles, and so on) to attain a more in-depth understanding of those subjects. Don't just rely on what you think you need to know to pass the CySA+ exam.

- If you are not familiar with a specific tool, download the tool (if open source) or a trial version (if commercial) and play with it a bit. Since we cover dozens of tools, you should prioritize them based on how unfamiliar you are with them.

Using the Objective Map

The table in Appendix A has been constructed to help you cross-reference the official exam objectives from CompTIA with the relevant coverage in the book. Each objective is listed along with the corresponding chapter number and heading that provides coverage of that objective.

Online Practice Exams

This book includes access to practice exams that feature the TotalTester Online exam test engine, which enables you to generate a complete practice exam or to generate quizzes by chapter module or by exam domain. See Appendix B for more information and instructions on how to access the exam tool.

PART I

Threat and Vulnerability Management

The Importance of Threat Data and Intelligence

In this chapter you will learn:

- The foundations of threat intelligence
- Common intelligence sources and the intelligence cycle
- Effective use of indicators of compromise
- Information sharing best practices

Every battle is won before it is ever fought.

—Sun Tzu

Modern networks are incredibly complex entities whose successful and ongoing defense requires a deep understanding of what is present on the network, what weaknesses exist, and who might be targeting them. Getting insight into network activity allows for greater agility in order to outmaneuver increasingly sophisticated threat actors, but not every organization can afford to invest in the next-generation detection and prevention technology year after year. Furthermore, doing so is often not as effective as investing in quality analysts who can collect and quickly understand data about threats facing the organization.

Threat data, when given the appropriate context, results in the creation of *threat intelligence*, or the knowledge of malicious actors and their behaviors; this knowledge enables defenders to gain a better understanding of their operational environments. Several products can be used to provide decision-makers with a clear picture of what's actually happening on the network, which makes for more confident decision-making, increases the cost to the adversary, improves operator response time, and in the worst case, reduces recovery time for the organization in the event of an incident. Sergio Caltagirone, coauthor of *The Diamond Model of Intrusion Analysis* (2013), defines cyber threat intelligence as "actionable knowledge and insight on adversaries and their malicious activities enabling defenders and their organizations to reduce harm through better security decision-making." Without a doubt, a good threat intelligence program is a necessary component of any modern information security program. Formulating a brief definition of so broad a term as "intelligence" is a massive challenge, but fortunately there are decades of studies on the primitives of intelligence analysis that we'll borrow from for the purposes of defining threat intelligence.

Discipline	Description
SIGINT	*Signals intelligence* is intelligence-gathering done via intercepts of communications, electronic, and/or instrumentation transmissions.
HUMINT	*Human intelligence* is derived from human sources through overt, covert, or clandestine methods.
OSINT	*Open source intelligence* is the collection and analysis of publicly available information appearing in print or electronic form.
MASINT	*Measurement and signature intelligence* is derived intelligence from data other than imagery and SIGINT.
GEOINT	*Geospatial intelligence* is the analysis of imagery and geospatial data concerning security-related activities on the earth.
All Source	This intelligence is derived from every available source on a subject or topic.

Table 1-1   A Sample of Intelligence Disciplines

Foundations of Intelligence

Traditional intelligence involves the collecting and processing of information about foreign countries and their agents. Usually conducted on behalf of a government, intelligence activities are carried out using nonattributable methods in foreign areas to further foreign policy goals and in support of national security. Another key aspect is the protection of the intelligence actions, the people and organizations involved, and the resulting intelligence products against unauthorized disclosure. Classically, intelligence is divided into the areas from which it is collected, as shown in Table 1-1. Functionally, a government may align one or several agencies with an intelligence discipline.

Intelligence Sources

Threat intelligence teams outside of the government do not often have the luxury of on-call intelligence assets available to those within government. Energy companies and Internet service providers (ISPs) do not deploy HUMINT agents or have SIGINT operations, but they will focus on using whatever public, commercial, or in-house resources are available. Fortunately, there are a number of free and paid sources to help teams meet their intelligence requirements, including commercial threat intelligence providers, industry partners, and government organizations. And, of course, there's also broad monitoring of social media and news for relevant data.

Open Source Intelligence

There are many ways to acquire free data associated with actor activity, be it malicious or benign. A common way to do this without interacting with those parties is by gathering open source intelligence (OSINT), free information that's collected in legitimate ways from public sources such as news outlets, libraries, and search engines. Using OSINT, practitioners can begin to answer questions critical to the intelligence process. How can

I create actionable threat intelligence from a variety of data sources? How can I share this threat information with the broader community, and what mechanism should I use? How can I use public information to limit the organization's exposure, while enabling my understanding of what adversaries know about us?

There are several additional benefits of developing OSINT skills. Threat analysts often use OSINT sources to help them keep pace with security industry trends and discussions in near real-time. This is a useful way for practitioners to understand what may be coming around the corner, even if the content comes in an irregular or inconsistent manner. Additionally, many security analysts rely on publicly available data sets to perform research on common threat indicators and mitigating controls. The information gleaned may also help in post-incident forensics efforts or in support of penetration testing activities.

From an adversary point of view, it is almost always preferable to get information about a target without directly touching it. Why? Because the less it is touched, the fewer fingerprints (or log entries) are left behind for the defenders and investigators to find. In an ideal case, adversaries gain all the information they need to compromise a target successfully without once visiting it, using OSINT techniques. *Passive reconnaissance*, for example, is the process by which an adversary acquires information about a target network without directly interacting with it. These techniques can be focused on individuals as well as companies. Just like individuals, many companies maintain a public face that can give outsiders a glimpse into their internal operations. In the sections that follow, we describe some of the most useful sources of OSINT with which you should be familiar.

Google

Google's vision is to organize all of the data in the world and make it accessible for everyone in a useful way. It should therefore not be surprising that Google can help an attacker gather a remarkable amount of information about any individual, organization, or network. The use of this search engine for target reconnaissance purposes drew much attention in the early 2000s, when security researcher Johnny Long started collecting and sharing examples of search queries that revealed vulnerable systems. These queries made use of advanced operators that are meant to allow Google users to refine their searches. Though the list of operators is too long to include in this book, Table 1-2 lists some of the ones we've found most useful over the years. Note that many others are available from a variety of online sources, and some of these operators can be combined in a search.

Operator	Restricts Search Results to	Example
site:	The specified domain or site	site:apache.org
inurl:	Having the specified text in the URL	inurl:/administrator/index.php
filetype:	The indicated type of file	filetype:xls
intitle:	Pages with the indicated text in their title	intitle:vitae
link:	Pages that contain a link to the indicated site or URL	link:www.google.com
cache:	Google's latest cached copies of the results	cache:www.eff.org

Table 1-2 Useful Google Search Operators

Index of /download/win64 - Wireshark
https://www.wireshark.org/download/win64/ ▾
Parent Directory, -. [DIR], all-versions/, 14-De[Cached] . [ ] · Wireshark-win64-2.0.8.exe, 16-Nov-
2016 20:55, 45M. [ ], Wireshark-win64-2.0.9.ex[Similar]

Figure 1-1   Using Google cached pages

Suppose your organization has a number of web servers. A potentially dangerous misconfiguration would be to allow a server to display directory listings to clients. This means that instead of seeing a rendered web page, the visitor could see a list of all the files (HTML, PHP, CSS, and so on) in that directory within the server. Sometimes, for a variety of reasons, it is necessary to enable such listings. More often, however, they are the result of a misconfigured and potentially vulnerable web server. If you wanted to search an organization for such vulnerable server directories, you would type the following into your Google search box, substituting the actual domain or URL in the space delineated by angle brackets:

```
site:<targetdomain or URL> intitle:"index of" "parent directory"
```

This would return all the pages in your target domain that Google has indexed as having directory listings.

You might then be tempted to click one of the links returned by Google, but this would directly connect you to the target domain and leave evidence there of your activities. Instead, you can use a page cached by Google as part of its indexing process. To see this page instead of the actual target, look for the downward arrow immediately to the right of the page link. Clicking it will give you the option to select Cached rather than connecting to the target (see Figure 1-1).

 EXAM TIP   You will not be required to know the specific symbols and words required for advanced Google searches, but it's useful as a security analyst to understand the various methods of refining search engine results, such as Boolean logic, word order, and search operators.

Internet Registries

Another useful source of information about networks is the multiple registries necessary to keep the Internet working. Routable Internet Protocol (IP) addresses as well as domain names need to be globally unique, which means that there must be some mechanism for ensuring that no two entities use the same IP address or domain. The way we, as a global community, manage this deconfliction is through the nonprofit corporations described next. They offer some useful details about the footprint of an organization in cyberspace.

Regional Internet Registries   As Table 1-3 shows, five separate corporations control the assignment of IP addresses throughout the world. They are known as the *regional Internet registries* (RIRs), and each has an assigned geographical area of responsibility.

Table 1-3 The Regional Internet Registries	Registry	Geographic Region
	AFRINIC	Africa and portions of the Indian Ocean
	APNIC	Portions of Asia and portions of Oceania
	ARIN	Canada, many Caribbean and North Atlantic islands, and the United States
	LACNIC	Latin America and portions of the Caribbean
	RIPE NCC	Europe, the Middle East, and Central Asia

Thus, entities wishing to acquire an IP address in Canada, the United States, or most of the Caribbean would deal (directly or through intermediaries) with the American Registry for Internet Numbers (ARIN). The activities of the five registries are coordinated through the Number Resource Organization (NRO), which also provides a detailed listing of each country's assigned RIR.

EXAM TIP You do not need to know the RIRs, but you do need to understand what information is available through these organizations together with the Internet Corporation for Assigned Names and Numbers (ICANN).

Domain Name System The Internet could not function the way it does today without the Domain Name System (DNS). Although DNS is a vital component of modern networks, many users are unaware of its existence and importance to the proper functionality of the Web. DNS is the mechanism responsible for associating domain names, such as www.google.com, with their server's IP address(es), and vice versa. Without DNS, you'd be required to memorize and input the full IP address for any website you wanted to visit instead of the easy-to-remember uniform resource locator (URL). Using tools such as nslookup, host, and dig in the command line, administrators troubleshoot DNS and network problems. Using the same tools, an attacker can interrogate the DNS server to derive information about the network. In some cases, attackers can automate this process to reach across many DNS servers in a practice called *DNS harvesting*.

In some cases, it may be necessary to replicate a DNS server's contents across multiple DNS servers through an action called a *zone transfer*. With a zone transfer, it is possible to capture a full snapshot of what the DNS server's records hold about the domain; this includes name servers, mail exchange records, and hostnames. Zone transfers are a potential vulnerable point in a network because the default behavior is to accept any request for a full transfer from any host on the network. Because DNS is like a map of the entire network, it's critical to restrict leakage to prevent DNS poisoning or spoofing.

NOTE DNS zone transfers are initiated by clients—whether from a secondary DNS server or network host. Because DNS data can be used to map out an entire network, it's critical that only authorized hosts be allowed to request full transfers. This is accomplished by implementing access control lists (ACLs). Zone transfers to unrecognized devices should *never* be allowed.

Showing results for: google.com
Original Query: google.com

Contact Information

Registrant Contact
Name: Dns Admin
Organization: Google Inc.
Mailing Address: Please contact
contact-admin@google.com, 1600
Amphitheatre Parkway, Mountain
View CA 94043 US
Phone: +1.6502530000
Ext:
Fax: +1.6506188571
Fax Ext:
Email:dns-admin@google.com

Admin Contact
Name: DNS Admin
Organization: Google Inc.
Mailing Address: 1600
Amphitheatre Parkway, Mountain
View CA 94043 US
Phone: +1.6506234000
Ext:
Fax: +1.6506188571
Fax Ext:
Email:dns-admin@google.com

Tech Contact
Name: DNS Admin
Organization: Google Inc.
Mailing Address: 2400 E.
Bayshore Pkwy, Mountain View CA
94043 US
Phone: +1.6503300100
Ext:
Fax: +1.6506181499
Fax Ext:
Email:dns-admin@google.com

Figure 1-2   A report returned from ICANN's WHOIS web-based service

Whenever a domain is registered, the registrant provides details about the organization for public display. This may include name, telephone, and e-mail contact information; domain name system details; and mailing address. This information can be queried using a tool called WHOIS (pronounced *who is*). Available in both command-line and web-based versions, WHOIS can be an effective tool for incident responders and network engineers, but it's also a useful information-gathering tool for spammers, identity thieves, and any other attacker seeking to get personal and technical information about a target. For example, Figure 1-2 shows a report returned from ICANN's WHOIS web-based service. You should be aware that some registrars (the service that you go through to register a website) provide *private* registration services, in which case the registrar's information is returned during a query instead of the registrant's. Although this may seem useful to limit an organization's exposure, the tradeoff is that in the case of an emergency, it may be difficult to reach that organization.

Job Sites
Sites offering employment services are a boon for information gatherers. Think about it: The user voluntarily submits all kinds of personal data, a complete professional history, and even some individual preferences. In addition to providing personally identifiable characteristics, these sites often include indications for a member's role in a larger network. Because so many of these accounts are often identified by e-mail address, it can often be the common link in the course of investigating a suspicious entity.

In understanding what your public exposure is, you must realize that it's trivial for attackers to automate the collection of artifacts about a target. They may perform activities to broadly collect e-mail addresses, for example, in a practice called *e-mail harvesting*. An attacker can use this to his benefit by taking advantage of business contacts to craft a more convincing phishing e-mail.

Beyond the social engineering implications for the users of these sites, companies themselves can be targets. If a company indicates that it's in the market for an administrator of a particular brand of firewall, then it's likely that the company is using that brand of firewall. This can be a powerful piece on information, because it provides clues about the makeup of the company's network and potential weak points.

Social Media

Social media sites can be rich sources of threat data. Twitter and Reddit, for example, are two platforms that often provide useful artifacts during high-impact events. As a vehicle to quickly spread news of major emergencies, these platforms can also be leveraged as a source for indicators about cyberattacks. They can also be highly targeted sources for personal information. As with employment sites, defenders can learn quite a bit about an attacker's social network and tendencies, should the attacker have a public persona. Conversely, an attacker can gain awareness about an individual or company using publicly available information. The online clues captured from personal pages enable an attacker to conduct *social media profiling,* which uses a target's preferences and patterns to determine their likely actions. Profiling is a critical tool for online advertisers hoping to capitalize on highly targeted ads. This information is also useful for an attacker in identifying which users in an organization may be more likely to fall victim to a *social engineering* attack, in which the perpetrator tricks the victim into revealing sensitive information or otherwise compromising the security of a system.

Many attackers know that the best route into a network is through a careless or untrained employee. In a social engineering campaign, an attacker uses deception, often influenced by the profile they've built about the target, to manipulate the target into performing an act that might not be in their best interest. These attacks come in many forms—from advanced phishing e-mails that seem to originate from a legitimate source, to phone calls requesting additional personal information. Phishing attacks continue to be a challenge for network defenders because they are becoming increasing convincing, fooling recipients into divulging sensitive information with regularity. Despite the most advanced technical countermeasures, the human element remains the most vulnerable part of the network.

OSINT in the Real World

One of the authors was asked to teach a class in an allied country to members of its nascent cyberspace workforce. The goal of the one-week course was to expose students to some open domain offensive techniques that they would have to master as a prerequisite to building their own capabilities. The first block of instructions was on reconnaissance, and the teachers of the class were given authorization to be fairly aggressive as long as we didn't actually compromise any systems. In preparation for the class, the author performed a fairly superficial OSINT-gathering exercise and found a remarkable amount of actionable information.

(continued)

Starting from the regional Internet registry, we were able to identify an individual named Daniel who appeared to be a system administrator for the target organization. We then looked him up on LinkedIn and confirmed his affiliation, but we were also able to learn all his experience, skills, and accomplishments. We then looked up the organization in a handful of prominent job sites and were able to confirm (and even refine) the tools the organization was using to manage its networks. We noted that one of the tools was notorious for having vulnerabilities. Finally, we looked up Daniel on Facebook and found a recent public post from his mother wishing him a happy birthday. At this point, we could have sent him an e-mail with a PDF resume attachment or an e-mail with a very convincing message from his "mother" with references to his three siblings and a link to a video of the birthday party. Either way, the probability of Daniel opening a malware-laden attachment or clicking a link would have been fairly high—and all it took was about 15 minutes on the Web.

Proprietary/Closed Source Intelligence

One of the key tenets of intelligence analysis is never relying on a single source of data when attempting to confirm a hypothesis. Ideally, analysts should look at multiple artifacts from multiple sources that support a hypothesis. Similarly, open source data is best used with corroborating data acquired from closed sources. Closed source data is any data collected covertly or as a result of privileged access. Common types of closed source data include internal network artifacts, dark web communications, details from intelligence-sharing communities, and private banking and medical records. Since closed source data tends to be higher quality, an analyst can confidently assess and verify findings using any number of intelligence analysis methods and tools. An added benefit of using multiple sources is that the practice reduces the effect of confirmation bias, or the tendency for an analyst to interpret information in a way that supports a prior strongly held belief.

Internal Network

Your organization's network will inherently have the most relevant threat data available of all the sources we'll discuss. Despite being completely germane to every security function, many organizations eschew internal threat data in favor of external data feeds. As an analyst, you must remember that by leveraging threat data from your own network, you can identify potential malicious activity with far greater speed and confidence than generic threat data. The most common sources for raw threat-related data include events, DNS, virtual private networks (VPNs), firewalls, and authentication system logs. By establishing a baseline of normal activity, analysts can use historic knowledge of past incident responses to improve awareness of emerging threats or ongoing malicious activity.

Classified Data

When handling closed source data, you must consider several important factors. In some cases, the mere disclosure of the data may jeopardize access to the source information, or worse, the individuals involved in the collection and analysis process. Since closed source data is sometimes not meant to be openly available to the public, there may be some legal stipulations regarding its handling. *Classified data*, or data whose unauthorized disclosure may cause harm to national security interests, is protected by several statutes that restrict its handling and sharing to trusted individuals. Those individuals undergo formal security screening to achieve clearance or the minimum eligibility required for them to handle or access classified data. Accordingly, leaks of classified data may result in steep administrative or criminal penalties. It will be your responsibility, should you encounter classified data, to handle it appropriately and to safeguard against unauthorized disclosure.

Traffic Light Protocol

The Traffic Light Protocol (TLP) was created by UK government's National Infrastructure Security Coordination Centre (NISCC) to enable greater threat information sharing between organizations. It includes a set of color-coded designations that are used to guide responsible sharing of sensitive information to the appropriate audience, while also protecting the information's sources. As shown in Table 1-4, the four-color designations are meant to be easily understood among participants, since its usage closely mirrors the

Color	When should it be used?	How may it be shared?
TLP:RED Not for disclosure, restricted to participants only	Use when information cannot be effectively acted upon by additional parties, and could lead to impacts on a party's privacy, reputation, or operations if misused.	Recipients may not share TLP:RED information with any parties outside of the specific exchange, meeting, or conversation in which it was originally disclosed. In the context of a meeting, for example, TLP:RED information is limited to those present at the meeting. In most circumstances, TLP:RED information should be exchanged verbally or in person.
TLP:AMBER Limited disclosure, restricted to participants' organizations	Use TLP:AMBER when information requires support to be effectively acted upon, yet carries risks to privacy, reputation, or operations if shared outside of the organizations involved.	Recipients may share TLP:AMBER information only with members of their own organization, and with clients or customers who need to know the information to protect themselves or prevent further harm. Sources are at liberty to specify additional intended limits of the sharing: these must be adhered to.
TLP:GREEN Limited disclosure, restricted to the community	Use TLP:GREEN when information is useful for the awareness of all participating organizations as well as with peers within the broader community or sector.	Recipients may share TLP:GREEN information with peers and partner organizations within their sector or community, but not via publicly accessible channels. Information in this category can be circulated widely within a particular community. TLP:GREEN information may not be released outside of the community.
TLP:WHITE Unlimited disclosure	Use TLP:WHITE when information carries minimal or no foreseeable risk of misuse, in accordance with applicable rules and procedures for public release.	Subject to standard copyright rules, TLP:WHITE information may be distributed without restriction.

Table 1-4 Traffic Light Protocol Designations (source: US Department of Homeland Security CISA website, https://www.us-cert.gov/tlp)

colors of traffic lights around the world. Despite its usage as a data-sharing guideline, TLP is not a classification or control scheme, nor is it designed to use as a mechanism to enforce intellectual property terms or how the data is to be used by the recipient.

Characteristics of Intelligence Source Data

Despite the many claims from vendors, there is no one solution for every organization when it comes to threat data and intelligence. Organizations must be able to map the threat intelligence products they acquire or produce to some distinct aspect of their threat profile. In short, analysts must prioritize data that is most relevant to their specific environment to ensure that they are not bogged down in unmanageable noise and that they can produce actionable, timely, and consistent results. Generic threat intelligence developed for environments that are too dissimilar from those in which the team is operating will not satisfy the organization's unique requirements. Furthermore, developing intelligence for environments that are not specific enough may result in a waste of resources. Security analysts working in a manufacturing environment, for example, must understand their environment and seek out and obtain threat intelligence products specific to manufacturing networks, in addition to general threat information. Intelligence must include context and provide recommendations if it is to have maximum value for decision-makers.

Good threat intelligence provides three critical elements to analyst so that they can appropriately provide answers to decision-makers. The intelligence must describe the threat using consistent and clear language, illustrate the impact to the business in terms that are relevant to the business, and provide a clear set of recommended actions. In addition to being complete, good threat intelligence often has three characteristics: timeliness, relevancy, and accuracy. Teams that can effectively use quality threat intelligence are able to address known gaps in their security posture and apply this data and context to nearly every aspect of their security operations; this will improve detection, enhance response, and strengthen prevention. Furthermore, the rate of noise generation and intelligence failure is inversely proportional to the timeliness, relevancy, and accuracy of the threat intelligence information.

Timeliness

All intelligence, whether in a traditional military operation or as it applies to information security, has a temporal dimension. After all, intelligence considers environmental conditions as a part of context, so it makes sense that it is most useful given the time-related stipulations within the intelligence requirements. You may find that intelligence may be extremely useful at one time and completely useless at another. Accordingly, intelligence that is not delivered in a timely manner is not as useful to decision-makers.

Relevancy

As discussed earlier, internal network data invariably yields the most useful threat intelligence, because it reflects the nuances of an organization. Additionally, relevancy varies based on the levels of operation, even within the same organization. It is therefore important to prepare threat intelligence products for the correct audience. Details about the exact nature of an adversary's technical capabilities, for example, may not be as useful for a strategic audience as it may be for a detection team analyst. This is an easy point to overlook, but intelligence requires human attention to consume and process, so irrelevant information is costly in terms of time and resources. Providing inconsequential intelligence is distracting, but it may be counterproductive as well.

Accuracy

Although it may seem self-evident in a field such as information security, accuracy is critical to enable a decision-maker to draw reliable conclusions and pursue a recommended course of action. The information must be factually correct. Acknowledging that it may be unrealistic to an exact understanding of the operational environment at any given time, an analyst must at minimum convey facts as they exist.

 NOTE Bias is a reality of human nature and is a deeply complicated matter. As humans, we attempt to make the world simpler by introducing assumptions to allow for quick decisions, particularly in a dangerous or confusing situation. As security analysts, it's important for us to recognize that although it may be impossible to remove all of our biases, we must account for it in our analysis process. We can limit bias by surrounding ourselves with teammates who bring a diversity of thought and backgrounds during the analysis process, and by using structured analytical techniques.

Confidence Levels

In a continuous effort to apply more rigorous standards to analytical assessment, intelligence providers often use three levels of analytic confidence made using *estimative language*. Estimative language aims to communicate intelligence assessments while acknowledging the existence of incomplete or fragmented information. The statements should not be seen as fact or proof, but as judgments based on analysis of collected information. As shown in Table 1-5, *confidence levels* reflect the scope and quality of the information supporting its judgments.

Level	Description
High	Assessments are based on high-quality information, and/or the nature of the issue makes it possible to render a solid judgment. The judgement is not to be interpreted as fact and still has a possibility of being incorrect.
Moderate	Information is credibly sourced and plausible but not of sufficient quality or corroboration to warrant a higher level of confidence.
Low	Information acquired is questionable or implausible and may be too fragmented or poorly corroborated to make solid analytic inferences. Additionally, significant concerns about sources may exist.

Table 1-5 Confidence Levels and Their Descriptions

Indicator Management

In discussing threat data, the term *indicator* will come up often to describe some observable artifact encountered on the network. Note that an indicator is not just data on its own. Indicators must include some context describing an aspect of an event, indicating *something* related to the intrusion of concern. Think about a time that someone has asked for your help in answering a question related to something you knew a lot about. If the question has ever begun with "what can you tell me about...," you're almost certainly going to ask follow-up questions to try to get as much information and context as possible before you provide an answer. Analyzing threat data is similar, in that context matters: a domain name is not an indicator on its own, but a domain name with the added context that is it used for phishing *is* an indicator.

Indicator Lifecycle

In working toward completing the picture of what's happened during an incident, an analyst will want to turn a newly discovered indicator into something actionable for remediation or detection at a later point. Your first step is to vet the indicator, a process of deciding whether the indicator is valid, researching the originating signal, and determining its usefulness in actually detecting the malicious activity you expect to find in your environment. Characteristics you may want to consider during vetting include the reliability of the indicator source and additional details about the artifact that you're able to uncover with follow-up research.

Indicators often have value beyond just your organization, even when they are derived from internal network data. Many financial service organizations use the same supporting technologies, have similar internal operations processes, and face the same kind of threat actors on a regular basis. It follows, therefore, that these organizations would be keenly interested in how others in the same space are detecting and responding to malicious activity. Sharing threat data is a key component of success of any security operations effort. Threat intelligence shared among partners, peers, and other trusted groups often helps focus detection efforts and prioritize the use of limited resources.

Structured Threat Information Expression

The *Structured Threat Information Expression,* or STIX, is a collaborative effort led by the MITRE Corporation to communicate threat data using a standardized lexicon. To represent threat information, the STIX 2.0 framework uses a structure consisting of twelve key STIX Domain Objects (SDOs) and two STIX Relationship Objects (SROs). In describing an event, an analyst may show the relationship between one SDO and another using an SRO. The goal is to allow for flexible, extensible exchanges while providing both human- and machine-readable data. STIX information can be represented visually for analysts or stored as JSON (JavaScript Object Notation) for use in automation.

Attack Pattern

Attack patterns are a class of tactics, techniques, and procedures (TTPs) that describe attacker tendencies in how they employ capabilities against their victims. This SDO is helpful in generally categorizing types of attacks, such as spear phishing. Additionally, these objects may be used to add insight into exactly how the attacks are executed. Using the spear phishing example, an attacker would identify a high-value target, craft a relevant phishing message, attach a malicious document, and send the message with the hope that the target downloads and opens the attachment. Each of these individual actions viewed together make up an attack pattern.

Campaign

A *campaign* is a collection of malicious actor behaviors against a common target over a finite timeframe. Campaign SDOs are often identified through various attributions methods to tie them to specific threat actors, but more important than the *who* behind the campaign is the identification of the unique use of tools, infrastructure, techniques, and targeting used by the actor. Tying this in with our phishing example, a campaign could be used to describe a threat actor group's attack that used a specially tailored phish kit to target specific executives of a multinational energy company over the course of eight weeks in the past winter.

Course of Action

A *course of action* is a preventative or response action taken to address an attack. This SDO describes any technical changes, such as a modification of a firewall rule, or policy changes, such as mandatory annual security training. To address the effectiveness of spear phishing, a security team may choose a course of action that includes enhanced phishing detection, automated attachment scanning, and link sanitization.

Identity

An *identity* is an SDO that represents individuals, organizations, or groups. They can be specific and named, such as John Doe or Acme Corporation, or broader, to refer to an entire sector, for example. In the act of collecting identifying information about the individuals and organizations involved in an incident, we may derive previously unseen patterns. The individuals that were the target of the specially crafted phishing message would be represented as identity objects using this framework.

Indicator

Similar to the previously defined indicator, the indicator SDO describes an observable that can be used to detect suspicious activity on a network or an endpoint. Again, this specific observable or pattern of observables must be accompanied with contextual data for it to be truly useful in communicating interesting aspects of a security event. Indicators found in spear phishing messages often include links with phishing domains, generic form language, or the use of ASCII homographs.

Intrusion Set

An *intrusion set* is a compilation of behaviors, TTPs, or other properties shared by a single entity. As with campaigns, the focus of an intrusion set is on identifying common resources and behaviors rather than just trying to figure out who's behind the activity. Intrusion set SDOs differ from campaigns in that they aren't necessarily restricted to just one timeframe. An intrusion set may include multiple campaigns, creating an entire attack history over a long period of time. If a company has identified that it was the target of multiple phishing campaigns over the past few years from the same threat actor, that activity may be considered an intrusion set.

Malware

Malware is any malicious code and malicious software used to affect the integrity or availability of a system or the data contained therein. Malware may also be used against a system to compromise its confidentiality, enabling access to information not otherwise available to unauthorized parties. In the context of this framework, malware is considered a TTP, most often introduced into a system in a manner that avoids detection by the user. As a STIX object, the malware SDO identifies samples and families using plain language to describe the software's capabilities and how it may affect an infected system. Examples of how these objects may be linked include connections to other malware objects to demonstrate similarities in its operations, or connections to identities to communicate the targets involved in an incident. The specific malicious code delivered in a phishing attempt can be described using the malware object.

Observed Data

The *observed data* SDO is used to describe any observable collected from a system or network device. This object may be used to communicate a single observation of an entity or the aggregate of observations. Importantly, observed data is not intelligence or even information, but is raw data, such as the number of times a connection is made or the summation of occurrences over a specified timeframe. The observed data object could be used in a phishing event to highlight the number of requests made to a particular phishing domain over the course of an hour.

Report

Reports are finished intelligence products that cover some specific detail of a security event. Reports can give relevant details about threat actors believed to be connected to an incident, the malware they may have used, or the methodologies used during a campaign. For example, a narrative that describes the details of the phish kits used in targeting energy company executives may be included in a report SDO. The report could include references to any of the other objects previously described.

Threat Actor

The *threat actor* SDO defines individuals or groups believed to be behind malicious activity. They conduct the activities described in campaign and intrusion set objects using the TTPs described in malware and attack pattern objects against the targets identified using identity objects. Their level of technical sophistication, personally identifiable information (PII), and assertions about motives can all be used in this object. If a determination can be made about the goals of the actors behind our fictional series of phishing messages, it could be included in the threat actor object along with any information about the individuals or groups involved.

Tool

The *tool* SDO describes software used by a threat actor in conducting a campaign. Unlike software described using the malware object, tools are legitimate utilities such as PowerShell or the terminal emulator. Tool objects may be connected to other objects describing TTPs to provide insight into levels of sophistication and tendencies. Understanding how and when actors use these tools can provide defenders with the knowledge necessary for developing countermeasures. Since the software described in tool objects is also used by power users, system administrators, and sometimes regular users, the challenge moves from simply detecting the presence of the software to detecting unusual usage and determining malicious intent. A caveat in using the tool object, in addition to avoiding using it to describe malware, is that this object is not meant to provide details about any software used by defenders in detecting or responding to a security event.

Vulnerability

A *vulnerability* SDO is used to communicate any mistake in software that may be exploited by an actor to gain unauthorized access to a system, software, or data. Although malware objects provide key characteristics about malicious software and when they are used in the course of an attack, vulnerability objects describe the exact flaw being leveraged by the malicious software. The two may be connected to show how a particular malware object targets a specific vulnerability object.

Relationship

The *relationship* SRO is can be thought of as the connective tissue between SDOs, linking them together and showing how they work with one another. In the previous description of vulnerability, we highlighted that there may be a connection made to a malware object to show how the malicious software may take advantage of a particular flaw. Using the relationship SRO relationship type *target*, we can show how a source and target are related using this framework. Table 1-6 highlights some common associations between a source SDO and target SDO using a relationship SRO.

Sighting

A *sighting* SRO provides information about the occurrence of an SDO such as indicator or malware. It's effectively used to convey useful information about trends and can be instrumental in developing intelligence about how an attacker's behavior may evolve or respond to mitigating controls. The sighting SRO differs primarily from the relationship SRO in that it can provide additional properties about when an object was first or last seen, how many times it was seen, and where it was observed. The sighting SRO is similar to the observed data SDO in that they both can be used to provide details about observations on the network. However, you may recall that an observed data SDO is not intelligence and provides only the raw data associated with the observation. While you would use the observed data SDO to communicate that you observed the presence of a particular piece of malware on a system, you'd use the sighting SRO to describe that a threat actor is likely to be behind the use of this malware given additional context.

Table 1-6
A Sample of Commonly Used Relationships

Source SDO	Relationship SRO	Target SDO
campaign	attributed to	threat actor
malware	targets	identity
attack pattern	uses	malware
course of action	mitigates	vulnerability
indicator	indicates	tool

Trusted Automated Exchange of Indicator Information

Trusted Automated Exchange of Intelligence Information (TAXII) defines how threat data may be shared among participating partners. It specifies the structure for how this information and accompanying messages are exchanged. Developed in conjunction with STIX, TAXII is designed to support STIX data exchange by providing the technical specifications for the exchange API.

TAXII 1.0 was designed to integrate with existing sharing agreements, including access control limitations, using three primary models: hub and spoke, source/subscriber, and peer-to-peer, as shown in Figure 1-3.

Figure 1-3
Primary models
for TAXII 1.0

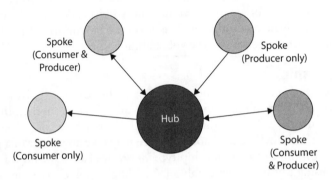

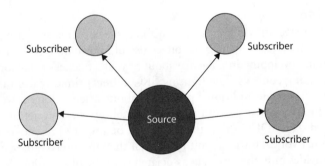

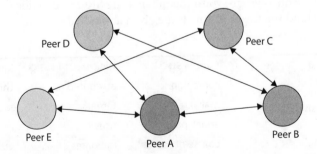

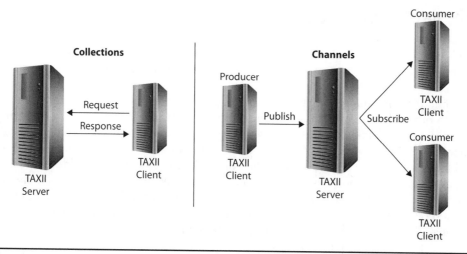

Figure 1-4   Collections and channel service architecture for TAXII 2.0

TAXII 2.0 defines two primary services, collections and channels, to facilitate exchange models. Collections are an interface to a logical store of threat data objects hosted by a TAXII server. Channels, maintained by the TAXII server, provide the pathway for TAXII clients to subscribe to the published data. Figure 1-4 illustrates the relationship between the server and clients for the collection and channel services.

OpenIOC

OpenIOC is a framework designed by Mandiant, an American cybersecurity firm that is now part of FireEye. The goal of the framework is to organize information about an attacker's TTPs and other indicators of compromise in a machine-readable format for easy sharing and automated follow-up. The OpenIOC structure is straightforward, consisting three of main components: IOC metadata, references, and the definition. The metadata component provides useful indexing and reference information about the IOC including author name, the IOC name, and a description of the IOC. The reference component is primarily meant to enable analysts to describe how the IOC fits in operationally with their specific environments. As a result, some of the information may not be appropriate to share, because it may refer to internal systems or sensitive ongoing cases. Analysts should be particularly careful to verify that reference information is suitable for sharing before sharing it externally. Finally, the definition component provides the indicator content most useful for investigators and analysts. The definition often contains Boolean logic to communicate the conditions under which the IOC is valid. For example, the requirement for an MD5 hash AND a file size attribute above a certain threshold would have to be fulfilled for the indicator to match.

Threat Classification

Before we go too deeply into the technical details regarding threats you may encounter while preparing for or responding to an incident, you need to understand the term *incident*. We use the term to describe any action that results in direct harm to your system or increases the likelihood of unauthorized exposure of your sensitive data. Your first step in knowing that something is harmful and out of place is to understand what normal looks like. In other words, establishing a baseline of your systems is the first step in preparing for an incident. Without your knowing what normal is, it becomes incredibly difficult for you to see the warning signs of an attack. Without a baseline for comparison, you will likely know that you've been breached only when your systems go offline. Making a plan for incident response isn't just a good idea—it may be compulsory, depending on your operating environment and line of business. As your organization's security expert, you will be entrusted to implement technical measures and recommend policy that keeps personal data safe while keeping your organization out of court.

Known Threats vs. Unknown Threats

In attempting to find malicious activity, antivirus software and more sophisticated security devices work by using signature-based and anomaly-based methods of detection. Signature-based systems rely on prior knowledge of a threat, which means that these systems are only as good as the historical data companies have collected. Although these systems are useful for identifying threats that already exist, they don't help much with regard to threats that constantly change form or have not been previously observed; these will slip by, undetected.

The alternative is to use a solution that looks at what the executable is doing, rather than what it looks like. This kind of system relies on *heuristic analysis* to observe the commands the executable invokes, the files it writes, and any attempts to conceal itself. Often, these heuristic systems will sandbox a file in a virtual operating system and allow it to perform what it was designed to do in that separate environment.

As malware is evolving, security practices are shifting to reduce the number of assumptions made when developing security policy. A report that indicates that no threat is present just means that the scanning engine couldn't find a match, and a clean report isn't worth much if the methods of detection aren't able to detect the newest types of threats. In other words, the absence of evidence is not evidence of absence. Vulnerabilities and threats are being discovered at a rate that outpaces what traditional detection technologies can spot. Because threats still exist, even if we cannot detect them, we must either evolve our detection techniques or treat the entire network as an untrusted environment. There is nothing inherently wrong about the latter; it just requires a major shift in thinking about how we design our networks.

Zero Day

The term *zero day*, once used exclusively among security professionals, is quickly becoming part of the public dialect. It refers to either a vulnerability or exploit never before seen in public. A *zero-day vulnerability* is a flaw in a piece of software that the vendor is unaware of and thus has not issued a patch or advisory for. The code written to take advantage of this flaw is called the *zero-day exploit.* When writing software, vendors often focus on providing usability and getting the most functional product out to market as quickly as possible. This often results in products that require numerous updates as more users interact with the software. Ideally, the number of vulnerabilities decreases as time progresses, as software adoption increases, and as patches are issued. However, this doesn't mean that you should let your guard down because of some sense of increased security. Rather, you should be more vigilant; even if an environment has protective software in place, it's defenseless should a zero-day exploit be used against it.

The Emergence of the Exploit Marketplace

Zero-day exploits were once extremely rare, but the security community has observed a significant uptick in their usage and discovery. As security companies improve their software, malware writers have worked to evolve their products to evade these systems, creating a malware arms race of sorts. Modern zero-day vulnerabilities are viewed as extremely valuable to some malicious users and criminal groups, and as with anything else of perceived value, markets have formed. Black markets for zero-day exploits exist with ample participation from criminal groups. On the opposite end of the spectrum, vendors have used *bug bounty* programs to supplement internal vulnerability discovery, inviting researchers and hackers to actively probe their software for bugs in exchange for money and prizes. Even the US Pentagon, a traditionally bureaucratic and risk-averse organization, saw the value in crowdsourcing security in this way. In March 2016, it launched the "Hack the Pentagon" challenge, a pilot program designed to identify security vulnerabilities on public-facing Department of Defense sites.

Preparation

Preparing to face unknown and advanced threats like zero-day exploits requires a sound methodology that includes technical and operational best practices. The protection of critical business assets and sensitive data should never be trusted to a single solution. You should be wary of solutions that suggest they are one-stop shops for dealing with these threats, because you are essentially placing the entire organization's fate in a single point of failure. Although the word "response" is part of your incident response plan, your team should develop a methodology that includes proactive efforts as well. This approach should involve active efforts to discover new threats that have not yet impacted the organization. Sources for this information include research organizations and threat intelligence providers. The SANS Internet Storm Center and the CERT Coordination Center at Carnegie Mellon University are two great resources for discovering the latest software bugs. Armed with new knowledge about attacker trends and techniques, you may be able to detect malicious traffic before it has a chance to do any harm. Additionally, you will give your security team time to develop controls to mitigate security incidents, should a countermeasure or patch not be available.

Advanced Persistent Threat

In 2003, analysts discovered a series of coordinated attacks against the Department of Defense, Department of Energy, NASA, and the Department of Justice. Discovered to have been in progress for at least three years by that point, the actors appeared to be on a mission and took extraordinary steps to hide evidence of their existence. These events, known later as "Titan Rain," would be classified as the work of an advanced persistent threat (APT), which refers to any number of stealthy and continuous computer hacking efforts, often coordinated and executed by an organization or government with significant resources. The goal for an APT is to gain and maintain persistent access to target systems while remaining undetected. Attack vectors often include spam messages, infected media, social engineering, and supply-chain compromise. The support infrastructure behind their operations, their TTPs during operations, and the types of targets they choose are all part of what makes APTs stand out. It's useful to analyze each word in the APT acronym to identify the key discriminators between APT and other actors.

Advanced

The operators behind these campaigns are often well equipped and use techniques that indicate formal training and significant funding. Their attacks indicate a high degree of coordination between technical and nontechnical information sources. These threats are often backed with a full spectrum of intelligence support, from digital surveillance methods to traditional techniques focused on human targets.

Persistent

Because these campaigns are often coordinated by government and military organizations, it shouldn't be surprising that each operator is focused on a specific task rather than rooting around without direction. Operators will often ignore opportunistic targets and remain focused on their piece of the campaign. This behavior implies strict rules of engagement and an emphasis on consistency and persistence above all else.

Threat

APTs do not exist in a bubble. Their campaigns show capability and intent, aspects that highlight their use as the technical implementation of a political plan. Like a military operation, APT campaigns often serve as an extension of political will. Although their code might be executed by machines, the APT framework is designed and coordinated by humans with a specific goal in mind. Because of the complex nature of APTs, it may be difficult to handle them alone. The concept of automatic threat intelligence sharing is a recent development in the security community. Because speed is often the discriminator between a successful and an unsuccessful campaign, many vendors provide solutions that automatically share threat data and orchestrate technical countermeasures for them.

NOTE Advanced persistent threats, regardless of affiliation, are characterized by resourcing, consistency, and a military-like efficiency during their actions to compromise systems, steal data, and cover their tracks.

Threat Actors

Threat actors are not equal in terms of motivation and capability; neither are they all necessarily overtly malicious. You will learn that the term "threat actor" is wide-ranging and can be categorized by sophistication as well as intent. We'll describe threat actors using several groups, but it's important for you to understand that these classifications are not mutually exclusive. Threat actors and threat actor groups may span across multiple classifications, usually depending on the targets and timeframe of the activity we're considering.

Nation-State Threat Actors

Nation-state threat actors are frequently among the most sophisticated adversaries, with dedicated infrastructure, training resources, and operational support behind their activities. Their activities are characterized by extensive planning and coordination and often reflect the strong government or military influence behind them. Like many government-supported operations, nation-state threat actor activities are often conducted to achieve political, economic, or strategic military goals. Identifying and tracking these actors can be difficult, since many of the individuals involved use common techniques across teams, operate behind robust infrastructure, and use methods to actively obfuscated their behavior. Alternatively, they many use toolsets that are not often seen or impossible to detect at the time of the security event, such as a zero-day exploit.

There are a few interesting notes about nation-state operations that make them unique. The first is that, depending on the countries involved, businesses can quickly become a part of the activity in either a direct or supporting capacity. Second, more sophisticated threat actors may incorporate false flag techniques, performing activities that lead defenders to falsely attribute their activity to another. Given the high degree of coordination that some nation-state actor activities require, this is becoming a frequent challenge for defenders to address. Finally, there's an aspect of perspective worth noting here: one nation's intelligence apparatus is another nation's malicious actor.

Hacktivists

Hacktivists are threat actors that typically operate with less resourcing than their nation-state counterparts, but nonetheless work to coordinate efforts to bring light to an issue or promote a cause. They often rely on readily available tools and mass participation to achieve their desired effects against a target. Though not always the case, their actions often have little lasting damage to their targets. Hacktivists are also known to use social media and defacement tactics to affect the reputation of their targets, hoping to erode public trust and confidence in their targets. Unlike other threat actors, hacktivists rarely seek to operate with stealth and look to bring attention to their cause along with notoriety for their own organization. As defenders, knowing that hacktivists frequently employ techniques to affect the availability of a system, we can use defensive techniques to mitigate denial-of-service (DoS) attacks. Furthermore, we can reduce the likelihood of successful social engineering efforts or unauthorized access to services and applications by enforcing multifactor authentication on system and social media accounts.

Organized Crime

Threat actors operating on behalf of organized crime groups are becoming an increasingly visible challenge for enterprise defenders to confront. Whether targeting theft of intellectual property or personal user data, these criminals' primary objective is to make money by selling stolen data. When compared to nation-state actors, organized crime may have a more moderate sophistication level, but as financial gain is often the goal, attacks will many times include the use of cryptojacking, ransomware, and bulk data exfiltration techniques. Despite having a well-understood operational model, organized crime threat actors still contribute to a significant percentage of security incidents. This is due in part to the comparatively low-risk, high-reward nature of their activities and the ease with which they can hide their activities online. Moreover, the rise in usage of digital currencies worldwide has allowed for these criminals to launder vast sums of money more easily.

Insider Threat Actors

Insider threat actors work within an organization and represent a particularly high risk of causing catastrophic damage due to their privileged access to internal resources. Because access is often an early goal for insider threat actors, having access as result of role or position in a company often means that traditional perimeter-focused security mechanisms are not effective in detecting and stopping their destructive activity. To address internal threats, it's critical that the security program is designed in a way that adheres to the principle of least privilege, as it relates to access. Furthermore, network security devices should be configured to allow or deny access based on robust access control rules and not simply as a result of a device's location within the network. Insider threats are unique in that they may be influenced by a combination of factors, from personal, to organizational, to technical. Accordingly, the solution to address them cannot be one-dimensional and must include policies, procedures, and technologies to mitigate the overall threat. For example, mandating annual training on cybersecurity awareness along with implementing technical controls to prevent unauthorized file access have been shown to reduce the occurrence and impact of insider threat events.

Intentional

Intentional insider threat actors may be employees, contractors, or any other business partners with established access to internal service, or any of these who have severed ties with the organization but have not lost access. Intentional actors behave in a manner that may be damaging to the organization, through data theft, data deletion, or vandalism. In many cases, these malicious insiders look to acts against the organization with the goal of personal financial gain, revenge, or both. Malicious insiders looking to steal intellectual property in order to facilitate a secondary income source will typically remove data slowly to avoid detection. A disgruntled employee, on the other hand, may work deliberately to sabotage an organization's critical systems. As defenders, we should be aware of anomalous activity such as high-volume network activity or indiscriminate file access attempts, especially following an employee's resignation or firing.

Unintentional

It's easy to think of insider threats as actors with malicious intent, but other factors may lead to an increased insider threat risk. Lack of security education, negligence, and human error are among the top contributors to unintentional insider security events. Such actions, though unwitting, may cause as much harm as those done intentionally by other threat actors. Hanlon's razor famously expresses that one shouldn't attribute to malice that which can be adequately explained by ignorance (or stupidity, per the actual adage). Mistakes happen, and it may be counterproductive for a security team to treat every user who is responsible for a security incident as a willful malicious actor. Not only would this ensure that the user would have a difficult learning experience, but it may lull the security team into the perception that the root issue is sufficiently addressed. If, for example, there are ten occurrences of data spillage that all point to a similar type of user error, perhaps the problem is less about the user and more about the usability of the system.

Intelligence Cycle

The *intelligence cycle* is a core process used by most government and business intelligence and security teams to process raw signals into finished intelligence for use in decision-making. Depending on the environment, the process is a five- or six-step method of adding clarity to a dynamic and ambiguous environment. Figure 1-5 shows the five-step cycle. Among the many benefits of its application is the increased situational awareness about the environment and the delineation of easily understood work efforts. Importantly, the intelligence cycle is also continuous and does not require perfect knowledge

Figure 1-5
The five-step
intelligence cycle

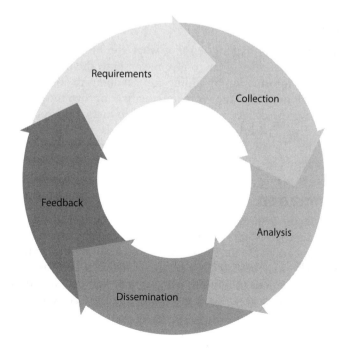

Requirements

Collection

Analysis

Dissemination

Feedback

about one phase to begin the next. In fact, the cycle is best used when output from one phase is used to feed the next while also refining the previous. For example, you may discover a new bit of information in a later stage than can be used to improve the inputs into the overall cycle.

Requirements

The requirements phase involves the identification, prioritization, and refinement of uncertainties about the operational environment that the security team must resolve to accomplish its mission. It includes key tasks related to the planning and direction of the overall intelligence effort. In simple terms, requirements are steps that are *needed*. The results of this phase are not always derived from authority, but are determined by aspects of the customer's operations as well as the capabilities of the intelligence team. As gaps in understanding are identified and prioritized, analysts will move on to figuring out ways to close these gaps, and a plan is set forth as to how they will get the data they need. This in turn will drive the collection phase.

Collection

At this phase, the plan that was previously defined is executed, and data is collected to fill the intelligence gap. Unlike a collection effort at a traditional intelligence setting, this effort at a business will likely not involve dispatching of HUMINT or SIGINT assets, but will instead mean the instrumentation of technical collection methods, such as setting up a network tap or enabling enhanced logging on certain devices. The sources of the raw data arrive from outside of the network, from news reports, social media, and public documents, or from closed and proprietary sources.

Analysis

Analysis is the act of making sense of what you observe. With the use of automation, highly trained analysts will try to give meaning to the normalized, decrypted, or otherwise processed information by adding context about the operational environment. They then prioritize it against known requirements, improve those requirements, and potentially identify new collection sources. The product of this phase is finished, actionable intelligence that is useful to the customer. Analysis can be a difficult process, but there are many structured analytical techniques that we may use to mitigate the effects of biases, to challenge judgments, and to manage uncertainty.

Dissemination

Distributing the requested intelligence to the customer occurs at the dissemination phase. Intelligence is communicated in whichever manner was previous identified in the requirements phase and must provide a clear way forward for the customer. The customer, who may be the security team itself, can then use these analytical products and recommendations to improve defense, gain a greater understanding of an adversary's social or computer network for counterintelligence, or even move toward legal action.

This also highlights an important concept in intelligence: the product provided must be useful. The words "actionable intelligence" are often misused, because intelligence is always meant to be actionable.

Feedback

Once intelligence is disseminated, more questions may be raised, which leads to additional planning and direction of future collection efforts. At each phase of the cycle, analysts are evaluating the quality of their input and outputs, but explicitly requesting feedback from consumers is extremely important to enable the security team to improve its activities and better align products to meet consumers' evolving intelligence needs. This phase also enables analysts to review their own analytical performance and to think about how to improve their methods for soliciting information, interacting with internal and external partners, and communicating their findings to the decision-makers.

Commodity Malware

Commodity malware includes any pervasive malicious software that's made available to threat actors via sale. Often made available in underground communities, this type of malware enables criminals to focus less on improving their technical sophistication and more on optimizing their illegal operations. There's a well-known military axiom that states that great organizations do routine things routinely well. Correspondingly, while commodity malware may not always be the most advanced or stealthy software, good security teams will know how to handle this malware quickly and effectively.

Malware-as-a-Service

Economic theory dictates that when there is a high demand for something, enterprising individuals will work toward creating the supply. This concept applies to malware in the rise of a phenomenon known as malware-as-a-service. This is malware designed, built, and sold to customers based on their individual specifications. Like any other piece of software, this malware software may offer customer support and release periodic updates, complete with bug fixes and improvements. Moreover, many of these tools are offered as a subscription service and based in the cloud, making it more attractive for potential "customers" to acquire, access, and suspend the service.

Information Sharing and Analysis Communities

Information sharing communities were created to make threat data and best practices more accessible by lowering the barrier to entry and standardizing how threat information is shared and stored between organizations. While information sharing occurs frequently between industry peers, it's usually in an informal and ad hoc fashion. One of the most effective formal methods of information sharing comes through information

ISAC	Description
Automotive (Auto-ISAC)	Founded in 2015, Auto-ISAC is the primary mechanism for global car manufacturers to share information about threats, vulnerabilities, and best practices related to connected vehicles.
Aviation (A-ISAC)	Open to trusted global private aviation companies, A-ISAC works with public aviation entities to ensure resilience of the shared global air transportation network.
Communications (NCC)	Also known as the National Coordinating Center for Communications, NCC facilitates the sharing of threat and vulnerability information among communications carriers, ISPs, satellite providers, broadcasters, vendors, and other stakeholders.
Electricity (E-ISAC)	Working with the US Department of Energy and the Electricity Subsector Coordinating Council (ESCC), the E-ISAC establishes awareness of incidents and threats relevant to the electricity sector.
Elections Infrastructure (EI-ISAC)	Established in 2018, EI-ISAC enables election bodies, from local municipalities to the federal government, to ensure the security and integrity of elections. The EI-ISAC is spearheaded by the nonprofit Center for Internet Security (CIS) and routinely collaborates with the Department of Homeland Security's Cybersecurity and Infrastructure Security Agency and the Election Infrastructure Subsector Government Coordinating Council (GCC).
Financial Services (FS-ISAC)	As one of the oldest ISACs, the FS-ISAC has existed in support of the resilience and continuity of the global financial services infrastructure with information sharing, education, and collaboration initiatives between private firms and government agencies.
Health (H-ISAC)	With membership comprising patient care providers, health IT companies, pharmaceutical companies, medical device manufacturers, and labs, H-ISAC exists to maintain the continuity of the health sector against cyber and physical threats.
Information Technology (IT-ISAC)	Operating since 2001, IT-ISAC has provided a forum for members of the IT sector to continuously share high-volume indicators related to their sector.
MultiState (MS-ISAC)	The MS-ISAC provides resources for information sharing for the nation's state, local, tribal, and territorial governments focused on response to and recovery from security events.

Table 1-7   A List and Description of Several ISACs

sharing and analysis centers (ISACs). ISACs, as highlighted in Table 1-7, are industry-specific bodies that facilitate sharing of threat information and best practices relevant to the specific and common infrastructure of the industry.

Another mechanism to achieve similar goals has been made possible via a 2015 executive order by then US President Barack Obama in the creation of Information Sharing and Analysis Organizations (ISAOs). These public information sharing organizations are similar to those of ISACs but without the alignment to a specific industry. ISAOs, by definition, are designed to be voluntary, transparent, inclusive, and flexible. Additionally, they are strongly encouraged to provide actionable products.

Chapter Review

Threat intelligence is not a "one-size-fits-all" solution, and it is not meant to be a perfect guide as to what to do next. If it were, we'd all be out of jobs, since security would be solved in short order. The usefulness of threat intelligence to your business depends on how well integrated business requirements are with intelligence requirements and the security operations effort. Threat actors of all kinds are constantly looking to gain advantages where ever they can find it. As security professionals, we're looking to set the conditions to disrupt their decision cycles with efficient and well-informed choices about how we prepare and defend our organizations. Developing an intelligence program, sharing indicators with peer organizations, and placing effective controls in defense of a complex network environment are all things we can do to help us achieve these goals.

Questions

1. Which of the following is *not* considered a form of passive or open source intelligence reconnaissance?

    A. Google hacking

    B. nmap

    C. ARIN queries

    D. nslookup

2. Which of the following is the term for collection and analysis of publicly available information appearing in print or electronic form?

    A. Signals intelligence

    B. Covert intelligence

    C. Open source intelligence

    D. Human intelligence

3. Information that may not be shared with parties outside of the specific exchange, meeting, or conversation in which it was originally disclosed is designated by which of the following?

    A. TLP:RED

    B. TLP:AMBER

    C. TLP:GREEN

    D. TLP:WHITE

4. Which of the following sources will most often produce intelligence that is most relevant to an organization?

    A. Open source intelligence

    B. Deep and dark web forums and communications platforms

    C. The organization's network

    D. Closed source vendor data

5. Which of the following is not a characteristic of high-quality threat intelligence source data?

   A. Timeliness

   B. Transparency

   C. Relevancy

   D. Accuracy

6. In the STIX 2.0 framework, which object may be used to represent individuals, organizations, or groups?

   A. Campaign

   B. Persona

   C. Intrusion set

   D. Identity

7. In which phase of the five-step intelligence cycle would an analyst communicate his or her findings to the customer?

   A. Communication

   B. Dissemination

   C. Collection

   D. Feedback

8. Threat actors whose activities lead to increased risk as a result of their privileged access or employment are best described by what term?

   A. Unwilling participant

   B. Nation-state actor

   C. Hacktivist

   D. Insider threat

Answers

1. **B.** nmap is a scanning tool that requires direct interaction with the system under test. All the other responses allow a degree of anonymity by interrogating intermediary information sources.

2. **C.** Open source intelligence, or OSINT, is free information that's collected in legitimate ways from public sources such as news outlets, libraries, and search engines. It should be used alongside intelligence gathered from closed sources to answer key intelligence questions.

3. **A.** TLP:RED information is limited to those present at a particular engagement, meeting, or joint effort. In most circumstances, TLP:RED should be exchanged verbally or in person. The use of TLP:RED outside of approved parties could lead to impacts on a party's privacy, reputation, or operations if misused.

4. **C.** An organization's network will inherently have the most relevant threat data available of all the sources listed.

5. **B.** Transparency is not a characteristic of high-quality threat intelligence source data. In addition to being complete, good threat intelligence often has three characteristics: timeliness, relevancy, and accuracy.

6. **D.** An identity is a STIX domain object that represents individuals, organizations, or groups. The object may be a specific and may include the names of the person or organization referenced, or it may be used to identify an entire industry sector, such as transportation.

7. **B.** Distributing the requested intelligence to the customer occurs at the dissemination phase. The product may be used to gain a greater understanding of an adversary's motivation or strengthen internal defenses, or to support legal action.

8. **D.** Insider threat actors are those who work within an organization and represent a particularly high risk of causing catastrophic damage due to their privileged access to internal resources.

Threat Intelligence in Support of Organizational Security

In this chapter you will learn:

- Types of threat intelligence
- Attack frameworks and their use in leveraging threat intelligence
- Threat modeling methodologies
- How threat intelligence is best used in other security functions

> *By "intelligence" we mean every sort of information about the enemy and his country—the basis, in short, of our own plans and operations.*
>
> —Carl von Clausewitz

Depending on the year and locale, fire departments may experience around a 10 percent false alarm rate resulting from accidental alarm tripping, hardware malfunction, and nuisance behavior. Given the massive amount of resources required for a response and the scarce and specialized nature of the responders, this presents a significant issue for departments to manage. After all, they cannot respond to a perceived emergency with anything less that their full attention. As with fire departments, false alarms are more than just an annoyance for security operations teams. As the volume of data that traverses the network increases, so do the alerts and logs that have to be triaged, interpreted, and actioned—and the chance for any of those alerts being a false alarm also rises. Unfortunately, growth of specialized security teams that work endlessly to protect an enterprise from threats isn't growing at the same pace.

Although organizations invest in new types of threat detection technologies, they may only add to the already overwhelming amount of noise that exists, resulting in *alert fatigue*. Not only are analysts simply unable to assess, prioritize, and act upon every alert that comes in, but they may often ignore some of them because of the high rate of false positives. Much like the townspeople who responded to the boy crying "wolf!" only to learn that there was no wolf, security responders may learn to ignore alarm bells over time if no malicious activity exists.

One effective way to mitigate the dangers of overwhelming alerts and the often-associated alert fatigue is to integrate a threat intelligence program into all aspects of security operations. As covered in Chapter 1, threat intelligence is about adding context to internal signals to make risk-based decisions. Whether these choices occur at the tactical level in the security operations center or at the strategic level in the boardroom, good threat intelligence makes for a far better-informed security professional. In this chapter we'll explore frameworks and best practices for integrating threat intelligence into your security program to improve operations at every level.

Levels of Intelligence

In his doctrine on characterizing the adversary, famed Prussian general Carl von Clausewitz states that the adversary is a thinking, animate entity that reacts to the decisions of his enemy. The essence of developing a strong operational plan is to discover the enemy's strategy, develop your own plan to confront the enemy, and execute it with precision. The delineation of *levels of war* has become a keystone in the military decision-making process for armies throughout the world. The key concepts of providing intelligence at these levels can also be applied to cybersecurity.

At the highest level, organizations think about their conduct in a *strategic* manner, the results of which should impair adversaries' abilities to carry out what they're trying to do. Strategic effects should aim to disrupt the enemy's ability to operate by neutralizing its centers of gravity or key resource providers. Strategic threat intelligence therefore should support decisions at this level by delivering products that are anticipatory in nature. These products will provide a comprehensive view of the environment, identify the key actors, and offer a glimpse into the future based on recommended courses of action or inaction. This type of intelligence is often designed to inform the decisions of senior leaders in an organization and is accordingly not overtly technical. It's aimed at addressing the concerns of that particular audience, covering topics such as regulatory and financial impacts to the organization.

The application of a company's cybersecurity strategy occurs at the *operational* level, and planning will address concepts such as what the organization is trying to defend, from whom, for what duration, and with what capabilities. Before defining exactly how all this is to happen, defenders must determine what major efforts need to be in place to accomplish strategic goals, what resources might be needed to accomplish them, and what defines the nature of the problem. Without this direction, organizations will fail to adequately confront the challenges posed by the adversary, possibly squandering resources and frustrating security analysts. Operational threat intelligence products will provide insight into conditions particular to the environment that the organization is looking to defend. Products of this type will inform decision-makers about how best to allocate resources to defend against specific threats.

Finally, at the lowest level of war are the engagements between attacker and defender. Decisions at the *tactical* level are focused on how, exactly, a defender will engage with the adversary, to include any technical countermeasures. The results of these activities may sometimes ripple out broadly to affect operational and strategic decision-making.

For example, a decision to block a service on a network may be required in response to a particularly damaging ongoing attack. If that decision then affects legitimate operations, or how the organization shares data with a strategic partner, it clearly extends beyond the immediate engagement at hand, and its second- and third-order effects will need to be considered moving forward. Tactical threat intelligence focuses on attacker tactics, techniques, and procedures (TTPs) and relates to the specific activity observed. These products are highly actionable and the results of the follow-up decisions they inform will be used by operational staff to ensure that technical controls and processes are in place to protect the organization moving forward.

Attack Frameworks

Security teams use frameworks as analytical tools to add structure when thinking about the lifecycle of a security incident and the actors involved. Frameworks allow for broad understanding of the key concepts, timelines, and motivations of attackers by providing consistent language and syntax to communicate most aspects of an attack. With knowledge of how attackers think, the tools they use, and where they might employ a particular technique, defenders are in a better position to make a decision earlier and can reduce the opportunity for an attacker to cause disruption.

MITRE ATT&CK

The MITRE Corporation manages a significant number of federally funded research groups throughout the United States. It has developed a number of systems and frameworks that are very important for the security industry, to include the Cyber Observable eXpression (CybOX) framework, the Common Vulnerabilities and Exposures (CVE) system, the Trusted Automated Exchange of Intelligence Information (TAXII), and Structured Threat Information Expression (STIX), some of which were covered in Chapter 1.

Beginning in 2013, MITRE began development on a model that would allow US government agencies and industry security teams to share information about attacker TTPs with one another in an effective manner. The model would come to be known as the Adversarial Tactics, Techniques, and Common Knowledge (ATT&CK) framework. It's currently one of the most effective methods of tracking adversarial behavior over time, based on observed activity shared from the security community. Within ATT&CK are three flavors: Enterprise ATT&CK, PRE-ATT&CK, and Mobile ATT&CK. As you may be able to guess from the names, PRE-ATT&CK models the TTPs that an attacker may use before conducting an attack, while Mobile ATT&CK models the TTPs an attacker may use to gain access to mobile platforms. We'll take some time to explore Enterprise ATT&CK, the most widely used and relevant model of the three to enterprise cybersecurity analysts.

The framework serves as an encyclopedia of previously observed tactics from bad actors. These behaviors are divided into 12 categories based on real-world observations:

- **Initial Access**   The steps the adversary takes to get into your network
- **Execution**   The methods the adversary uses to run malicious sode on a local or remote system

- **Persistence**   The means by which the adversary maintains a presence on your systems
- **Privilege Escalation**   The techniques the adversary uses to gain positions of higher privilege
- **Defense Evasion**   The maneuvers the adversary uses to avoid detection
- **Credential Access**   The way the adversary gathers credentials such as account names, passwords and tokens
- **Discovery**   The way in which the adversary gains an understand of your network layout and its technologies
- **Lateral Movement**   The techniques the adversary uses to pivot and gain access to other systems on the network
- **Collection**   The methods the adversary uses to capture artifacts on hosts and servers
- **Command and Control**   The techniques the adversary uses to communicate with systems on a victim network
- **Exfiltration**   The methods the adversary employs to move data out of the network
- **Impact**   The steps an adversary takes to prevent access to, damage or destroy data on your network

These activities are communicated using a common language so that different teams and organizations worldwide can use the same reference to describe activities that might otherwise have other names. This aspect of the framework makes it particularly useful with prioritizing behaviors to focus on. What makes this framework so useful for defenders is that the behaviors are not meant to include strictly red-teaming or theoretical attacks, but rather those that adversaries have conducted and that they are likely to do based on observations in the wild. Furthermore, the model itself is free and built for global accessibility, reducing the barriers to entry for participation and usage, such as team maturity or organizational resources.

Although there are only 12 tactics in the Enterprise ATT&CK framework, there are hundreds of techniques that offer details about activity related to specific operating systems including Windows, macOS, and Linux, and several cloud platforms. Many of the techniques listed under the tactic categories can be used in other categories. A useful way to determine when and how a particular technique can be used is by browsing MITRE's open source web app, MITRE ATT&CK Navigator (https://mitre.github.io/attack-navigator/enterprise/). It allows for basic navigation of all of the framework's components from one pane. A user can select a particular technique and see it highlighted across multiple categories, as shown in Figure 2-1.

Accompanying each of the techniques are details that will be very useful for defenders. Using the group FIN7 as an example, shown in Figure 2-2, you can see that the information provided includes a unique identification number, dates the group was first observed, and a brief description that often includes alternate group names.

Initial Access	**Execution**	**Persistence**	**Privilege Escalation**	**Defense Evasion**	**Credential Access**	**Discovery**
11 items	34 items	62 items	32 items	69 items	21 items	23 items
Drive-by Compromise	AppleScript	.bash_profile and .bashrc	Access Token Manipulation	Access Token Manipulation	Account Manipulation	Account Discovery
Exploit Public-Facing Application	CMSTP	Accessibility Features	Accessibility Features	Binary Padding	Bash History	Application Window Discovery
External Remote Services	Command-Line Interface	Account Manipulation	AppCert DLLs	BITS Jobs	Brute Force	Browser Bookmark Discovery
Hardware Additions	Compiled HTML File	AppCert DLLs	AppInit DLLs	Bypass User Account Control	Credential Dumping	Domain Trust Discovery
Replication Through Removable Media	Component Object Model and Distributed COM	AppInit DLLs	Application Shimming	Clear Command History	Credentials from Web Browsers	File and Directory Discovery
Spearphishing Attachment	Control Panel Items	Application Shimming	Bypass User Account Control	CMSTP	Credentials in Files	Network Service Scanning
Spearphishing Link	Dynamic Data Exchange	Authentication Package	DLL Search Order Hijacking	Code Signing	Credentials in Registry	Network Share Discovery
Spearphishing via Service	Execution through API	BITS Jobs	Dylib Hijacking	Compile After Delivery	Exploitation for Credential Access	Network Sniffing
Supply Chain Compromise	Execution through Module Load	Bootkit	Elevated Execution with Prompt	Compiled HTML File	Forced Authentication	Password Policy Discovery
Trusted Relationship	Exploitation for Client Execution	Browser Extensions	Emond	Component Firmware	Hooking	Peripheral Device Discovery
Valid Accounts	Graphical User Interface	Change Default File Association	Exploitation for Privilege Escalation	Component Object Model Hijacking	Input Capture	Permission Groups Discovery
	InstallUtil	Component Firmware	Extra Window Memory Injection	Connection Proxy	Input Prompt	Process Discovery
	Launchctl	Component Object Model Hijacking	File System Permissions	Control Panel Items	Kerberoasting	Query Registry
	Local Job Scheduling	Create Account		DCShadow	Keychain	Remote System Discovery
	LSASS Driver	DLL Search Order Hijacking		Deobfuscate/Decode Files or Information	LLMNR/NBT-NS Poisoning and Relay	Security Software Discovery
		Dylib Hijacking		Disabling Security Tools	Network Sniffing	Software Discovery
				DLL Search Order Hijacking		
				DLL Side-Loading		

Figure 2-1 MITRE ATT&CK Navigator with technique highlights

The information page goes on to list all of the techniques associated with that group. An analyst can then select a technique to find examples of the technique as seen in the wild, detection methods, and mitigations.

Using the framework has quantitative benefits for the security team as well. Outside of simply being aware of the most commonly used attack techniques, a security team can compare TTPs associated with groups that it believes are priority and highlight the TTPs associated with them, identifying overlaps, while providing an effective visual reference. For example, security analysts at a bank may use the framework to highlight the TTPs used by FIN7 and Cobalt Group, two threat actor groups known for targeting financial institutions. Using separate colors to identify the groups, let's say red for FIN7 and blue for Cobalt Group, analysts can overlay the color-coded TTPs on a single matrix to determine which techniques to prioritize defenses against, in this case identified in purple.

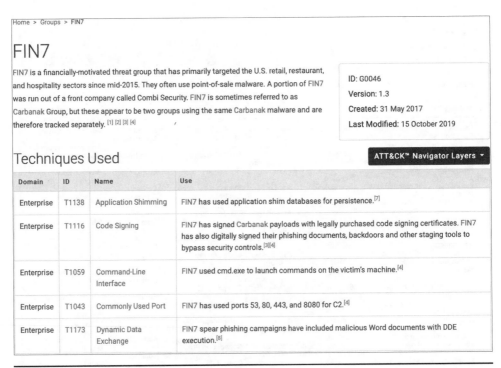

Figure 2-2   Snapshot of FIN7 threat actor group

Additionally, structuring TTPs in this way easily enables defenders to measure the adversaries they're tracking and countermeasures they've put in place. In addition, presenting hard figures alongside trends observed over time to show progress in addressing specific TTPs often resonates well with senior leadership.

The Diamond Model of Intrusion Analysis

While looking for an effective way to analyze and track the characteristics of cyberintrusions by advanced threat actors, security professionals Sergio Caltagirone, Andrew Pendergast, and Christopher Betz developed the Diamond Model of Intrusion Analysis, shown in Figure 2-3, which emphasizes the relationships and characteristics of four basic components: Adversary, Capability, Victim, and Infrastructure. The components are represented as vertices on a diamond, with connections between them. Using the connections between these entities, you can use the model to describe how an adversary uses a capability in an infrastructure against a victim. A key feature of the model is that enables defenders to pivot easily from one node to the next in describing a security event. As entities are populated in their respective vertex, an analyst will be able to read across the model to describe the specific activity (for example, FIN7 [the Adversary] uses phish kits [the Capability] and actor-registered domains [the Infrastructure] to target bank executives [the Victim]).

Figure 2-3
The Diamond
Model

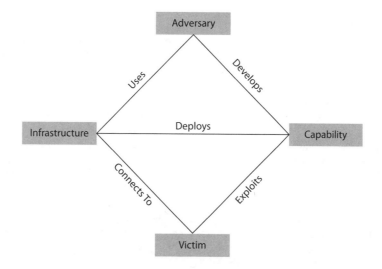

It's important to note that the model is not static, but rather adjusts as the adversary changes TTPs, infrastructure, and targeting. Because of this, this model requires a great deal of attention to ensure that the details of each component are updated as new information is uncovered. If the model is used correctly, analysts can easily pivot across the edges that join vertices in order to learn more about the intrusion and explore new hypotheses.

As we use this model to capture and communicate details about malicious activity we may encounter, we should also consider the seven axioms that Caltagirone describes as capturing the nature of all threats. Table 2-1 lists the axioms and what they mean for defenders.

Axiom	What it means for defenders
For every intrusion event there exists an adversary taking a step toward an intended goal by using a capability over infrastructure against a victim to produce a result.	Every security incident has the four components of adversary, infrastructure, capability, and victim. We use the Diamond Model to build strategies for detecting, tracking, and mitigating malicious activity.
There exists a set of adversaries (insiders, outsiders, individuals, groups, and organizations) that seek to compromise computer systems or networks to further their intent and satisfy their needs.	Threat actors are constantly trying to gain access to systems for specific reasons. If we can determine those reasons, we may be able to defend our systems more effectively.
Every system, and by extension every victim asset, has vulnerabilities and exposures.	Assume that no technology, and by extension no system, is safe from vulnerabilities and exploits.

Table 2-1 Diamond Model Axioms *(continued)*

Axiom	What it means for defenders
Every malicious activity contains two or more phases that must be successfully executed in succession to achieve the desired result.	Dependencies need to be fulfilled for an attack to be successful. The kill chain is one model that describes the phases of an attack.
Every intrusion event requires one or more external resources to be satisfied prior to success.	If we determine the adversary's external requirements, we may develop methods to deny the adversary access to them to frustrate their efforts.
A relationship always exists between the adversary and their victim(s), even if distant, fleeting, or indirect.	Because gaining access can be a difficult undertaking, adversaries will always have a reason to dedicated time and resources to a particular victim.
There exists a subset of the set of adversaries that have the motivation, resources, and capabilities to sustain malicious effects for a significant length of time against one or more victims while resisting mitigation efforts. Adversary-victim relationships in this subset are called persistent adversary relationships.	Depending on the nature of the operations, some adversaries need long-term access to their victims to be successful. Determining the nature of these operations will be helpful in developing techniques to deter or prevent persistence.

Table 2-1 Diamond Model Axioms

Kill Chain

The *kill chain* is a phase-based model that categorizes the activities that an enemy may conduct in a kinetic military operation. One of the first and most commonly used kill chain models in the military is the 2008 F2T2EA, or Find, Fix, Track, Target, Engage, and Assess. Driven by the need to improve response times for air strikes, then Air Force Chief of Staff General John Jumper pushed for the development of an agile and responsive framework to achieve his goals, and thus F2T2EA was born. Similar to the military kill chain concept, the cyber kill chain defines the steps used by cyberattackers in conducting their malicious activities. The idea is that by providing a structure that breaks an attack into stages, defenders can pinpoint where along the lifecycle of an attack an activity is and deploy appropriate countermeasures. The kill chain is meant to represent the deterministic phases adversaries need to plan and execute in order to gain access to a system successfully.

The Lockheed Martin Cyber Kill Chain is perhaps the most well-known version of the kill chain as applied to cybersecurity operations. It was introduced in a 2011 whitepaper authored by security team members Michael Cloppert, Eric Hutchins, and Rohan Amin. Using their experience from many years on intelligence and security teams, they describe a structure consisting of seven stages of an attack. Figure 2-4 shows the progression from stage to stage in the Cyber Kill Chain. Since the model is meant to approach attacks in a linear manner, if defenders can stop an attack early on at the exploitation stage, they can have confidence that the attacker is far less likely to have progressed to further stages. Defenders can therefore avoid conducting a full incident response plan.

Figure 2-4
The Lockheed
Martin Cyber
Kill Chain

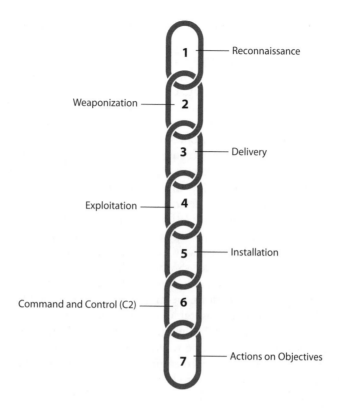

1 — Reconnaissance

Weaponization — 2

3 — Delivery

Exploitation — 4

5 — Installation

Command and Control (C2) — 6

7 — Actions on Objectives

Furthermore, understanding the phase progression, typical behavior expected at each phase, and inherent dependencies in the overall process allows for defenders to take appropriate measures to disrupt the kill chain.

The kill chain begins with *Reconnaissance*, the activities associated with getting as much information as possible about a target. Reconnaissance will often show the attacker focusing on gaining understanding about the topology of the network and key individuals with system or specific data access. As described in Chapter 1, reconnaissance actions (sometimes referred to as recon) can be passive or active in nature. Often an adversary will perform passive recon to acquire information about a target network or individual without direct interaction. For example, an actor may monitor for new domain registration information about a target company to get technical points of contact information. Active reconnaissance, on the other hand, involves more direct methods of interacting with the organization to get a lay of the land. An actor may scan and probe a network to determine technologies used, open ports, and other technical details about the organization's infrastructure. The downside (for the actor) is that this type of activity may trigger detection rules designed to alert defenders on probing behaviors. It's important for the defender to understand what types of recon activities are likely to be leveraged against the organization and develop technical or policy countermeasures to mitigate those threats. Furthermore, detecting reconnaissance activity can be very useful in revealing the intent of an adversary.

With the knowledge of what kind of attack may be most appropriate to use against a company, an attacker would move to prepare attacks tailored to the target in the *Weaponization* phase. This may mean developing documents with naming schemes similar to those used by the company, which may be used in a social engineering effort at a later point. Alternatively, an attacker may work to create specific malware to affect a device identified during the recon. This phase is particularly challenging for defenders to develop mitigations for, because weaponization activity often occurs on the adversary side, away from defender-controlled network sensors. It's nonetheless an essential phase for defenders to understand, because it occurs so early in the process. Using artifacts discovered during the Reconnaissance phase, defenders may be able to infer what kind of weaponization may be occurring and prepare defenses for those possibilities. Even after discovery, it may be useful for defenders to reverse the malware to determine how it was made. This can inform detection efforts moving forward.

The point at which the adversary goes fully offensive is often at the *Delivery* phase. This is the stage when the adversary transmits the attack. This can happen via a phishing e-mail or Short Message Service (SMS), by delivery of a tainted USB device, or by convincing a target to switch to an attacker-controlled infrastructure, in the case of a rogue access point or physical man-in-the-middle attack. For defenders, this can be a pivotal stage to defend. It's often measurable since the rules developed by defenders in the previous stages can be put into use. The number of blocked intrusion attempts, for example, can be a quick way to determine whether previous hypotheses are likely to be true. It's important to note that technical measures combined with good employee security awareness training continually proves to be the most effect way to stop attacks at this stage.

In the unfortunate event that the adversary achieves successful transmission to his victim, he must hope to somehow take advantage of a vulnerability on the network to proceed. The *Exploitation* phase includes the actual execution of the exploit against a flaw in the system. This is the point where the adversary triggers the exploit against a server vulnerability, or when the user clicks a malicious link or executes a tainted attachment, in case of a user-initiated attack. At this point, an attack can take one of two courses of action. The attacker can install a *dropper* to enable him to execute commands, or he can install a *downloader* to enable additional software to be installed at a later point. The end goal here is often to get as much access as possible to begin establishing some permanence on the system. Hardening measures are extremely important at this stage. Knowing what assets are present on the network and patching any identified vulnerabilities improves resiliency against such attacks. This, combined with more advanced methods of determining previously unseen exploits, puts defenders in a better position to prevent escalation of the attack.

For the majority of attacks, the adversary aims to achieve *persistence*, or extended access to the target system for future activities. The attacker has taken a lot of steps to get to this point, and would likely want to avoid going through them every time he wants access to the target. *Installation* is the point where the threat actor attempts to emplace a backdoor or implant. This is frequently seen during insertion of a web shell on a compromised web server, or a remote access Trojan (RAT) on a compromised machine. Endpoint detection is frequently effective against activities in the stage; however, security analysts may sometimes need to use more advanced logging interpretation techniques to identify clever or obfuscated installation techniques.

In the *Command and Control* (C2) phase, the attacker creates a channel in order to facilitate continued access to internal systems remotely. C2 is often accomplished through periodic beaconing via a previously identified path outside of the network. Correspondingly, defenders can monitor for this kind of communication to detect potential C2 activity. Keep in mind that many legitimate software packages perform similar activity for licensing and update functionality. The most common malicious C2 channels are over the Web, Domain Name System (DNS), and e-mail, sometimes with falsified headers. For encrypted communications, beacons tend to use self-signed certificates or custom encryption to avoid traffic inspection. When the network is monitored correctly, it can reveal all kinds of beaconing activity to defenders hunting for this behavior. When looking for abnormal outbound activities such as this, we must think like our adversary, who will try to blend in with the scene and use techniques to cloak his beaconing. To complicate things more, beaconing can occur at any time or frequency, from a few times a minute to once or twice weekly.

The final stage of the kill chain is the Actions on Objectives, or the whole reason the attacker wanted to be there in the first place. It could be to exfiltrate sensitive intellectual property, encrypt critical files for extortion, or even sabotage via data destruction or modification. Defenders can use several tools at this stage to prevent or at least detect these actions. Data loss prevention software, for example, can be useful in preventing data exfiltration. In any case, it's critical that defenders have a reliable backup solution that they can restore from in the worst-case scenario. Much like the Reconnaissance stage, detecting activity during this phase can give insight into attack motivations, albeit much later than is desirable.

While the Cyber Kill Chain enables organizations to build defense-in-depth strategies that target specific parts of the kill chain, it may fail to capture attacks that aren't as dependent on all of the phases to achieve end goals. One such example is modern phishing attacks in which attackers rely on victims to execute an attached script. Additionally, the kill chain is very malware-focused and doesn't capture the full scope of other common threat vectors such as insider threats, social engineering, or any intrusion in which malware wasn't the primary vehicle for access.

Threat Research

Thanks to the increasing acceptance of intelligence concepts across security teams and the use of attack frameworks, threat research is gaining analytic rigor and completeness required to be a repeatable and scalable practice. Though threat research can be used to enrich alerts raised with existing detection technologies, it can also be used to uncover novel attacker TTPs within an environment not already discovered by detection rules. As an analyst coming across an artifact, you should work to answer a few initial questions about it: Is this benign? Has anyone seen this before, and if so, what do they have to say about it? Why is this present in my system? There are many different approaches we can use to get answers to these questions, but often we must conduct some sort of initial enrichment to determine the next steps. Conducting threat research is a critical part of the threat intelligence process, and we'll step through a few effective methods to getting answers to those questions.

Reputational

Security team members need valid malware signatures and reputation data about IPs and domains to help filter the vast amounts of data that flow through the network. They use this information to enable firewalls, gateways, and other security devices to make decisions that prevent attacks while maintaining access for legitimate traffic. There are many free and commercial services that assign reputation scores to URLs, domains, and IP addresses across the Internet. Scores that correlate to the highest risk are associated with malware, spyware, spam, phishing, C2, and data exfiltration servers. Reputation scores can also be assigned to computers and websites that have already been identified as compromised. Google's Safe Browsing is a useful service that enables users to check the status of a website manually. The service also allows for automatic URL blacklisting for users of Chrome, Safari, and Firefox web browsers, helping users identify whether they are attempting to access web resources that contain malware or phishing content.

Cisco's threat intelligence team, Talos, provides excellent reputational lookup features in a single dashboard in its Reputation Center service. Figure 2-5 is a snapshot of the report page generated for a suspicious IP address. Included in the report are details about location, blacklist status, and IP address owner information. For the Reputation

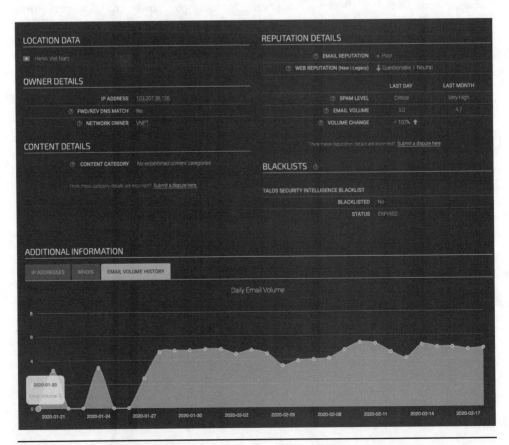

Figure 2-5 Talos Reputation Center report

Figure 2-6 VirusTotal homepage

Details section, Talos breaks the reputational assessments out by e-mail, malware, and spam, with historical information to give a sense of trends associated with the IP address. Finally, at the bottom of the report is a section for additional information. In this case, the IP address carried "critical" and "very high" spam levels from the previous day and month, respectively. The Additional Information section in this case provides daily e-mail volume information to give some historical context to the rating.

For high-volume reputational information, VirusTotal, a security-focused subsidiary of Google, is one of the most reliable services. VirusTotal aggregates the results of submitted URLs and samples from more than 70 antivirus scanners and URL/domain blacklisting services to return a verdict about the likelihood of content being malicious. In addition to the web-based submission method shown in Figure 2-6, users may submit samples programmatically using any number of scripting options via VirusTotal's application programming interface (API).

Behavioral

Sometimes we are unable or unwilling to invest the effort into reverse engineering a binary executable, but we still want to find out what it does. This is where an isolation environment, or sandbox, comes in handy. Unlike endpoint protection sandboxes, this tool is usually instrumented to assist the security analyst in understanding what a running executable is doing as samples of malware are executed to determine their behaviors.

Cuckoo Sandbox is a popular open source isolation environment for malware analysis. It uses either VirtualBox or VMware Workstation to create a virtual computer on which to run the suspicious binary safely. Unlike other environments, Cuckoo is just as capable in Windows, Linux, macOS, or Android virtual devices. Another tool with which you may want to experiment is REMnux, which is a Linux distribution loaded with malware reverse engineering tools.

Using these tools, we may get insight into how a particular piece of software behaves when executed in the target environment. Be aware that malware writers are increasingly leveraging techniques to detect sandboxes and control various malware activities in the presence of a virtualized environment.

Indicator of Compromise

An indicator of compromise (IOC) is an artifact that indicates the possibility of an attack or compromise. As covered in Chapter 1, IOCs need two primary components: data and context. Of the countless commercial feeds available to security teams, the most appropriate one to use depends on the industry and specific organizational requirements. As for free sources of IOC, there are a few high-quality sources that we'll highlight. In addition to the ISACs covered earlier, the Computer Incident Response Center Luxembourg (CIRCL) operates several open source malware information sharing platforms, or MISPs, to facilitate automated sharing of IOCs across private and public sectors. Domestically, the FBI's InfraGard Portal provides historical and ongoing threat data relevant to 16 sectors of critical infrastructure.

Security architect David Bianco developed a great model to categorize IOCs. His model, the Pyramid of Pain, shown in Figure 2-7, is used to show how much cost we can impose on the adversary when security teams address indicators at different levels. Hashes are easy to alert upon with high confidence; however, they are also easy to change and can therefore cause a trivial amount of pain for the adversary if detected and actioned. Changing IP addresses is more difficult than changing hashes, but most

Figure 2-7
Bianco's Pyramid
of Pain

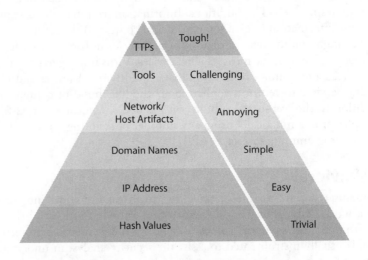

adversaries have disposable infrastructure and can also change the IP addresses of their hop points and command and control nodes once they are compromised. The bottom half of the pyramid contains the indicators that are most likely uncovered using highly automated solutions, while the top half includes more behavioral based indicators.

As security professionals, we want to operate right at the top whenever possible, where the TTPs, if identified, require that attackers change nearly every aspect of how they operate. As you can probably guess, it is more difficult to alert on network and host artifacts and TTPs as well, but if we can address these high-confidence indicators, it will have a lasting impact on the security of our networks.

Common Vulnerability Scoring System

A well-known standard for quantifying severity is the Common Vulnerability Scoring System (CVSS). As a framework designed to standardize the severity ratings for vulnerabilities, this system ensures accurate quantitative measurement so that users can better understand the impact of these weaknesses. With the CVSS scoring standard, members of industries, academia, and governments can communicate clearly across their communities.

 EXAM TIP   The Common Vulnerability Scoring System is the de facto standard for assessing the severity of vulnerabilities. Therefore, you should be familiar with CVSS and its metric groups: base, temporal, and environmental. These groups represent various aspects of a vulnerability. Base are those characteristics that do not change over time, temporal describes those that do, and environmental represents those that are unique to a user's environment.

Threat Modeling Methodologies

Threat modeling promotes better security practices by taking a procedural approach to thinking like the adversary. At their core, threat modeling techniques are used to create an abstraction of the system, develop profiles of potential attackers, and bring awareness to potential weaknesses that may be exploited. Exactly how the threat modeling activity is conducted depends on the goals of the model. Some threat models may be used to gain a general understanding about all aspects of security, while others may be focused on related aspects such as user privacy.

To gain the greatest benefit from threat modeling, it should be performed early and continuously as an input directly into the software development lifecycle (SDLC). Not only might this prevent a catastrophic security issue in the future, but it may lead to architectural decisions that help reduce vulnerabilities without sacrificing performance. For some systems, such as industrial control systems and cyber-physical systems, threat modeling may be particularly effective, because the costs of failure are not just monetary. By promoting development with security considerations, rather than security as an afterthought, defenders should have an easier time with incident response because of their increased awareness of the software architecture.

Adversary Capability

Understanding what a potential attacker is capable of can be a daunting task, but it is often a key competence of any good threat intelligence team. The first step in understanding adversary capability is to document the types of threat actors that would likely be threats, what their intent might be, and what capabilities they might bring to bear in the event of a security incident. As previously described, we can develop an understanding of adversary TTPs with the help of various attack frameworks and resources such as MITRE ATT&CK.

Total Attack Surface

The attack surface is the logical and physical space that can be targeted by an attacker. Logical areas include infrastructure and services, while physical areas include server rooms and workstations. Mapping out what parts of a system need to be reviewed and tested for security vulnerabilities is a key part of understanding the total attack surface. As each component is addressed, defenders need to keep track of how the overall attack might change as compensating controls are put into place. Analysis of the attack surface is usually conducted by penetration testers, software developers, and system architects, but as a security analyst, you will have significant influence in the architecture decisions as the local expert on security operations in the organization.

Attack Vector

With a potential adversary in mind and critical assets identified, the natural next step is to determine the most likely path for the adversary to get their hands on the goods. This can be done using visual tools, as part of a red-teaming exercise, or even as a tabletop exercise. The goals of mapping out attack vectors consist of identifying realistic or likely paths to critical assets and identifying which security controls are in place to mitigate specific TTPs. If no mitigation exists, security teams can put compensating controls in place while they work out a long-term remediation plan.

Impact

Impact is simply the potential damage to an organization in the case of a security incident. Impact types can include but aren't limited to physical, logical, monetary, and reputational. Impact is used across the security field to communicate risk to an organization. We'll cover various aspects in impact later the in the book when discussing risk analysis.

Likelihood

The possibility of a threat actor successfully exploiting a vulnerability that results in a security incident is referred to as likelihood. The Nation Institute of Standards and Technology, or NIST, provides a formal definition of likelihood as it relates to security operations as "a weighted factor based on a subjective analysis of the probability that a given threat is capable of exploiting a given vulnerability or a set of vulnerabilities."

Threat	Property Affected	Definition	Example
Spoofing	Authentication	Impersonating someone or something else	A outside sender pretending to be HR in an e-mail
Tampering	Integrity	Modifying data on disk, in memory, or elsewhere	A program modifying the contents of a critical system file
Repudiation	Nonrepudiation	Claiming to have not performed an action or have knowledge of who performed it	A user claiming that she did not receive a request
Information disclosure	Confidentiality	Exposing information to parties not authorized to see it	An analyst accidentally revealing the inner details of the network to outside parties
Denial of service	Availability	Denying or degrading service to legitimate users by exhausting resources needed for a service	Users flooding a website with thousands of requests a second, causing it to crash
Elevation of privilege	Authorization	Gaining capabilities without the proper authorization to do so	A user bypassing local restrictions to gain administrative access to a workstation

Table 2-2 STRIDE Threat Categories

STRIDE

STRIDE is a threat modeling framework that evaluates a system's design using flow diagrams, system entities, and events related to a system. The framework name is a mnemonic, referring to the security threats in six categories shown in Table 2-2. Invented in 1999 and developed over the last 20 years by Microsoft, STRIDE is among the most used threat modeling method, suitable for application to logical and physical systems alike.

PASTA

PASTA, or the Process for Attack Simulation and Threat Analysis, is a risk-centric threat modeling framework developed in 2012. Focused on communicating risk to strategic-level decision-makers, the framework is designed to bring technical requirements in line with business objectives. Using seven stages show in Table 2-3, PASTA encourages analysts to solicit input from operations, governance, architecture, and development.

Stage	Key Tasks
Define Objectives	Identify business objectives Identify security and compliance requirements Perform business impact analysis
Define Technical Scope	Record infrastructure, application, and software dependencies Record scope of the technical environment
Application Decomposition	Identify use cases Identify actors, assets, services, roles, and data sources Create data flow diagrams
Threat Analysis	Analyze attack scenarios Perform threat intelligence correlation and analytics
Vulnerability and Weaknesses Analysis	Catalog vulnerability reports and issues Map existing vulnerabilities Perform design flaw analysis
Attack Modeling	Analyze complete attack surface
Risk and Impact Analysis	Qualify and quantify business impact Catalog mitigating strategies and techniques Identify residual risk

Table 2-3 PASTA Stages

Threat Intelligence Sharing with Supported Functions

Organizations dedicate significant resources to attracting, training, and retaining security professionals. Integrating threat intelligence concepts enables responders to act more quickly in the face of uncertainty and frees them up to deal with new and unexpected threats when they arise.

Incident Response

Incident responders have some of the most sought-after skills required by any organization because of their ability to rapidly and accurately address potentially wide-ranging issues on a consistent basis. Although many incident responders thrive in stressful environments, the job can be challenging for even the most seasoned security professionals. Looking at the upward trend of security event volume and complexity, there will likely be a constant demand for responders well into the future. Incident response is not usually an entry-level security function because it requires such a diverse skill set, from malware analysis, to forensics, to network traffic analysis. Furthermore, at the core of a responder's modus operandi is speed—speed to confirm a potential incident, speed to remediate, and speed to address the root cause. When we take a look at what's required across the skill spectrum from an incident responder, and combine that with the need for speed, it becomes clear why reducing response time and moving toward proactive measures are so important.

As security teams attempt to move away from a reactive nature, they must do whatever they can to prepare themselves for the possibility of a security event. Many teams are using playbooks, or predefined sets of automated actions, more and more in their response efforts. Although it may take some time to understand the company's IT environment, the entities involved, and the external and internal threats posed against the company, preparation pays off in several ways. Threat intelligence information is a critical part of the preparation phase because it enables teams to more accurately develop strong, consistent processes to cope with issues should they arise. These not only dramatically reduce the time needed to respond, but as repeatable and scalable processes, they reduce the likelihood of an analyst error.

Vulnerability Management

Vulnerability management teams are all about making risk-based decisions. Thinking back to one of the axioms of the Diamond Model of Intrusion Analysis, we're reminded that "every system, and by extension every victim asset, has vulnerabilities and exposures." Vulnerabilities seem to be a fact of life, but that doesn't mean that a team can forsake its responsibility to identify vulnerable assets and deploy patches.

Throughout the book so far, we've referred to useful sources for threat data, many of them providing vulnerability information. Another database, NIST's National Vulnerability Database (NVD), makes it easy for organizations to determine whether they are likely to be affected by disclosed vulnerabilities. But the NVD and other databases miss a key feature that threat intelligence adds. Threat intelligence takes vulnerability management concepts a step further and provides awareness about vulnerabilities in an operational context—that is, answering the question, "Are these vulnerabilities actively being exploited?" Table 2-4 provides a short list of some of the authors' favorite types of free sources for intelligence related to vulnerabilities.

Threat intelligence communicates exploitation relevant to the organization instead of general exploitability. This is important, because thinking back to the scale of vulnerabilities across the enterprise, a vulnerability management team's core function is really prioritization. By identifying what *is* being exploited versus what *can be* exploited, these teams can make better decisions about where to place resources.

Type	Description
Information security sites	Vendor blogs and official vendor disclosure notices are a great source for up-to-date disclosure information.
Social media	"Security Twitter," the community of researcher and security-adjacent personas, is one of the best sources for vulnerability information and observations of exploitation in the wild.
Code repositories	GitHub is a popular platform for sharing proof-of-concept exploit code.
Paste sites	Pastebin, Ghostbin, and other free and anonymous services often host lists of exploitable vulnerabilities.

Table 2-4   Sources of Intelligence Related to Vulnerabilities

NOTE Many times, vulnerabilities with high scores may not be the ones actually being exploited in the wild. There are a number of reasons for this, from technical complexity preventing the creation of a practical exploit to the fact that threat actors are going to use whatever works as long as they can, regardless of CVSS score.

Risk Management

As with vulnerability management teams, risk management teams speaking the language of risk in terms of impact and probability. If we understand risk to mean the impact to an asset by a threat actor exploiting a vulnerability, we see that the presence of a threat actor is necessary in communicating risk accurately. Drilling down further, three components need to be present for a threat to be accurately described: capability, intent, and opportunity. Providing answers to these three components as they appear at present is exactly what threat intelligence is designed to do. Furthermore, good threat intelligence will also be able to predict what the threat will likely be in the future, or if there are likely to be more. Figure 2-8 highlights the various components necessary to describe a threat and how it all fits in with defining risk.

Predicting the future is what all risk team members want to do, and though that's not really possible, threat intelligence does provide answers to questions that risk managers and security leaders ask. Primary among these are identifying what type of attacks are becoming more or less prevalent and what assets are attackers likely to target in the future. In terms of quantifying the cost to an organization, risk teams may also want to know which of these attacks are likely to be most costly to the business. The logistics sector, for example, has a completely different cost of downtime than does the automotive industry.

Figure 2-8
Relationship of vulnerability, impact, threat, and risk

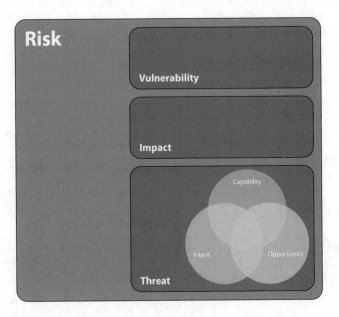

Security Engineering

Security engineers of all flavors, whether on a product security or corporate security team, regularly benefit from threat intelligence data. Threat intelligence gathered from security research or criminal communities can offer insight into the effectiveness of security measures across a company. While the motivations of these two communities are very different, they both can provide a unique outlook on your organization's security posture. This feedback can then be analyzed and operationalized by your organization's security engineers.

Detection and Monitoring

As discussed throughout Chapter 1, threat intelligence as applied to security operations is all about enriching internal alerts with the external information and context necessary to make decisions. For analysts working in a security operations center (SOC) to interpret incoming detection alerts, context is critical in enabling them to triage quickly and move on to scoping potential incidents. Because a huge part of a detection analyst's time is spent interpreting the output of dashboards, identifying relevant inputs early on reduces the cognitive load for the analyst down the line.

Let's explore a typical workflow for a detection analyst, where threat intelligence can significantly speed up the decision-making cycle. When an analyst receives an alert, she is getting only a headline of the activity and a few artifacts to support the alert condition. Attempting to triage this initial alert without access to enough context will not give the analyst a sense of what the true story is. Even if a repository exists and is made available to the analyst with all information, it would be impractical to perform the manual steps necessary to assimilate and correlate with other data related to the alert. Automated threat intelligence tailored to the needs of the detection team improves the analyst workflow by providing timely details. A great example is enrichment around suspicious domains. A detection team can easily leverage automation techniques to query threat intelligence data to extract reputation information, passive DNS details, and malware associations linked to that domain. Joining this information with what's already in the alert content provides so much more awareness for the analyst to make a call.

Chapter Review

Threat intelligence enables organizations to anticipate, respond to, and remediate threats. In some cases, organizations can use threat intelligence to speed up its decision-making cycle, causing increased cost to attackers conducting malicious activity with continual improvements in its security posture and response efforts. Useful threat intelligence can be generated only after establishing a clear understanding of the organization's goals, the role of the security operations team within the organization, and the role of threat intelligence in supporting those security teams. Although the threat intelligence team may often reside within or adjacent to a security operations center, it serves customers throughout the organization. Risk managers, vulnerability managers, incident responders, financial analysts, and C-suite members can all benefit from threat intelligence products. Key to providing the best products possible is to have a baseline understanding of adversaries that may target the organization, what their capabilities are, and what motivates them. Understanding these aspects focuses the threat intelligence effort and ensures not only that the value delivered by the security team is in line with organizational goals, but that it is also relevant and actionable.

Questions

1. Which of the following is a commonly utilized four-component framework used to communicate threat actor behavior?

   A. Kill chain

   B. STIX 2.0

   C. The Diamond Model

   D. MITRE ATT&CK

2. Developed by Microsoft, which threat modeling technique is suitable for application to logical and physical systems alike?

   A. PASTA

   B. STRIDE

   C. The Diamond Model

   D. MITRE ATT&CK

3. Defining what a threat actor might be able to achieve in the event of an attack is also known as determining the actor's _____?

   A. Means

   B. Skillset

   C. Intent

   D. Capability

4. Details about domains that may include scoring information, blacklist status, and association with malware are also known as what kind of data?

   A. Behavioral

   B. Threat

   C. Reputational

   D. Malware

5. Accurately describing a threat includes all but which of the following components?

   A. Intent

   B. Capability

   C. Opportunity

   D. Operations

6. What is the term for a predefined set of automated actions that incident responders and security operations center analysts use to enhance their operations?

   A. CVSS

   B. Playbook

   C. Repudiation

   D. Detection set

7. Consisting of the two components of data and context, what term describes an artifact that indicates the possibility of an attack?

   A. Indicator of compromise

   B. Security indicators and event monitor

   C. Security information and event management

   D. Simulations, indications, and environmental monitors

8. Which of the following is an open framework for communicating the characteristics and severity of software vulnerabilities?

   A. STRIDE

   B. NVD

   C. CVSS

   D. CVE

Answers

   1. C. The Diamond Model of Intrusion Analysis describes how an adversary uses a capability in an infrastructure against a victim.

   2. B. STRIDE is a threat modeling framework that evaluates a system's design using flow diagrams, system entities, and events related to a system.

   3. D. Defining a threat actor's capability, the ability to use skills and tools to perform an attack, helps indicate what the actor might be able to achieve in the event of an attack.

   4. C. Reputational services often assign scores to URLs, domains, and IP addresses across the Internet that are generated based on the entity's links with malware, spyware, spam, phishing, C2, and data exfiltration servers.

5. **D.** Operations is not one of the three components that need to be present for a threat to be accurately described. The components are capability, intent, and opportunity.

6. **B.** Security playbooks are customized, scalable security workflows that use integration with software and hardware platforms, usually via APIs, to automate parts of security operations.

7. **A.** An indicator of compromise (IOC) is an artifact consisting of context applied to observable data that indicates the possibility of an attack.

8. **C.** The Common Vulnerability Scoring System (CVSS) is the de facto standard for assessing the severity of vulnerabilities.

Vulnerability Management Activities

In this chapter you will learn:

- The requirements for a vulnerability management process
- How to determine the frequency of vulnerability scans to meet your needs
- The types of vulnerabilities found in various systems
- Considerations when configuring tools for scanning

*Of old, the expert in battle would first make himself invincible
and then wait for his enemy to expose his vulnerability.*

—Sun Tzu

Like many other areas in life, vulnerability management involves a combination of things we want to do, things we should do, and things we have to do. Assuming you don't need help with the first, we'll focus our attention in this chapter on the latter two. First of all, we identify the requirements that we absolutely have to satisfy. Broadly speaking, these come from external authorities (such as laws and regulations), internal authorities (such as organizational policies and executive directives), and best practices. This last source may be a bit surprising to some, but keep in mind that we are required to display due diligence in our application of security principles to protecting our information systems. To do otherwise risks liability issues and even our very jobs.

Vulnerability Identification

Vulnerability scanning, the practice of automating security checks against your systems, is a key part of securing a network. These checks help focus your efforts on protecting the network by pointing out the weak parts of the system. In many cases, scanning tools even suggest options for remediation.

Although we promote the regular use of vulnerability scanners, you should consider some important limitations of this practice before you use them. Many vulnerability scanners do a tremendous job of identifying weaknesses, but they are often single-purpose tools. Specifically, they often lack the functionality of capitalizing on a weakness and elevating to the exploit stage automatically. As a defender, you must understand that an actual attacker will combine the results from his own vulnerability scan, along with other intelligence about the network, to formulate a smart plan on how to get into the network. What's more, scanning tools will usually not be able to perform any type of advanced correlation on their own, and you'll likely require additional tools and processes to determine the overall risk of operating the network. This is due not only to the large variety of network configurations, but also to other nontechnical factors such as business requirements, operational requirements, and organizational policy.

Consider a simple example—a discovery of several low-risk vulnerabilities across the network. Although the vulnerability scanner may classify each of these occurrences as "low risk," without the context of the security posture in other areas of the network, it's impossible for you to truly understand the cumulative effect of these low-risk vulnerabilities. It may even turn out that the cumulative effect is beyond what the additive effect may be. In other words, the combined effect of several relatively low severity vulnerabilities is amplified by the fact that they may be exploited together. A weak password used by an administrator or a normal user are both problematic, but the potential effect of a compromise of an administrator's account can have far greater impact on the organization. No single tool is going to be able to provide such a depth of insight automatically, so it's important for you to understand the limitations of your tools as their capabilities.

Regulatory Environments

A *regulatory environment* is an environment in which an organization exists or operates that is controlled to a significant degree by laws, rules, or regulations implemented by government (federal, state, or local), industry groups, or other organizations. In a nutshell, when you have to play by someone else's rules or else risk serious consequences, you're dealing with a regulatory environment. Regulatory environments commonly have enforcement groups and procedures in place to deal with noncompliance.

You, as a cybersecurity analyst, may have to take action in a number of ways to ensure compliance with one or more regulatory requirements. A sometimes overlooked example is the type of contract that requires one of the parties to ensure that certain conditions are met with regard to information system security. It is not uncommon, particularly when dealing with the government, to be required to follow certain rules, such as preventing access by foreign nationals to certain information or ensuring that everyone working on the contract is trained on proper information-handling procedures.

In this section, we discuss some of the most important regulatory requirements with which you should be familiar in the context of vulnerability management. The following three standards cover the range of examples, from those that are completely optional to those that are required by law.

ISO/IEC 27001 Standard

The International Organization for Standardization (ISO) and the International Electrotechnical Commission (IEC) jointly maintain a number of standards, including ISO/IEC 27001, which covers Information Security Management Systems (ISMS). ISO/IEC 27001 is arguably the most popular voluntary security standard in the world and covers every important aspect of developing and maintaining good information security. One of its provisions, covered in control A.12.6.1, deals with vulnerability management in particular. Named "Management of Technical Vulnerabilities," this control, whose implementation is required for certification, essentially states that the organization has a documented process in place for timely identification and mitigation of known vulnerabilities.

ISO/IEC 27001 certification, which is provided by an independent certification body, is performed in three stages. First, a desk-side audit verifies that the organization has documented a reasonable process for managing its vulnerabilities. The second stage is an implementation audit aimed at ensuring that the documented process is actually being carried out. Finally, surveillance audits confirm that the process continues to be followed and improved upon.

Payment Card Industry Data Security Standard

The Payment Card Industry Data Security Standard (PCI DSS) applies to any organization involved in processing credit card payments using cards branded by the five major issuers: Visa, MasterCard, American Express, Discover, and JCB. Each of these organizations had its own vendor security requirements, so in 2006 they joined efforts and standardized these requirements across the industry. The PCI DSS is periodically updated and, as of this writing, is in version 3.2.

Requirement 11 of the PCI DSS deals with the obligation to "regularly test security systems and processes," and Section 2 describes the requirements for vulnerability scanning. Specifically, the requirement states that the organization must perform two types of vulnerability scans every quarter: internal and external. Internal scans must use qualified members of the organization, whereas external scans must be performed by approved scanning vendors (ASVs). It is important to know that the organization must be able to show that the personnel involved in the scanning have the required expertise to do so. Requirement 11 also states that both internal and external vulnerability scans must be performed whenever significant changes are made to the systems or processes.

Finally, PCI DSS requires that any "high-risk" vulnerabilities uncovered by either type of scan be resolved. After resolution, another scan is required to demonstrate that the risks have been properly mitigated.

Health Insurance Portability and Accountability Act

The Health Insurance Portability and Accountability Act (HIPAA) establishes penalties (ranging from $100 to $1.5 million) for covered entities that fail to safeguard protected health information (PHI). Though HIPAA does not explicitly call out a requirement to conduct vulnerability assessments, Section 164.308(a)(1)(i) requires organizations to conduct accurate and thorough vulnerability assessments and to implement security measures that are sufficient to reduce the risks presented by those assessed vulnerabilities to a reasonable level. Any organization that violates the provisions of this act, whether willfully or through negligence or even ignorance, faces steep civil penalties.

EXAM TIP   You do not have to memorize the provisions of ISO/IEC 27001, PCI DSS, or HIPAA, but you need to know that there are regulatory environments that require vulnerability management. Although these examples are intended to be illustrative of the exam requirement, being somewhat familiar with them will be helpful.

Corporate Security Policy

A *corporate security policy* is an overall general statement produced by senior management (or a selected policy board or committee) that dictates what role security plays within the organization. Security policies can be organizational, issue-specific, or system-specific. In an organizational security policy, management establishes how a security program will be set up, lays out the program's goals, assigns responsibilities, shows the strategic and tactical values of security, and outlines how enforcement should be carried out. An issue-specific policy, also called a *functional policy,* addresses specific security issues identified by management that need more detailed explanation and attention to make sure a comprehensive structure is built and all employees understand how they are to comply with these security issues. A system-specific policy presents the management's decisions that are specific to the actual computers, networks, and applications.

Typically, organizations will have an issue-specific policy covering vulnerability management, but it is important to note that this policy is nested within the broader corporate security policy and may also be associated with system-specific policies. The point is that it is not enough to understand the vulnerability management policy (or develop one if it doesn't exist) in a vacuum. We must understand the organizational security context within which this process takes place.

Data Classification

An important item of metadata that should be attached to all data is a classification level. This classification tag is important in determining the protective controls we apply to the information. The rationale behind assigning values to different types of data is that the values enable a company to gauge the resources that should go toward protecting each type of data, because not all of it has the same value to the company. There are no hard-and-fast rules on the classification levels an organization should use. Typical classification levels include the following:

- **Private**   Information whose improper disclosure could raise personal privacy issues
- **Confidential**   Data that could cause grave damage to the organization
- **Proprietary (or sensitive)**   Data that could cause some damage, such as loss of competitiveness to the organization
- **Public**   Data whose release would have no adverse effect on the organization

Each classification should be unique and separate from the others and should not have any overlapping effects. The classification process should also outline how information is controlled and handled throughout its lifecycle, from creation to termination. The following list shows some criteria parameters an organization might use to determine the sensitivity of data:

- The level of damage that could be caused if the data were disclosed
- The level of damage that could be caused if the data were modified or corrupted

- Lost opportunity costs that could be incurred if the data is not available or is corrupted
- Legal, regulatory, or contractual responsibility to protect the data
- Effects the data has on security
- The age of data

Asset Inventory

You cannot protect what you don't know you have. Though inventorying assets is not what most of us would consider glamorous work, it is nevertheless a critical aspect of managing vulnerabilities in your information systems. In fact, this aspect of security is so important that it is prominently featured at the top of the Center for Internet Security (CIS) list of Critical Security Controls (CSC). Critical control number 1 applies to the inventory of authorized and unauthorized devices, and number 2 deals with the software running on those devices.

Keep in mind, however, that an asset is anything of worth to an organization. Apart from hardware and software, this includes people, partners, equipment, facilities, reputation, and information. For the purposes of the CySA+ exam, we focus on hardware, software, and information. Determining the value of an asset can be difficult and is oftentimes subjective.

 NOTE Network appliances are computers that are specifically designed to perform one or more functions such as proxying network traffic or serving files. They will normally exhibit the same vulnerabilities as other servers, though it may be easier for IT staff to overlook the need to secure or patch them.

Servers

Perhaps the most common vulnerability on servers stems from losing track of a server's purpose on the network and allowing it to run unnecessary services and open ports. The default installation of many servers includes hundreds, if not thousands, of applications and services, most of which are not really needed for a server's main purpose. If this extra software is not removed, disabled, or at the very least hardened and documented, it may be difficult to secure the server.

Another common vulnerability is the misconfiguration of services. Most products offer many more features than what we actually need, but many of us simply ignore the "bonus" features and focus on configuring the critical ones. This can come back to haunt us if these bonus features allow attackers to gain a foothold easily by exploiting legitimate system features that we were not even aware of. The cure to this problem is to ensure that we know the full capability set of anything we put on our networks, and disable anything we don't need.

NOTE The term *shadow IT* is used to describe unmonitored and unsupported devices and services on a network. They can be problematic, because defenders often have no idea they exist on the network. Since a vulnerability management effort aims to identify and quantify risk, shadow IT and rogue devices must be considered in designing a scanning strategy.

Endpoints

Endpoints are almost always end-user devices (mobile or otherwise). They are the most common entry point for attackers into our networks, and the most common vectors are e-mail attachments and web links. In addition to the common vulnerabilities (especially updates/patches), the most common problem with endpoints is lack of up-to-date malware protection. This, of course, is the minimum standard. We should really strive to have more advanced, centrally managed, host-based security systems.

Another common vulnerability at the endpoint is system misconfiguration or default configurations. Though most modern operating systems pay attention to security, they oftentimes err on the side of functionality. The pursuit of a great user experience can sometimes come at a high cost. To counter this vulnerability, you should have baseline configurations that can be verified periodically by your scanning tools. These configurations, in turn, are driven by your organizational risk management processes as well as any applicable regulatory requirements.

Critical Assets

A *critical asset* is anything that is absolutely essential to performing the primary functions of your organization. If you work at an online retailer, for example, critical assets would include your web platforms, data servers, and financial systems, among other things. They probably wouldn't include the workstations used by your web developers or your printers. Critical assets clearly require a higher degree of attention when it comes to managing vulnerabilities. This attention can be expressed in a number of different ways, but you should focus on at least two: the thoroughness of each vulnerability scan and the frequency of each scan.

 EXAM TIP In the context of vulnerability management, the CySA+ exam will require only that you decide how you would deal with critical and noncritical assets.

Noncritical Assets

A *noncritical asset,* though valuable, is not required for the accomplishment of your main mission as an organization. You still need to include these assets in your vulnerability management plan, but given the limited resources with which we all have to deal, you would give them a lower priority than critical assets.

 EXAM TIP Every security decision—including how, when, and where to conduct vulnerability assessments—must consider the implications of these controls and activities on the core business of the organization.

Active vs. Passive Scanning

Scanning is a method used to get more details about the target network or device by poking around and taking note of the target's responses. Attackers have scanned targets attempting to find openings since the early days of the Internet, starting with a technique called *war dialing*. By using a device to dial sequentially and automatically through a list of phone numbers and listen to the response, an attacker could determine whether it was a human or a machine on the other side. Back when many computers were connected to the Web via unprotected modems, war dialing was the easiest and most effective method for gaining access to these systems. *Host scanning* remains an effective way to inventory and discover details of a system by sending a message and, based on the response, either classifying that system or taking further exploratory measures. Scanners generally come in three flavors: network mappers, host (or port) scanners, and web app vulnerability scanners.

Mapping/Enumeration

The goal of *network mapping* is to understand the topology of the network, including perimeter networks, demilitarized zones, and key network devices. The process used during network mapping is referred to as *topology discovery*. The first step in creating a network map is to find out what devices exist by performing a "sweep." As with the example of war dialing, network sweeping is accomplished by sending a message to each device and recording the response. A popular tool for this is Network Mapper, more commonly referred to as nmap. In executing a network sweep, nmap's default behavior is to send an ICMP echo request, a TCP SYN to port 443, a TCP ACK to port 80, and an ICMP timestamp request. A successful response to any of these four methods is evidence that the address is in use. nmap also has a traceroute feature that enables it to map out networks of various complexities using the clever manipulation of the time-to-live values of packets. After mapping a network, an adversary may have an inventory of the network but may want to fill in the details.

Port Scanning

Port scanners are programs designed to probe a host or server to determine what ports are open. They are an important tool for administrators. This method of enumerating the various services a host offers is one means of *service discovery*. It enables an attacker to add details to the broad strokes by getting insight into what services are running on a target. Because network-connected devices often run services on well-known ports such as 80 and 25, port scanning is a reliable source of information on these devices. Depending on the response time, the response type, and other criteria, the scanning software can identify what services are running—and some software can even provide *OS fingerprinting* to identify the device's operating system. However, identifying the OS isn't perfect, because the values that the software relies on for detection can change depending on the network configuration and other settings. With the information provided by the scanner, the attacker is in a better position to choose what kind of attack may be most effective.

 NOTE OS fingerprinting is not an exact science. You should not conclude that a host is running a given OS simply because the scanner identified it as such.

Web App Vulnerability Scanning

A *web application vulnerability scanner* is an automated tool that scans web applications to determine security vulnerabilities. Included in popular utilities are common tests, such as those for SQL injection, command injection, cross-site scripting, and improper server configuration. As usage of these applications has increased over the years, so has the frequency of attacks by their exploitation. These scanners are extremely useful because they automatically check against many types of vulnerabilities across many systems on a network. The scans are often based on a preexisting database of known exploits, so it's important to consider this when using these types of scanners. Although vulnerability scanners in the strictest definition don't offer anything beyond identification of existing vulnerabilities, some scanners offer additional correlation features or can extend their functionality using plug-ins and APIs.

Scanning Parameters and Criteria

With all these very particular vulnerabilities floating around, how often should we be checking for them? As you may have guessed, there is no one-size-fits-all answer to that question. The important issue to keep in mind is that the *process* is what matters. If you haphazardly do vulnerability scans at random intervals, you will have a much harder time answering the question of whether or not your vulnerability management is being effective. If, on the other hand, you do the math up front and determine the frequencies and scopes of the various scans given your list of assumptions and requirements, you will have much more control over your security posture.

Just as you must weigh a host of considerations when determining how often to conduct vulnerability scans, you also need to think about different but related issues when configuring your tools to perform these scans. Today's tools typically have more power and options than most of us will sometimes need. Our information systems may also impose limitations or requirements on which of these features can or should be brought to bear. When configuring scanning tools, you have a host of different considerations, but here we focus on the main ones you will be expected to know for the CySA+ exam. The list is not exhaustive, however, and you should probably grow it with issues that are specific to your organization or sector.

Risks Associated with Scanning Activities

The *risk appetite* of an organization is the amount of risk that its senior executives are willing to assume. You will never be able to drive risk down to zero, because there will always be a possibility that someone or something causes losses to your organization. What's more, as you try to mitigate risks, you will rapidly approach a point of diminishing returns. When you start mitigating risks, you will go through a stage in which a great many risks can be reduced with some common sense and inexpensive controls. After you start running out of low-hanging fruit, the costs (for example, financial and opportunity) will start rapidly increasing. You will then reach a point where further mitigation is fairly expensive. How expensive is "too expensive" is dictated by your organization's risk appetite.

When it comes to the frequency of vulnerability scans, it's not as simple as doing more if your risk appetite is low, or vice versa. Risk is a deliberate process that quantifies the likelihood of a threat being realized and the net effect it would have in the organization. Some threats, such as hurricanes and earthquakes, cannot be mitigated with vulnerability scans. Neither can the threat of social engineering attacks or insider threats. So the connection between risk appetite and the frequency of vulnerability scans requires that we dig into the risk management plan and see which specific risks require scans and then how often they should be done to reduce the residual risks to the agreed-upon levels.

Regulatory Requirements

If you thought the approach to determining the frequency of scans based on risk appetite was not very definitive, the opposite is true of regulatory requirements. Assuming you've identified all the applicable regulations, the frequencies of the various scans will be given to you. For instance, requirement 11.2 of the PCI DSS requires vulnerability scans (at least) quarterly as well as after any significant change in the network. HIPAA, on the other hand, imposes no such scanning frequency requirements. Still, to avoid potential problems, most experts agree that covered organizations should run vulnerability scans at least semiannually.

Technical Constraints

Vulnerability assessments require resources such as personnel, time, bandwidth, hardware, and software, many of which are likely limited in your organization. Of these, the top technical constraints on your ability to perform these tests are qualified personnel and technical capacity. Here, the term *capacity* is used to denote computational resources expressed in cycles of CPU time, bytes of primary and secondary memory, and bits per second (bps) of network connectivity. Because any scanning tool you choose to use will require a minimum amount of such capacity, you may be constrained in both the frequency and scope of your vulnerability scans.

If you have no idea how much capacity your favorite scans require, quantifying it should be one of your first next steps. It is possible that in well-resourced organizations such requirements are negligible compared to the available capacity. In such an environment, it is possible to increase the frequency of scans to daily or even hourly for high-risk assets. It is likelier, however, that your scanning takes a noticeable toll on assets that are also required for your principal mission. In such cases, you want to balance the mission and security requirements carefully so that one doesn't unduly detract from the other.

Workflow

Another consideration when determining how often you conduct vulnerability scanning is established workflows of security and network operations within your organization. As mentioned earlier, qualified personnel constitute a limited resource. Whenever you run a vulnerability scan, someone will have to review and perhaps analyze the results to determine what actions, if any, are required. This process is best incorporated into the workflows of your security and/or network operations centers personnel.

A recurring theme in this chapter has been the need to standardize and enforce repeatable vulnerability management processes. Apart from well-written policies, the next best way to ensure this happens is by writing it into the daily workflows of security and IT personnel. If you work in a security operation center (SOC) and know that every Tuesday morning your duties include reviewing the vulnerability scans from the night before and creating tickets for any required remediation, then you're much more likely to do this routinely. The organization, in turn, benefits from consistent vulnerability scans with well-documented outcomes, which, in turn, become enablers of effective risk management across the entire system.

Sensitivity Levels

Earlier in this chapter, we discussed the different classifications we should assign to our data and information, as well as the criticality levels of our other assets. We return now to these concepts as we think about configuring our tools to do their jobs while appropriately protecting our assets. When it comes to the information in our systems, we must take great care to ensure that the required protections remain in place at all times. For instance, if we are scanning an organization covered by HIPAA, we should ensure that nothing we do as part of our assessment in any way compromises PHI. We have seen vulnerability assessments that include proofs such as sample documents obtained by exercising a security flaw. Obviously, this is not advisable in the scenario we've discussed.

Besides protecting the information, we also need to protect the systems on which it resides. Earlier we discussed critical and noncritical assets in the context of focusing attention on the critical ones. Now we'll qualify that idea by saying that we should scan them in a way that ensures they remain available to the business or other processes that made them critical in the first place. If an organization processes thousands of dollars each second and our scanning slows that down by an order of magnitude, even for a few minutes, the effect could be a significant loss of revenue that might be difficult to explain to the board. Understanding the nature and sensitivity of these assets can help us identify tool configurations that minimize the risks to them, such as scheduling the scan during a specific window of time in which no trading occurs.

Vulnerability Feed

Unless you work in a governmental intelligence organization, odds are that your knowledge of vulnerabilities mostly comes from commercial or community feeds. These services have update cycles that range from hours to weeks and, though they eventually tend to converge on the vast majority of known vulnerabilities, one feed may publish a threat significantly before another. If you are running hourly scans, you would obviously benefit from the faster services and may be able to justify the higher cost. If, on the other hand, your scans are weekly, monthly, or even quarterly, the difference may not be as significant. As a rule of thumb, you want a vulnerability feed that is about as frequent as your own scanning cycle.

If your vulnerability feed is not one with a fast update cycle, or if you want to ensure you are absolutely abreast of the latest discovered vulnerabilities, you can (and perhaps should) subscribe to alerts in addition to those of your provider. The National Vulnerability Database (NVD) maintained by the National Institute of Standards and Technologies (NIST) provides two Rich Site Summary (RSS) feeds, one of which will alert you to any new vulnerability reported, and the other provides only those that have been analyzed. The advantage of the first feed is that you are on the bleeding edge of notifications. The advantage of the second is that it provides you with specific products that are affected as well as additional analysis. A number of other organizations provide similar feeds that you should probably explore as well.

Assuming you have subscribed to one or more feeds (in addition to your scanning product's feed), you will likely learn of vulnerabilities in between programmed scans. When this happens, you will have to consider whether to run an out-of-cycle scan that looks for that particular vulnerability or to wait until the next scheduled event to run the test. If the flaw is critical enough to warrant immediate action, you may have to pull from your service provider or, failing that, write your own plug-in to test the vulnerability. Obviously, this would require significant resources, so you should have a process by which to make decisions like these as part of your vulnerability management program.

Scope

Whether you are running a scheduled or a special scan, you have to define its scope carefully and configure your tools appropriately. Though it would be simpler to scan everything at once at set intervals, the reality is that this is oftentimes not possible simply because of the load this places on critical nodes, if not the entire system. What may work better is to have a series of scans, each with a different scope and parameters.

Whether you are doing a global or targeted scan, your tools must know which nodes to test and which ones to leave alone. The set of devices that will be assessed constitutes the scope of the vulnerability scan. Deliberately scoping these events is important for a variety of reasons, but one of the most important ones is the need for credentials, which we discuss next.

Noncredentialed vs. Credentialed

A noncredentialed vulnerability scan evaluates the system from the perspective of an outsider, such as an attacker just beginning to interact with a target. This is a sort of black-box test in which the scanning tool doesn't get any special information or access into the target. The advantage of this approach is that it tends to be quicker, while still being fairly realistic. It may also be a bit more secure, because there is no need for additional credentials on all tested devices. The disadvantage, of course, is that you will most likely not get full coverage of the target.

 NOTE Noncredentialed scans look at systems from the perspective of the attacker but are not as thorough as credentialed scans.

To know all that is vulnerable in a host, you typically need to provide the tool with credentials so it can log in remotely and examine the inside as well as the outside. Credentialed scans will always be more thorough than noncredentialed ones, simply because of the additional information that a login provides the tool. Whether or not this additional thoroughness is important to you is for you and your team to decide. An added benefit of credentialed scans is that they tend to reduce the amount of network traffic required to complete the assessment.

NOTE It is very rare to need full domain admin credentials to perform a vulnerability scan. If you are doing credentialed scans, you should avoid using privileged accounts unless you are certain you cannot otherwise meet your requirements.

Server Based vs. Agent Based

Vulnerability scanners tend to fall into two classes of architectures: those that require a running process (agent) on every scanned device, and those that do not. The difference is illustrated in Figure 3-1. A server-based (or agentless) scanner consolidates all data and processes on one or a small number of scanning hosts, which depend on a fair amount of network bandwidth to run their scans. It has fewer components, which could make maintenance tasks easier and help with reliability. Additionally, it can detect and scan devices that are connected to the network but do not have agents running on them (for example, new or rogue hosts).

Agent-based scanners have agents that run on each protected host and report their results back to the central scanner. Because only the results are transmitted, the bandwidth required by this architectural approach is considerably less than a server-based solution. Also, because the agents run continuously on each host, mobile devices can still be scanned even when they are not connected to the corporate network.

NOTE While both agent-based and agentless scanners are suitable for determining patch levels and missing updates, agent-based (or serverless) vulnerability scanners are typically better for scanning mobile devices.

Figure 3-1
Server-based and agent-based vulnerability scanner architectures

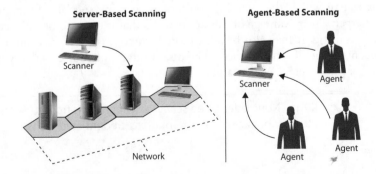

Internal vs. External

Many modern vulnerability scanning services allow for scanning of IP addresses on your network from an outside location or from within the corporate network. Internal scanners usually use appliances located within the network to perform the scanning activity, taking advantage of their privileged position to gain visibility of devices across the network. With an external scan, it's possible to get a sense of what vulnerabilities exist from an outsider's point of view. In some cases, a larger organization may use external scanners combined with special access control rules to allow for elevated visibility into internal assets. In this way, they act as internal scanners, but there is no need to carve out internal resources.

Types of Data

As you configure your scanning tool, you must consider the information that should or must be included in the report, particularly when dealing with regulatory compliance scans. This information will drive the data that your scan must collect, which in turn affects the tool configuration. Keep in mind that each report (and there may be several as outputs of one scan) is intended for a specific audience. This affects both the information in the report as well as the manner in which it is presented.

Tool Updates and Plug-Ins

Vulnerability scanning tools work by testing systems against lists of known vulnerabilities. These flaws are frequently being discovered by vendors and security researchers. It stands to reason that if you don't keep your vulnerabilities lists up-to-date, whatever tool you use will eventually fail to detect vulnerabilities that are known by others, especially your adversaries. This is why it is critical to keep an eye on current vulnerabilities lists.

A vulnerability scanner plug-in is a simple program that looks for the presence of one specific flaw. In Nessus, plug-ins are coded in the Nessus Attack Scripting Language (NASL), which is a very flexible language able to perform virtually any check imaginable. Figure 3-2 shows a portion of a NASL plug-in that tests for FTP servers that allow anonymous connections.

SCAP

How do you ensure that your vulnerability management process complies with all relevant regulatory and policy requirements *regardless of which scanning tools you use?* Each tool, after all, may use whatever standards (such as rules) and reporting formats its developers desire. This lack of standardization led NIST to team up with industry partners to develop the Security Content Automation Protocol (SCAP). SCAP uses specific standards for the assessment and reporting of vulnerabilities in the information systems of an organization. Currently in version 1.3, SCAP incorporates about a dozen different components that standardize everything from an asset reporting format (ARF), to Common Vulnerabilities and Exposures (CVE), to the Common Vulnerability Scoring System (CVSS).

At its core, SCAP leverages baselines developed by NIST and its partners that define minimum standards for vulnerability management. If, for instance, you want to ensure that your Windows 10 workstations are complying with the requirements of the Federal Information Security Management Act (FISMA), you would use the appropriate SCAP

```
#
# The script code starts here :
#

include("ftp_func.inc");

port = get_kb_item("Services/ftp");
if(!port)port = 21;

if (get_kb_item('ftp/'+port+'/backdoor')) exit(0);

state = get_port_state(port);
if(!state)exit(0);
soc = open_sock_tcp(port);
if(soc)
{
 domain = get_kb_item("Settings/third_party_domain");
 r = ftp_log_in(socket:soc, user:"anonymous", pass:string("nessus@", domain));
 if(r)
 {
  port2 = ftp_get_pasv_port(socket:soc);
  if(port2)
  {
   soc2 = open_sock_tcp(port2, transport:get_port_transport(port));
   if (soc2)
   {
    send(socket:soc, data:'LIST /\r\n');
    listing = ftp_recv_listing(socket:soc2);
    close(soc2);
    }
  }

  data = "
This FTP service allows anonymous logins. If you do not want to share data
with anyone you do not know, then you should deactivate the anonymous account,
since it may only cause troubles.
```

Figure 3-2 NASL script that tests for anonymous FTP logins

module that captures these requirements. You would then provide that module to a certi-fied SCAP scanner (such as Nessus), and it would be able to report this compliance in a standard language. As you should be able to see, SCAP enables full automation of the vulnerability management process, particularly in regulatory environments.

Special Considerations

Apart from the considerations in a credentialed scan already discussed, the scanning tool must have the correct permissions on whichever hosts it is running, as well as the necessary access across the network infrastructure. It is generally best to have a dedicated account for the scanning tool or, alternatively, to execute it within the context of the user respon-sible for running the scan. In either case, minimally privileged accounts should be used to minimize risks (that is, do not run the scanner as root unless you have no other choice).

Network access is also an important configuration, not so much of the tool as of the infrastructure. Because the vulnerability scans are carefully planned beforehand, it should be possible to examine the network and determine what access control lists (ACLs), if any, need to be modified to allow the scanner to work. Similarly, network intrusion detection systems (IDSs) and intrusion prevention systems (IPSs) may trigger on the scanning activity unless they have been configured to recognize it as legitimate. This may also be true for host-based security systems (HBSSs), which might attempt to mitigate the effects of the scan.

Finally, the tool is likely to include a reporting module for which the right permissions must be set. It is ironic that some organizations deploy vulnerability scanners but fail to secure the reporting interfaces properly. This allows users who should be unauthorized to access the reports at will. Although this may seem like a small risk, consider the consequences of adversaries being able to read your vulnerability reports. This ability would save them significant effort, because it would enable them to focus on the targets you have already listed as vulnerable. As an added bonus, they would know exactly how to attack the hosts.

Intrusion Prevention System, Intrusion Detection System, and Firewall Settings

Here's a tricky situation: the purpose of a security device is to block or redirect unwanted network traffic, while the purpose of a vulnerability scanner is to detect weaknesses in a system, sometimes using aggressive techniques that resemble attacks. See the problem here? Firewalls, IDS, and IPS devices can pose a serious problem for network vulnerability assessment efforts if they are not correctly configured to allow incoming traffic from those devices. It's generally good practice for the vulnerability management team to publish its schedule of scans along with source device information so that the detection team can whitelist those devices.

Generating Reports

Report generation is an important part of the incident response process and is particularly critical for vulnerability management. All vulnerability scanners perform reporting functions of some kind, but they don't all come with customization options. Nessus provides its reports in common formats such as PDF, HTML, and CSV. Additionally, you can also use Nessus's own formats.

As an administrator, it's important that you consider what kinds of reporting your utility is capable of and how you may automate the reporting process. Getting the pertinent information to the right people in a timely fashion is the key to capitalizing successfully on vulnerability scans.

Automated vs. Manual Distribution

Creating reporting templates enables you to rapidly prepare customized reports based on vulnerability scan results, which can then be forwarded to the necessary points of contact. For example, you can have all the web server vulnerabilities automatically collected and sent to the web server administrator. Similarly, you can have occurrences of data storage violations sent to your spillage team for faster action. Unless there is only one administrator, it may make sense to automate the report delivery process to alleviate the primary administrator from having to manage every report manually. The service administrators can be more efficient because they are getting the reports that are most relevant to their role.

Validation

No automated vulnerability report is ever perfectly accurate. It is up to the analyst to review and make sense of it before passing it on to others in the organization. The two most important outcomes of the review process are to determine the validity of reported vulnerabilities and to determine exceptions to policies. The aim is to have the most accurate information about your network, because it means more confidence in the decisions made by your technical staff and company leadership. With vulnerabilities accurately identified and the most appropriate courses of action developed and refined through open lines of communication, you can prioritize responses that have minimal impact throughout the company.

True Positives

Armed with the feedback from vulnerability scan reports, you may find verifying the results a straightforward process. Figure 3-3 shows the output for an uncovered vulnerability on a Windows host located at 10.10.0.115 that's related to the Remote Desktop functionality.

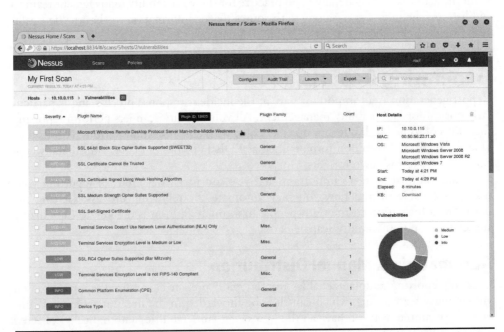

Figure 3-3   Details on a vulnerability in the Remote Desktop protocol on a Windows host

Figure 3-4
The Remote
Desktop options
in the System
Properties dialog
for a Windows 10
host

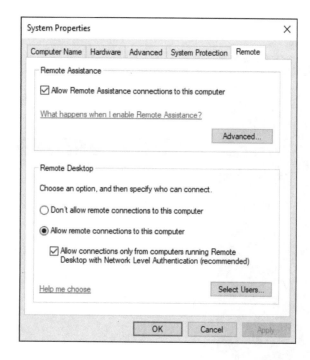

The protocol was found to have a weakness in its implementation of the cryptographic exchange during identity verification. As a solution, Nessus suggests that we either force the use of SSL for the service or enable Network Level Authentication. When we check the System properties of the same Windows host identified in Figure 3-4, we can see that the option for the use of Network Level Authentication is available for us to select.

We see that the vulnerability scanner successfully identified the less-secure state of the Windows host. Fortunately for us, we don't have to manually verify and adjust for every occurrence; this can all be automated by enforcing a new group policy or by using any number of automated remediation solutions.

NOTE The goal of major databases such as the Open Source Vulnerability Database (OSVDB) and the NVD is to publish CVEs for public awareness. These databases are incredibly useful but do not always have complete information on vulnerabilities, because many are still being researched. Therefore, you should use supplemental sources in your research, such as the Bugtraq, OWASP, and CERT.

Compare to Best Practices or Compliance

Several benchmarks for industry, academia, and government are available for you use to improve your network's security. On military networks, the most widely used set of standards is developed by the Defense Information Systems Agency (DISA). Its Security Technical Implementation Guides (STIGs), combined with the National Security Agency

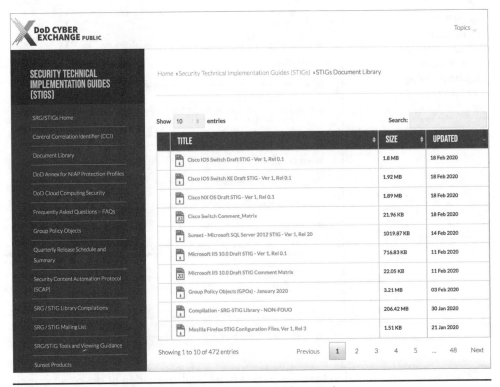

Figure 3-5 DISA Security Technical Implementation Guides portal

(NSA) guides, are the configuration standards used on US Department of Defense (DoD) information systems. Figure 3-5 shows the DISA STIG public portal, with a small sample of the latest guides.

STIGs provide the technical steps required to harden network devices, endpoints, and software. Note that although many STIG benchmarks are available to the public, some require a DoD Public Key Infrastructure (PKI) certificate, such as those found in Common Access Cards (CACs), for access. Using the SCAP specification, you can apply these standards or monitor for compliance across your network.

Reconcile Results

If there's one thing that's certain in incident response and forensic analysis, it's that taking thorough notes will make your job much easier in the end. These notes include the steps you take to configure a device, validate its configuration, verify its operation, and of course test vulnerabilities. Taking notes on how you uncovered and dealt with a vulnerability will aid in continuity, and it might be required based on the industry in which you operate. Both Nessus and OpenVAS provide ways to track how the corrective action performs on network devices. Should your network activity be examined by an investigation, it's also good to know you've taken thorough notes about every action you performed to make the network safer.

Review Related Logs and/or Other Data Sources

When reviewing the report, you should also review event logs and network data. You can compare running services, listening ports, and open connections against a list of authorized services to identify any abnormal behavior. Correlating the vulnerability scan output with historical network and service data serves several functions. First, it verifies that your logging mechanism is capturing the activities related to the vulnerability scans, because these scans will often trigger logging. Second, you should be able to see changes in the network based on the patches or changes you've made because of compensating controls. Finally, the logs may give insight into whether any of the uncovered vulnerabilities have been acted upon already. Security information and event management (SIEM) tools can assist tremendously with validation, because they will likely be able to visualize all the scanning activity. And because you are likely already ingesting other log data, these tools provide a useful place to begin correlation.

Determine Trends

Using either the built-in trending functionality or with help from other software, you can track how vulnerabilities in the network have changed over time. Trending improves context and enables your security response team to tailor its threat mitigation strategies to its efforts more efficiently. Additionally, you can also determine whether any of your solutions are taking hold and are effective. Linking the vulnerability scanners with existing SIEM platforms isn't the only option; you can also track progress on fixing problems using existing trouble ticket software. This helps with the internal tracking of the issues and allows for visibility from leadership, in case outside assistance is required to enforce a policy change.

False Positives

False positives with vulnerability scanners are particularly frustrating, because the effort required to remediate a suspected issue can be resource intensive. A false positive rate of 2 percent may not be a problem for smaller organizations, but the same rate on a large network with thousands of endpoints will cause significant problems for the security staff. Although it's important that you quickly produce a solution to an uncovered vulnerability, you should take a moment to consider the reasons why a scanner might cry wolf. Sometimes, the logic that a check, NVT, or plug-in uses is flawed, resulting in a report of a vulnerability that does not exist.

Understand that these tests are authored with certain assumptions about a system, because it is impossible to write logic in such a way that it applies perfectly to every system. There is no way for the authors of a vulnerability test to know the details of your network, so they must create rules that are sometimes less granular, which may lead to false positives. In this case, it may be useful for you to customize your own test after a false positive has been discovered. Another reason for a false positive could be that you've already determined the appropriate compensating control for an issue but have not correctly disposed of the alert.

NOTE Although vulnerability scanners have improved over the years, OS and software detection in vulnerability scanners isn't perfect. This makes detection in environments with custom operating systems and devices particularly challenging. Many devices use lightweight versions of Linux and Apache web server that are burned directly onto the device's read-only memory. You should expect to get a higher number of alerts in these cases because the vulnerability scanner may not be able to tell exactly what kind of system it is. However, you should also take care not to dismiss alerts immediately on these systems either. Sometimes a well-known vulnerability may exist in unexpected places because of how the vulnerable software packages were ported over to the new system.

True Negatives

Despite how it may sound, a *true negative* is actually a good thing for vulnerability scanners. In short, a true negative outcome indicates that there is no vulnerability reported and no vulnerability exists. It's *very* important, however, that you understand that proving a true negative is almost impossible. Without getting into a deep discussion about the intricacies of informal logic, we can look to the old aphorism for assistance: "Absence of evidence is not evidence of absence."

False Negatives

A false negative, also referred to as a Type II error, is a result that indicates that no vulnerability is present when, in fact, a vulnerability does actually exist. Reasons behind this outcome include a lack of technical capability to detect the vulnerability. It could well be that a vulnerability is too new and that detection rules for the scanner do not yet exist. Or perhaps an incorrect type of vulnerability scan was initiated by the analyst. As troublesome as false positives are in terms of the effort expended to prove they are false, a false negative is far worse, because there a vulnerability actually exists but is undetected, and therefore it will not be remediated and will be possibly exploitable.

Remediation

When the scanner uncovers a vulnerability, it provides as much information about it as possible. Figure 3-6 shows the detail screen provide by OpenVAS on an uncovered vulnerability. This screen shows a summary of the vulnerability, the location of the resource of concern, the vulnerability's impact, how it was discovered, and any solutions or workarounds.

Remediation of network vulnerabilities should be done as quickly as possible after discovery, but not so haphazardly as to make the situation worse. Effective remediation requires a continuous examination for vulnerabilities combined with a thoughtful process to remedy problems and keep the organization's resources confidential and accessible. The simplest mechanism for verifying remediation is to compare consecutive

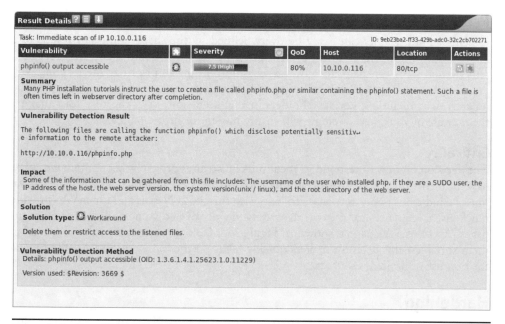

Figure 3-6 OpenVAS vulnerability detail screenshot

vulnerability scans to determine that vulnerabilities were addressed via software patches or upgrades. Many vendors offer solutions to manage the whole process of patching in a separate practice, patch management. Should a compensating control be used, you can provide feedback into the vulnerability management system by adding a note to the report, or you can override the alert. This helps document the actions you took to address the problem if you are unable to apply an update.

Patching

Patching is a necessary evil for administrators. You don't want to risk network outages due to newly introduced incompatibilities, but you also don't need old software being exploited because of your reservations about patching. For many years, vendors tried to ease the stress of patching by regularly releasing their updates in fixed intervals. Major vendors such as Microsoft and Adobe got into a rhythm with issuing updates for their products. Admins could therefore make plans to test and push updates with some degree of certainty. Microsoft recently ended its practice of "Patch Tuesday," however, in part due to this predictability. A downside of the practice emerged as attackers began reverse engineering the fixes as soon as they were released to determine the previously unknown vulnerabilities. Attackers knew that many administrators wouldn't be able to patch all their machines before they figured out the vulnerability, and thus the moniker "Exploit Wednesday" emerged. Although darkly humorous, it was a major drawback that convinced the company to focus instead on improving its automatic update features.

Prioritizing

Systems administrators can easily be overwhelmed by the sheer volume of results of a vulnerability scan. This is where prioritization of the vulnerabilities and the associated remediation steps can help. Ideally, a discussion on how to prioritize the response will include the capabilities of the technical staff as well as the overall business goals of the organization. Including key stakeholders in the discussion, or at the very least making them aware of your methodology, will ensure buy-in for future policy changes.

Criticality

The decision on how to respond to the results of a scan is driven by economics; we either have limited time, money, or personnel that can be used to remediate an issue, so we must be judicious in our response. To help with this, both Nessus and OpenVAS provide quick visual references for the overall severity of a discovered vulnerability on their result pages. An OpenVAS page is shown in Figure 3-7. Color-coding the results and making them sortable helps decision-makers quickly focus on the most critical issues and make the best use of limited resources.

Hardening

The application of the practices in this chapter is part of an effort to make the adversary work harder to discover what our network looks like, or to make it more likely that his efforts will be discovered. This practice of hardening the network is not static and requires constant monitoring of network and local resources. It's critical that we take

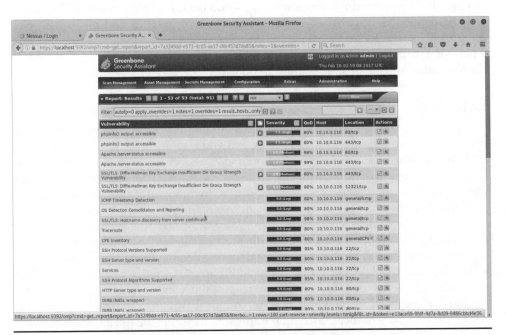

Figure 3-7   Results of an OpenVAS vulnerability scan, sorted by severity

a close look at what we can do at the endpoint to avoid only relying on detection of an attack. Hardening the endpoint requires careful thought about the balance of security and usability. We cannot make the system so unusable that no work can be done. Thankfully, there are some simple rules to follow:

- Resources should be accessed only by those who need them to perform their duties. Administrators also need to have measures in place to address the risk associated with a requirement if they are not able to address the requirement directly.

- Hosts should be configured with only the necessary applications and services necessary for their role. Does the machine need to be running a web server, for example? Using policies, these settings can be standardized and pushed to all hosts on a domain.

- Updates should be applied early and often. In addition to proving functional improvements, many updates come with patches for recently discovered flaws.

Usability vs. Security

Sometimes, the more secure we try to make a system, the less secure it becomes in practice. Think about it: If you make password requirements too difficult to remember for users, they'll start writing their passwords on sticky notes and storing them under keyboards and on monitors for easy access. Without other factors to assist in the authentication process, this process hardly makes for a more secure system. It may be hard to imagine a world without passwords: no complex string of letters and numbers to remember and almost instant access to systems can mean an immensely usable but woefully insecure system. However, usability and security are not mutually exclusive concepts. An entire field of information security is dedicated to improving the usability of systems while maintaining a high level of confidentiality, integrity, and availability. As we work toward improving usability while maintaining high security, it's important that we understand the tradeoffs in our current systems so that we can make smart decisions about our security policy.

We're always amazed at how many consumer products designed to be connected to a network leave unnecessary services running, sometimes with default credentials. Some of these services are shown in Figure 3-8. If you do not require a service, it should be disabled. Running unnecessary services not only means more avenues of approach for an adversary, but it also creates more work for the administrator. Services that aren't used are essentially wasted energy because the computers will still run the software, waiting for a connection that is not likely to happen by a legitimate user.

 EXAM TIP The UDP and TCP ports between 0 and 1023 are known as the *well-known ports* because they are used for commonly used services. Some notable well-known ports are 20 (FTP), 22 (SSH), 25 (SMTP), and 80 (HTTP). Ports 1024 to 49151 are *registered* ports, and ports above 49151 are *ephemeral* or *dynamic* ports.

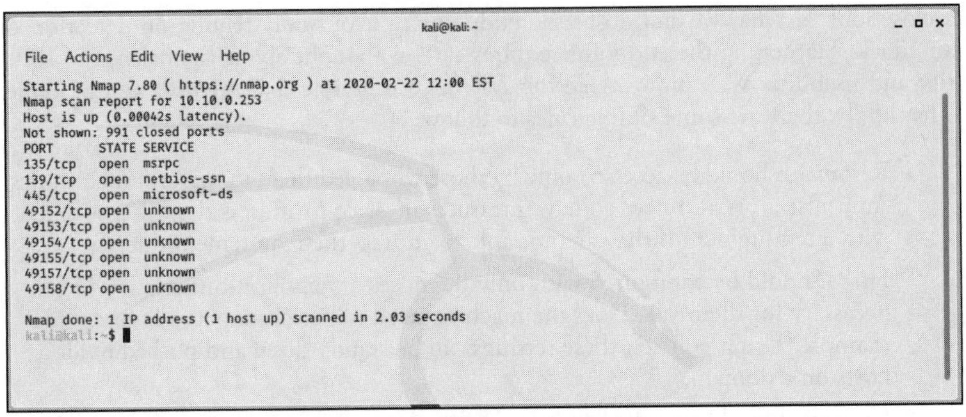

```
                                           kali@kali: ~                              _ □ ×

 File  Actions  Edit  View  Help

 Starting Nmap 7.80 ( https://nmap.org ) at 2020-02-22 12:00 EST
 Nmap scan report for 10.10.0.253
 Host is up (0.00042s latency).
 Not shown: 991 closed ports
 PORT       STATE SERVICE
 135/tcp    open  msrpc
 139/tcp    open  netbios-ssn
 445/tcp    open  microsoft-ds
 49152/tcp open  unknown
 49153/tcp open  unknown
 49154/tcp open  unknown
 49155/tcp open  unknown
 49157/tcp open  unknown
 49158/tcp open  unknown

 Nmap done: 1 IP address (1 host up) scanned in 2.03 seconds
 kali@kali:~$ ▉
```

Figure 3-8　Listing of open ports on an unprotected Windows 8 machine

Compensating Controls

You may encounter challenges in remediation because of the difficulty in implementing a solution. If there is a significant delay in implementation because of technical reasons or cost, you should still work to achieve the goals of a security requirement using some compensating control. Compensating controls are any means used by organizations to achieve the goals of a security requirement even if they were unable to meet the goals explicitly because of some legitimate external constraint or internal conflict. Here's an example: A small business processes credit card payments on its online store and is therefore subject to PCI DSS. This business uses the same network for both sensitive financial operations and external web access. Although best practices would dictate that physically separate infrastructures be used, with perhaps an air gap between the two, this may not be feasible because of cost. A compensating control would be to introduce a switch capable of VLAN management and to enforce ACLs at the switch and router levels. This alternative solution would "meet the intent and rigor of the original stated requirement," as described in the PCI DSS standard.

Risk Acceptance

Every vulnerability management plan should include a strategy for dealing with accepted vulnerabilities. Business priorities, resource restrictions, or other issues may dictate that risk mitigation wait for another time. Although such risk acceptance is often reasonable and a fact of security life, accepted vulnerabilities should not simply be ignored. The security team should lead the process for regularly inventorying open vulnerabilities and the systems that are impacted by them and understand how these risks may change over time. Should a vulnerability somehow become exposed to an attacker who has the intent and the means to launch an attack, addressing the issue becomes a priority and may transition to a response effort.

Verification of Mitigation

Security controls may fail to protect information systems against threats for a variety of reasons. If the control is improperly installed or configured, or if you chose the wrong control to begin with, the asset will remain vulnerable. (You just won't know it.) For this reason, you should create a formal procedure that describes the steps by which your organization's security staff will verify and validate the controls they use. Verification is the process of ensuring that the control was implemented correctly. Validation ensures that the (correctly installed) control actually mitigates the intended threat.

Inhibitors to Remediation

Even a solid plan for remediation that has stakeholder buy-in sometimes faces obstacles. Many of the challenges arise from processes that have major dependencies on the IT systems or from a stale policy that fails to address the changing technological landscape adequately. In this section, we cover some common obstacles to remediation and how we can avoid them.

Memorandum of Understanding

The *memorandum of understanding* (MOU) outlines the duties and expectations of all concerned parties. As with a penetration test, a vulnerability scan should have a clearly defined scope, along with formal rules of engagement that dictate above all else what can be done during the assessment and in the event of a vulnerability discovery. For example, conducting a scan on production systems during times of high usage would not be suitable. There might also be a situation that, without a formal MOU in place, would leave too much ambiguity. It wouldn't be hard to imagine that the discovery of a vulnerability on your network may have implications on an adjacent network not controlled by you. Also, a misconfiguration of an adjacent network may have a direct impact on your organization's services. In either case, a MOU that covers such conditions will clarify how to proceed in a way that's satisfactory for everyone involved.

Service Level Agreement

Many IT service providers perform their services based on an existing service level agreement (SLA) between them and the service recipient. An SLA is a contract that can exist within a company (say, between a business unit and the IT staff) or between the organization and an outside provider. SLAs exist to outline the roles and responsibilities for the service providers, including the limits of the services they can perform. Unless remediation is explicitly part of an SLA, providers cannot be compelled to perform those steps.

Organizational Governance

Corporate governance is the system of processes and rules an organization uses to direct and control its operations. Corporate governance aims to strike a sensible balance among the competing priorities of company stakeholders. In some cases, governance may interrupt the application of remedial steps, because those actions may negatively affect other business areas. This highlights the importance of communicating your actions with corporate leadership so that they can factor in the effects of remedial action with other issues to make a decision. Strong communication enables timely decision-making in the best interest of the company.

Business Process Interruption

There's never a good time to apply a patch or take other remedial actions. Highly efficient business and industrial processes such as just-in-time manufacturing have enabled businesses to reduce process time and increase overall efficiency. Underpinning these systems are production IT systems that themselves are optimized to the business. A major drawback, however, is that some systems may be more susceptible to disruption because of their optimized states. This fear of unpredictably or instability in the overall process is often enough for company leadership to delay major changes to production systems, or to avoid them altogether.

Degrading Functionality

Although there's no equivalent to the Hippocratic Oath in network administration, we must always try to "do no harm" to our production systems. Sometimes the recommended treatment, such as quarantining key systems after critical vulnerabilities are discovered, may be deemed unacceptable by leadership. How much risk you and your leadership are willing to underwrite is a decision for your organization, but you should aim to have the most accurate information possible about the state of every vulnerability. If you discover that an important remedial action breaks critical applications in a test environment, the alternative isn't to avoid patching. Rather, you must devise other mitigating controls to address the vulnerabilities until a suitable patch can be developed.

Legacy and Proprietary Systems

Legacy and proprietary system are a unique challenge to manage. Often, the equipment is older than many of the analysts on staff or may have been built in such a complicated manner that businesses choose to keep them on board instead of replacing them completely. The equipment still works, but an upgrade is out of the question because of complex upgrade procedures, unique connectivity requirements, or the immense cost of making any kind of changes. What's more, some specialized systems have their own proprietary communication protocols, and keeping tabs on these is a challenge in itself. Such systems often leave vulnerabilities unpatched or lack modern security features such as Secure Socket Layer (SSL) or Transport Layer Security (TLS). Though it's easy to think "if it ain't broke don't fix it," simply repeating this does nothing to reduce risk. Addressing the challenges of having these systems on the network takes a focused and comprehensive approach that includes constant network monitoring, compensating controls, and multiple layered security mechanisms.

Ongoing Scanning and Continuous Monitoring

Where feasible, you should schedule automated vulnerability scanning to occur daily. Depending on the types of networks you operate and your security policies, you may opt to perform scans more often, always using the most updated version of the scanning tool. You should pay extra attention to critical vulnerabilities and aim to remediate them within 48 hours. Recognizing that maintaining software, libraries, and reports can be tedious for administrators, some companies have begun to offer web-based scanning solutions. Qualys and Tenable, for example, both provide cloud-enabled web application security scanners that can be run from any number of cloud service providers. Promising increased scalability and speed across networks of various sizes, these companies provide several related services based on subscription tiers.

Chapter Review

This chapter has focused on developing deliberate, repeatable vulnerability management processes that satisfy all the internal and external requirements. The goal is that you, as a cybersecurity analyst, will be able to ask the right questions and develop appropriate approaches to managing the vulnerabilities in your information systems. Vulnerabilities, of course, are not all created equal, so you have to consider the sensitivity of your information and the criticality of the systems on which it resides and is used. As mentioned repeatedly, you will never be able to eliminate every vulnerability and drive your risk to zero. What you can and should do is assess your risks and mitigate them to a degree that is compliant with applicable regulatory and legal requirements and consistent with the risk appetite established by your executive leaders.

You can't do this unless you take a holistic view of your organization's operating environment and tailor your processes, actions, and tools to your particular requirements. Part of this involves understanding the common types of vulnerabilities associated with the various components of your infrastructure. You also need to understand the internal and external requirements to mitigating the risks of flaws. Finally, you need to consider the impact on your organization's critical business processes—that is, the impact of both the vulnerabilities and the processes of identifying and correcting them. After all, no organization exists for the purpose of running vulnerability scans on its systems, but these assessments are required in support of the organization's mission.

Questions

1. What popular framework aims to standardize automated vulnerability assessment, management, and compliance level?

   A. CVSS

   B. SCAP

   C. CVE

   D. PCAP

2. An information system that might require restricted access to, or special handling of, certain data as defined by a governing body is referred to as a what?

   A. Compensating control

   B. International Organization for Standardization (ISO)

   C. Regulatory environment

   D. Production system

3. Which of the following are parameters that organizations should *not* use to determine the classification of data?

   A. The level of damage that could be caused if the data were disclosed

   B. Legal, regulatory, or contractual responsibility to protect the data

   C. The age of data

   D. The types of controls that have been assigned to safeguard it

4. What is the term for the amount of risk an organization is willing to accept in pursuit of its business goals?

   A. Risk appetite

   B. Innovation threshold

   C. Risk hunger

   D. Risk ceiling

5. Insufficient storage, computing, or bandwidth required to remediate a vulnerability is considered what kind of constraint?

   A. Organizational

   B. Knowledge

   C. Technical

   D. Risk

6. What is a reason that patching and updating occur so infrequently with ICS and SCADA devices?

   A. These devices control critical and costly systems that require constant uptime.

   B. These devices are not connected to networks, so they do not need to be updated.

   C. These devices do not use common operating systems, so they cannot be updated.

   D. These devices control systems, such as HVAC, that do not need security updates.

7. All of the following are important considerations when deciding the frequency of vulnerability scans *except*:

   A. Security engineers' willingness to assume risk

   B. Senior executives' willingness to assume risk

   C. HIPAA compliance

   D. Tool's impact on business processes

Use the following scenario to answer Questions 8–10:

A local hospital has reached out to your security consulting company because it is worried about recent reports of ransomware used on hospital networks across the country. The hospital wants to get a sense of what weaknesses exist on its network and wants your guidance on the best security practices for its environment. The hospital has asked you to assist with its vulnerability management policy and provided you with some information about its network. The hospital provides laptops to staff members, and each device can be configured using a standard baseline. However, the hospital is not able to provide a smartphone to everyone and allows user-owned devices to connect to the network.

8. What kind of vulnerability scanner architecture do you recommend be used in this environment?

    A. Zero agent

    B. Server-based

    C. Agent-based

    D. Network-based

9. Which vulnerabilities would you expect to find mostly on the hospital's laptops?

    A. Misconfigurations in IEEE 802.1X

    B. Fixed passwords stored in plaintext in the file shares

    C. Lack of VPN clients

    D. Outdated malware signatures

10. Which of the following is *not* a likely reason you would use to prohibit user-owned devices from accessing the network?

    A. The regulatory environment explicitly prohibits these kinds of devices.

    B. There are concerns about staff recruiting and retention.

    C. There is no way to enforce who can have access to the device.

    D. The organization has no control over what else is installed on the personal device.

Answers

1. **B.** The Security Content Automation Protocol (SCAP) is a method of using open standards called components to identify software flaws and configuration issues.

2. **C.** In a regulatory environment, the way an organization exists or operates is controlled by laws, rules, or regulations put in place and required by a governing body.

3. **D.** Although there are no fixed rules regarding the classification levels that an organization uses, some common criteria parameters used to determine the sensitivity of data include the level of damage that could be caused if the data were disclosed; legal, regulatory, or contractual responsibility to protect the data; and the age of data. The classification should determine the controls used and not the other way around.

4. **A.** Risk appetite is a core consideration when determining your organization's risk management policy and guidance and will vary based on factors such as criticality of production systems, impact to public safety, and financial concerns.

5. **C.** Any limitation on the ability to perform a task on a system due to limitations of technology is a technical constraint and must have acceptable compensating controls in place.

6. **A.** The cost involved and potential negative effects of interrupting business and industrial processes often dissuade these device managers from updating and patching these systems.

7. **A.** An organization's risk appetite, or amount of risk it is willing to take, is a legitimate consideration when determining the frequency of scans. Risk appetite is not determined by security engineers; only executive leadership can make that determination.

8. **C.** Because every laptop has the same software baseline, an agent-based vulnerability scanner is a sensible choice. Agent-based scanners have agents that run on each protected host and report their results back to the central scanner. The agents can also scan continuously on each host, even when not connected to the hospital network.

9. **D.** Malware signatures are notoriously problematic on endpoints, particularly when the devices are portable and not carefully managed. Although VPN client problems may be an issue, they would not be as significant a vulnerability as outdated malware signatures. IEEE 802.1X problems would be localized at the network access points and not on the endpoints.

10. **B.** Staff recruiting and retention are frequently quoted by business leaders as reasons to allow personal mobile devices on their corporate networks. Therefore, staffing concerns would typically not be a good rationale for prohibiting these devices.

Vulnerability Assessment Tools

In this chapter you will learn:

- When and how you can use different tools and technologies
- How to choose among similar tools and technologies
- How to review and interpret results of vulnerability scan reports
- How to use vulnerability assessment tools for specialized environments

We shape our tools and thereafter our tools shape us.

—Father John Culkin

The purpose of this chapter is to introduce you to (or perhaps reacquaint you with) vulnerability assessment tools with which you will need to be familiar for the CySA+ exam. We are not trying to provide a full review of each tool or even cover all the features. Instead, we give you enough information to help you understand the purpose of each tool and when you may want to use one over another. If you read about a tool for the first time in this chapter, you may want to spend more time familiarizing yourself with it before you take the exam. Given the diversity and scale of the modern network, making sense of the output of a vulnerability scan may be a daunting task. Fortunately, many tools deliver their comprehensive reports with visual tools and technical details behind the vulnerabilities they uncover. Understanding why vulnerabilities exist and how they can be exploited will assist you in analyzing the final scan report.

Some of the tools within a single category do pretty much the same thing, albeit in different ways. For each tool category, we provide an overview before comparing notable products in that class. We then provide an illustrative scenario that offers a more detailed description of each tool and how it fits in that scenario.

 EXAM TIP　You do not need to know about these tools in detail for the exam, but you do need to know when and how you would use the different classes of tools.

Web Application Scanners

Web application vulnerability scanners are automated scanning tools that scan web applications, normally from the outside, and from the perspective of a malicious user. Like other vulnerability scanners, they will scan only for vulnerabilities and malware for which plug-ins have been developed. Some of the most common types of vulnerabilities that scanners look for are listed here:

- Outdated server components
- Misconfigured server
- Secure authentication of users
- Secure session management
- Information leaks
- Cross-site scripting (XSS) vulnerabilities
- Improper use of HTTPS

Many commercial and open source web application vulnerability scanners are available. Most of them enable you to develop customized tests for your specific environment, so if you have some unique policies or security requirements that must be satisfied, it is worthwhile to learn how to write plug-ins or tests for your preferred scanner.

OWASP Zed Attack Proxy

The OWASP Zed Attack Proxy (ZAP) is one of the Open Web Application Security Project (OWASP) Foundation's flagship projects, popular because of its powerful features. ZAP uses its position between the user's browser and the web application to intercept and inspect user requests, modify the contents if required, and then forward them to a web server. This is exactly the same process that occurs during a man-in-the-middle attack. ZAP is designed to be used by security practitioners at all levels. Figure 4-1 shows the results of an attack conducted with nothing more than a target specified.

In this example, ZAP rapidly fabricates a list of GET requests using known directories and files to test the site. These locations should not be publicly viewable, so the results here will inform an administrator of misconfigurations in the server and unexpected data exposure.

Burp Suite

Burp Suite is an integrated web application testing platform. Often used to map and analyze a web application's vulnerabilities, Burp offers seamless use of automated and manual functions when finding and exploiting vulnerabilities. In proxy mode, Burp will enable the user to manually inspect every request passing through from the user to the server. Options to forward or drop the request are shown in Figure 4-2.

Although Burp is designed for a human to be in the decision loop, it does offer a point-and-click web-scanning feature in the paid version that may be useful.

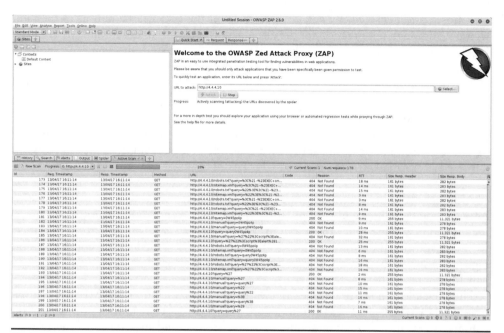

Figure 4-1   ZAP screen showing progress in an active scanning attack

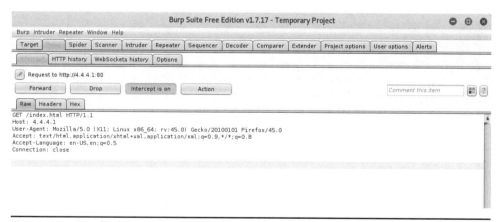

Figure 4-2   Burp Suite intercept screen and options in proxy mode

Nikto

Nikto is a web server vulnerability scanner whose main strength is finding vulnerabilities such as SQL and command-injection susceptibilities, XSS, and improper server configuration. Although Nikto lacks a graphical interface as a command-line–executed utility, it's able to perform thousands of tests very quickly and provides details on the nature of the weaknesses it finds. To conduct a scan against a web server, you specify the IP with the **–host** option enabled. By default, the results of the scan will be output to the same window. Although not practical for detailed analysis, Nikto is useful for quickly confirming the status of a host. By using other options in the command line, you can export the results to an output file for follow-up evaluation. Note that the output includes the type of vulnerability, a short description, and any reference information about the vulnerability.

 NOTE We focus on specific vulnerability scanners in this chapter, but the workflow of vulnerability scanning execution, report generation, and report distribution is similar with nearly all other types of vulnerability scanners on the market.

Although its utility is limited to web servers, Nikto's strength is its speed in assessing software vulnerabilities and configuration issues. As a command-line utility, it's not as user friendly as some other tools. Nikto requires at least a target host to be specified, with any additional options, such as nonstandard ports, added in the command line. Figure 4-3 shows the command issued to perform a scan against a web team's new site, which operates on port 3780.

Nikto enables reports to be saved in a variety of ways, including HTML. An example of Nikto HTML test files is shown in Figure 4-4. This report includes a summary of the command issued, information about the servers tested, and hyperlinks to the relevant resources and their vulnerability data. Although this is good for technical teams to act on, it may not be as useful for nontechnical decision-makers.

```
root@kali:~# nikto -host 4.4.4.28 -port 3780
- Nikto v2.1.6
---------------------------------------------------------------------
+ Target IP:        4.4.4.28
+ Target Hostname:  4.4.4.28
+ Target Port:      3780
---------------------------------------------------------------------
+ SSL Info:         Subject:  /CN=CompanyX/O=bl
                    Ciphers:  ECDHE-RSA-AES256-GCM-SHA384
                    Issuer:   /CN=CompanyX/O=bl
+ Start Time:       2017-04-10 12:14:28 (GMT-7)
---------------------------------------------------------------------
+ Server: Product Information
+ The site uses SSL and the Strict-Transport-Security HTTP header is not defined.
+ Root page / redirects to: https://4.4.4.28:3780/login.jsp
+ No CGI Directories found (use '-C all' to force check all possible dirs)
+ Hostname '4.4.4.28' does not match certificate's names: CompanyX
+ Allowed HTTP Methods: GET, HEAD, POST, PUT, DELETE, OPTIONS
+ OSVDB-397: HTTP method ('Allow' Header): 'PUT' method could allow clients to save files
on the web server.
+ OSVDB-5646: HTTP method ('Allow' Header): 'DELETE' may allow clients to remove files on
the web server.
+ OSVDB-67: /_vti_bin/shtml.dll/_vti_rpc: The anonymous FrontPage user is revealed through
a crafted POST.
+ /login.html: Admin login page/section found.
+ 7499 requests: 0 error(s) and 7 item(s) reported on remote host
+ End Time:         2017-04-10 12:20:26 (GMT-7) (358 seconds)
---------------------------------------------------------------------
+ 1 host(s) tested
```

Figure 4-3 Nikto command-line output for a test against a web server located at IP 4.4.4.28

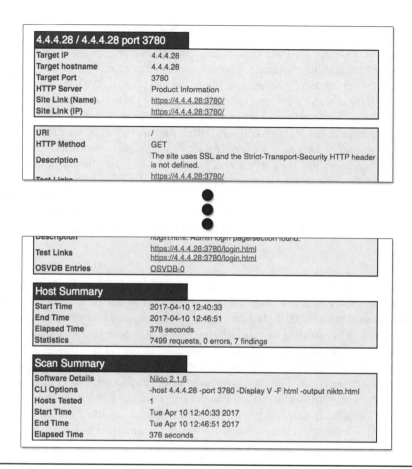

4.4.4.28 / 4.4.4.28 port 3780	
Target IP	4.4.4.28
Target hostname	4.4.4.28
Target Port	3780
HTTP Server	Product Information
Site Link (Name)	https://4.4.4.28:3780/
Site Link (IP)	https://4.4.4.28:3780/

URI	/
HTTP Method	GET
Description	The site uses SSL and the Strict-Transport-Security HTTP header is not defined.
Test Links	https://4.4.4.28:3780/

Description	/login.html: Admin login page/section found.
Test Links	https://4.4.4.28:3780/login.html https://4.4.4.28:3780/login.html
OSVDB Entries	OSVDB-0

Host Summary	
Start Time	2017-04-10 12:40:33
End Time	2017-04-10 12:46:51
Elapsed Time	378 seconds
Statistics	7499 requests, 0 errors, 7 findings

Scan Summary	
Software Details	Nikto 2.1.6
CLI Options	-host 4.4.4.28 -port 3780 -Display V -F html -output nikto.html
Hosts Tested	1
Start Time	Tue Apr 10 12:40:33 2017
End Time	Tue Apr 10 12:46:51 2017
Elapsed Time	378 seconds

Figure 4-4 Nikto HTML output from a test against a web server located at IP 4.4.4.28

Arachni

Arachni is a Ruby-based, modular web app scanner with a special focus on speed. Unlike many other scanners, Arachni performs many of its scans in parallel, enabling the app to scale to large test jobs without sacrificing performance. The tool's developer, Tasos Laskos, designed Arachni to train itself throughout its audits using several techniques. Among them is the tool's ability to incorporate the feedback it gets from initial responses during the test to inform which new techniques should be used moving forward. In presenting the final results, Arachni is also able to reduce the occurrence of false positives through a process the author refers to as *meta-analysis,* which considers several factors from the tests responses. Arachni can perform audits for vulnerabilities including SQL injection, cross-site request forgery (CSRF), code injection, LDAP injection, path traversal, file inclusion, and XSS.

Despite being fairly performant, Arachni has few requirements to operate. It needs only a target URL to get started and can be initiated via a web interface or command line.

Infrastructure Vulnerability Scanners

Modern scanners cannot find weaknesses they're not aware of or do not understand. The most popular vulnerability scanners have amassed enormous libraries of vulnerabilities that cover the vast majority of flaws most likely to be exploited. We'll discuss a few popular vulnerability scanners in the next few pages. Many of these tools enable analysts to get a picture of the network from the perspective of an outsider as well as from a legitimate user. In the latter case, the scanner will most often perform an *authenticated scan* in one of two primary ways. The first method is to install local agents on the endpoints to synchronize with the vulnerability scan server and provide analysis on the endpoint during the course of the scan. The second method is to provide administrative credentials directly to the scanner, which it will invoke as necessary during the scan. It's good practice to use both authenticated and unauthenticated scans during an assessment, because the use of one type may uncover vulnerabilities that would not be found by the other.

Nessus

Nessus, a popular and powerful scanner, began its life as an open source and free utility in the late 1990s and has since become a top choice for conducting vulnerability scans. With more than 80,000 plug-ins, Tenable's Nessus enables users to schedule and conduct scans across multiple networks based on custom policies. Nessus includes basic port-scanning functionality. Its real power, however, lies in its multitude of features for vulnerability identification, misconfiguration detection, default password exposure, and compliance determination. The standard installation includes the Nessus server, which will coordinate the vulnerability scan, generate reports, and facilitate the vulnerability management feature. It can reside on the same machine as the Nessus web client or can be located elsewhere on the network. The client is designed to be run from the web interface, which enables the administrator to manipulate scan settings using any browser that supports HTML5.

Figure 4-5 shows the Nessus architecture being used against several targets on the network. Located on the Nessus server are the various plug-ins used in conducting assessments against the targets. With registration, Tenable provides updates of plug-ins and the server software often, usually once a day.

Assuming the server is running on the same local machine as the client, as is often the case, you can access the Nessus web interface by pointing your browser to http://localhost:8834, as shown in Figure 4-6. When you start Nessus for the first time, there is

Figure 4-5
Nessus
client/server
architecture
shown against
several network
targets

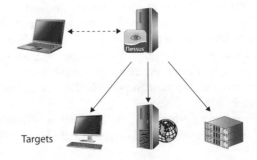

Targets

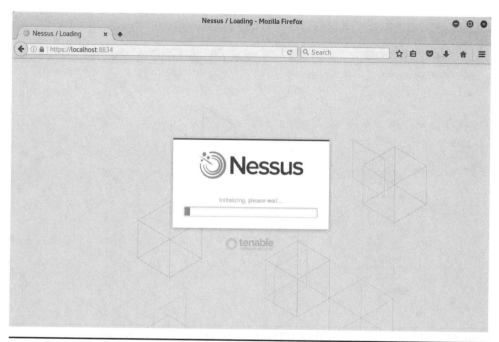

Figure 4-6 View of the initial Nessus loading from a standard web browser

a bit of a delay for initial configuration, registration, and updating. Be patient, and you'll soon be ready to conduct your first scan.

Once the initial setup is complete, you can specify the details for any type of scan from the same interface. By default, the most popular scans are already enabled, but it's good practice to walk through all the settings to learn what exactly will be happening on your network. Figure 4-7 shows the general Settings page, which provides space for a name and description of the scan. Scans created here can be used for immediate or scheduled action. Targets can be specified in one of several ways, including via a single IPv4 address, a single IPv6 address, a range of IPv4 or IPv6 addresses, or a hostname. In addition, the server will also correctly interpret classless inter-domain routing (CIDR) or netmask notation to specify IPv4 subnets. Nessus also provides a space to upload groups of specific target machines in ASCII text format, making it easy to reuse pre-populated lists. In this Settings page, you can also set schedules for scans, adjust notification preferences, and define certain technical limits for the scan. Nessus classifies some plug-ins as dangerous, meaning that their use may cause damage to some systems in certain conditions. When you're preparing to execute a scan, it may be useful to use the Nessus "safe checks" option to avoid launching potentially destructive attacks. We'll step through setting up for a basic Nessus scan over the next few pages.

Nessus allows for great flexibility and depth in the scanning process. You can configure Nessus to pass along any credentials that may be useful. In the Credentials page, shown in Figure 4-8, there is space to configure credentials for Windows hosts. Nessus supports passing authentication for a wide range of cloud services, databases, hosts, network devices, and hypervisors.

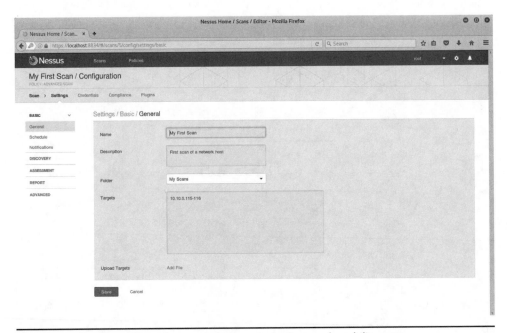

Figure 4-7   Nessus configuration screen before conducting a vulnerability scan

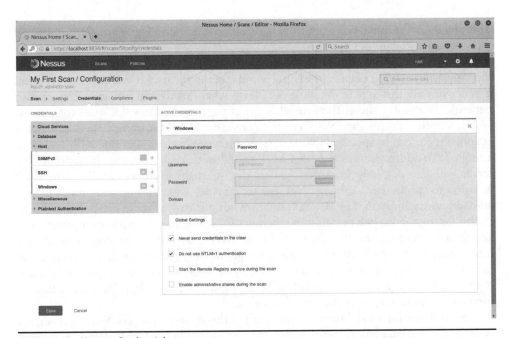

Figure 4-8   Nessus Credentials page

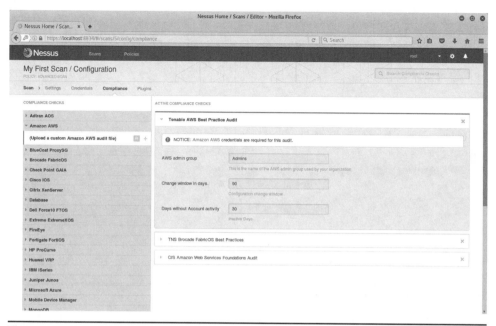

Figure 4-9 Nessus compliance checks with the Tenable AWS Best Practice Audit options displayed

In the Compliance page, shown in Figure 4-9, you can configure compliance checks for the scan. Included in the default installation are many preconfigured checks developed in-house or based on industry best practices and benchmarks. As an admin, you can also develop and upload your own custom configuration, called an *audit file,* for use in the compliance check. The audit file gives instructions used to assess the configuration of endpoints and network device systems against a compliance policy or for the presence of sensitive data.

When using compliance checks, you should be aware of some of the tradeoffs. Enabling these checks may slow down the scan, because many more aspects of the target system will be checked, and potentially at a deeper level. In some cases, active scanning may reduce functionality of both the client and target machines. Furthermore, these compliance checks may also be interpreted as intrusive by the target, which may trigger alerts on intermediate security devices and endpoint software.

On the next page, Plugins, you can see the status of all the plug-ins available for scanning. Nessus maintains a library of these small programs, which checks for known flaws. Plug-ins are written in the Nessus Attack Scripting Language (NASL) and contain information about the vulnerability, its remediation steps, and the mechanism that the plug-in uses to determine the existence of the vulnerability. Usually released within 24 hours of a public disclosure, plug-ins are constantly updated as part of the Nessus subscription. As shown in Figure 4-10, you can activate (or deactivate) any plug-ins required for the scan, or just get details into what exactly is performed by the plug-in during the assessment.

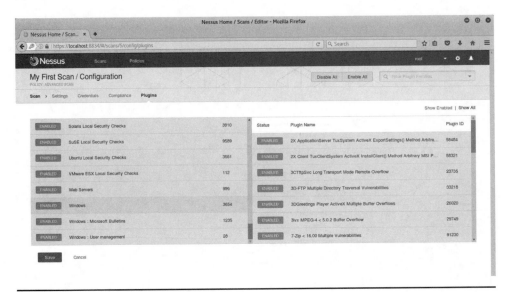

Figure 4-10 Nessus plug-in selection interface

With all the necessary settings saved, you can begin scanning targets for vulnerabilities. When Nessus discovers a vulnerability, it assigns a severity level to it in the scan results, as shown in Figure 4-11.

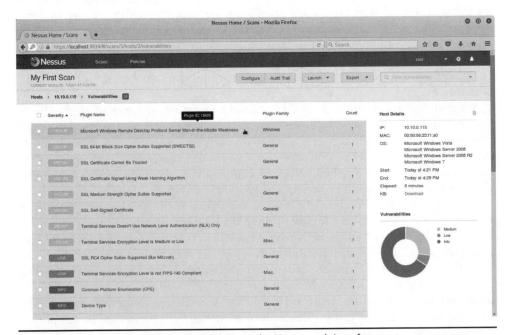

Figure 4-11 List of discovered vulnerabilities in the Nessus web interface

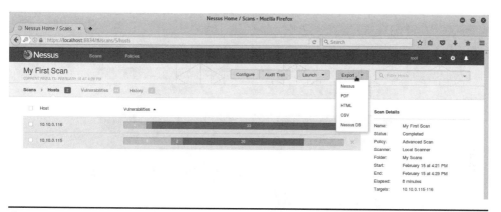

Figure 4-12 Nessus export options for report generation after a vulnerability scan

Technical details for each vulnerability, the method used in identifying it, and any database references are displayed here. Nessus is particularly strong at assessing compliance using its library of compliance checks. These compliance checks, or any other type of scan, can be scheduled to occur as desired, fulfilling your CISO's desire to conduct periodic scans automatically. As for reports, Nessus offers several export options, as shown in Figure 4-12.

Nessus can generate reports that list only vulnerabilities with an associated exploit alongside suggested remediation steps. For an audience such as company leaders, explaining in plain language the concrete steps that may be taken to improve organizational security is key. Nessus provides these suggestions for individual hosts as well as for the network at large.

OpenVAS

The Open Vulnerability Assessment System, or OpenVAS, is a free framework that consists of several analysis tools for both vulnerability identification and management. OpenVAS is a fork of the original Nessus project that began shortly after Tenable closed development of the Nessus framework. OpenVAS is similar to Nessus in that it supports browser-based access to its OpenVAS Manager, which uses the OpenVAS Scanner to conduct assessments based on a collection of more than 47,000 network vulnerability tests (NVTs). Results of these NVTs are then sent back to the Manager for storage.

You can access OpenVAS's interface by using a standard browser to access http://localhost:9392. Figure 4-13 shows the welcome screen from which an admin can access all settings for both the OpenVAS Manager and OpenVAS Scanner. There is also an empty field on the right side of the screen that can be used to launch quick scans.

OpenVAS also provides details on active NVTs used in the scan, as shown in Figure 4-14. You can see the status of each of the tests and, as with Nessus, get details on the test itself. In addition to the summary of the NVT, a vulnerability score is given, plus a level of confidence assigned to the detection method.

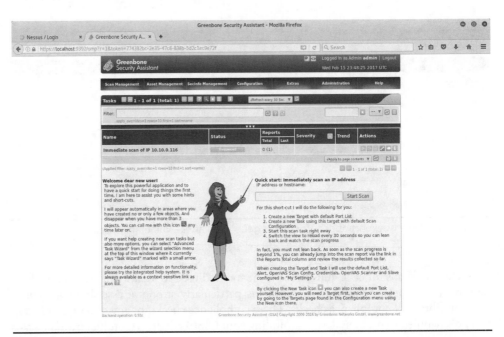

Figure 4-13 OpenVAS welcome screen

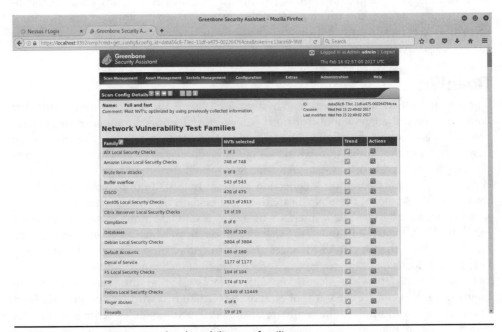

Figure 4-14 OpenVAS network vulnerability test families

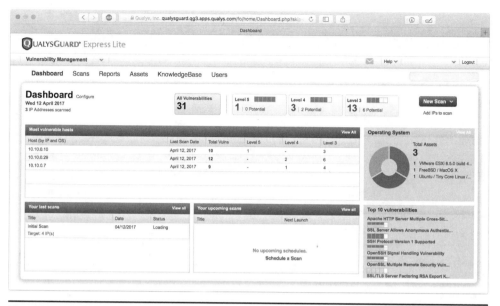

Figure 4-15 The QualysGuard dashboard for managing its cloud-based vulnerability assessment and management tasks

Qualys

QualysGuard is a product of the California-based security company Qualys, an early player in the vulnerability management market. The company currently provides several cloud-based vulnerability assessment and management products through a Software as a Service (SaaS) model. For internal scans, a local virtual machine conducts the assessment and reports to the Qualys server. Figure 4-15 shows a QualysGuard dashboard with various options under the vulnerability management module.

All network discovery, mapping, asset prioritization, scheduling, vulnerability assessment reporting, and remediation tracking tasks can be accessed via the web-based UI. The platform can generate detailed reports using several templates. Included in the default installation is a template called "Executive Report," which provides just the type of data the CISO needs for her discussion. A portion of results of this report is shown in Figure 4-16.

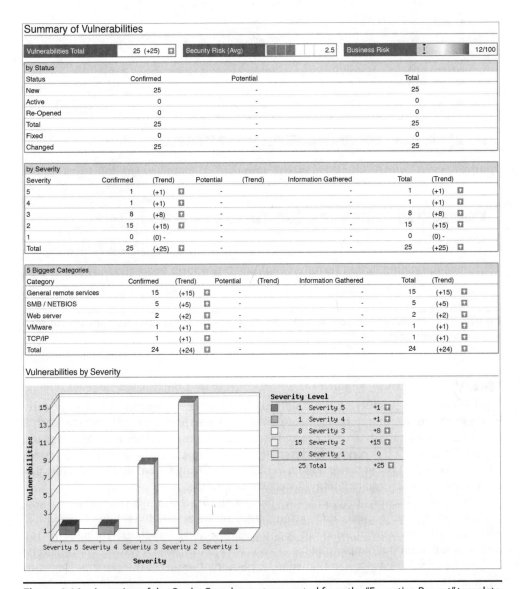

Figure 4-16 A portion of the QualysGuard report generated from the "Executive Report" template

Software Assessment Tools and Techniques

As software becomes larger and complex, the potential for vulnerabilities increases. Software vulnerabilities are part of the overall attack surface, and attackers waste no time discovering what flaws exist—in many cases using the techniques we'll describe shortly. Software vulnerability detection methods usual fall into the categories of static analysis, dynamic analysis, reverse engineering, or fuzzing. These methods can be done separately or, ideally, together and as a part of a comprehensive software assessment process.

Static Analysis

Static code analysis is a technique meant to help identify software defects or security policy violations and is carried out by examining the code without executing the program (hence the term *static*). The term *static analysis* is generally reserved for automated tools that assist analysts and developers, whereas manual inspection by humans is generally referred to as *code review*. Because it is an automated process, static analysis enables developers and security staff to scan their source code quickly for programming flaws and vulnerabilities.

Figure 4-17 shows an example of a tool called Lapse+, which was developed by OWASP to find vulnerabilities in Java applications. This tool is highlighting an instance wherein user input is directly used, without sufficient validation, to build a SQL query against a database. In this particular case, this query is verifying the username and password. This insecure code block would allow a threat actor to conduct a SQL injection attack against this system. This actor would likely gain access by providing the string `'foo' OR 1==1 --` if the database was on a MySQL server.

Automated static analysis like that performed by Lapse+ provides a scalable method of security code review and ensures that secure coding policies are being followed. There are numerous manifestations of static analysis tools, ranging from tools that simply consider the behavior of single statements to tools that analyze the entire source code at once. However, you should keep in mind that static code analysis cannot usually reveal logic errors or vulnerabilities (that is, behaviors that are evident only at runtime) and therefore should be used in conjunction with manual code review to ensure a more thorough evaluation.

Figure 4-17   Code analysis of a vulnerable web application (Source: www.owasp.org)

Dynamic Analysis

This method doesn't really care about what the binary *is*, but rather what the binary *does*. Referred to as *dynamic analysis,* this method often requires a sandbox in which to execute the malware. This sandbox creates an environment that looks like a real operating system to the software and provides such things as access to a file system, network interface, memory, and anything else the software might need. Each request is carefully documented to establish a timeline of behavior that enables us to understand what it does. The main advantage of dynamic analysis is that it tends to be significantly faster and requires less expertise than alternatives. It can be particularly helpful for code that has been heavily obfuscated or is difficult to interpret. The biggest disadvantage is that dynamic analysis doesn't reveal all that the software does, but simply all that it did during its execution in the sandbox.

Reverse Engineering

Reverse engineering is the detailed examination of a product to learn what it does and how it works. In this approach to understanding what software is doing, a highly skilled analyst will either disassemble or decompile the binary code to translate its 1's and 0's into either assembly language or whichever higher level language it was created in. This enables a reverse engineer to see all possible functions of the software, not just those that it exhibited during a limited run in a sandbox. It is then possible, for example, to understand what kind of input is expected, discover dependencies, and highlight inefficient or dangerous techniques. It can be argued that static analysis is a subset, or supporting function, of an overall reverse engineering effort.

Engineering and Reversing Software

Computers can understand only sequences of 1's and 0's (sometimes represented in hexadecimal form for our convenience), which is why we call this representation of software *machine language*. It would be tedious and error-prone to write complex programs in machine language, which is why we invented *assembly language* many decades ago. In this language, the programmer uses operators (such as push and add) and operands (such as memory addresses, CPU registers, and constants) to implement an algorithm. The software that translates assembly language to machine language is called an *assembler.* Though assembly language was a significant improvement, we soon realized that it was still rather ineffective, which is why we invented higher level programming languages such as C/C++. This higher level source code is translated into assembly language by a compiler before being assembled into binary format, as shown here.

(continued)

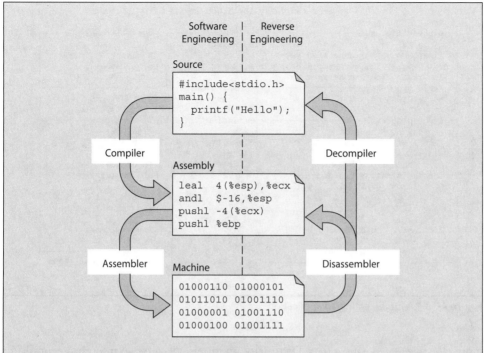

When reverse engineering binary code, we can translate it into assembly language using a tool called a *disassembler*. This is the most common way of reversing a binary. In some cases, we can also go straight from machine language to a representation of source code using a *decompiler*. The problem with using decompilers is that there are infinitely many ways to write source code that will result in a given binary. The decompiler makes educated guesses as to what the original source code looked like, but it's unable to replicate it exactly.

Fuzzing

Fuzzing is a technique used to discover flaws and vulnerabilities in software by sending large amounts of malformed, unexpected, or random data to the target program in order to trigger failures. Attackers could manipulate these errors and flaws to inject their own code into the system and compromise its security and stability. Fuzzing tools are commonly successful at identifying buffer overflows, denial of service (DoS) vulnerabilities, injection weaknesses, validation flaws, and other activities that can cause software to freeze, crash, or throw unexpected errors. Figure 4-18 shows a popular fuzzer called American Fuzzy Lop (AFL) crashing a targeted application.

Fuzzers don't always generate random inputs from scratch. Purely random generation is known to be an inefficient way to fuzz systems. Instead, they often start with an input that is pretty close to normal and then make lots of small changes to see which ones seem

```
                        american fuzzy lop 1.86b (test)
 ┌─ process timing ──────────────────────┐ ┌─ overall results ──────┐
 │        run time : 0 days, 0 hrs, 0 min, 2 sec │   cycles done : 0
 │   last new path : none seen yet              │   total paths : 1
 │  last uniq crash : 0 days, 0 hrs, 0 min, 2 sec │ uniq crashes : 1
 │   last uniq hang : none seen yet             │    uniq hangs : 0
 ├─ cycle progress ──────────────┐ ┌─ map coverage ─────────┐
 │   now processing : 0 (0.00%)  │ │    map density : 2 (0.00%)
 │  paths timed out : 0 (0.00%)  │ │ count coverage : 1.00 bits/tuple
 ├─ stage progress ──────────────┘ ├─ findings in depth ────┘
 │     now trying : havoc            │  favored paths : 1 (100.00%)
 │    stage execs : 1464/5000 (29.28%) │ new edges on : 1 (100.00%)
 │    total execs : 1697             │ total crashes : 39 (1 unique)
 │    exec speed : 626.5/sec         │   total hangs : 0 (0 unique)
 ├─ fuzzing strategy yields ─────────┐ ├─ path geometry ────────┐
 │     bit flips : 0/16, 1/15, 0/13  │ │     levels : 1
 │    byte flips : 0/2, 0/1, 0/0     │ │    pending : 1
 │   arithmetics : 0/112, 0/25, 0/0  │ │   pend fav : 1
 │    known ints : 0/10, 0/28, 0/0   │ │  own finds : 0
 │    dictionary : 0/0, 0/0, 0/0     │ │   imported : n/a
 │        havoc : 0/0, 0/0           │ │   variable : 0
 │         trim : n/a, 0.00%         │ └────────────────────────┘
 └───────────────────────────────────┘          [cpu: 92%]
```

Figure 4-18 A fuzzer testing an application

more effective at exposing a flaw. Eventually, an input will cause an interesting condition in the target, at which point the security team will need a tool that can determine where the flaw is and how it could be exploited (if at all). This observation and analysis tool is often bundled with the fuzzer, because one is pretty useless without the other.

untidy

untidy is a popular Extensible Markup Language (XML) fuzzer. Used to test web application clients and servers, untidy takes valid XML and modifies it before inputting it into the application. The untidy fuzzer is now part of the Peach Fuzzer project.

Peach Fuzzer

The Peach Fuzzer is a powerful fuzzing suite that's capable of testing a wide range of targets. Peach uses XML-based modules, called *pits,* to provide all the information needed to run the fuzz. These modules are configurable based on the testing needs. Before conducting a fuzz test, the user must specify the type of test, the target, and any monitoring settings desired.

Microsoft SDL Fuzzers

As part of its Security Development Lifecycle (SDL) toolset, Microsoft released two types of standalone fuzzers designed to be used in the verification phase of the SDL: the Mini-Fuzz File Fuzzer and the Regex Fuzzer. However, Microsoft has dropped support and no longer provides these applications for download.

Enumeration Tools and Techniques

Network enumeration is the interrogation of a set of hosts to look for specific information. A horizontal network scan, for instance, sends messages to a set of host addresses asking the question, "Are you there?" Its goal is to determine which addresses correspond to active (responding) systems. A vertical scan, on the other hand, sends messages to a set of protocol/port combinations (for example, UDP 53 or TCP 25) asking the same question, with the goal of determining which ports are listening for client connection attempts. It is possible to combine both horizontal and vertical scans, as shown in Figure 4-19. The point of a network scan of any flavor is to find out who is listening (and responding) on a network. This is useful in finding systems that are not behaving as they ought to.

Consider the case in which a software development team sets up a test web server to ensure that the web app on which the team members are working is functioning properly.

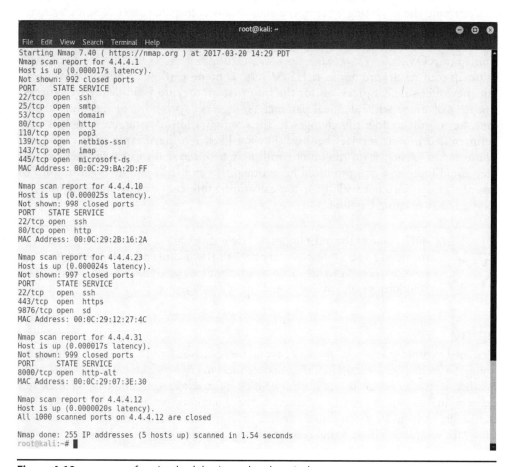

Figure 4-19   nmap performing both horizontal and vertical scans

There is no malice, but not knowing that this server is running (and how) could seriously compromise the integrity of the defenses. On the other hand, we can sometimes find evidence that a host has been compromised and is now running a malicious service. Though these situations should be rare in a well-managed network, scanning is one of the quintessential skills of any security professional, because it is one of very few ways to know and map what is really on our networks.

nmap

nmap, one of the most popular tools used for the enumeration of a targeted host, is synonymous with network scanning. The name is shorthand for "network mapper," which is an apt description of what it does. nmap works by sending specially crafted messages to the target hosts and then examining the responses. This can tell not only which hosts are active on the network and which of their ports are listening, but it can also help us determine the operating system, hostname, and even patch level of some systems. Though nmap is a command-line interface (CLI) tool, a number of front ends provide a graphical user interface (GUI), including Zenmap (Windows), NmapFE (Linux), and Xnmap (macOS).

nmap scan results are not as rich in diverse content as those of other data sources, but they offer surgical precision for the information they do yield. Moreover, successive runs of nmap with identical parameters, together with a bit of scripting, enable the user to quickly identify changes to the configuration of a target. Attackers may be interested in new services because they are likelier to have exploitable configuration errors. Defenders, on the other hand, may be interested in new services because they could indicate a compromised host. nmap is even used by some organizations to inventory assets on a network by periodically doing full scans and comparing hosts and services to an existing baseline.

 NOTE One of the most important defensive measures you can take is to maintain an accurate inventory of all the hardware and software on your network. nmap can help with this, but various open source and commercial solutions are also available for IT asset management.

---

hping

Designed to provide advanced features beyond those offered from the built-in ping utility, hping is a useful enumeration tool that enables users to craft custom packets to assist with the discovery of network flaws, or it can be used by attackers to facilitate targeted exploit delivery. In addition to providing packet analysis functionality of TCP, UDP, and ICMP traffic, hping includes a traceroute mode and supports IP fragmentation. Figure 4-20 shows the output of a basic hping command targeting the loopback address.

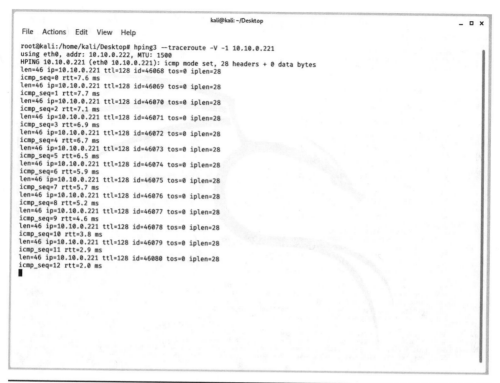

Figure 4-20 Output from hping performed against a host on the LAN

Passive vs. Active Enumeration Techniques

Passive enumeration techniques are used to gain information about the target without interfacing with or interrogating the target directly. In fact, access to the target isn't necessarily required since alternative sources for target information are queried. Common examples include using a whois query, nslookup, or specialized tools like dnsrecon. A useful application of these tools for the purpose of vulnerability assessment could be to assist in discovery of what an asset looks like to outsiders. Figure 4-21 shows a dnsrecon lookup against google.com and provides a good deal of insight into the records associated with the company.

Active enumeration techniques involve interfacing directly with the target system. Port scanning is one of the most widely used active scanning techniques and uses any number of tools and techniques directly against a target to discover open ports and available services.

```
                                    kali@kali:~/Desktop                                    _ □ ×

  File   Actions   Edit   View   Help

root@kali:/home/kali/Desktop# dnsrecon -d google.com
[*] Performing General Enumeration of Domain: google.com
[-] DNSSEC is not configured for google.com
[*]     SOA ns1.google.com 216.239.32.10
[*]     NS ns1.google.com 216.239.32.10
[*]     NS ns1.google.com 2001:4860:4802:32::a
[*]     NS ns2.google.com 216.239.34.10
[*]     NS ns2.google.com 2001:4860:4802:34::a
[*]     NS ns3.google.com 216.239.36.10
[*]     NS ns3.google.com 2001:4860:4802:36::a
[*]     NS ns4.google.com 216.239.38.10
[*]     NS ns4.google.com 2001:4860:4802:38::a
[*]     MX alt4.aspmx.l.google.com 173.194.175.26
[*]     MX aspmx.l.google.com 64.233.179.26
[*]     MX alt1.aspmx.l.google.com 172.253.112.27
[*]     MX alt2.aspmx.l.google.com 173.194.77.26
[*]     MX alt3.aspmx.l.google.com 64.233.177.27
[*]     MX alt4.aspmx.l.google.com 2607:f8b0:400d:c0b::1a
[*]     MX aspmx.l.google.com 2607:f8b0:4003:c12::1b
[*]     MX alt1.aspmx.l.google.com 2607:f8b0:4023::1b
[*]     MX alt2.aspmx.l.google.com 2607:f8b0:4023:401::1a
[*]     MX alt3.aspmx.l.google.com 2607:f8b0:4002:c08::1a
[*]     A google.com 216.58.217.14
[*]     AAAA google.com 2001:4860:4802:38::75
[*]     TXT google.com docusign=05958488-4752-4ef2-95eb-aa7ba8a3bd0e
[*]     TXT google.com docusign=1b0a6754-49b1-4db5-8540-d2c12664b289
[*]     TXT google.com facebook-domain-verification=22rm551cu4k0ab0bxsw536tlds4h95
[*]     TXT google.com globalsign-smime-dv=CDYX+XFHUw2wml6/Gb8+59BsH31KzUr6c1l2BPvqKX8=
[*]     TXT google.com v=spf1 include:_spf.google.com ~all
[*] Enumerating SRV Records
[*]     SRV _ldap._tcp.google.com ldap.google.com 216.239.32.58 389 0
[*]     SRV _ldap._tcp.google.com ldap.google.com 2001:4860:4802:32::3a 389 0
[*]     SRV _carddavs._tcp.google.com google.com 216.58.217.14 443 0
[*]     SRV _carddavs._tcp.google.com google.com 2001:4860:4802:38::75 443 0
[*]     SRV _caldav._tcp.google.com calendar.google.com 216.58.217.14 80 0
[*]     SRV _caldav._tcp.google.com calendar.google.com 2607:f8b0:4000:804::200e 80 0
[*]     SRV _caldavs._tcp.google.com calendar.google.com 216.58.217.14 443 0
[*]     SRV _caldavs._tcp.google.com calendar.google.com 2607:f8b0:4000:804::200e 443 0
[*]     SRV _xmpp-server._tcp.google.com alt4.xmpp-server.l.google.com 173.194.68.125 5269 0
[*]     SRV _xmpp-server._tcp.google.com xmpp-server.l.google.com 64.233.179.125 5269 0
```

Figure 4-21   Output from dnsrecon performed against google.com

nslookup

The name server lookup, or nslookup, utility can be thought of as the user interface for Domain Name System (DNS). It enables us to resolve the IP address corresponding to a fully qualified domain name (FQDN) of a host. Depending on the situation, it is also possible to do the inverse (that is, resolve the IP address of an FQDN). The tool allows for the specification of a DNS server to be used, or it can use the system default. Lastly, it is possible to fully interrogate the target server and obtain other record data, such as Mail Exchange (MX) for e-mail or Canonical Name (CNAME).

responder

In the Windows environment, the Link-Local Multicast Name Resolution (LLMNR) protocol or NetBIOS Name Service (NBT-NS) can be used to query local computers on a LAN if a host is unable to resolve a hostname using DNS. If either of these two fallback protocols does not work, any host can respond with an answer that may be viewed as authoritative by the requesting machine. responder is a powerful tool that can be used

Figure 4-22 responder usage and options screen

to gain remote access by poisoning name services to gather hashes and credentials from systems within a LAN. Figure 4-22 is a snapshot of the responder reference page from the command line–based tool.

Wireless Assessment Tools

It is difficult to run any kind of corporate network without considering the implications of wireless networks. Even if you don't allow wireless devices at all (not even mobile phones) in your building, how would you know that the policy is being followed by all employees, all the time? Admittedly, most organizations will not (and often cannot) implement such draconian measures, which makes wireless local area network (WLAN) auditing and analysis particularly important.

To conduct a WLAN analysis, you must first capture data. Normally, when a WLAN interface card connects to a wireless access point (WAP), the client device will be in managed mode and the WAP will be in master mode. Master mode (also known as *infrastructure* mode) means that the interface will be responsible for managing all aspects of the

WLAN configuration (such as channel and service set ID [SSID]). The client in managed mode is being managed by the master and thus will change channels or other settings when told to do so. Wireless interfaces can also communicate directly in *mesh* mode, which enables the interface to negotiate directly with another interface in a relationship that does not have master and managed nodes (also known as *ad hoc* mode). In each of these three modes, the interface will be limited to one connection to one network. To monitor multiple networks simultaneously, we need a fourth mode of operation. In *monitor* mode, the wireless interface will be able to see all available WLANs and their characteristics without connecting to any of them. This is the mode we need in order to perform a WLAN audit. Fortunately, WLAN analyzers, such as Kismet, take care of these details and enable us simply to run the application and see what is out there.

The most important step to a security analysis of your WLANs is to know your devices. Chief among these, of course, are the WAPs, but you must also keep track of wireless clients. How would you know that something odd is going on? Quite simply, by keeping track of what "normal" looks like. When analyzing the structure of WLANs, you must start from a known-good list of access points and client devices. Because, presumably, your organization installed (or had someone install) all the WAPs, you should have a record of their settings (for example, protocol, channel, and location) as well as their Media Access Control (MAC) and IP addresses. As you conduct your periodic audits, you will be able to tell if a new WAP shows up in your scan, potentially indicating a rogue access point.

Looking for rogue or unauthorized clients is a bit tricky because it is not difficult to change the MAC address on many networked devices. Indeed, all major operating systems have built-in tools that enable you to do just that. Because the main indicator of an end device's identity is so susceptible to forgery, you may not be able to detect unauthorized nodes unless you implement some form of authentication, such as implementing WPA Enterprise and IEEE 802.1x. Absent authentication, you will have a very difficult time identifying all but the most naïve intruders connected to your WLAN.

Aircrack-ng

Aircrack-ng is the most popular open source wireless network security tool. Despite the name, the tool is actually a full-featured suite of wireless tools. It's used primarily for its ability to audit the security of WLANs through attacks on WPA keys, replay attacks, deauthentication, and the creation of fake access points, as shown in Figure 4-23. In support of all of this functionality, Aircrack-ng also allows for indiscriminate wireless monitoring, packet captures, and, if equipped with the correct hardware, wireless injection.

Reaver

It's well known that many wireless security protocols have significant flaws associated with them, which may allow for an attacker to gain foothold into a network that's within reach of a wireless signal. Reaver takes advantage of a vulnerability that exists in access points that use the Wi-Fi Protected Setup (WPS) feature. The WPS protocol

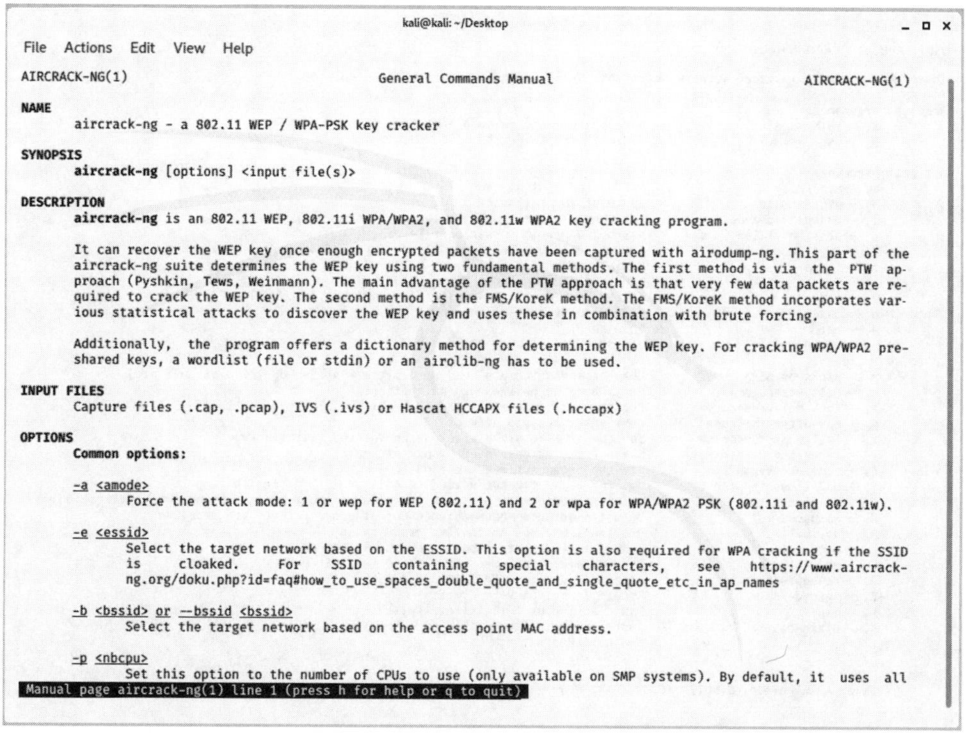

```
                                    kali@kali: ~/Desktop                              _ □ ×

 File   Actions   Edit   View   Help

 AIRCRACK-NG(1)                     General Commands Manual                    AIRCRACK-NG(1)

 NAME
        aircrack-ng - a 802.11 WEP / WPA-PSK key cracker

 SYNOPSIS
        aircrack-ng [options] <input file(s)>

 DESCRIPTION
        aircrack-ng is an 802.11 WEP, 802.11i WPA/WPA2, and 802.11w WPA2 key cracking program.

        It can recover the WEP key once enough encrypted packets have been captured with airodump-ng. This part of the
        aircrack-ng suite determines the WEP key using two fundamental methods. The first method is via the PTW ap-
        proach (Pyshkin, Tews, Weinmann). The main advantage of the PTW approach is that very few data packets are re-
        quired to crack the WEP key. The second method is the FMS/KoreK method. The FMS/KoreK method incorporates var-
        ious statistical attacks to discover the WEP key and uses these in combination with brute forcing.

        Additionally, the program offers a dictionary method for determining the WEP key. For cracking WPA/WPA2 pre-
        shared keys, a wordlist (file or stdin) or an airolib-ng has to be used.

 INPUT FILES
        Capture files (.cap, .pcap), IVS (.ivs) or Hascat HCCAPX files (.hccapx)

 OPTIONS
        Common options:

        -a <amode>
               Force the attack mode: 1 or wep for WEP (802.11) and 2 or wpa for WPA/WPA2 PSK (802.11i and 802.11w).

        -e <essid>
               Select the target network based on the ESSID. This option is also required for WPA cracking if the SSID
               is     cloaked.    For    SSID    containing    special    characters,    see    https://www.aircrack-
               ng.org/doku.php?id=faq#how_to_use_spaces_double_quote_and_single_quote_etc_in_ap_names

        -b <bssid> or --bssid <bssid>
               Select the target network based on the access point MAC address.

        -p <nbcpu>
               Set this option to the number of CPUs to use (only available on SMP systems). By default, it uses all
 Manual page aircrack-ng(1) line 1 (press h for help or q to quit)
```

Figure 4-23 Aircrack-ng description and option screen

emerged in the 2000s as a way for users to get set up on a network without having to remember complicated passwords or configuration settings. Unfortunately, a major flaw lies in the how WPS handles validation of the PIN. After splitting up the PIN into two halves, each portion is validated separately. This method significantly reduces the sample space, which Reaver leverages to guess the full password. As you can see from Figure 4-24, usage is straightforward. The analyst simply calls the command from the CLI, specifies the interface over which the attack will be delivered, and identifies the target access point. From a vulnerability management point of view, the solution is to disable WPS altogether.

oclHashcat

Recent developments in graphics card technology have allowed for incredibly detailed graphics on home PC and dedicated gaming rigs. The same array of processors powering those graphics cards has also enabled users to perform similar computationally intensive operations, such as password cracking, on the cheap. A few years ago, an analyst might

```
                                    kali@kali: ~/Desktop                              _ ☐ ✕

 File  Actions  Edit  View  Help

 Copyright (c) 2011, Tactical Network Solutions, Craig Heffner <cheffner@tacnetsol.com>

 Required Arguments:
       -i, --interface=<wlan>          Name of the monitor-mode interface to use
       -b, --bssid=<mac>               BSSID of the target AP

 Optional Arguments:
       -m, --mac=<mac>                 MAC of the host system
       -e, --essid=<ssid>              ESSID of the target AP
       -c, --channel=<channel>         Set the 802.11 channel for the interface (implies -f)
       -s, --session=<file>            Restore a previous session file
       -C, --exec=<command>            Execute the supplied command upon successful pin recovery
       -f, --fixed                     Disable channel hopping
       -5, --5ghz                      Use 5GHz 802.11 channels
       -v, --verbose                   Display non-critical warnings (-vv or -vvv for more)
       -q, --quiet                     Only display critical messages
       -h, --help                      Show help

 Advanced Options:
       -p, --pin=<wps pin>             Use the specified pin (may be arbitrary string or 4/8 digit WPS pin)
       -d, --delay=<seconds>           Set the delay between pin attempts [1]
       -l, --lock-delay=<seconds>      Set the time to wait if the AP locks WPS pin attempts [60]
       -g, --max-attempts=<num>        Quit after num pin attempts
       -x, --fail-wait=<seconds>       Set the time to sleep after 10 unexpected failures [0]
       -r, --recurring-delay=<x:y>     Sleep for y seconds every x pin attempts
       -t, --timeout=<seconds>         Set the receive timeout period [10]
       -T, --m57-timeout=<seconds>     Set the M5/M7 timeout period [0.40]
       -A, --no-associate              Do not associate with the AP (association must be done by another application)
       -N, --no-nacks                  Do not send NACK messages when out of order packets are received
       -S, --dh-small                  Use small DH keys to improve crack speed
       -L, --ignore-locks              Ignore locked state reported by the target AP
       -E, --eap-terminate             Terminate each WPS session with an EAP FAIL packet
       -J, --timeout-is-nack           Treat timeout as NACK (DIR-300/320)
       -F, --ignore-fcs                Ignore frame checksum errors
       -w, --win7                      Mimic a Windows 7 registrar [False]
       -K, --pixie-dust                Run pixiedust attack
       -Z                              Run pixiedust attack

 Example:
       reaver -i wlan0mon -b 00:90:4C:C1:AC:21 -vv
```

Figure 4-24   Reaver standard and advanced options screen

be able to perform a successful password-cracking operation using one or two processors and the Hashcat utility in a matter of hours or days. Nowadays, the same cracking operation performed on a graphics card–optimized version called oclHashcat enables that analyst to complete the same task orders of magnitude faster. For the user, this means the time to recover passwords successfully goes from days to hours or minutes.

 NOTE   For the rest of the section, we'll refer to both variants of the utility as just Hashcat since the usage is the same in the modern version.

Hashcat supports many operating modes, character sets, and formats. A small sample of the operating modes that may be most useful to a security analyst are listed in Table 4-1.

	Name	Type
Table 4-1 Selection of Operating Modes for Hashcat/ oclHashcat	MD4	Raw Hash
	MD5	Raw Hash
	SHA1	Raw Hash
	SHA2-256	Raw Hash
	SHA2-512	Raw Hash
	NetNTLMv2	Network Protocols
	Kerberos 5 AS-REQ Pre-Auth etype 23	Network Protocols
	DNSSEC (NSEC3)	Network Protocols
	PostgreSQL CRAM (MD5)	Network Protocols
	MySQL CRAM (SHA1)	Network Protocols
	SIP digest authentication (MD5)	Network Protocols
	TACACS+	Network Protocols
	PostgreSQL	Database Server
	MySQL4.1/MySQL5	Database Server
	LM	Operating Systems
	NTLM	Operating Systems
	bcrypt $2*$, Blowfish (Unix)	Operating Systems
	7-Zip	Archives
	VeraCrypt	Full-Disk Encryption (FDE)
	FileVault 2	Full-Disk Encryption (FDE)
	Plaintext	Plaintext

In addition to the robust mode support, Hashcat also supports the following attack modes:

- Brute-force attack
- Combinator attack
- Dictionary attack
- Hybrid attack
- Mask attack
- Rule-based attack

```
●  ●  ●                            🖥 Desktop — -bash — 128×56
hashcat (v5.1.0) starting...

OpenCL Platform #1: Apple
========================
* Device #1: Intel(R) Xeon(R) CPU E5-1680 v2 @ 3.00GHz, skipped.
* Device #2: AMD Radeon HD - FirePro D700 Compute Engine, 1536/6144 MB allocatable, 32MCU
* Device #3: AMD Radeon HD - FirePro D700 Compute Engine, 1536/6144 MB allocatable, 32MCU

Hashes: 5 digests; 5 unique digests, 1 unique salts
Bitmaps: 16 bits, 65536 entries, 0x0000ffff mask, 262144 bytes, 5/13 rotates
Rules: 1

Applicable optimizers:
* Optimized-Kernel
* Zero-Byte
* Precompute-Init
* Precompute-Merkle-Demgard
* Meet-In-The-Middle
* Early-Skip
* Not-Salted
* Not-Iterated
* Single-Salt
* Raw-Hash

Minimum password length supported by kernel: 0
Maximum password length supported by kernel: 31

Watchdog: Hardware monitoring interface not found on your system.
Watchdog: Temperature abort trigger disabled.

Dictionary cache hit:
* Filename..: rockyou.txt
* Passwords.: 14344384
* Bytes.....: 139921497
* Keyspace..: 14344384


Session.........: hashcat
Status..........: Cracked
Hash.Type.......: MD5
Hash.Target.....: test.txt
Time.Started....: Tue Feb 18 20:31:39 2020 (1 sec)
Time.Estimated..: Tue Feb 18 20:31:40 2020 (0 secs)
Guess.Base......: File (rockyou.txt)
Guess.Queue.....: 1/1 (100.00%)
Speed.#2........: 44116.2 kH/s (1.62ms) @ Accel:256 Loops:1 Thr:256 Vec:1
Speed.#3........:        0 H/s (0.00ms) @ Accel:256 Loops:1 Thr:256 Vec:1
Speed.#*........: 44116.2 kH/s
Recovered.......: 5/5 (100.00%) Digests, 1/1 (100.00%) Salts
Progress........: 4195254/14344384 (29.25%)
Rejected........: 950/4195254 (0.02%)
Restore.Point...: 0/14344384 (0.00%)
Restore.Sub.#2..: Salt:0 Amplifier:0-1 Iteration:0-1
Restore.Sub.#3..: Salt:0 Amplifier:0-1 Iteration:0-1
Candidates.#2...: 123456 -> rogabac
Candidates.#3...: SALCIDO -> roeiend
```

Figure 4-25 Hashcat in operation against a list of MD5 hashes using two GPUs

Figure 4-25 shows the output after initiating a dictionary brute-force attack on a list of hashes. Upon starting, Hashcat identifies suitable processor candidates to perform the password cracking. It then lists the unique digests, or hashes, on which the utility will operate. For the duration of the cracking operation, Hashcat will output useful statistics about its performance. This includes the estimated time of completion, cracking speed, and successfully recovered passwords.

In this example, we've specified that the successfully cracked passwords be outputted to a text file, shown in Figure 4-26. Note that the hash and its plaintext equivalent are placed side by side.

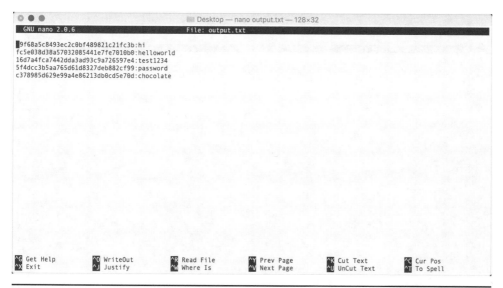

Figure 4-26   Contents of Hashcat output file

Cloud Infrastructure Assessment Tools

Although the elastic nature of cloud computing and storage makes for rapid scalability and efficiency, its dynamic nature also means you need to think about tracking and prioritizing vulnerabilities in a different way. Thankfully, with the rise of cloud services, several fantastic tools are now available to assess cloud host vulnerabilities caused by misconfigurations, access flaws, and custom deployments. Many of these tools automate monitoring against industry standards, compliance checklists, regulatory mandates, and best practices to prevent common issues such as data spillage and instance takeover.

Scout Suite

Scout Suite is an open source auditing tool developed by NCC Group, a security consulting company that specializes in verification, auditing, and managed services. With support for Amazon Web Services (AWS), Microsoft Azure, and Google Cloud Platform, the tool enables security teams to determine the security posture of their cloud assets. It works by managing the interactions with could assets via the platform API, gathering information to make determinations of potentially vulnerable configurations. The results can be easily prepared for manual inspection or follow-up orchestration because of its structured format.

The tool requires a bit of preparation in the way of terms of service agreements and environmental setup, but usage is straightforward once configuration is in place. While Scout Suite ships with a default set of rules, the framework enables an analyst to provide custom rules to change the breadth of coverage during a scan. As shown the Figure 4-27, the rules are captured in a JSON file, which can be adjusted to the needs of the vulnerability team and specified during the utility's execution.

```
                                          kali@kali: ~/Desktop                                    _ □ X

  File   Actions   Edit   View   Help

  GNU nano 4.5                              detailed.json
    "about": "This ruleset consists of numerous rules that are considered standard by NCC Group. The rules enabled rang▶
    "rules": {
        "certificate-with-transparancy-logging-disabled.json": [
            {
                "enabled": true,
                "level": "warning"
            }
        ],
        "cloudformation-stack-with-role.json": [
            {
                "enabled": true,
                "level": "danger"
            }
        ],
        "cloudtrail-duplicated-global-services-logging.json": [
            {
                "enabled": true,
                "level": "warning"
            }
        ],
        "cloudtrail-no-data-logging.json": [
            {
                "enabled": true,
                "level": "warning"
            }
        ],
        "cloudtrail-no-global-services-logging.json": [
            {
                "enabled": true,
                "level": "danger"
            }
        ],
        "cloudtrail-no-log-file-validation.json": [
            {
                "enabled": true,
                            [ Read 1204 lines ]
  ^G Get Help    ^O Write Out   ^W Where Is    ^K Cut Text    ^J Justify    ^C Cur Pos     M-U Undo    M-A Mark Text
  ^X Exit        ^R Read File   ^\ Replace     ^U Paste Text  ^T To Spell   ^_ Go To Line  M-E Redo    M-6 Copy Text
```

Figure 4-27   JSON file containing Scout Suite ruleset

Prowler

Similar to Scout Suite, Prowler is a framework designed as a scalable and repeatable method of acquiring measurable data related to the security readiness of your organization's cloud infrastructure. Prowler assesses cloud assets based in part on the Center for Internet Security (CIS) best practices for configuring security options for various AWS resources. Like Scout Suite, Prowler offers a fair amount of configuration to allow the tool programmatic access via the Amazon API. Additionally, the AWS CLI will have to be installed on the machine or instance running the tool. Prowler allows for the definition of custom checks, but by default it ships with the following best practice test configurations, or "groups":

- Identity and Access Management
- Logging
- Monitoring
- Networking
- CIS Level 1

- CIS Level 2
- Forensics
- GDPR
- HIPAA

Pacu

Rhino Security Labs, a boutique security services firm made famous by its cutting-edge cloud penetration testing services, released the AWS exploitation framework Pacu in 2018. Pacu is open source and has a modular architecture based on common syntax and data structure to allow for simple expansion of its features. A key difference between Pacu and other cloud-focused assessment tools is that it's meant to be used in penetration tests and not just compliance checks. Accordingly, the framework makes it easy to document and export results in a manner familiar to those who frequently perform security engagements. Pacu is available via Rhino Lab's GitHub page, along with all the documentation and initial configurations needed to get started testing instances and services. Pacu ships with more than 30 modules that enable a range of attacks, including user privilege escalation, enumeration, and attacking vulnerable Lambda functions. Each of the modules has comprehensive details about the module's usage, as shown in Figure 4-28.

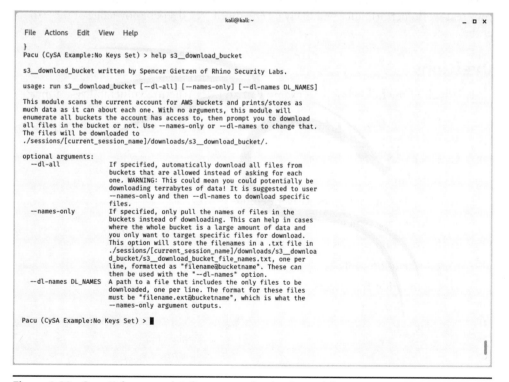

Figure 4-28   Pacu Help screen detailing usage of a plug-in module

Chapter Review

Vulnerability scanning is a key responsibility of any security team. Your taking the steps to understand and track the vulnerabilities your network faces is important in determining the best mitigation strategies. Keeping key stakeholders involved in the effort will also enable you to make decisions that are in the best interests of the organization. When you increase the visibility of the vulnerability status of your network, you ensure that your security team can focus its efforts in the right place and that leadership can devote the right resources to keeping the network safe. We covered several tools, including Nessus, OpenVAS, and Nikto, all of which provide vulnerability scan information using continuously updated libraries of vulnerability information. Some of these tools also offer the ability to automate the process and output in formats for ingestion in other IT systems. Vulnerability scanning is a continuous process, requiring your security team to monitor the network regularly to determine changes in detected vulnerabilities, gauge the efficacy of the patches and compensating controls, and adjust its efforts accordingly to stay ahead of threats.

As a CySA+, you should at least be familiar with every tool we described here, and, ideally, you should be proficient with each. Our goal was not to provide you with the depth of knowledge we believe you should possess on these tools, but rather to give a high-level survey of these essential tools of the trade. We hope that you will use this as a springboard for your own self-study into any products with which you are not familiar. Though the exam will require that you have only a general familiarity with these tools, your real-world performance will likely be enhanced by a deeper knowledge of these tools of the trade.

Questions

1. Which of the following is an example of an infrastructure vulnerability scanner?

   A. Bro

   B. Aircrack-ng

   C. OpenVAS

   D. Burp Suite

2. To what class of tools does Reaver belong?

   A. Wireless assessment tools

   B. Interception proxies

   C. Enumeration tools

   D. Exploitation frameworks

Use the following scenario to answer Questions 3–5:

Your company's internal development team just developed a new web application for deployment onto the public-facing web server. You are trying to ensure that it conforms to best security practices and does not introduce any vulnerabilities into your systems.

3. Which would be the best tool to use if you want to ensure that the web application is not transmitting passwords in cleartext?

   A. Nikto

   B. Hashcat

   C. Burp Suite

   D. Aircrack-ng

4. Having tested the web application against all the input values you can think of, you decide to try random data to see if you can force instability or crashes. Which is the best tool for this purpose?

   A. Untidy

   B. hping

   C. Nessus

   D. Qualys

5. Your developer team is preparing to roll out a beta version of the software tomorrow and wants to quickly test for vulnerabilities including SQL injection, path traversal, and cross-site scripting. Which of the following tools do you recommend to them?

   A. nmap

   B. Arachni

   C. Hashcat

   D. nslookup

6. Which utility provides advanced features beyond those offered by the built-in network tools?

   A. OpenVAS

   B. Zed Attack Proxy

   C. hping

   D. nslookup

7. You have a list of password hashes, and you want to quickly assess and recover a password from each. Which is the most appropriate tool to use for this operation?

   A. Hashcat

   B. ifconfig

   C. nslookup

   D. MD5

8. Of the cloud infrastructure assessment tools we covered in this chapter, which of the following is designed as both a compliance assessment tool and penetration testing framework?

   A. Pacu

   B. Scout Suite

   C. Reaver

   D. Prowler

Answers

1. **C.** OpenVAS (in addition to Qualys, Nessus and Nikto) is an infrastructure vulnerability scanner with which you should be familiar.

2. **A.** Reaver is a wireless assessment tool that takes advantage of a vulnerability that exists in access points that use the Wi-Fi Protected Setup (WPS) feature.

3. **C.** Burp Suite is an integrated web application testing platform often used to map and analyze a web application's vulnerabilities. It is able to intercept web traffic and enables analysts to examine each request and response.

4. **A.** The class of tools that tests applications by bombarding them with random values is called fuzzing tools. Untidy and Peach Fuzzer are all examples covered in this chapter.

5. **B.** Arachni has the ability to incorporate the feedback it gets from initial responses during the test to inform what new techniques should be used moving forward. In presenting the final results, Arachni is also able to reduce the occurrence of false positives through a process its author refers to as *meta-analysis*, which considers several factors from the tests responses.

6. **C.** hping is a free packet generator and network analyzer that supports TCP, UDP, and ICMP, and it includes a traceroute functionality.

7. **A.** Hashcat and its graphics card–optimized version oclHashcat are advanced password recovery tools that support a variety of operating modes, character sets, and formats.

8. **A.** Cloud infrastructure assessment tools assess cloud host vulnerabilities caused by misconfigurations, access flaws, and custom deployments, many times checking against industry standards and compliance checklists. Pacu performs these assessments and also enables penetration testers to exploit configuration flaws that are discovered.

Threats and Vulnerabilities Associated with Specialized Technology

In this chapter you will learn:

- How to identify vulnerabilities associated with unique systems
- Most common threat vectors for specialized technologies
- Vulnerability assessment tools for specialized technologies
- Best practices for protecting cyber-physical systems

To kill an error is as good a service as, and sometimes even better than, the establishing of a new truth or fact.

—Charles Darwin

Most threat actors don't want to work any harder than they absolutely have to. Unless they are specifically targeting your organization, cutting off the usual means of exploitation is often sufficient to encourage them to move on to lower hanging fruit elsewhere. Fortunately, we know a lot about the mistakes that many organizations make in securing their systems, because, sadly, we see the same issues time and again. Before we delve into common flaws on specific types of platforms, here are some that are applicable to most, if not all, systems:

- **Missing patches/updates**   A system could be missing patches or updates for numerous reasons. If the reason is legitimate (for example, an industrial control system cannot be taken offline), then this vulnerability should be noted, tracked, and mitigated using an alternate control.

- **Misconfigured firewall rules**   Whether or not a device has its own firewall, the ability to reach it across the network, which should be restricted by firewalls or other means of segmentation, is oftentimes lacking.

- **Weak passwords**   Even when default passwords are changed, it is not uncommon for users to choose weak passwords if they are allowed to do so. Our personal favorite was an edge firewall that was deployed for an exercise by a highly skilled team of security operators. The team, however, failed to follow its own checklist and was so focused on hardening other devices that it forgot to change the default password on the edge firewall.

Access Points

Perhaps the most commonly vulnerable network infrastructure components are the wireless access points (WAPs). Particularly in environments where employees can bring (and connect) their own devices, it is challenging to strike the right balance between security and functionality. It bears pointing out that the Wired Equivalent Privacy (WEP) protocol has been known to be insecure since at least 2004 and has no place in our networks. (Believe it or not, we still see them in smaller organizations.) For best results, use the Wi-Fi Protected Access 2 (WPA2) protocol.

Even if your WAPs are secured (both electronically and physically), anybody can connect a rogue WAP or any other device to your network unless you take steps to prevent this from happening. The IEEE 802.1X standard provides port-based Network Access Control (NAC) for both wired and wireless devices. With 802.1X, any client wishing to connect to the network must first authenticate itself. With that authentication, you can provide very granular access controls and even require the endpoint to satisfy requirements for patches/upgrades.

Virtual Private Networks

Virtual private networks (VPNs) connect two or more devices that are physically part of separate networks and enable them to exchange data as if they were connected to the same LAN. These virtual networks are encapsulated within the other networks in a manner that segregates the traffic in the VPN from that in the underlying network. This is accomplished using a variety of protocols, including the Internet Protocol Security (IPSec) Layer 2 Tunneling Protocol (L2TP), Transport Layer Security (TLS), and the Datagram Transport Layer Security (DTLS) used by many Cisco devices. These protocols and their implementations are, for the most part, fairly secure.

 NOTE   In considering VPN vulnerabilities, we focus exclusively on the use of VPNs to connect remote hosts to corporate networks. We do not address the use of VPNs to protect mobile devices connecting to untrusted networks (for example, coffee shop WLANs) or to ensure the personal privacy of network traffic in general.

The main vulnerability in VPNs lies in what they potentially allow us to do: connect untrusted, unpatched, and perhaps even infected hosts to our networks. The first risk comes from which devices are allowed to connect. Some organizations require that VPN client software be installed only on organizationally owned and managed devices. If this is not the case and any user can connect any device, provided they have access credentials, then the risk of exposure increases significantly.

Another problem, which may be mitigated for official devices but not so much for personal ones, is the patch/update state of the device. If we do a great job at developing secure architectures but then let unpatched devices connect to them, we are providing adversaries a convenient way to render many of our controls moot. The best practice for mitigating this risk is to implement a Network Access Control (NAC) solution that actively checks the device for patches, updates, and any required other parameter *before* allowing it to join the network. Many NAC solutions enable administrators to place noncompliant devices in a "guest" network so they can download the necessary patches/ updates and eventually be allowed in.

Finally, with devices that have been "away" for a while and show back up on our networks via VPN, you have no way of knowing whether they are compromised. Even if they don't spread malware or get used as pivot points for deeper penetration into your systems, any data these devices acquire would be subject to monitoring by unauthorized third parties. Similarly, any data originating in such a compromised host is inherently untrustworthy.

Mobile Devices

Smartphones and mobile technologies have overtaken any other form of computing in terms of its accessibility. There's no question that these devices and the applications that run on them have become a fixture in most people's daily routine. The integration of mobile technologies into nearly every aspect of life means that these devices have a far greater density of stored personal information that any other media in history. The ability to interact with service providers that have traditionally operated in a strictly brick-and-mortar sense, such as financial service and health service providers, demonstrates the increasing criticality of these devices. Additionally, improvements in infrastructure, distributed systems performance, and high-speed communications have enabled service providers to offer their services reliably. But as mobile technologies and platforms evolve, there are an accompanying set of vulnerabilities targeting these systems, devices, and apps connected to the mobile ecosystem.

Given the potential for impactful and persistent access, it's no wonder that mobile devices have become a lucrative target for malicious programmers. We'll describe the spectrum of mobile vulnerabilities and threat vectors using three categories: network, device, and application.

 NOTE There is a well-known adage that says that if I can gain physical access to your device, there is little you can do to secure it from me. This applies to any computer or network device and reminds us of a common vulnerability of all mobile devices: theft. We have to assume that every mobile device will at some point be taken (permanently or temporarily) by someone who shouldn't have access to it. What happens then? In Part V, we'll discuss the importance of privacy and the role it plays in security.

Network Vulnerabilities

We can look at mobile network vulnerabilities from two angles: flaws of the telecommunications backbone that the device relies on, and flaws of the device's communications interface to these networks. An example of the first kind was brought to light as early as 2008 and again throughout the next decade related to the Signaling System No. 7 (SS7) protocols. SS7 is a set of protocols that enable large communications networks to exchange the information needed to pass calls and texts seamlessly and to ensure correct delivery to the user and correct billing information to the carrier. If you travel to a foreign country, for example, SS7 allows your calls to be transparently forwarded without the need for additional action from you. However, if given access to these exchange points, actors can record or listen in on calls, read SMS messages sent between phones, and collect location data of the communicating parties. While the likelihood of gaining that access is extremely low, it's not out of the question, especially for sophisticated and well-resourced actors.

The Heartbleed Bug is an example of the second kind of mobile network vulnerability that we'll cover. Heartbleed was a serious vulnerability in the popular OpenSSL cryptographic library and was disclosed in 2014. It impacted the security and privacy of a wide number of apps and services, such as web, e-mail, and messaging platforms, as well as connected devices such as VOIP phones, routers, and network-attached storage (NAS). The bug's name is derived from the flaw of the OpenSSL heartbeat function, which helps keep network connections alive by exchanging data periodically between the client and server. As explained by researchers from Codenomicon and Google, the flaw could enable an attacker to recover up to 64K of data, which could include session cookies, passwords, and cryptographic private keys of the communicating parties. Filed as CVE-2014-0160, it was believed to have impacted almost 20 percent of web servers around the world.

Device Vulnerabilities

Device vulnerabilities can be devastating, because a successful exploitation means that the device's entire functionality, contents, and associated personal user data are often at risk. For many years, researchers theorized of an attack on device memory involving the manipulation of the electrical charges that are responsible for keeping the memory elements of a device in state—that is, on or off. The challenge with these types of attacks is that they exploit fundamental properties in the operation of the device, so patching is difficult if not impossible.

When a processor accesses dynamic random-access memory (DRAM), it will attempt to read an element, or bit, of memory that is held in state by an electric charge. These small charges are sent to the memory elements to encode data in 1's and 0's. For performance reasons, DRAM will access the state of a memory bit by reading the whole row, modifying a bit, and replacing the contents of the row. Occasionally, this operation will cause charges to leak and make changes to neighboring bits, causing them to change their state.

Figure 5-1
Depiction of
Rowhammer
attack

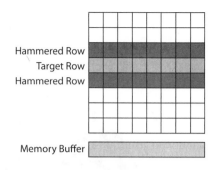

Hammered Row
Target Row
Hammered Row

Memory Buffer

These routine and random occurrences are addressed through various correction mechanisms, but if an attacker can intentionally force this charge spillage in a predictable way, he may be able to cause specific changes to bit states that have an upstream affect to the software in use. This is exactly what attackers hope to achieve using the Rowhammer technique. By repeatedly accessing the adjacent rows of a target row—the row immediately above and the row immediately below, as shown in Figure 5-1—the attacker may cause a specific bit flip to change in the target row with all the "hammering."

This bit change may be exactly what's necessary to initiate a cascading effect, which eventually grants an attacker access to a resource he previously did hot have. Fortunately, an attack like Rowhammer is exceedingly rare and difficult to execute; nevertheless, it demonstrates the relative difficulty in defending against attacks of this kind.

Operating System Vulnerabilities

Operating system makers regularly publish system and firmware updates that include improvements to operating system functionality and security. An interesting phenomenon occurs after these bulletins are released. As with most software security notifications, the authors hope that adoption can be as complete and rapid as possible, while attackers hope to maximize the *window of vulnerability*, or the amount of time they have to exploit these recently disclosed flaws. For a variety of reasons, Android devices typically have longer vulnerability windows than iOS systems. This is primarily because of the challenge of coordinating multiple manufacturers and carriers to develop an operating system patch for Android devices. Figure 5-2 shows a chart of Android distribution worldwide, according the Google Developers portal. Though version 10 is the latest major release as of the writing of this book, it doesn't have nearly the adoption rate as the four previous versions.

In contrast, iOS versions see rapid adoption and no fragmentation, in part because of the existence of a centrally controlled ecosystem. Figure 5-3 shows the latest iOS version as of the writing of this book, iOS 13, as the clear dominant shareholder for iOS installs across all devices. According to the Apple App Store, iOS 13 accounts for 77 percent of the active install base, iOS 12 accounts for around 17 percent, and previous versions account for 6 percent.

Figure 5-2
Android version
adoption
(May 2019)

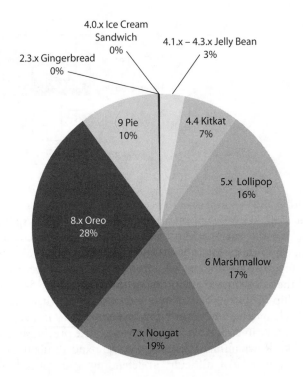

Figure 5-3
iOS version
adoption
(January 2020)

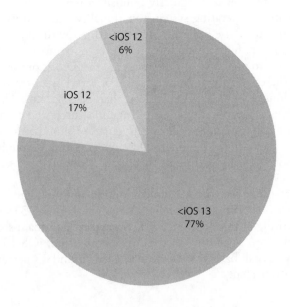

App Vulnerabilities

While security engineers and architects work tirelessly to improve core security protocols, user security and privacy issues regularly occur at the app level, where data is often collected and stored for the first time. Even while security controls around data in transit and at rest improve, when a human enters or reads personal data, malicious actors routinely take advantage of the opportunity.

Apps experience increasingly short development cycles as creators hope to ship their products as quickly as possible to the end user. Unfortunately, this often means that security concerns are second to functionality. We'll cover this larger concept of secure software development in depth in later Chapter 9, but with quick software development cycles come vulnerabilities. Of the top ten risks to mobile platforms identified by the Open Web Application Security Project (OWASP) in its 2016 list, eight of them are directly related to applications. We'll discuss some of these categories and their potential for impact in the following sections.

Improper Platform Usage

In addition to building in technical controls to prevent abuse, platform creators publish documentation that is designed to be consistent and well understood. Despite this, some developers take shortcuts that introduce vulnerabilities that may allow attackers to circumvent existing controls. Although not all of this behavior is intentional, occasionally security researchers will discover developer behaviors that blatantly contradict the best practices recommended by platform guidelines or that are obviously designed to circumvent security controls. In these cases, the developers' digital code-signing signatures may be revoked, or they may be banned from the platform altogether.

Many developers attempt to do the right thing, but sometimes they may allow in errors related to how the app interfaces with the platform. This may occur in the form of incorrectly crafted API calls or incorrect implementation of built-in security controls.

Insecure Data Storage

Vulnerabilities related to data storage often have as much impact on privacy as they do on security. Such vulnerabilities usually occur when developers fail to handle data transfer and storage in a manner that protects confidentiality and integrity. World readable files, improper logging, and surreptitiously collected analytics data fall into this group.

Insecure Authentication

Authentication mechanisms can be difficult to implement, and sometime developers may not fully grasp all the requirements for correct implementation of these mechanisms. There are a few ways that applications may be vulnerable from insecure authentication. First, there is the unfortunately common occurrence of locally stored passwords and secrets without the use of built-in security systems, such as iCloud Keychain. Authenticating a user locally often introduces the possibility of bypassing an authentication routine completely, especially on jailbroken or rooted devices. Second, an app may not sufficiently enforce password policies, allowing for weak passwords to be entered into a system. Although this doesn't have immediate impact, it nonetheless introduces a vulnerability that may be exploited at a later point.

Insecure Authorization

As with authentication vulnerabilities, authorization vulnerabilities may allow for access to resources on a mobile platform that an actor would not otherwise have. The main difference, of course, is that authorization vulnerabilities affect the mechanism that ensures that a user gains access to the correct resource, rather than verifying that the user is who he claims to be. Mitigating these types of vulnerabilities involves verifying that the roles and permissions associated with a user are correct and appropriate, and that they are verified using only back-end systems that are not accessible by the user. Sometimes authorization vulnerabilities are also authentication vulnerabilities, such as the case of a resource request in which the authentication occurs out of order—for example, if an API request occurs and is somehow fulfilled before any validation occurs.

Code Quality Vulnerabilities

To this day, some code practices are technically legal, but they still introduce the possibility for issues such as buffer overflow vulnerabilities. Code-level mistakes continue to be an issue, especially when older code is reused. Sometimes, when the only solution is to completely rewrite a major portion of the app to remove a flaw, some developers simply will not do so.

When executed, code runs within an environment that the original developer is not in control of, so the possibility of modification after the fact is real. Attackers will often attempt to make binary changes directly to the application package or its core dependences. Additionally, developers are more reliant on third-party libraries and frameworks, which themselves may be subject to targeting and compromise.

One of the best ways to prevent vulnerabilities of this kind is via manual secure code review using subject matter experts. Although this may be laborious, it helps security teams understand what the app is trying to do before it's executed. Furthermore, an app should be written so that it is able detect at runtime any code that may have been manipulated. The app must be able to react appropriately to ensure integrity and to prevent malicious actors from injecting functionality into otherwise trusted code.

Messaging Platforms

In late 2019, WhatsApp owner Facebook revealed that a security researcher had discovered a critical flaw in the app that could lead to remote code execution (RCE). The flaw, reported as CVE-2019-11932, affected users of Android versions 8.1 and 9.0 and could enable a malicious actor to achieve full access with root privileges, giving an attacker complete access to all the files on that device. Classified as a double-free memory vulnerability, the vulnerability affected the WhatsApp image preview library. An attacker could take advantage of the flaw by sending a user a specially crafted image that would trigger an automatic preview of the image and then execute the malicious code.

What's noteworthy about this vulnerability and subsequent exploitation is that it affected not only the app, but the entire mobile device. It came at a time that the app was already in the spotlight for several other high-profile security and privacy vulnerabilities that year, including a zero-day vulnerability that could allow spyware injection onto victims' phones, another vulnerability that allowed interception of media files, and another that allowed an attacker to remotely upload malicious code onto a phone by sending packets of data masquerading as phone calls.

Internet of Things

The *Internet of Things* (IoT), the broad term for Internet-connected nontraditional computing devices, is expanding rapidly in terms of adoption. IoT devices, which include televisions and fridges, now offer features not usually associated with them—or, in some cases, needed by their users. Nevertheless, they are here to stay, so we must contend with the security implications they introduce. For the enterprise, the most common IoT devices are those responsible for adding efficiencies to the daily operations of an organization while lowering cost, since these specialized devices reduce the requirement of having dedicated computer systems to perform the same functionality. Your physical security team, for example, may rely on a network of closed-circuit television cameras to monitor the organization's property. The facilities team may use smart thermostats and lighting, which will adjust their operations based on sensors designed to detect occupancy. More and more, printer companies are adding functionality not only to access print services from the desktop network, but to allow for the printer to send messages to administrators in the case of low ink and paper.

For the security analyst, it's important to understand the massive impact these devices may have on increasing an organization's attack surface. As organizations put more faith in these connected devices to give us visibility over their assets, they must take extra care to protect their networks and maintain the confidentiality, integrity, and availability of the data that flows through them. As mentioned several times throughout this book, your understanding what devices are connected is the beginning of your understanding how best to protect the organization.

Patching, as you know, is a necessary part of the vulnerability management process. Some devices, while Internet-connected, do not have straightforward mechanisms to update their software. In many cases, users may have to download updates manually and then interface directly with the device to apply the change. These additional steps are costly, especially if the IoT footprint is hundreds or thousands of devices, and they often lead to administrators deprioritizing maintenance of these devices, which results in them running outdated software, and thus being more vulnerable to exploit.

Manufacturers often release connected devices that contain default, static, or easily decipherable passwords. When we combine this with the fact that many of them ship with services open to remote access, these devices are a target for attackers running automated scripts for bulk exploitation. Several times in the last decade, IoT devices with default passwords and services in place have been targeted, exploited, and used in massive botnets. After they're initially accessed, attackers upload malicious code and maintain connectivity with them to

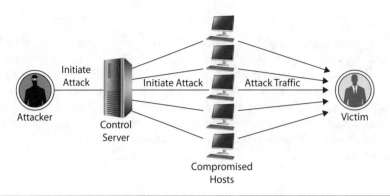

Figure 5-4 Typical botnet structure

form large clusters of centrally controlled devices capable of performing distributed denial-of-service (DDoS) attacks, data theft, and large-scale spamming, as shown in Figure 5-4.

The Mirai Botnet

Mirai is a malware strain that infects IoT devices and was behind one of the largest and most effective botnets in recent history. The Mirai botnet took down major websites via massive DDoS attacks against several sites and service providers using hundreds of thousands of compromised IoT devices. In October 2016, a Mirai attack targeted the popular DNS provider Dyn, which provided name resolution to many popular websites such as Airbnb, Amazon, GitHub, HBO, Netflix, PayPal, Reddit, and Twitter. After taking down Dyn, Mirai left millions of users unable to access these sites for hours.

 NOTE Strong passwords are incredibly effective in protecting IoT devices, but along with changing defaults passwords, you should always minimize the number of open ports and disable unnecessary services such as Telnet and Universal Plug and Play (UPnP). As with any network hardware or servers, IoT devices should not be physically exposed to unauthorized access.

Medical Devices

IoT devices are now commonplace in the medical industry as a means to acquire up-to-date medical information or to deliver lifesaving functionality. Pacemakers and heart monitors, for example, enable doctors to keep track of patient health and adjust device settings remotely. The exploitation of devices like these, either through acquisition of legitimate credentials or via an exploitation of a hardware flaw, can have fatal consequences. Although there have been several proofs of concept demonstrating the possibility for remote takeover and execution of code on medical devices, it's exceedingly rare to see exploits in the wild. In any case, a strategy for securing such IoT hardware needs to be a part of an organization's overall security strategy if its employees use these devices.

Embedded Systems

Embedded systems are characterized by lightweight software running specialized tasks on low-power microprocessors. These instructions are written to the device's firmware, or they are hardwired instructions, which govern every aspect of the device's operation and interface with the rest of the world. Despite the firmware being written using standard programming languages, these programs often pose a special challenge to monitor for and remediate vulnerabilities given the way the programs are executed on the hardware.

Real-Time Operating Systems

A *real-time operating system* (RTOS) is a crucial underlying technology of many high-performance, specialized, connected devices. It is an operating system designed to provide low-latency responses on input. Usually found in vehicle electronics, manufacturing hardware, and aircraft electronics, RTOS solutions excel at scheduling tasks to avoid any possibility for delays in processing or delivery. The vulnerabilities that affect RTOS are not unique, but the vast deployment of RTOS technology in high-performance devices makes them especially important.

There's no greater example than the recent disclosure of vulnerabilities affecting VxWorks, one of the most popular RTOS implementations in the world. In 2019, Armis Labs discovered and published details on 11 zero-day vulnerabilities in VxWorks that allowed for complete control of affected devices. These devices ranged from medical equipment, such as MRI scanners, to elevator controllers. The vulnerabilities stemmed from IPnet, VxWorks' TCP/IP stack implementation. As a result of the severity and potential for impact of the vulnerabilities, named URGENT/11, the US Food and Drug Administration (FDA) issued a warning about the potential for an attacker to "remotely take control of the medical device and change its function, cause denial of service, or cause information leaks or logical flaws, which may prevent device function." The unusual nature of the FDA's involvement highlights not only the increasing dependency on connected technology from organizations of all kinds, but also the importance of federal agencies, manufacturers, and security researchers in working together to identify, communicate, and prevent abuse of vulnerabilities when they are discovered.

System on a Chip

A *system on a chip* (SoC) is the integration of software and hardware onto a single integrated circuit and a processor. Similar to microcontrollers in terms of their minimalist approach to tackling special tasks, SoCs usually involve more complicated circuitry. The integration of all the necessary functions offers multiple advantages related to cost and power consumption, but this tight integration means a higher likelihood for system-wide impact in the case of an exploit. There are also considerations of trusted hardware when it comes to SoC that do not necessarily apply to traditional operating systems. (We'll discuss the concept of trusted hardware more in Chapter 6.) Because the software and hardware are so tightly coupled with SoC solutions, it's critical to ensure that hardware verification is performed alongside any software vulnerability assessments.

Field Programmable Gate Array

Specialized chips often offer increased performance and savings at the cost of device flexibility. One possible solution to the inflexibility of integrated circuit functionality is the use of a *field-programmable gate array* (FPGA). A FPGA is a kind of programmable chip that is widely used in many areas, including automotive electronics, medical devices, and consumer electronics. The design of FPGAs enables programmers to reconfigure the hardware itself to accommodate new software functionality.

The range of vulnerabilities to FPGAs is fairly broad and includes those that apply to both hardware and software. Like other hardware, systems based on FPGAs remain vulnerable to flaws found in higher level protocols used in the system's operations, along with the challenges of proper implementation. Additionally, FPGAs are a target for very specialized attacks that rely on compromising the device functionality at the level it's written. FPGAs use a hardware description language (HDL) to define the operation of the chip. Should an actor manage to gain access to the HDL process, it would be possible for him to introduce malware, which might be extremely difficult to detect in upstream operations.

Common Vulnerabilities and Exposures (CVEs) are occasionally issued for vulnerabilities related to FPGAs in commercial products. CVE-2019-1700, for example, describes a vulnerability related to the FPGA used in a series of Cisco firewall devices. In this case, the FPGA vulnerability could allow for an attacker to cause a denial-of-service (DoS) condition by taking advantage of a logic error in the FPGA implementation. Because of the way the FPGA processes certain packets, an attacker could initiate the condition from an adjacent subnet by sending a series of specially crafted packets to the device for processing. The resulting *queue wedge condition* would cause the device to stop processing subsequent packets.

Physical Access Control

Physical access control systems usually rely on a combination of hardware and logical artifacts to make a determination about physical access for a user presenting the credentials. Radiofrequency identification (RFID) badges and readers are among the most common types of physical access control. Every complicated system has vulnerabilities up and down the stack. With physical access control systems, attackers often rely on taking advantage of one or more flaws somewhere between the hardware and presentation layers to gain access to decision systems.

First, it's often possible for an attacker to acquire a hardware reader and perform a teardown to determine whether any mechanical or electrical flaws exist. If, for example, there is logic performed on the device between the RFID card and the reader that simply results in a 50ms 5-volt pulse from the device reader to the door mechanism, then gaining access could be as simple as removing the reader from the wall and creating that pulse using another tool. This is an extremely simplified way to remind you of the point that physical systems are vulnerable to physical vulnerabilities and physical attacks.

One of the more publicized vulnerabilities of physical access systems is their susceptibility to spoofing attacks by way of replay or cloning. *Replay* is a method of capturing and emulating the signals exchanged between legitimate components of a physical access control system. Early systems were particularly vulnerable to attacks in which threat actors captured the radio signals emitted from a card and replayed them at a later point to the reader device to gain access. Many of these methods have since been defeated using encryption and the use of unique client/server exchanges. *Cloning* is a method of capturing the information from a card and emulating its functionality, or copying it to a similar device.

Connected Vehicles

For organizations in the automotive sector, or those that use a massive fleet of connected vehicles, the vulnerabilities associated with connected cars are high priorities in their security strategy. Threat modeling realistic attack vectors and attacker methodologies will absolutely be part of their vulnerability assessment plans. The key questions that these organizations will hope to answer are as follows:

- What kinds of vulnerabilities most commonly affect a modern connected car?
- What vectors are attacks likely to use to compromise a vehicle?
- How can we mitigate these vulnerabilities?
- How do we prioritize our resources to address these challenges?

Most times, when there is a discussion about a vulnerability related to a vehicle, it's in the context of some kind of tethered, or physically connected, compromise. However, with the emergence of connected vehicles, this can no longer be the assumption for security professionals, anyone working in the automotive industry, or even the general public.

Imagine for a moment that a security system is installed on a vehicle's computer system. Suppose an attack that gains access to the vehicle's communication and control systems to wreak havoc is indistinguishable from the system's normal operation. But even if a detection system did alert you to the malicious behavior that intended to create an immediately hazardous situation in the vehicle, the system's priority would unquestionably be to preserve life first, and then react to the alert. An attacker trying to cause chaos on such a system would probably be less concerned about being detected and more interested in making sure the attack was successful.

CAN Bus

The Controller Area Network (CAN bus) vehicle standard defines the ways in which independent components of vehicles control systems, usually in the form of electronic control units (ECUs), communicate with one another without having a central controller. The primary security issues with the CAN bus standard lie in the fact that, as a low-level protocol developed in the 1980s, it includes no inherent security features to protect the signals that are transmitted across the network. As a result, mechanisms that work to maintain the security of the data are applied often at the component level, with no measures for consistency. Signals that pass along the CAN bus are also not protected by encryption by default, so the chance of a man-in-the-middle attack is far greater than an attack on other systems with protection. A typical CAN bus configuration is shown in Figure 5-5.

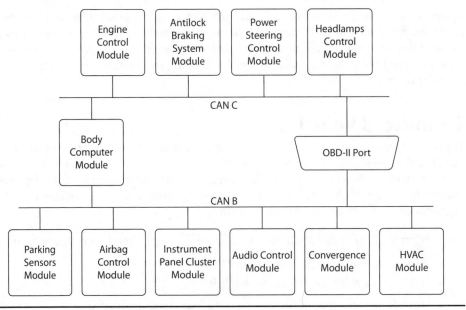

Figure 5-5   A typical CAN bus configuration

A few years ago, security researchers Charlie Miller and Chris Valasek presented a demonstration of a vehicle takeover initiated from the vehicle's Uconnect cellular connection. After gaining access to the vehicle's network via a flaw in the Uconnect system, the two pivoted to take advantage of a vulnerability in the vehicle's entertainment system. After replacing the radio unit's functionality with their own, they were able to send commands through the car's CAN bus to its physical components, including the engine and the wheels. Both Miller and Valasek worked closely with the car's manufacturer to ensure that all aspects of their research would be correctly addressed, in a great example of responsible disclosure. Their research advanced the public discussion about connected devices and even helped move forward legislation regarding connected vehicles.

Drones

Hobby and professional quadcopters, often called drones, are at their core flying robots capable of a wide range of activities. Photographers and government organizations take advantage of the unique access and perspective that these small and maneuverable devices offer when taking photos or performing inspections. Drone platforms have seen rapid adoption, in part because of the plummeting cost of the hardware, and also because of improvements in the user interface and integration with other familiar technologies such as smartphones and tablets.

Drones are, however, a growing concern for the law enforcement, physical security, and aviation communities because of their potential use as tools of disruptive and dangerous activities. (The forced grounding of hundreds of flights at Gatwick Airport in December 2018, for example, occurred following reports of drone sightings close to the runway.)

Many of those expressing concern are also concerned about drone use in low-cost surveillance, enabling controllers to capture high-quality audio and video at a fraction of the cost of higher end tools. Furthermore, skilled operators can modify these drones to perform signals collection, tracking, and identification functionality in much the same way that some intelligence organizations might.

From a security point of view, regardless of the intent of the user behind the controls, we must remember that attackers often use vulnerabilities to take over and abuse the legitimate access of well-meaning users on the network. It's no different with drones— a malicious actor may take advantage of vulnerabilities in the drone's communication systems, its hardware, or the manufacturer's web portal to steal information, perform surveillance, or cause disruption.

Hardware Security

As we've stated throughout the book, physical access is often the best kind of access from an adversary's perspective. Many low-cost drones use easy-to-find components, standard storage media, and common connectors. If your organization uses drones for monitoring or inspection purposes, their security begins with preventing the devices from getting into the wrong hands to begin with. Recovering media and photos from these devices often just involves removing the storage media and connecting it to another computer.

Communications Channels Security

In the interest of cost savings and ease of use, early drones often operated as Wi-Fi hotspots that users would connect to via a mobile phone or tablet to control the device. In many cases, these hotspots were not protected by any kind of password or encryption. The security community quickly discovered this problem, along with the fact that some drones, such as those in the popular Parrot AR line, left many network services such as telnet wide open, as shown in Figure 5-6.

Additionally, many of these devices are susceptible to DoS attacks via Wi-Fi; an attacker could overload the drone with at a high volume of connection requests until its hardware is overwhelmed and subsequently shuts down. The result would almost certainly involve damage to the drone and whatever is below it at the time.

Drone manufacturers have since improved drone security following early lessons of shipping the devices without any kind of protection against hijacks and disruptions via the drone's communications channels. Although modern drones are far more resilient to these kinds of attacks, researchers are constantly working to discover new vulnerabilities and improve the security of these devices.

```
Nmap scan report for 192.168.1.1
Host is up (0.033s latency).
Not shown: 997 closed ports
PORT      STATE SERVICE
21/tcp    open  ftp
23/tcp    open  telnet
5555/tcp  open  freeciv

Nmap done: 1 IP address (1 host up) scanned in 13.54 seconds
```

Figure 5-6   nmap scan results of probe performed on Parrot AR.Drone 2.0

Web Portal Security

In late 2018, a security research team from Check Point Software Technologies, an Israeli security software and appliance company, demonstrated the grave effects that could result from a drone web portal takeover. The researchers were able to determine that drone vendor DJI's website login process contained a vulnerability that allowed anyone to authenticate to any account without legitimate credentials. Leveraging the vulnerability, team members were able to log into a target account and gain access to the DJI FlightHub, the web-based app that gives users complete awareness and management over their drones. The researchers worked with DJI to patch the issue quickly, but their research illustrates a successful exploit via vector that might be completely out of control of the user. This dependency on third parties has to be considered when developing your security strategy.

Industrial Control Systems

Industrial control systems (ICS) are cyber-physical systems that enable specialized software to control their physical behaviors. For example, ICSs are used in automated automobile assembly lines, building elevators, and even HVAC systems. A typical ICS architecture is shown in Figure 5-7. At the bottom layer (level 0), are the actual physical

Figure 5-7
Simple industrial control system

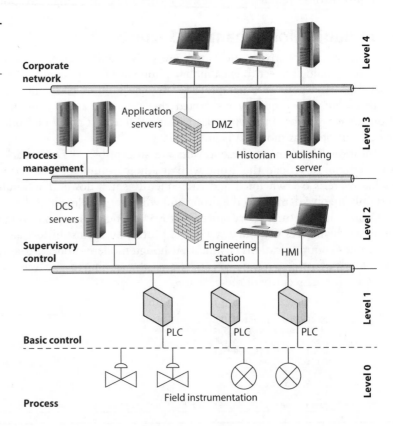

devices such as sensors and actuators that control physical processes. These are connected to remote terminal units (RTUs) or programmable logic controllers (PLCs), which translate physical effects to binary data, and vice versa. These RTUs and PLCs at level 1 are, in turn, connected to database servers and human–machine interaction (HMI) controllers and terminals at level 2. These three lower levels of the architecture are known as the operational technology (OT) network. The OT network was traditionally isolated from the IT network that inhabits levels 3 and 4 of the architecture. For a variety of functional and business reasons, this gap between OT (levels 0 through 2) and IT (levels 3 and 4) is now frequently bridged, providing access to physical processes from anywhere on the Internet.

Much of the software that runs an ICS is burned into the firmware of devices such as programmable logic controllers (PLCs), such as those that run the uranium enrichment centrifuges targeted by the Stuxnet worm in 2010. This is a source of vulnerabilities because updating the ICS software cannot normally be done automatically or even centrally. The patching and updating, which is pretty infrequent to begin with, typically requires that the device be brought offline and manually updated by a qualified technician. Between the costs and efforts involved and the effects of interrupting business processes, it should not come as a surprise to learn that many ICS components are never updated or patched. To make matters worse, vendors are notorious for not providing patches at all, even when vulnerabilities are discovered and made public. In its "2016 ICS Vulnerability Trend Report," FireEye Cybersecurity described how 516 of the 1552 known ICS vulnerabilities had no patches available.

Another common vulnerability in ICSs is the password. Unlike previous mentions of this issue in this chapter, the issue here is not the users choosing weak passwords, but the manufacturer of the ICS device setting a trivial password in the firmware, documenting it so all users (and perhaps abusers) know what it is, and sometimes making it difficult if not impossible to change. In many documented cases, these passwords are stored in plaintext. Manufacturers are getting better at dealing with these issues, but many devices still have unchangeable passwords controlling critical physical systems around the world.

Vulnerabilities in Interconnected Networks

In late 2013, the consumer retail giant Target educated the entire world on a major vulnerability of interconnected networks. One of the largest data breaches in recent history was accomplished not by attacking the retailer's data systems directly, but by using a heating, ventilation, and air conditioning (HVAC) vendor's network as an entry point. The vendor had access to the networks at Target stores to monitor and manage their HVAC systems, but the vulnerability induced by the interconnection was not fully considered by Target's security personnel. In a world that grows increasingly interconnected and interdependent, we should all take stock of which of our partners may present our adversaries with a quick way into our systems.

SCADA Devices

A supervisory control and data acquisition (SCADA) system is a specific type of ICS characterized by its ability to monitor and control devices throughout large geographic regions. Whereas an ICS typically controls physical processes and devices in one building or a small campus, a SCADA system is used for pipelines and transmission lines covering hundreds or thousands of miles. SCADA is most commonly associated with energy (for example, petroleum or power) and utility (for example, water or sewer) applications. The general architecture of a SCADA system is depicted in Figure 5-8.

SCADA systems introduce two more types of common vulnerabilities in addition to those normally found in ICS. The first of these is induced by the long-distance communications links. For many years, most organizations using SCADA systems relied on the relative obscurity of the communications protocols and radio frequencies involved to provide a degree (or at least an illusion) of security. In one of the

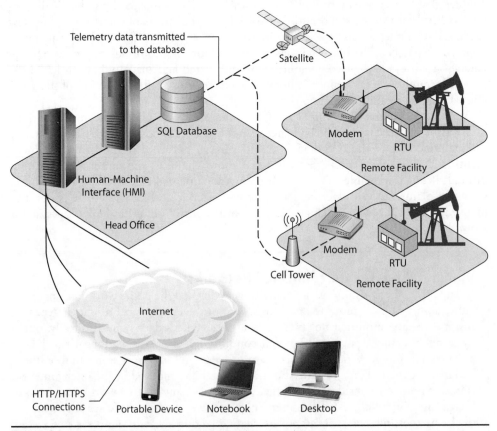

Figure 5-8 Typical architecture of a SCADA system

first cases of attacks against SCADA systems, an Australian man apparently seeking revenge in 2001 connected a rogue radio transceiver to a remote terminal unit (RTU) and intentionally caused millions of gallons of sewage to spill into local parks and rivers. Though these wireless systems have mostly been modernized and hardened, they still present potential vulnerabilities.

The second weakness, particular to a SCADA system, is its reliance on isolated and unattended facilities. These remote stations provide attackers with an opportunity to gain physical access to system components. Though many of these stations are now protected by cameras and alarm systems, their remoteness makes responding significantly slower compared to that of most other information systems.

 EXAM TIP Although the exact protocols used may vary from standard networked devices, the concepts of network scanning, identification of vulnerable targets, and exploit delivery remain the same. For the CySA+ exam, you should focus on broad mitigation strategies for these types of activity.

Modbus

Like CAN bus, the Modbus system was developed to prioritize functionality over security. A communications system created in the late 1970s by Modicon, now Schneider Electric, Modbus enables communications among SCADA devices quickly and easily. Since its inception, Modbus has quickly become the de facto standard for communications between programmable logic controllers (PLCs). But as security was not built in, Modbus offers little protection against attacks. An attacker residing on the network can simply collect traffic using a tool like Wireshark, find a target device, and issue commands directly to the device.

Process Automation Systems

Process automation systems (PASs), which are sometimes called workflow automation systems (WASs), are technologies used by organizations to automate their day-to-day business processes. They help increase efficiency and accuracy in functions that are well established and don't have too many weird edge cases. For example, when new employees are hired, specific applications need to be provisioned, usually with different permissions, depending on their work roles. Many organizations handle this by running a script or tool that accurately takes care of this in seconds instead of relying on a systems administrator to manually configure each account. PAS/WAS technologies are used in virtually all sectors, from automating the processing of sales orders to controlling assembly line robots or even the flow of electricity through a power grid in SCADA systems.

But problems can arise when the processes and workflows that get automated are complex, subject to edge conditions, or just poorly understood. For example, in 2012 a financial services firm called Knight Capital Group unintentionally bought $7 billion worth of stocks in one hour due to a flaw in its automated trading software, a mistake that nearly ended the company. Threat actors can deliberately exploit vulnerabilities in a PAS, as was the case in the attack on the Bangladesh Bank in 2016 that resulted in the theft of $81 million. These attackers exploited the automated workflows used by the Society for Worldwide Interbank Financial Telecommunication (SWIFT) systems. These examples show the importance of thoroughly testing PASs and protecting them as we would any other system.

Chapter Review

Your ability to protect specialized technologies begins with understanding the roles they play in the organization, mapping their attack surfaces, and identifying likely attack vectors. Security vulnerabilities in specialized technologies include many of the same vulnerabilities that affect technology at large, including failures to adequately define system data sensitivities, insufficient security measures against common attacks, poor implementation of appropriate authentication and authorization systems, and lack of mitigation that adheres to the defense-in-depth concept. Sometimes, however, there are special considerations beyond those for standard servers and endpoints. Trusted hardware becomes much more of a concern when you're dealing with highly customized systems that work in sensitive spaces, for example. Security teams must also consider the effect that a successful exploitation of any system dealing with a physical dimension may have. Whether it is allowing physical access to an area or controlling vehicles or heavy equipment, the impact of a security event may have effects that extend into the real world.

Questions

1. Which of these wireless protocols is considered insecure and should be avoided?

   A. RADIUS

   B. WEP

   C. WPA2

   D. ICS

2. World readable files, improper logging, and surreptitiously collecting analytics data are examples of what kind of vulnerability?

   A. Insecure authorization

   B. Insecure authentication

   C. Insecure data storage

   D. Replay vulnerability

3. Which communications standard defines the ways in which independent components of vehicles control systems?

   A. CAN bus

   B. Modbus

   C. SS7

   D. RCE

4. Of the following, which is a type of specialized system whose design is optimized for high-speed, low-latency processing of live data?

   A. FPGA

   B. SoC

   C. RFID

   D. RTOS

5. Which of the following is *not* among the challenges with dealing with IoT devices on your network?

   A. Increased attack surface

   B. Difficulties of patching

   C. Static credentials

   D. Difficulties of integration

6. What is a reason that patching and updating occur so infrequently with ICS and SCADA devices?

   A. These devices control critical and costly systems that require constant uptime.

   B. These devices are not connected to networks, so they do not need to be updated.

   C. These devices do not use common operating systems, so they cannot be updated.

   D. These devices control systems, such as HVAC systems, that do not need security updates.

7. When describing RFID attacks, which of the following best describes the act of capturing signals between card and reader, and emulating those signals at a later point using specialized hardware?

   A. Replay attack

   B. Brute-force attack

   C. Spoofing attack

   D. Man-in-the-middle attack

8. The distributed computational power and network access provided by a collection of infected and hijacked IoT devices can be used to perform distributed denial-of-service attacks. This collection of infected and hijacked devices is referred to as what?

   A. SoC

   B. Parallel computer

   C. Botnet

   D. SCADA

Answers

1. **B.** Introduced in the late 1990s, the Wired Equivalent Privacy (WEP) is a security algorithm for 802.11 wireless networks. It's known to be susceptible to attacks that can reveal keys in a matter of minutes.

2. **C.** Insecure data storage occurs when developers fail to handle data transfer and storage in a manner than protects confidentiality and integrity.

3. **A.** The Controller Area Network (CAN bus) standard defines the ways in which independent components of vehicles control systems, usually in the form of electronic control units (ECUs), communicate with each other without having a central controller.

4. **D.** Often found in vehicle and manufacturing electronics, a real-time operating system (RTOS) is a crucial underlying technology of many high-performance, specialized connected devices. It is an operating system designed to provide low-latency responses on input.

5. **D.** Difficulties of integration are not a challenge with dealing with Internet of Things devices. IoT is the broad term for Internet-connected nontraditional computing devices that add significantly to an organization's attack surface. Many of these devices ship with default or static credentials and may be difficult to update.

6. **A.** The cost involved and potential negative effects of interrupting business and industrial processes often dissuade device managers from updating and patching these systems.

7. **A.** In a Replay attack, valid communications transmissions are captured and retransmitted, often to impersonate a legitimate user to gain access to a system or steal information.

8. **C.** Botnets are large clusters of centrally controlled devices capable of performing distributed denial-of-service (DDoS) attacks, data theft, and large-scale spamming. Attackers will often upload malicious code and maintain connectivity with various IoT devices to form botnets.

Threats and Vulnerabilities Associated with Operating in the Cloud

In this chapter you will learn:

- How to identify vulnerabilities associated with common cloud service models
- Current and emerging cloud deployment models
- Challenges of operating securely in the cloud
- Best practices for securing cloud assets

Bronze Age. Iron Age. We define entire epics of humanity by the technology they use.

—Reed Hastings

Cloud solutions offer many benefits to an organization looking to develop new applications and services rapidly or to save costs in the setup and maintenance of an IT infrastructure. To make the best cloud solution decisions, however, an organization must be fully informed of the vulnerabilities that exist and the myriad financial, technical, and compliance risks that they may assume by using this technology. Cloud environments, for the most part, are subject to the same types of vulnerabilities faced by any other computing platform. Clouds run software and hardware and use networking devices in much the same way that traditional data centers do. Although the targeting and specific exploit techniques used in the cloud may vary, we can approach defending cloud environments as we would in any other environment. Key requirements in forming a defense plan include understanding what your organization's exposure is as a result of using cloud technologies, what types of threat actors exist, what capabilities they may leverage against your organization, and what steps you can take to shore up defenses.

Cloud Service Models

Cloud computing enables organizations to access on-demand network, storage, and compute power, usually from a shared pool of resources. Cloud services are characterized by their ease of provisioning, setup, and teardown. There are three primary service models for cloud resources, listed next. All of these services rely on the shared

143

responsibility model of security that identifies the requirements for both the service provider and server user for mutual security.

- **Software as a Service (SaaS)** A service offering in which the tasks for providing and managing software and storage are undertaken completely by the vendor. Google Apps, Office 365, iCloud, and Box are all examples of SaaS.
- **Platform as a Service (PaaS)** A model that offers a platform on which software can be developed and deployed over the Internet.
- **Infrastructure as a Service (IaaS)** A service model that enables organizations to have direct access to their cloud-based servers and storage as they would with traditional hardware.

Shared Responsibility Model

The shared responsibility model covers how organizations should approach designing their security strategy as it relates to cloud computing, as well as the day-to-day responsibilities involved in executing that strategy. Although your organization may delegate some technical responsibility to your cloud provider, your company is always responsible for the security of your deployments and the associated data, especially if the data relates to customers. You should take the approach of *trust but verify*. To complement your cloud provider's policies and security operations, your organization must have the right tools and procedures in place to manage and secure the risks effectively in order to keep company and user data safe. In each of the following service models, the parts of the technology stack the vendor is primarily responsible for and those for which your organization has full control and oversight over are different, but all aspects are important, and all offer various benefits your company may choose to adopt.

Software as a Service

SaaS is among the most commonly used software delivery models on the market. In the SaaS model, organizations access applications and functionality directly from a service provider with minimal requirements to develop custom code in-house; the vendor takes care of maintaining servers, databases, and application code. To save money on licensing and hardware costs, many companies subscribe to SaaS versions of critical business applications such as graphics design suites, office productivity programs, customer relationship managers, and conferencing solutions. As Figure 6-1 shows, those applications reside at the highest level of the technology stack. The vendor provides the service and all of the supporting technologies beneath it.

SaaS is a cloud subscription service designed to offer a number of technologies associated with hardware provision, all the way up to the user experience. A full understanding of how these services protect your company's data at every level will be a primary concern for your security team. Given the popularity of SaaS solutions, providers such as Microsoft, Amazon, Cisco, and Google often dedicate large teams to securing all aspects of their service infrastructure. Increasingly, any security problems that arise occur at the data-handling level, where these infrastructure companies do not have the responsibility

Figure 6-1
Technology stack
highlighting areas
of responsibility
in a SaaS service
model

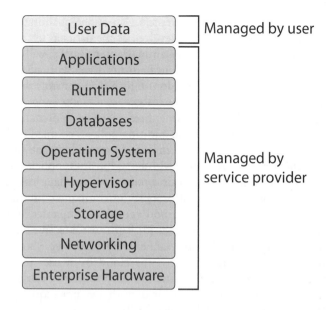

User Data — Managed by user

Applications

Runtime

Databases

Operating System — Managed by
service provider

Hypervisor

Storage

Networking

Enterprise Hardware

or visibility required to take action. This leaves the burden primarily on the customer to design and enforce protective controls. The most common types of SaaS vulnerabilities exist in one or more of three spaces: visibility, management, and data flow.

As organizations use more and more distributed services, simply gaining visibility on what's being used at any time becomes a massive challenge. The "McAfee 2019 Cloud Adoption and Risk Report" describes the disconnect between the number of cloud services IT teams believe are being accessed by their users, verses what are actually being accessed. The discrepancy, according to the report, is several orders of magnitude, with the average company using nearly 2000 unique services. Given that IT teams expect service usage to be a number in the dozens or hundreds, they are often not prepared to fully map and audit usage. This is where you, as a security analyst, can assist in bringing to light the reality of your company's usage, what it means to the attack surface, and some thoughtful courses of action.

A second major challenge to using SaaS solution security is the complexity of identity and access management (IAM). IAM in practice is a perfect example of the friction between security and usability. IAM services enable data owners or security teams to set access rights across services and provide centralized auditing reports for user privileges, provisioning, and access activity. While robust IAM systems are great for an organization because they offer granular control over every aspect of data access, they are often extremely difficult to understand and manage. If the security team finds the system too difficult to manage, the risk increases for a misconfiguration or failure to implement a critical control.

The final major SaaS challenge concerns data, the thing your organization hoped that the SaaS solution would enable you to focus on by removing the need to manage infrastructure. SaaS solutions often offer the ability for you to share data widely, both within an organization and with trusted partners. Although collaboration controls can be set on an organizational, directory, or file level, once the data has left the system, it's impossible to bring it back. Employees, inadvertently or intentionally, will share documents through messaging apps, e-mail, and other mechanisms. Even with clear indication of a sensitive or confidential document, if there are no technical controls to keep it from leaving the network, it *will* leave the network.

Addressing these vulnerabilities requires a layered approach that includes policy and technical controls. Compliance and auditing policies coupled with the features provided by SaaS offerings will give the security team reporting capabilities that may help them determine whether they are in line with larger business policies or with government and industry regulations. From a technical point of view, data loss prevention (DLP) solutions will help protect intellectual property in company communications when used in SaaS products. A number of DLP products enable companies to easily define the boundaries at which DLP may be enforced. DLP is most effective when used in conjunction with a strong encryption method. Data encryption protects both data at rest in the SaaS application and during transit across the cloud from user to user. Some government regulations require encryption for financial information, healthcare data, and personally identifiable information. But even for data that is not required to be encrypted, it's a good idea to use encryption whenever feasible.

Security as a Service

Security as a service (SECaaS) is a cloud-based model for service delivery by a specialized security service provider. SECaaS providers usually offer services such as authentication, antivirus, intrusion detection, and security assessments. Similar to managed security service providers (MSSPs), SECaaS providers bring value to organizations primarily based on cost savings and access to on-demand specialized expertise. The primary difference between the two models is that SECaaS, like other cloud service models, enables organizations to pay only for what they use. Unlike most MSSP agreements, SECaaS models do not necessarily follow an annual contract-based payment agreement.

Platform as a Service

PaaS shares a similar set of functionalities as SaaS and provides many of the same benefits in that the service provider manages the foundational technologies of the stack in a manner transparent to the end user. PaaS differs primarily from SaaS in that it offers direct access to a development environment to enable organizations to build their own solutions on a cloud-based infrastructure rather than providing their own infrastructures, as shown in Figure 6-2. PaaS solutions are therefore optimized to provide value focused on software development. PaaS, by its very nature, is designed to provide organizations with tools that interact directly with what may be the most important company asset: its source code.

Figure 6-2
Technology stack
highlighting areas
of responsibility
in a PaaS service
model

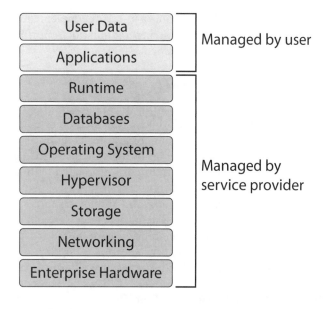

PaaS offers effective and reliable role-based controls to control access to your source code, which is a primary security feature of this model. Auditing and account management tools available in PaaS may also help determine inappropriate exposure. Accordingly, protecting administrative access to the PaaS infrastructure helps organizations avoid loss of control and potential massive impacts to an organization's development process and bottom line.

At the physical infrastructure, in PaaS, service providers assume the responsibility of maintenance and protection and employ a number of methods to deter successful exploits at this level. This often means requiring trusted sources for hardware, using strong physical security for its data centers, and monitoring access to the physical servers and connections to and from them. Additionally, PaaS providers often highlight their protection against DDoS attacks using network-based technologies that require no additional configuration from the user.

At the higher levels, your organization will have to understand any risks that you explicitly inherit and take steps to address them. You can use any of the threat modeling techniques we discussed earlier to generate solutions. In Table 6-1, you can see that

Threat	Security Property	Mitigation
Spoofing	Authentication	Require HTTPS connections
Tampering	Integrity	Validate SSL certificates
Repudiation	Nonrepudiation	Enable monitoring and diagnostics
Information disclosure	Confidentiality	Encrypt sensitive data
Denial of service	Availability	Implement filtering and blocking
Elevation of privilege	Authorization	Use an identity management solution

Table 6-1 STRIDE Model as Applied to PaaS

a PaaS provider may offer mitigation, or it can be implemented by your security team, based on a STRIDE (spoofing, tampering, repudiation, information disclosure, denial of service, elevation of privilege) assessment.

Infrastructure as a Service

Moving down the technology stack from the application layer to the hardware later, we next have the cloud service offering of IaaS. As a method of efficiently assigning hardware through a process of constant assignment and reclamation, IaaS offers an effective and affordable way for companies to get all of the benefits of managing their own hardware without incurring the massive overhead costs associated with acquisition, physical storage, and disposal of the hardware. In this service model, shown in Figure 6-3, the vendor would provide the hardware, network, and storage resources necessary for the user to install and maintain any operating system, dependencies, and applications they want.

The previously discussed vulnerabilities associated with PaaS and SaaS may exist for IaaS solutions as well, but there are additional opportunities for flaws at this level, because we're now dealing with hardware resource sharing across customers. Any of the vulnerabilities that could take advantage of flaws in hard disks, RAM, CPU caches, and GPUs can affect IaaS platforms. One attack scenario affecting IaaS cloud providers could enable a malicious actor to implant persistent backdoors for data theft into bare-metal cloud servers. A vulnerability in either the hypervisor supporting the visualization of various tenant systems or a flaw in the firmware of the hardware in use could introduce a vector for this attack. This attack would be difficult for the customer to detect because it would be possible for all services to appear unaffected at a higher level of the technology stack.

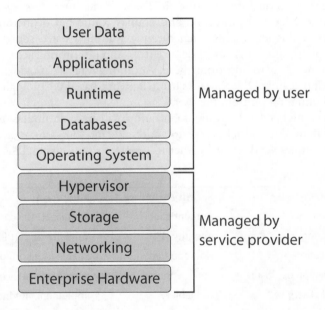

Figure 6-3
Technology stack highlighting areas of responsibility in an IaaS service model

In 2019, researchers at Eclypsium, an enterprise firmware security company, described such a scenario using a flaw they uncovered and named Cloudborne. The vulnerability existed in the baseboard management controller (BMC) of the device, the mechanism responsible for managing the interface between system software and platform hardware. Usually embedded on the motherboard of a service, BMCs grant total management power over the hardware. Among the most noteworthy part of Eclypsium's research efforts was the demonstration of long-term persistence even after the infrastructure had been reclaimed. The security team got initial access to a bare-metal server, verified the correct BMC, initiated the exploit, and made a few innocuous changes to the firmware. Later, after they had given up access to that original server, they attempted to locate and reacquire the same server from the provider. Though surface changes had been made, the team was able to note that the BMC remained modified, indicating that any malicious code implanted into the firmware would have remained unchanged throughout the reclamation process.

Though the likelihood of a successful exploit of this kind of vulnerability is quite low, there may still be significant cost to vulnerabilities detected at this level not related to an actual exploit. Take, for example, the 2014 hypervisor update performed by Amazon Web Services (AWS), which essentially forced a complete restart of a major cloud offering, the Elastic Compute Cloud (EC2). In response to the discovery of a critical security flaw in the open source hypervisor Xen, Amazon forced EC2 instances globally to restart to ensure the patch would take correctly and that customers remained unaffected. In most cases, though, as with many other cloud services, attacks against IaaS environments are possible because of misconfiguration on the customer side.

Cloud Deployment Models

Although the public cloud is often built on robust security principles to support the integrity of customer data, the possibility of spillage may rise above the risk appetite of an organization, particularly organizations operating with loads of sensitive data. Additionally, concerns about the permanence of data, or the possibility that data might remain after deletion, can push an organization to adopt a cloud deployment model that is not open to all. Before implementing or adopting a cloud solution, your organization should understand what it's trying to protect and what model works best to achieve that goal. We'll discuss three models, their security benefits, and their disadvantages.

Public

In public cloud solutions, the cloud vendor owns the servers, infrastructure, network, and hypervisor used in the provision of service to your company. As tenants to the cloud solution provider, your organization borrows a portion of that shared infrastructure to perform whatever operations are required. Services offered by Google, Amazon, and Microsoft are among the most popular cloud-based services worldwide. So, for example, although the data you provide to Gmail may be yours, the hardware used to store and process it is all owned by Google.

Private

In private cloud solutions, organizations accessing the cloud service own all the hardware and underlying functions to provide services. For a number of regulatory or privacy reasons, many healthcare, government, and financial firms opt to use a private cloud in lieu of shared hardware. Using private cloud storage, for example, enables organizations that operate with sensitive data to ensure that their data is stored, processed, and transmitted by trusted parties, and that these operations meet any criteria defined by regulatory guidelines. In this model, the organization is wholly responsible for the operation and upkeep of the physical network, infrastructure, hypervisors, virtual network, operating systems, security devices, configurations, identity and access management, and, of course, data.

Community

In a community cloud, the infrastructure is shared across organizations with a common interest in how the data is stored and processed. These organizations usually belong to the same larger conglomerate, or all operate under similar regulatory environments. The focus for this model may be the consistency of operation across the various organizations as well as lowered costs of operation for all concerned parties. While the community cloud model helps address common challenges across partnering organizations, including issues related to information security, it adds more complexity to the overall operations process. Many vulnerabilities, therefore, are likely related to processes and policy rather than strictly technical. Security teams across these organizations should have, at minimum, standard operation procedures that outline how to communicate security events, indicators of compromise, and remediation efforts to other partners participating in the community cloud infrastructure.

Hybrid

Hybrid clouds combine on-premises infrastructure with a public cloud, with a significant effort placed in the management of how data and applications leverage each solution to achieve organizational goals. Organizations that use a hybrid model can often see benefits offered by both public and private models while remaining in compliance with any external or internal requirements for data protection. Often organizations will also use a public cloud as a failover to their private cloud, allowing the public cloud to take increased demand on resources if the private cloud is insufficient.

Serverless Architecture

Hosting a service usually means setting up hardware, provisioning and managing servers, defining load management mechanisms, setting up requirements, and running the service. In a serverless architecture, the services offered to end users, such as compute, storage, or messaging, along with their required configuration and management, can be performed without a requirement from the user to set up any server infrastructure.

The focus is strictly on functionality. These models are designed primarily for massive scaling and high availability. Additionally, from a cost perspective, they are attractive, because billing occurs based on what cycles are actually used versus what is provisioned in advance. Integrating security mechanisms into serverless models is not as simple as ensuring that the underlying technologies are hardened. Because there is limited visibility into host infrastructure operations, implementing countermeasures for remote code execution or modifying access control lists isn't as straightforward as it would be with traditional server design. In this model, security analysts are usually restricted to applying controls at the application or function level.

Function as a Service

Function as a Service (FaaS) is a type of cloud-computing service that further abstracts the requirements for running code, well beyond that of even PaaS solutions. FaaS lets developers focus on the action to be performed and scale it in a manner than works best for them. In the FaaS model, logic is executed at the time it's required and no sooner. Accordingly, cloud service providers invoice based on the time the function operates and the resources required in completing the task. The basic requirement for FaaS is that the code is presented in a manner that is compliant with the service provider's platform. FaaS promotes code that is simple, highly scoped, and efficiently written. Because it is oriented around functions, code is written to execute a specific task in response to a specific trigger.

Amazon Lambda is currently the most popular example of this event-driven, serverless computing framework. As shown in Figure 6-4, the only requirement for initial Lambda function setup is to define the runtime. Amazon supports several programming

Figure 6-4   Amazon Lambda runtime requirements setup screen showing supported runtimes

languages by default, although you can implement your own runtime in any programming language, provided the code can be compiled and executed in the Amazon Linux environment.

Infrastructure as Code

The pace at which developers can create code and iterate through various virtual environments for configuration and testing purposes has accelerated, in large part because of the ability to provision hardware and system configurations automatically with a great deal of granularity. The repeatable nature of the provision process through human- and machine-readable code is the core of what is offered by Infrastructure as Code (IaC).

IaC aims to reduce the impact of environment drift on a development cycle. As small changes are made to environments, they transform over time to a state that cannot be easily replicated. In deploying an IaC solution, developers can rely on the fact that their predefined environments will be the same every time, regardless of external or internal factors. Modern continuous integration and continuous delivery (CI/CD), as the standard practice governing how we ship software today, relies heavily on automation as part of the software development lifecycle (SDLC). Using IaC, developers can orchestrate the creation of these standard platforms and environments to deliver software products more quickly and with fewer problems.

IaC tools generally take one of two approaches: *imperative* or *declarative*. In terms of software engineering, these are two ways that a programmer can get a computer to perform functions to achieve an end state. Imperative approaches involve the programmer specifying exactly how a computer is to reach the goal with explicit step-by-step instructions. Declarative approaches, on the other hand, involve describing the desired end state and relying on the computer to perform whatever is necessary to get there. Regardless of which method is used, IaC tools will invoke the intermediate functions and create the abstraction layer necessary to interface with the infrastructure to achieve the end state.

Figure 6-5 is a snapshot of an AWS CloudFormation template used to create an EC2 instance with a network interface that has both public and private IPs. CloudFormation is among the well-supported IaC solutions on the market right now.

Though IaC is often used to improve a development team's security posture by ensuring consistent environments that are up-to-date on patches, you should be aware of some issues when using the technology. The vulnerabilities associated with IaC solutions don't lie in the templates used to create the environments, but in the processes used to spin up services. The templates don't cause the issue per se, but if a developer trusts a template that then calls an insecure service, or a developer establishes a service with insecure qualities, she unwittingly introduces risk into her operations. So with better processes in place, this potential vulnerability can actually turn into a power tool to standardize security practices across the software development process at an organization.

```
1      [
2        "AWSTemplateFormatVersion": "2010-09-09",
3        "Description": "Template Creates a single EC2 instance with a single ENI which has
       multiple private and public IPs",
4        "Parameters":{
5        "Subnet": {
6          "Description": "ID of the Subnet the instance should be launched in, this will link
       the instance to the same VPC.",
7          "Type": "List<AWS::EC2::Subnet::Id>"
8
9        }
10     },
11       "Resources": {
12         "EIP1": {
13           "Type": "AWS::EC2::EIP",
14           "Properties": {
15             "Domain": "VPC"
16           }
17         },
18         "EIP2": {
19           "Type": "AWS::EC2::EIP",
20           "Properties": {
21             "Domain": "VPC"
22           }
23         },
24         "Association1":
25         {
26         "Type": "AWS::EC2::EIPAssociation",
27         "DependsOn" : ["ENI","EIP1"],
28         "Properties": {
29           "AllocationId": { "Fn::GetAtt" : [ "EIP1", "AllocationId" ]},
30           "NetworkInterfaceId": {"Ref":"ENI"},
31           "PrivateIpAddress": {"Fn::Select" : [ "0", {"Fn::GetAtt" : [ "ENI" ,
       'SecondaryPrivateIpAddresses"]} ]}
32         }
33       },
34       'Association2":
35         {
36         "Type": "AWS::EC2::EIPAssociation",
37         "DependsOn" : ["ENI","EIP2"],
38         "Properties": {
39           "AllocationId": { "Fn::GetAtt" : [ "EIP2", "AllocationId" ]},
40            "NetworkInterfaceId": {"Ref":"ENI"},
41           "PrivateIpAddress": {"Fn::Select" : [ "1", {"Fn::GetAtt" : [ "ENI" ,
       'SecondaryPrivateIpAddresses"]} ]}
42         }
43       },
44       "ENI":
45       [
46         "Type" : "AWS::EC2::NetworkInterface",
47         "Properties" : {
48           "SecondaryPrivateIpAddressCount" : 2,
49           "SourceDestCheck" : true,
50           "SubnetId" : { "Fn::Select" : [ "0", {"Ref" : "Subnet"} ] }
51       }
52
```

Figure 6-5   Amazon CloudFormation template example

Insecure Application Programming Interface

Nearly all cloud service providers offer a means for users to interface with their services by way of an exposed application programming interface (API). With the correct credentials, users can interact via the API with the service to initiate actions such as provisioning, management, orchestration, and monitoring. Since the API is a potential avenue for unwanted action, it must be protected from abuse, and any API credentials must be protected in the same way as any other secret.

In addition to securing API credentials, you must assess the security of the interfaces themselves. Among the many web security projects run by the Open Web Application Security Project (OWASP) foundation is the API Security Project, an effort to identify and provide solutions to mitigate vulnerabilities unique to APIs. We'll explore the top ten vulnerabilities in the following sections.

 EXAM TIP For the CySA+ exam, it's important that you are able to identify high-level problems with implementation and how they may lead to compromise. You don't need to know the details of every specific type of flaw.

Broken Object Level Authorization

When exposing services via API, some servers fail to authorize on an object basis, potentially creating the opportunity for attackers to access resources without the proper authorization to do so. In some cases, and attacker can simply change a URI to reflect a target resource and gain access. Object level authorization checks should always be implemented and access granted based on the specific role of the user. Additionally, universally unique identifiers (UUIDs) should be implemented in a random manner to avoid the chance that an attacker can successfully enumerate them.

Broken User Authentication

Authentication mechanisms exist to give legitimate users access to resources, provided they are able to present the correct credentials. When these mechanisms fail to validate credentials correctly, allow credentials that are too weak, accept credentials in an insecure manner, or allow brute-forcing of credentials, they create conditions that an attacker can take advantage of. To prevent potential issues related to broken authentication, analysts should determine and test all the ways a user can authenticate to all APIs, use short-lived tokens, mandate two-factor authentication where practical, and enforce rate-limiting and password strength policies.

Excessive Data Exposure

Many APIs return the full contents of an object and rely on the client to do further filtering. This may create a situation where the response contains irrelevant data—or worse yet, it exposes the user to too much data he or she should not have access to. API responses should be built to be as specific as possible while also being as complete as possible. This will require thought and effort into creating complete descriptions of what each API schema looks like, and constant monitoring of the contents of the response data. If developers expect that an API return may include protected health information (PHI), personally identifiable information (PII), or other sensitive information, they should consider enforcing additional checks or require separate API calls for that subset of special data.

Lack of Resources and Rate Limiting

APIs do not always impose restrictions on the data returned to users after a legitimate request, which may make them vulnerable to resource exhaustion if abused. As with excessive data exposure, this occurs when the API developer expects the client to perform the heavy lifting. The most effective solution to prevent this is rate limiting or throttling. Rate limiting measures are often put in place to protect services from excessive use and to maintain service availability. Even though cloud services are built to be highly scalable and available, they are still susceptible to overconsumption of resources or bandwidth from malicious or unwitting users. Combining this with limitations on payload or the use of pagination will prevent responses from overwhelming clients.

Broken Function Level Authorization

As with object level authorization problems, function level authorization vulnerabilities arise because of the complex nature of setting up access policies. The simplest solutions are to take a more restrictive default deny approach to access and to apply permissions to roles and groups as they come in for evaluation.

Mass Assignment

Mass assignment is a vulnerability that may allow an attacker to guess fields that they might not normally have access to, or to modify server-side variables not intended by the web app. Back in 2012, GitHub experienced a security incident as a result of an exploited mass assignment vulnerability. An unauthorized attacker was able to upload his public key to the platform to facilitate read/write access to the Ruby on Rails organization code repository. While the attacker did not cause widespread problems, the vulnerability demonstrated the importance of validating inputs and restricting users to read-only as a default condition.

Security Misconfiguration

Default configurations, open access, open cloud storage, misconfigured headers, and verbose error messages all fall under the security misconfiguration umbrella. Misconfigurations are responsible for a significant number of security issues with APIs, but by following a checklist, you can mitigate vulnerabilities consistently and easily. Automation tools may also assist you in creating deployments that are hardened and well-defined, and that contain the minimal set of features required to provide a service.

Injection

Injection flaws occur when untrusted data is inputted into a system and executed blindly by the interpreter. An attacker's malicious input can trick the interpreter into executing commands, granting access to otherwise closed-off data. The most common types of

API call injection include SQL, LDAP, operating system commands, and XML. Input sanitization, validation, and filters are particularly effective against this technique. From the response point of view, developers should limit API outputs to prevent data leaks.

Improper Asset Management

At this point, you've no doubt realized that asset management plays as important a role as any technical implementation when it comes to managing vulnerabilities. Mismanaged API assets lead to all kinds of data exposure problems. Attackers will often find access to the data they're looking for in deprecated, testing, or staging API environments that developers have forgotten about or have failed to harden as they would production environments. Keeping an up-to-date inventory of all API endpoints and retiring those no longer in use will prevent many vulnerabilities related to improper asset management. For testing environments, limiting access to training data while maintaining the same level of rigor in the authentication process are effective controls.

Insufficient Logging and Monitoring

You cannot act on what you can't see. As in all security events, visibility is important in knowing when traffic, user behavior, and, in this case, API usage deviates from the norm. The lack of proper logging, monitoring, and alerting enables attackers to go unnoticed. This applies not just to the existence of logging capability, but to the protections around them as well. Security teams should log failed attempts, denied access, input validation failures, and any other relevant security-related events. Logs should be formatted and normalized so that they are easy to ingest into centralized monitoring systems or to hand off to other tools. Finally, logs should be protected from access and tampering, like any other sensitive data.

Improper Key Management

In many cases, data owners cannot physically reach out and touch the hard drives on which their data is stored, and this reality often brings to light the importance of finding a reliable encryption solution. Regardless of which service or deployment model the organization chooses, encryption provides the means to protect data at every phase of processing, storage, and transport. Even with customers with security or data residency policies that mandate that they use on-premises private cloud solutions, encryption is just as important. Whether it is used to protect against internal threats or unauthorized disclosure reliable encryption remains a priority.

The distributed and decentralized nature of cloud computing, however, often makes key management and encryption more complex for administrators and defenders. Cloud customers can often take advantage of service provider–provided encryption services. Often this means more reliable integration and superior performance than can be offered by homebrew solutions. Additionally, using a solution from a vendor reduces the requirement for your organization to perform any processing of the data on your end. Some others may offer hardware-based solutions such as a hardware security module (HSM) service to enable customers to generate and manage their own encryption keys easily in the cloud.

One common method to ensure that sensitive data is never exposed to cloud providers is to ship data that is already encrypted at the customer interface. This not only ensures that the data cannot be viewed in transit, but also protects the data in the case of a failure of tenancy controls. This method is frequently used by organizations that want to enjoy the benefits of the cloud but may not have the resources or personnel to manage their own private setup. The main drawbacks of this method include the upfront cost of architecting such a solution and the increased computational resource required on the customer side to perform the initial encryption.

As with credentials, if an attacker gains access to encryptions keys, the fallout can be major. Protecting keys from human error is also crucial. Although some cloud service providers may have a mechanism to recover from lost private keys, if an organization manages its own keys and somehow loses them, the data may be lost forever.

 NOTE Key management is a challenging component to an overall security program and is not restricted to whatever cloud solutions your organization employs. Cloud service provider solutions are often designed to integrate with other cloud and local services to reduce the amount of specialized knowledge and customer development resources required to protect and process data in the cloud.

Unprotected Storage

Along with malicious internal or external actor activities, misconfiguration of a server is often at the root of security issues. Enterprise data sets can easily be exposed if administrators aren't aware that they are marked for the public cloud. Cloud storage options such as Amazon Simple Storage Service (S3) are frequent targets of inadvertent exposure. Although these storage services are private by default, permissions may be altered in error, causing the data to become accessible to anyone.

The impact of such leaks can be quite high. Consider, for example, the 2017 discovery of a publicly accessible S3 bucket linked to the defense and intelligence contracting company Booz Allen Hamilton. Analysts at security firm UpGuard were able to identify that the unprotected bucket contained files connected to the US National Geospatial-Intelligence Agency (NGA), the US military's intelligence agency responsible for providing imagery intelligence. Later that year, the same team discovered data belonging to the US Army Intelligence and Security Command (INSCOM), an organization tasked with gathering intelligence for US military, intelligence, and political leaders. Data from that exposure included details on sensitive communications networks. Exposures of such sensitive information highlights in part the difficulty with using cloud storage solutions. Surely there is no lack of understanding as to the nature of this data and the impact of its unauthorized disclosure, so how could organizations so experienced with protecting data fall victim?

It's important to reiterate that these kinds of leaks aren't always the result of hackers or careless IT admins, but are more often a product of flaws in business processes that don't fully consider the access that cloud services provide. Though security through obscurity may have been a marginally viable practice years ago, it will absolutely not fly in today's highly connected and automated world. Protecting these assets requires full visibility into the real-time state cloud resources and the ability to remediate as quickly as these flaws are detected.

 TIP Along with Security Technical Implementation Guides, the US Department of Defense has created a "Cloud Computing Security Requirements Guide" (CCSRG) with checklists and best practices to help you assess various security aspects of cloud service providers, including those related to data exposure and protection.

Logging and Monitoring

When transitioning over to a cloud solution, an organization may lose visibility of certain points on the technology stack, particularly if it's subscribing to PaaS or SaaS solutions. Because the responsibility of protecting portions of the stack falls to the service provider, it does sometimes mean the organization loses monitoring capabilities, for better or worse. A major byproduct of the shared responsibility model in practice is that it's forced a paradigm shift for security teams. In a manner familiar to many security professionals, we now have to do more with less and learn to perform monitoring and analysis related to our data and users without using network-based methodologies.

Chapter Review

A good cloud security strategy begins with implementing the policies and technical controls to identify and authenticate users correctly, the ability to assign users correct access rights, and the ability to create and enforce access control policies for resources. It is important to remember that cloud service providers use a shared responsibility model for security. Although providers accept responsibility for some portion of security, usually as it relates to the part of infrastructure under the provider's purview, the subscribing organization is also responsible for taking the necessary steps to protect the rest. Throughout this chapter, we've provided awareness of the most common vulnerabilities related to cloud technologies and provided steps and resources to assist you in remediating them. Use these as a starting point in crafting your own cloud security strategy or in developing tactical guides on how to address cloud-related security issues that may come up.

Questions

1. Which cloud service model provides customers direct access to hardware, the network, and storage?

    A. SaaS

    B. IaaS

    C. PaaS

    D. SECaaS

2. Mechanisms that fail to provide access to resources on a granular basis may fall under which of the following categories?

   A. Improper key management

   B. Improper asset management

   C. Injection

   D. Broken object level authorization

3. Rate limiting is most often used as a mitigation to prevent exploitation of what kind of flaw?

   A. Excessive data exposure

   B. Broken function level authorization

   C. Lack of resources

   D. Mass assignment

Use the following scenario to answer Questions 4–8:

Your company is looking to cut costs by migrating several internally hosted services to the cloud. The company is primarily interested in providing all global employees access to productivity software and custom company applications, built by internal developers. The CIO has asked you to join the migration planning committee.

4. Which cloud service model do you recommend to enable access to developers to write custom code while also providing all employees access from remote offices?

   A. SaaS

   B. PaaS

   C. IaaS

   D. SECaaS

5. Pleased with the stability and access that the cloud solution has provided after six months, the CIO is now considering hiring more staff to manage more aspects of the cloud solution to include the underlying hardware, network, and storage resources required to support the company's custom code. What model would you suggest that the company consider using?

   A. SaaS

   B. PaaS

   C. IaaS

   D. SECaaS

6. Although the new team is having success managing cloud-based hardware and software resources, they are looking for a new way to create development environments more efficiently and consistently. What technology do you recommend they adopt?

   A. SECaaS

   B. FaaS

   C. IaC

   D. SaaS

7. What is a major security consideration as your organization invests in the cloud service model you recommend?

   A. Key management complexity

   B. Increased cost of maintenance

   C. Hypervisor update management

   D. Physical security of new company data center

8. As your organization expands into offering more financial services, it wants to ensure that there is sufficient redundancy to maintain service availability while also allowing for the possibility for on-premises data retention. Which cloud deployment model is most suitable for these goals?

   A. Public

   B. Private

   C. Community

   D. Hybrid

Answers

1. **B.** Infrastructure as a Service (IaaS) offers an effective and affordable way for companies to get all of the benefits of managing their own hardware without massive overhead costs associated with acquisition, physical storage, and disposal of the hardware.

2. **D.** When exposing services via API, some servers fail to authorize on an object-level basis, potentially creating the opportunity for unauthorized attackers to access resources.

3. **C.** In situations where resourced are lacking, rate limiting measures are often put in place to protect services from excessive use and to maintain service availability.

4. **B.** PaaS solutions optimized to provide value focused on software development, offering direct access to a development environment to enable an organization to build its own solutions on the cloud infrastructure, rather than providing its own infrastructure.

5. **C.** An IaaS solution provides the hardware, network, and storage resources necessary for the customer to install and maintain any operating system, dependencies, and applications they want.

6. **C.** IaC solutions enable developers to orchestrate the creation of custom development environments to deliver software products more quickly and with fewer problems.

7. **A.** The nature of cloud computing often makes key management and encryption more complex for administrators, especially as more stacks of the network have to be managed.

8. **D.** Hybrid clouds combine on-premises infrastructure with public clouds and offer the benefits of both models, while allowing for the flexibility to remaining in compliance with any external or internal requirements on data protection.

Mitigating Controls for Attacks and Software Vulnerabilities

In this chapter you will learn:

- How common attacks may threaten your organization
- Best practices for securing environments from commonly used attacks
- Common classes of vulnerabilities
- Mitigating controls for common vulnerabilities

The attacker only needs to be right once, but defenders must be right all the time.

—Unknown

The threat of a cyberattack is a fact of life in our connected world. As new vulnerabilities are discovered, attackers will often look to take advantage of them using custom and/or commodity tools. Given the scale of these threats, operating in a strictly response capacity, or not at all, is not a prudent option. Protecting your organization's operations and brand by establishing strong security habits combined with forward-looking practices will yield the best results. It's useful for you to understand both the mechanics of common attacks as well as the trends associated with their usage in preparing your proactive and reactive defense plans. Additionally, having a full understanding of how common vulnerabilities may be exploited will prepare you for defending against future attacks not yet observed in the wild.

Attack Types

There are seemingly countless ways for a bad actor to gain access to privileged systems, and often a single flaw can be exploited to cause a nightmare. Many times, the exploited vulnerabilities are already well known across the Web. Hackers know that despite companies' various disclosure mechanisms and security best practices, these targets may still be completely unaware of their malicious activities. In this section, we'll cover common classes of attacks and how they are used to create the conditions that enable attackers to get in easily.

Injection Attacks

In injection attacks, attackers execute malicious operations by submitting untrusted input into an interpreter that is then evaluated. Security flaws may enable system calls into the operating system, where they invoke scripts and outside functionality or requests to back-end databases via database languages. Any time a web application uses an interpreter to assess untrusted data, it is potentially vulnerable to injection attacks. Injection attacks can take many forms, depending on what the attacker is trying to do and what target system is available. Injection vulnerabilities, though sometimes obscure, can be very easy to discover and exploit. The results of a successful injection attack range from a nuisance denial of service to, under certain circumstances, complete system compromise.

Remote Code Execution

Remote code execution (RCE) describes an attacker's ability to execute malicious code on a target platform, often following an injection attack. RCEs are widely considered one of the most dangerous types of computer vulnerabilities, because they may allow for arbitrary command execution and do not require physical connectivity.

Here's an example: An attacker may achieve RCE using a PHP built-in function called **system()**. If the attacker, through some reconnaissance, can determine that a web server is vulnerable to an injection attack, she might be able to invoke a shell on the server by wrapping the shell command in what is otherwise a legitimate PHP command:

```
http://victimwebsite.com/?code=system('whoami');
```

RCE attacks can be incredibly costly to the victim. Over the last decade, the popular web development framework Apache Struts has been the subject of many vulnerability disclosures. One such disclosure, covered by CVE-2017-5638, described a flaw in a parsing function for certain versions of Struts 2 that would allow for execution of arbitrary commands. This Struts 2 vulnerability gained a lot of attention from attackers and was quickly weaponized throughout the world in a matter of days. Two months after initial disclosure by Apache, it was discovered that hackers had accessed and exposed personal data of 143 million customers in what's now known as the Equifax Breach. As of this writing, the estimated cost of remediation, clean up, penalties, and claims settlements is around $1.4 billion.

Extensible Markup Language Attack

Extensible Markup Language (XML) injection is a class of attacks that relies on manipulation or compromise of an application by abusing the logic of an XML parser to cause some unwanted action. XML injections that target vulnerable parsers generally take two forms: XML bombs and XML External Entity (XXE) attacks.

An XML bomb is an attack designed to cause the XML parser or the application it supports to crash by overloading it with data. A common XML bomb attack, the Billion Laughs attack, uses nested entities, each referring to another entity that itself may contain several others. When triggered, the parser will attempt to evaluate the input, expand the entities, and eventually run out of resources. The resulting crash can cause denial of service and application downtime. Figure 7-1 shows an example of code behind a Billion Laughs attack.

```
1    <?xml version="1.0"?>
2    <!DOCTYPE lolz [
3    <!ENTITY lol "lol">
4    <!ELEMENT lolz (#PCDATA)>
5    <!ENTITY lol1 "&lol;&lol;&lol;&lol;&lol;&lol;&lol;&lol;&lol;&lol;">
6    <!ENTITY lol2 "&lol1;&lol1;&lol1;&lol1;&lol1;&lol1;&lol1;&lol1;&lol1;&lol1;">
7    <!ENTITY lol3 "&lol2;&lol2;&lol2;&lol2;&lol2;&lol2;&lol2;&lol2;&lol2;&lol2;">
8    <!ENTITY lol4 "&lol3;&lol3;&lol3;&lol3;&lol3;&lol3;&lol3;&lol3;&lol3;&lol3;">
9    <!ENTITY lol5 "&lol4;&lol4;&lol4;&lol4;&lol4;&lol4;&lol4;&lol4;&lol4;&lol4;">
10   <!ENTITY lol6 "&lol5;&lol5;&lol5;&lol5;&lol5;&lol5;&lol5;&lol5;&lol5;&lol5;">
11   <!ENTITY lol7 "&lol6;&lol6;&lol6;&lol6;&lol6;&lol6;&lol6;&lol6;&lol6;&lol6;">
12   <!ENTITY lol8 "&lol7;&lol7;&lol7;&lol7;&lol7;&lol7;&lol7;&lol7;&lol7;&lol7;">
13   <!ENTITY lol9 "&lol8;&lol8;&lol8;&lol8;&lol8;&lol8;&lol8;&lol8;&lol8;&lol8;">
14   ]>
15   <lolz>&lol9;</lolz>
```

Figure 7-1   Example XML code designed to trigger a Billion Laughs attack

XXE can be used to initiate denial of service, conduct port scanning, and perform server-side request forgery (SSRF) attacks. The attack works though abuse of the XML parser to execute functions on behalf of the attacker, using a reference to an external entity. One feature of the XML standard is that it can be used to reference outside resources. By providing an input referencing an external entity, an attacker may be able to get a vulnerable XML parser to read and process normally protected data. In doing so, the parser could inadvertently leak sensitive information.

Structured Query Language Injection

SQL injection (SQLi) is a popular form of injection in which an attacker injects arbitrary SQL commands to extract data, read files, or even escalate to an RCE. To exploit this vulnerability, the attacker must find a legitimate input parameter that the web app passes through to the SQL database. The attacker then embeds malicious SQL commands into the parameter and hopes that it's passed on for execution in the back end. These attacks are not particularly sophisticated, but the consequences of their successful usage in an attack are particularly damaging, because an attacker can obtain, corrupt, or destroy database contents.

Cross-Site Scripting

Cross-site scripting (XSS) is a type of injection attack that leverages a user's browser to execute malicious code that can access sensitive information in the user's browser, such as passwords and session information. Because the malicious code resides on the site that the user accesses, it's often difficult for the user's browser to know that the code should not be trusted. XSS thus takes advantage of this inherent trust between browser and site to run the malicious code at the security level of the website.

XSS comes in two forms: persistent and nonpersistent. With persistent attacks, malicious code is stored on a site, usually via message board or comment postings. When other users attempt to use the site, they unwittingly execute the code hidden in the previously posted content. Nonpersistent attacks, also referred to as *reflected* XSS, take advantage of a flaw in the server software. If an attacker notices a XSS vulnerability on a site, he can craft a special link, which when passed to and clicked on by other users, would cause the browser to visit the site and reflect the attack back onto the victim. This could cause an inadvertent leak of session details or user information to whatever server the attacker specifies. These links are often passed along through e-mail and text messages and appear to be innocuous and legitimate.

Another type of attack, called a DOM-based XSS attack, occurs when an attacker injects a malicious script into the client-side HTML being parsed by a browser. The DOM, or Document Object Model, is the standard by which a client, or more specifically, the browser, interacts with HTML. If website code either does not properly validate input or encode data correctly, it creates the opportunity for an attacker to modify code en route to initiate a XSS attack. Importantly, no server resources are affected using the method. The page, its HTML contents, and the associated HTTP response do not change, but the modifications that result from the malicious script cause the client to execute abnormally. Detecting DOM-based attacks using server-side tools is impossible because the traffic will look identical to legitimate traffic. Defense against DOM-based XSS involves sanitizing client-side code through inspection of DOM objects and using intrusion prevention systems that are able to inspect inbound traffic to determine if unusual parameters are being passed.

Directory Traversal

A directory traversal attack enables an attacker to view, modify, or execute files in a system that he wouldn't normally be able to access. For web applications, these files normally reside outside of the web root directory and should not be viewable. However, if the server has poorly configured permissions, a user may be able to view other assets on the server. If an attacker determines a web application is vulnerable to directory traversal attack, he may use one or more explicit Unix-compliant directory traversal character sequences (*../*) or an encoded variation of it to bypass security filters and access files outside of the web root directory. Figure 7-2 shows the execution of a simple directory traversal attack against a vulnerable server. In this example, an attacker uses the wget utility to crawl up and down the file system to recover the unprotected /etc/passwd file.

 TIP Sometimes avoiding user-supplied input as a defense against directory traversal isn't possible, but there are other ways to mitigate the effects of abusive behavior that can be performed on the server side. At a minimum, input should be validated before it's processed. This ensures that only the expected type of content is moved on to interpretation by the server.

Buffer Overflow Attacks

Attackers often will write malware that takes advantage of some quality or operation of main memory. Memory is an extremely complex environment, and malicious activities are prone to disrupt the delicate arrangement of elements in that space. The temporary space that a program has allocated to perform operating system or application functions is referred to as the *buffer*. Buffers usually reside in main memory, but they may also exist

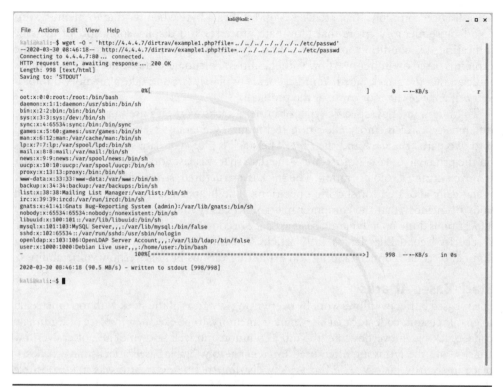

Figure 7-2 Output from a successful directory traversal

in hard drive and cache space. You may be familiar with the type of buffer used by video-streaming services. Streaming services often uses buffers to ensure smooth video playback and to mitigate the effects of an unexpected drop in connectivity. As the video is playing, upcoming portions are saved to memory in a sliding window. If the connection is dropped, the video can be played directly from the buffer until the connection resumes; the more buffer space the system has, the longer it is able to maintain the temporary streaming. Buffers in software enable operating systems to access data efficiently since RAM is among the fastest types of storage.

When the volume of data exceeds the capacity of the buffer, the result is *buffer overflow*. If this occurs, a system may attempt to write data past the limits of the buffer and into other memory spaces. Attackers will sometimes find ways to craft input into a program in a way that will cause the program to attempt to write too much data to a buffer. If the boundaries of data are well understood, an attacker may be able to overwrite legitimate executable data and replace it with malicious code.

Buffer overflows affect nearly every type of software and can result in unexpected results if not managed correctly. The Heartbleed Bug discussed in Chapter 5 resulted from a buffer overflow vulnerability that was not correctly addressed and resulted in exposure of information. Even though these types of attack are well understood, they are still fairly common. If an attacker is off by even a byte when writing to memory, this could cause memory errors that terminate processes and display some sort of message indicating this condition to the user. This type of symptom is particularly likely if the exploit is based on buffer overflow vulnerabilities. Fortunately, these messages sometimes indicate that the attack failed. Your best bet is to play it safe and take a memory dump so you can analyze the root cause of the problem.

To understand how specific types of buffer attacks work, it's useful to understand the different types of memory allocation techniques. Generally, system memory is divided into two parts, the *stack* and the *heap*. The stack is a type of data structure that operates on the principle of last-in, first-out—the data that was last added to the stack is the first removed during a read operation. The stack is a structured, sequential memory space that is statically allocated for specific operations. The heap, in contrast, is an unstructured body of memory that the computer may use to satisfy memory requests. Heap memory allocation is done on a dynamic basis as space becomes available to use. A useful resource is security legend Elias Levy's 1996 article, "Smashing the Stack for Fun and Profit," a seminal piece into understanding and exploiting various buffer overflow vulnerabilities.

Stack-Based Attacks

Stack-based buffer overflows work by overwriting key areas of the stack with too much data to enable custom code, located elsewhere in memory, to be executed in place of legitimate code. Stack-based overflows are the most common and well known of all buffer overflow attacks, and the terms are often used interchangeably. Stack-based attacks have a special place in security history. The first widely distributed Internet worm was made possible through a successful stack-based buffer attack. The Morris Worm, written by graduate student Robert Tappan Morris from Cornell University in the late 1980s, took advantage of a buffer overflow vulnerability in a widely used version of fingerd, a daemon for a simple network protocol used to exchange user information. In short, the worm's main purpose was to enumerate Internet-connected computers by connecting to a machine, replicating itself, and sending that copy on to neighboring computers. Morris's activity also resulted in the first conviction under the 1986 Computer Fraud and Abuse Act.

Heap-Based Attacks

Attacks targeting the memory heap are usually more difficult for attackers to implement because the heap is dynamically allocated. In many cases, heap attacks involve exhausting the memory space allocated for a program. To make things more complicated for attackers, the success of a heap-based overflow does not always mean a successful exploit, since data in the heap must be corrupted rather than just overwritten.

Integer Attacks

An integer overflow takes advantage of the fixed architecture-defined memory regions associated with integer variables. In the C programming language, the integer memory regions are capable of holding values of up to 4 bytes, or 32 bits. This means the integer value range is −2,147,483,648 to 2,147,483,647. If a value is submitted that is larger in size than the 4-byte limit, then the integer buffer may be exceeded. The challenge with an integer overflow lies in its difficulty of detection, since there may not be an inherent way for a process to determine whether a result it calculated is correct. Most integer overflows are not exploitable because memory is not being overwritten, but they can lead to other types of overflow conditions. Real-life examples of exploits are rare but not unheard of.

In 2001, the CERT Coordination Center released Vulnerability Note VU#945216, which described a flaw in the SSH1 protocol that could allow for remote code execution. Curiously, the fault existed in a function designed to detect cyclic redundancy check (CRC) attacks. The function makes use of a hash table to store connection information used in determining abuse patterns. With a specially crafted packet of a length greater than the function's integer buffer, an attacker could create a null-sized table, or one with no value, and allow for the attacker to modify arbitrary addresses within the daemon. What's worse is that this attack could be executed before any authentication occurred, meaning the system may essentially have no defenses against it.

Privilege Escalation

Privilege escalation is simply any action that enables a user to perform tasks she is not normally allowed to do. This often involves exploiting a bug, implementation flaw, or misconfiguration. Escalation can happen in a *vertical* manner, meaning that a user gains the privileges of a higher privilege user. Alternatively, *horizontal* privilege escalation can be performed to get the access of others in the same privilege level. Attackers will use these privileges to modify files, download sensitive information, or install malicious code.

Jailbreaking and Rooting

Jailbreaking, the act of bypassing Apple iOS restrictions, uses privilege escalation to enable users to perform functions that they normally could not. Jailbreaking enables Apple mobile device users to install custom software or modified operating systems. Similarly, "rooting" an Android device gives a user privileged access to the device's subsystem. Developing these kinds of exploits is a big deal, because mobile device manufacturers expend enormous resources to standardize their devices. While gaining freedom to install additional apps and modify a mobile device seems like a good idea, it makes the device less secure, because it's likely that the protections that could prevent malicious activity were removed to achieve the jailbreak or root in the first place.

Authentication Attacks

In securing authentication systems, the main challenge lies in identifying and communicating just the right amount of information to the authentication system to make an accurate decision. These are machines, after all, and they will never truly know who we are or what our intentions may be. They can form a decision based only on the information we give them and the clues about our behavior as we provide that data. If an attacker is clever enough to fabricate a user's information sufficiently, he is effectively the same person in the eyes of the authentication system.

Password Spraying

Password spraying is a type of brute-force technique in which an attacker tries a single password against a system, and then iterates though multiple systems on a network using the same password. In doing so, an attacker may avoid account lockouts from a single system. Detecting spraying attempts is much easier if there is a unified interface, such as that provided by a security information and event management (SIEM) solution, on which an analyst can visualize events from a range of security information sources. Common ways to detect password spraying include the following:

- High number of authentication attempts within a defined period of time across multiple systems
- High number of bad usernames, or usernames that don't match company standard
- High number of account lockouts over a defined period of time
- Multiple successful logins from a single IP in a short timeframe
- Multiple failed logins from a single IP in a short timeframe

Mitigation of spray techniques includes several technical and policy measures designed to add cost to the attackers, or to add visibility and response capabilities for analysts. The following is a short list of mitigations for attacks against password-spraying attacks:

- Implement and enforce multifactor authentication (MFA) on all systems.
- Enforce a password policy that requires complexity, password resets, and password rotation.
- Disallow use of the same password on multiple systems whenever possible.
- Increase alerting and monitoring through integration into a SIEM solution.
- Enforce IP blacklisting for abusive traffic.

Credential Stuffing

Credential stuffing is a type of brute-force attack in which credentials obtained from a data breach of one service are used to authenticate to another system in an attempt to gain access. Given the scale and rate of data breaches, credential stuffing continues to be a major problem for the security and safety of organizations and data. Recent data breaches have each exposed hundreds of thousands to millions of credentials. When aggregated as "collections," these numbers can easily top billions. If even a small fraction of a credentials list can be used to gain access to accounts, it's worth it from the attacker's vantage point, especially with the use of automation. For organizations, mandating MFA is effective in slowing the effectiveness of attacks, especially those that are automated.

Impersonation

Sometimes attackers will impersonate a service to harvest credentials or intercept communications. Fooling a client can be done one of several ways. First, if the server key is stolen, the attacker appears to be the server without the client possibly knowing about it.

Additionally, if an attacker can somehow gain trust as the certificate authority (CA) from the client, or if the client does not check to see if the attacker is actually a trusted CA, then the impersonation will be successful.

Man-in-the-Middle

Essentially, man-in-the-middle (MITM) attacks are impersonation attacks that face both ways: the attacker impersonates both the client to the real server and the server to the real client. Acting as a proxy or relay, the attacker will use her position in the middle of the conversation between parties to collect credentials, capture traffic, or introduce false communications. Even with an encrypted connection, it's possible to conduct an MITM attack that works similarly to an unencrypted attack. In the case of HTTPS, the client browser establishes a Secure Sockets Layer (SSL) connection with the attacker, and the attacker establishes a second SSL connection with the web server. The client may or may not see a warning about the validity of the client, as shown in Figure 7-3.

If a warning appears, it's very likely that the victim may ignore or click though the warning. The tendency for users to ignore warning highlights the importance of user training. The Firefox browser attempts to give as much information to the user about

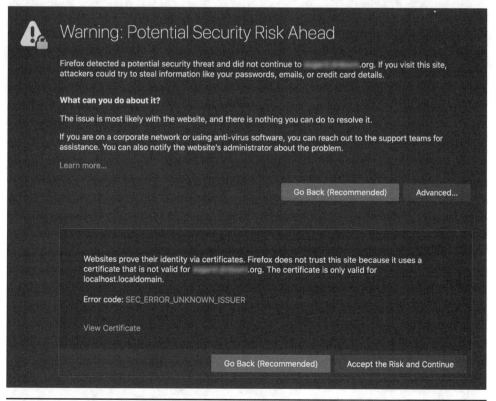

Figure 7-3 Firefox warning displaying information regarding a website's certificate validity

what's happening in as plainly as possible. Firefox offers detailed information about the certificate for more technical users to use in troubleshooting any potential issues. Note that the certificate details shown in Figure 7-4 are valid only for certain domains, which prompted the initial warning from the browser. It's possible for the warning not to appear at all, which would indicate that the attacker has managed to get a certificate signed by a trusted CA.

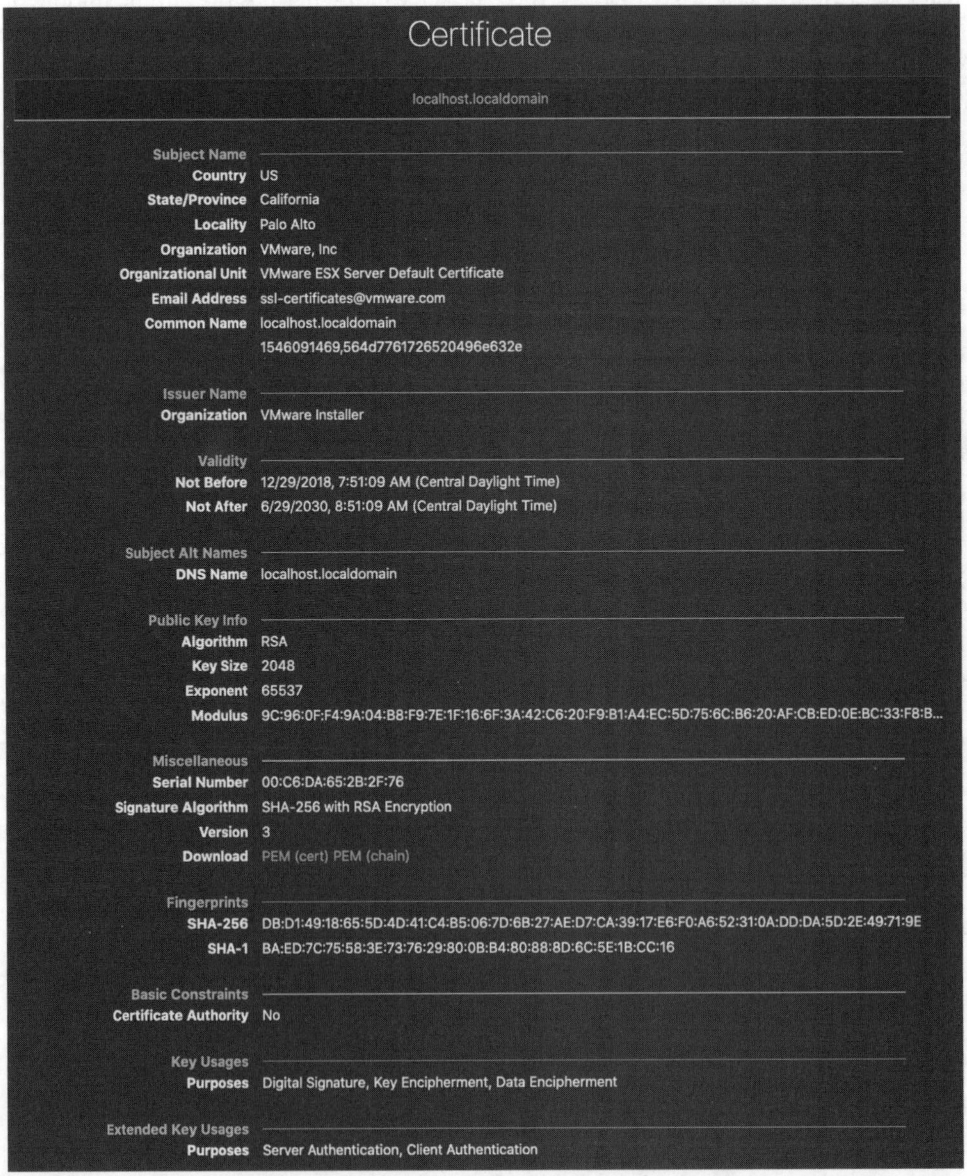

Figure 7-4 Firefox certificate details screen

Session Hijacking

Session hijacking is a class of attacks by which an attacker takes advantage of valid session information, often by stealing and replaying it. HTTP traffic is stateless and often uses multiple TCP connections, so it uses sessions to keep track of client authentication. Session information is just a string of characters that appears in a cookie file, the URL itself, or other parts of the HTTP traffic. An attacker can get existing session information through traffic capture, an MITM attack, or by predicting the session token information. By capturing and repeating session information, an attacker may be able to take over, or hijack, the existing web session to impersonate a victim.

Rootkits

Rootkits are among the most challenging types of malware because they are specially designed to maintain persistence and root-level access on a system without being detected. As with other types of malware, rootkits can be introduced by attacker leveraging vulnerabilities to achieve privilege escalation and clandestine installation. Alternatively, they may be presented to a system as an update to BIOS or firmware. Rootkits are difficult to detect because they sometimes reside in the lower levels of an operating system, such as in device drivers and in the kernel, or even in computer hardware itself, so the system cannot necessarily be trusted to report any modifications it has undergone.

Vulnerabilities

Attackers often use software vulnerabilities to get around an organization's security policies intended to protect its data. Sometimes the flaws exist in how the rules are written, how the mechanism is implemented, or what functions are called in the operation of the software. All computer systems have vulnerabilities, but depending on the number and type of compensating controls in place, they may or may not result in damage if exploited. As we describe these errors, it's important for you to remember that all of the vulnerabilities have some relevance to the secure operation of your company's software and network. They aren't just mistakes that inhibit usability or the look of the software, but rather enable attackers to do something they wouldn't otherwise be able to do. They will sometimes result in unexpected and often undesirable behavior, but in many cases, they can be addressed using secure coding practices and thoughtful review processes before code ever goes to production.

Improper Error Handling

Error handling is an important and normal function in software development. Although developers work to identify problems quickly, consistently, and early, errors related to memory, network, systems, and databases routinely pop up in a production environment. Improper error handling can be a security concern when too much information is disclosed about an exception to outside users. When internal error messages such as stack traces, database dumps, and error codes are displayed to the user, they may reveal details about internal network configuration or other implementation details that should never be revealed. For attackers, this kind of information can provide the situational awareness they need to craft very specific exploits.

Sometimes error messages don't reveal a lot of detail, but they can still provide clues to help attackers. Notices about file unavailability, timeout conditions, and access denial could reveal the presence or nonpresence of a resource or indications about a file system's directory structure. As part of a secure coding practice, policies on error handling should be documented, including what kind of information will be visible to users and how this information is logged. Messages should be simple, conveying only what's needed in a consistent way to avoid disclosing too much about the platform.

Dereferencing

A dereferencing vulnerability, or null point dereference, is a common flaw that occurs when software attempts to access a value stored in memory that does not exist. A null pointer is a practice in programming used to direct a program to a location in memory that does not contain a valid object. This vulnerability has legitimate uses when used as a special marker, but security issues arise when it is dereferenced or accessed. Often, dereferencing results in an immediate crash and subsequent instability of an application. Attackers will try to trigger a null pointer dereference in the hopes that the resulting errors enable them to bypass security measures or learn more about how the program works by reading the exception information.

Insecure Object Reference

Insecure object reference vulnerabilities occur when the object identifiers in requests are used in a way that reveals a format or pattern in underlying or back-end technologies, such as files, directories, database records, or URLs. If database records are stored in a sequential manner and referenced directly by a function, for example, an attacker may be able to use this predictable pattern to identify resources to target directly, thereby saving him time and effort.

Rather than referencing sources directly, developers can avoid exposing resources by using an indirect reference map. With this method, a random value is used in place of a direct internal reference, preventing inadvertent disclosure of internal asset locations. Enforcing access controls at the object level also addresses the primary issue with this vulnerability: insufficient or missing access check. While client-side validations for access may be useful, server-side will be more effective at verifying that the current user owns or is allowed to access the requested data.

Race Condition

A race condition vulnerability is a defect in code that creates an unstable quality in the operation of a program arising from timing variances produced by programming logic. These deviations from timing could mean that a program that relies on sequential execution of a series of actions may experience two actions attempting to complete at the same time, or actions that attempt to complete out of order. A subclass of race condition vulnerabilities, time-of-check to time-of-use (TOCTOU), is often leveraged by attackers to cause issues with data integrity. TOCTOU is used by software to check the state of a resource before its use. If the state of the resource is changed between the checking and usage window, it may result in unexpected behavior from the software.

All kinds of software are vulnerable to race condition attacks. One way that an attacker may target this vulnerability is to bypass any restrictions imposed by an access control list (ACL). CVE-2016-7098 refers to such a possibility in its description of a flaw in wget, a popular command-line tool that allows file retrieval from servers. This vulnerability affects the way that wget applies its ACLs during the download process. In this case, if a download is initiated with recursive options, the utility will wait until the completion of the download to apply the rules. An attacker may take advantage of this by inserting malicious code or altering a file mid-download, before the application has a chance to finish the transfer. Because the file gets deleted only after the connection closes, the attacker could keep the connection open and make use of the malicious file before its deletion.

Sensitive Data Exposure

Sensitive data exposure vulnerabilities occur when an application or system does not adequately protect data from access to unauthorized parties. Data can include authentication information, such as logins, passwords, and tokens, as well as personally identifiable information (PII), protected health information (PHI), or financial information. Recent advances in legislation and privacy laws, such as the European Union's General Data Protection Regulation (GDPR), attempt to enhance protections around this kind of sensitive data. To help identify vulnerabilities in the space, you may find it useful to go through a few questions related to the data your organization handles:

- Is any data transmitted in cleartext across your networks?
- Are any old, weak, or custom cryptographic algorithms and libraries used in your organization's code?
- Are default cryptographic keys in use?
- Are cryptographic keys rotated on a regular basis?
- Is encryption enforced across services?
- Do clients properly verify server certificates?

At the minimum, all organizations can follow a set of practices to ensure baseline protections against exploits of data exposure vulnerabilities. The process begins with identifying which data is sensitive according to privacy laws, regulatory requirements, or your organization's own definitions. As the data is introduced into various systems and applications for processing and transmission, there needs to be a way to identify these various levels of sensitivity by some technical means. From this point, it's straightforward for the data to be processed, stored, and transmitted in accordance with these standards. Furthermore, auditing can be made easier if the technical foundation is in place.

In terms of data storage, it's important to collect and store only the minimum amount of data needed to fulfill a business requirement. It's important that sensitive information be stored only as long as necessary and not beyond its usefulness. Stored data should be encrypted using strong standard algorithms and protocols. Private keys should be stored in a way that protects their disclosure and integrity. Strong encryption is also critical for all data in transit. When possible, your organization should promote usage of standards such as Transport Layer Security (TLS) with perfect forward secrecy (PFS) ciphers and HTTP Strict Transport Security (HSTS).

Insecure Components

Though modern software development practices make use of open source and external components to speed up the software development process, overreliance on these components, especially when they're not fully examined from a security aspect, can lead to dangerous exposure of your organization's data. Developers need to know what these components are doing, how they're interacting with internal assets and data, and what compensating controls can be put into place to address security issues related to their use. Some vulnerabilities related to insecure components may lead to minor impacts in performance, but others can lead to major security events. What's more, promises about the security of an external component may even lead to a false sense of security, making the impact potentially worse because there was some expectation of protection.

Insufficient Logging and Monitoring

Earlier in the chapter, we touched on the concept of errors as a way for developers to get feedback about their software and find ways to improve performance and stability. Like errors, logging gives critical feedback about the state of a system, enabling analysts to trace events back to their beginning as they work to determine root cause. The main vulnerability with logging is the nonexistence of logging mechanisms at key points within the network that would give insight into malicious and abusive behaviors. Nearly all network devices, operating systems, and other types of hosts provide one or more options for logging, often at various levels of detail (or verbosity). However, developers might disable or misconfigure logging during development and never turn it back on. Good logging practice is also much more than just having the feature enabled—it's about setting the conditions under which logs can be ingested, understood, and operationalized easily. This may include developing intermediate steps such as setting up specialized servers and normalizing the data. Security without logging is so much harder because, as with asset visibility, you cannot defend what you cannot see.

As best practice, the Open Web Application Security Project (OWASP) recommends that logging be configured for the following activities:

- Input validation failures such as invalid parameter names and values, and protocol violation
- Output validation failures such as invalid data encoding
- Authentication successes and failures
- Access control failures
- Session management failures
- Application and component syntax and runtime errors
- Connectivity problems
- Malware detection

- Configuration modifications
- Application and related system startups and shut-downs
- Logging functionality and state changes
- Actions by administrative accounts

It's also important to define what to exclude for logging. Data that is not legally sanctioned, or that may have serious privacy implications for users, should not be part of the logging effort. This usually includes some types of privileged communications, data collected without consent, payment card holder data, encryption keys, sensitive PII, access tokens, passwords, and intellectual property data. Keep in mind that local laws may explicitly forbid the collection of certain classes of data.

Weak or Default Configurations

Earlier we covered the concept of a *vulnerability window*, the time between the disclosure of a flaw from a vendor and the moment an organization is able to patch the flaw. Attackers will often attempt to exploit unpatched flaws during this time, or they may perform actions to cause a system to revert to a weakened state. Though Microsoft aimed to stop the cat-and-mouse game between attackers and defenders by modifying their predictable "Patch Tuesday" release model, attackers nevertheless pay close attention to disclosure announcements on a continual basis to maximize the vulnerability window.

Often attackers will attempt to take advantage of default accounts and settings in the hopes that administrators have not changed them. Security misconfigurations like this can exist at any level, from a factory password left on a wireless access point to static credentials left on enterprise devices. In some cases, despite a security team's best effort, these credentials may not be able to be changed. An example of this last condition is described in an advisory issued by Cisco in 2019 and covered in CVE-2019-1723. There was a flaw in a previous version of the Cisco Common Service Platform Collector (CSPC), a tool used for gathering information from Cisco devices on a network, that allowed anyone to access and control the tool using a static default password.

Downgrade Attack

Some kind of attacks, such as a downgrade attack, attempt to take advantage of the backward-compatibility features offered by certain services by causing them to operate in a mode that offers lower levels of security. One common example of a downgrade attack occurs when an attacker redirects a visitor from an HTTPS version of a website to an HTTP version. When this occurs, a user loses all of the benefit of strong encryption between the client and server. Downgrade attacks are often used to enable MITM attacks, in which the switch to a lower version of crypto enables an attacker to eavesdrop on communications. One such vulnerability existed in a previous version of the popular OpenSSL service that could allow for an attacker to negotiate a lower version of TLS between the client and server.

Use of Insecure Functions

Even with a major focus on secure coding practices across the development industry, the use of insecure and dangerous functions is still a common occurrence. This vulnerability occurs when a function is invoked by a program that introduces a weakness into a system based on its implementation or inherent qualities. These functions may be used to perform the job they were designed for, but at greater risk to the organization employing them. Sometimes, the function can never be guaranteed to be used safely. This is the case with **gets()**, an inherently unsafe C function that operates by blindly copying all input from STDIN to the buffer without checking size. This is problematic because it may enable the user, intentionally or unwittingly, to provide input that is larger than the buffer size, resulting in an overflow condition.

strcpy

The **strcpy()** function simply copies memory contents into the buffer, ignoring the size of the area allocated to the buffer. This attribute can easily be misused in a manner that could enable an attacker to cause a buffer overflow. A short example of how this may occur is shown here:

```
char str1[10];
char str2[]="AStringThatIsClearlyTooMuchForTheBuffer";
strcpy(str1,str2);
```

Although there are some alternatives to some of these functions advertised as safe, those functions may themselves be vulnerable to other types of attacks. The **strncpy()** function, for example, is said to be a safer version of **strcpy()**, because it enables a maximum size to be specified. However, the **strncpy()** function doesn't null terminate the destination if the buffer is completely filled, which may lead to stability problems in code. As a security analyst, it's important that you not take alternative recommendations for granted. Doing so can give you a false sense of security and may introduce additional vulnerabilities.

Chapter Review

There are effective ways to reduce your organization's exposure to the common types of attacks that we discussed in this chapter. Additionally, there are a number of controls that your organization can put into place at low cost to reduce the likelihood of the existence or exploitation of system flaws. With strong access controls, your security team will be able to defend both at the perimeter of the network and within its boundaries. Modern security appliances and software enable access control methods to be applied at a granular level, and it's worth taking advantage of these features. Web proxies and filters can be used to detect and block executable downloads, or to block access completely to known malicious domains. For commonly used malware, it's important for you to maintain a method to detect and respond to known malicious code using an endpoint and network-based tooling. Furthermore, preventing unknown software from being able to run at all using whitelisting rules is extremely effective in preventing exploits. To address software and implementation flaws, it's critical that you develop a vulnerability remediation plan to enable the team to detect and patch common vulnerabilities.

Questions

1. What kind of attack takes advantage of compatibility functionality provided by a program or protocol to force a user into operating in a less secure mode?

   A. Downgrade attack

   B. Side-channel attack

   C. Passthrough attack

   D. Hijacking

2. Attacks that rely on an interpreter to evaluate untrusted input and execute tasks on an attacker's behalf are known by what name?

   A. Injection attacks

   B. Authentication attacks

   C. Exhaustion attacks

   D. Overflow attacks

3. What is the primary difference between a memory heap and memory stack?

   A. The stack allows for memory reservations beyond the limits of the buffer, while the heap does not.

   B. The stack is a structured, statically allocated memory space, while the heap space is not reserved in advance.

   C. The heap allows for memory reservations beyond the limits of the buffer, while the stack does not.

   D. The heap is a structured, statically allocated memory space, while the stack space is not reserved in advance.

4. Software that checks the status of a resource before using it, but allows a change between initial check and final usage, suffers from what kind of vulnerability?

   A. Buffer overflow

   B. Race condition

   C. Injection

   D. Derefencing

Use the following scenario to answer Questions 5–8:

Your security team is made aware of a massive data breach of a popular social media platform. According to analysts of the breach, the records include details such as credit card information, e-mail addresses, language preference, and passwords. Although there is no connection between the social media platform and your organization, you are concerned about the security implications to your company given the popularity of the platform.

5. You are aware that password reuse is a common occurrence and urge the team to issue a mandatory password reset for all users in hopes of preventing which kind of attack?

   A. Session hijacking

   B. Downgrade attack

   C. Password spraying

   D. Credential stuffing

6. Your security dashboard sends an alert indicating that in the last hour, the same user had several unsuccessful login attempts across eight internal systems. What type of attack has most likely occurred?

   A. Session hijacking

   B. Man-in-the-middle attack

   C. Password spraying

   D. Firehose attack

7. Security analysts investigating the breach have now indicated that attackers were able to target their attack to a specific system based on information recovered from an exception that was raised during their reconnaissance. Based on this information, what type of vulnerability did the attacks likely take advantage of?

   A. Improper error handling

   B. Insecure object reference

   C. Use of insecure functions

   D. Use of default configurations

8. The information recovered from the previously identified reconnaissance effort indicated the use of a publicly facing security device known to have a static password in previous versions of its firmware. Analysts discovered that the firmware was never upgraded and static credentials were used to get into the device. What flaw is likely to have been exploited?

   A. Improper error handling

   B. Insecure object reference

   C. Use of insecure functions

   D. Use of default configurations

Answers

1. **A.** Downgrade attacks are characterized by techniques that cause services to operate in a mode that offers lower levels of security, such as HTTP instead of HTTPS.

2. **A.** In injection attacks, attackers execute undesirable operations by submitting untrusted input into an interpreter that is then evaluated.

3. **B.** The stack is a type of reserved, sequential memory space that operates on the principle of last-in, first-out, while the heap is an unstructured body of memory that the computer may use as needed.

4. **B.** A race condition vulnerability is a flaw in the operation of a program that may allow two actions to attempt completion at the same time or to perform actions out of order, resulting in integrity or stability issues.

5. **D.** Credential stuffing is a type of brute-force attack where credentials obtained from a data breach of one service are used to authenticate to another system.

6. **C.** Password spraying is a type of brute-force technique in which an attacker tries a single password against a system, and then iterates though multiple systems on a network using the same password.

7. **A.** Improper error handling is a vulnerability that allows for too much information to be disclosed about an exception to outside users.

8. **D.** Default configuration exist at many levels, from standard factory passwords to static credentials on enterprise devices. Attackers will take advantage of these settings to get around other security mechanisms.

PART II

Software and Systems Security

Security Solutions for Infrastructure Management

In this chapter you will learn:

- Common network architectures and their security implications
- How to manage assets and changes
- Technologies and policies used to identify, authenticate, and authorize users
- Different types of encryption and when to use them

You can't build a great building on a weak foundation.

—Gordon B. Hinckley

In Part I of this book, we looked at how adversaries operate against our systems. We now turn our attention to the architectural foundations of the systems they'll be attacking and we are defending. Though we would all agree that it is difficult to protect something we don't understand, a remarkable number of security professionals are not fully aware of all that is in and on their networks. This situation is further complicated by the fact that a rapidly shrinking number of organizations have anything resembling a physical perimeter. Most modern computing environments involve a mix of local and remote resources, some physical and some virtual, with many of them moving to various forms of cloud computing platforms.

As a cybersecurity analyst, you don't need to be an expert in all of the various technologies that make up your network. However, you do need to have a working knowledge of them, with an emphasis on those that impact security the most. You should be tracking not only these technologies, but also the manner in which they are implemented in your environment and, just as importantly, the way in which changes to assets and their configurations impact security. For example, many organizations that have carefully deployed strong defenses are experiencing them gradually eroding because of seemingly inconsequential configuration changes. The solution is to have good processes and programs around tracking your assets and deliberately managing changes in their configurations.

In addition to having a solid grip on your network architecture and an understanding of the importance of (and how to do) asset and change management, you also need to deepen your knowledge of how to identify, authenticate, and authorize system users.

This is true regardless of where users are located, what resources they're trying to use, and what they intend to do with them. A central topic in this regard is cryptography, so we'll spend a good amount of time reviewing the concepts that make it work.

Cloud vs. On-Premises Solutions

As we saw in Chapter 6, *cloud computing* refers to an extensive range of services enabled by high-performance, distributed computing. They generally require only a network connection to the service provider, which handles most of the computation and storage. By contrast, *on-premises* solutions involve hardware and software that are physically located within the organization. Although there has been much debate about which approach is intrinsically more secure, in reality the answer is complicated.

With most cloud solutions, service providers are responsible for some (or even most) of the security controls. They do this at scale, which means they make extensive use of automation to secure their infrastructure. If you go with well-known cloud providers, they'll probably have really good security teams and controls. Depending on the capabilities in your own staff, this may make it easier for you to achieve a better cybersecurity posture. Your team still has an important role to play, but the burden can be largely shifted to your provider.

It is important to keep in mind that some critical cloud security controls are different from those you'd use in on-premises environments. This means that you can't completely rely on your service provider to handle every control. For example, misconfigurations account for the lion's share of compromises in cloud environments. Misconfigured permissions, in particular, are behind most of the incidents you've heard or read about in the news. This vulnerability is exacerbated by some organizations' failure to encrypt sensitive data in the cloud. The underlying problem for both these commonly exploited vulnerabilities is not that the cloud is insecure, but that the customer was using the cloud insecurely.

On the other hand, there are situations in which your organization may need or want to keep systems on premises. This makes a lot of sense when you're dealing with cyber-physical systems (such as industrial control systems) or certain kinds of extremely sensitive data (such as national intelligence), or when you have the sunk costs (incurred costs that cannot be recovered) of an existing infrastructure and a well-resourced cybersecurity team. In these cases, your team will shoulder the entire burden of security unless you outsource some of that.

Most organizations are moving to cloud-based solutions primarily because of costs. Wherever your assets are stored, the security solutions you need for infrastructure management are fundamentally similar and yet different in subtle but important ways. The rest of this chapter will cover the infrastructure concepts you'll need to know as a cybersecurity analyst.

Network Architecture

A *network architecture* refers to the nodes on a computer network and the manner in which they are connected to one another. There is no universal way to depict this, but it is common to, at least, draw the various subnetworks and the network devices (such as routers, firewalls) that connect them. It is also better to list the individual devices

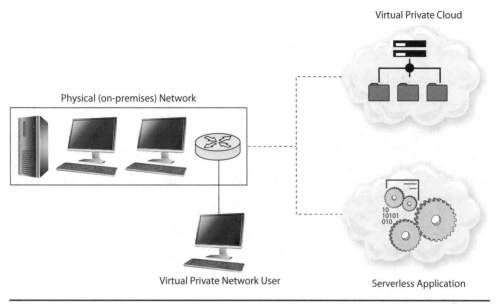

Figure 8-1   Hybrid network architecture

included in each subnet, at least for valuable resources such as servers, switches, and network appliances. The most mature organizations draw on their asset management systems to provide a rich amount of detail as well as helpful visualizations.

A network architecture should be prescriptive in that it should determine the way things shall be. It should not just be an exercise in documenting where things are but in placing them deliberately in certain places. Think of a network architecture as a military commander arraying her forces in defensive positions in preparation for an enemy assault. She wouldn't go to her subordinates and ask where they want to be. Instead, she'd study the best available intelligence about the enemy capabilities and objectives, combine that with an understanding of her own mission and functions, and then place her forces intentionally.

In most cases, network architectures are hybrid constructs incorporating physical, software-defined, virtual, and cloud assets. Figure 8-1 shows a high-level example of a typical hybrid architecture. The following sections describe each of the architectures in use by many organizations.

Physical Network

The most traditional network architecture is a physical one. In a *physical network architecture*, we describe the manner in which physical devices such as workstations, servers, firewalls, and routers relate to one another. Along the way, we decide what traffic is allowed from where to where and develop the policies that will control those flows. These policies are then implemented in the devices themselves, for example as firewall rules or access control lists (ACLs) in routers. Most organizations use a physical network architecture, or perhaps more than one. While physical network architectures are well known to every-body who's ever taken a networking course at any level, they are limited in that they typically require static changes to individual devices to adapt to changing conditions.

Software-Defined Network

Software-defined networking (SDN) is a network architecture in which software applications are responsible for deciding how best to route data (the control layer) and then for actually moving those packets around (the data layer). This is in contrast to a physical network architecture in which each device needs to be configured by remotely connecting to it and giving it a fixed set of rules with which to make routing or switching decisions. Figure 8-2 shows how these two SDN layers enable networked applications to communicate with each other.

One of the most powerful aspects of SDN is that it decouples data forwarding functions (the data plane) from decision-making functions (the control plane), allowing for holistic and adaptive control of how data moves around the network. In an SDN architecture, SDN is used to describe how nodes communicate with each other.

Virtual Private Cloud Network

One of the benefits of using cloud computing is that resources such as CPUs, memory, and networking are shared by multiple customers, which means we pay only for what we use, and we get to add resources seamlessly whenever we need them. The (security) challenge comes when we consider what those resources are doing when we are not using them: they're being used by someone else. Although cloud computing service providers go to great lengths to secure their customers' computing, shared resources can introduce a level of risk. For this reason, most enterprises that decide to host significant or sensitive parts of their assets on the cloud want more security than that.

Figure 8-2
Overview of a software-defined networking architecture

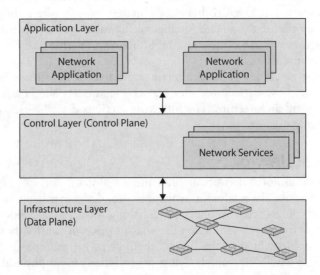

A *virtual private cloud* (VPC) is a private set of resources within a public cloud environment. Any systems within the VPC can communicate with other systems as though they were all in the same physical or software-defined network architecture, though VPCs are hosted in the cloud. The "virtual" part differentiates this approach from a private cloud, in which the hardware resources are owned by the enterprise (not by a cloud service provider) and are typically in a private data center.

So if we want to compare a public cloud to a VPC or to a physical or SDN architecture, we could say that the public cloud is like living in an apartment with (generally) well-behaved roommates who are not always home (giving you the illusion of having the place to yourself). A VPC would be like living in a condo where nobody else has a key and you know you can have the place to yourself while enjoying the convenience of having someone else take care of the common areas, maintenance, and so on. A physical or SDN architecture would be like buying a lot and building your own house away from the neighbors.

Virtual Private Network

A *virtual private network* (VPN) provides a secure tunnel between two endpoints over some shared medium. This enables your users, for instance, to take their mobile devices to another country and connect to your network with the same amount of security they'd have if they were working in the same building. A VPN is most commonly created on demand between an endpoint and a network device (typically a router or VPN concentrator). VPNs can also be persistently established between network devices to create a VPN backbone between corporate locations such as branch offices and corporate headquarters.

VPNs are important architectural components, especially when supporting remote workers. Without VPNs, your staff would not be protected by any security controls that do not reside in their devices. For example, a user surfing the Web directly could more easily be compromised if she is not protected by your network security appliances. If she connects to your VPN, her traffic could go through the same controls that protect anyone who was in the office.

Perhaps the biggest drawback of using VPNs is that the appliances that terminate remote user connections can become bottlenecks and slow things down. A common way to mitigate this is to use the VPN only for some of the user traffic. A *VPN split tunnel* is a configuration in which the VPN client sends some traffic to the corporate gateway and enables other traffic to flow directly into the Internet. This configuration would be helpful if you were a remote worker and wanted to access your protected corporate systems but still print on the printer in your home office. The split tunnel could be configured to send all traffic to the gateway except that intended for the printer. Conversely, the tunnel could be configured to send only traffic to the gateway that is destined for corporate URLs and leave everything else (including the printer) outside the tunnel. The first scenario still enables your corporate security stack to protect you as you surf the Web, while the second case does not. Which is best is typically a complex decision, but, generally speaking, it is more secure not to use split tunneling, and if you must, to very selectively whitelist the destinations you'll allow outside the tunnel.

Serverless Network

In a *serverless network architecture* (sorry, no acronym here), a relatively smart client (think single-page web application with a bunch of JavaScript or an applet) needs some specific functions provided by a cloud resource. Because of the focus on providing functions in the cloud, serverless architectures are sometimes called Function as a Service (FaaS) models.

A comparison may be helpful here. Suppose you have an online store that sells widgets. In a traditional client/server model, all the processing and data storage would be taken care of by one (or maybe two) servers hosted by your company. Whether you sell hundreds of widgets per hour or one every few weeks, you still need the server(s) standing by, waiting for orders, and this gets expensive in a hurry. Now, suppose you implement a serverless architecture in which your customers download a mobile app, which would be a fairly thick client with processing and storage functions. Whenever customers want to browse or search your catalog of widgets, their apps call functions on a cloud service, and those functions return the appropriate data, which is then further processed and rendered by the app. If they want to place a purchase, their app invokes a different set of functions that place the order and charge a credit card. Now you have no need to host web, application, or data servers; you just break it down into functions that are hosted by a cloud service provider, sit back, and rake in the money.

Virtualization

Virtualization is the creation and use of computer and network resources to allow for varied instances of operating systems and applications on an ad hoc basis. Virtualization technologies have revolutionized IT operations because they have vastly reduced the hardware needed to provide a wide array of service and network functions. Virtualization's continued use has enabled large enterprises to achieve a great deal of agility in their IT operations without adding significant overhead. For the average user, virtualization has proven to be a low-cost way to gain exposure to new software and training. Although virtualization has been around since the early days of the Internet, it didn't gain a foothold in enterprise and home computing until the 2000s.

Hypervisors

As previously described, virtualization is achieved by creating large pools of logical storage, CPUs, memory, networking, and applications that reside on a common physical platform. This is most commonly done using software called a *hypervisor*, which manages the physical hardware and performs the functions necessary to share those resources across multiple virtual instances. In short, one physical box can "host" a range of varied computer systems, or *guests*, thanks to clever hardware and software management.

Hypervisors are classified as either Type 1 or Type 2. Type-1 hypervisors are also referred to as *bare-metal* hypervisors, because the hypervisor software runs directly on the host computer hardware. Type-1 hypervisors have direct access to all hardware and manage guest operating systems. Today's more popular Type-1 hypervisors include

VMware ESXi, Microsoft Hyper-V, and Kernel-based Virtual Machine (KVM). Type-2 hypervisors are run from within an already existing operating system. These hypervisors act just like any other piece of software written for an operating system and enable guest operating systems to share the resources that the hypervisor has access to. Popular Type-2 hypervisors include VMware Workstation Player, Oracle VM VirtualBox, and Parallels Desktop.

Virtual Desktop Infrastructure

Traditionally, users log into a local workstation running an operating system that presents them resources available in the local host and, optionally, on remote systems. *Virtual desktop infrastructure* (VDI) separates the physical devices that the users are touching from the systems hosting the desktops, applications, and data. VDI comes in many flavors, but in the most common ones, the user's device is treated as a thin client whose only job is to render an environment that exists on a remote system and relay user inputs to that system. A popular example of this is Microsoft Remote Desktop Protocol (RDP), which enables a Windows workstation to connect remotely to another workstation desktop and display that desktop locally. Another example is the Desktop as a Service (DaaS) cloud computing solution, in which the desktop, applications, and data are all hosted in a cloud service provider's system to which the user can connect from a multitude of end devices.

VDI is particularly helpful in regulated environments because of the ease with which it supports data retention, configuration management, and incident response. If a user's system is compromised, it can quickly be isolated for remediation or investigation, while a clean desktop is almost instantly spawned and presented to the user, reducing the downtime to seconds. VDI is also attractive when the workforce is highly mobile and may log in from a multitude of physical devices in different locations. Obviously, this approach is highly dependent on network connectivity. For this reason, organizations need to consider carefully their own network speed and latency when deciding how (or whether) to implement it.

Containerization

As virtualization software matured, a new branch called *containers* emerged. Whereas operating systems sit on top of hypervisors and share the resources provided by the bare metal, containers sit on top of operating systems and share the resources provided by the host OS. Instead of abstracting the hardware for guest operating systems, container software abstracts the kernel of the operating system for the applications running above it. This allows for low overhead in running many applications and improved speed in deploying instances, because a whole virtual machine doesn't have to be started for every application. Rather, the application, services, processes, libraries, and any other dependencies can be wrapped up into one unit. Additionally, each container operates in a sandbox, with the only means to interact being through the user interface or application programming interface (API) calls. Containers have enabled rapid development operations, because developers can test their code more quickly, changing only the components necessary in the container and then redeploying.

Network Segmentation

Network segmentation is the practice of breaking up networks into smaller subnetworks. The subnetworks are accessible to one another through switches and routers in on-premises networks, or through VPCs and security groups in the cloud. Segmentation enables network administrators to implement granular controls over the manner in which traffic is allowed to flow from one subnetwork to another. Some of the goals of network segmentation are to thwart the adversary's efforts, improve traffic management, and prevent spillover of sensitive data. Beginning at the physical layer of the network, segmentation can be implemented all the way up to the application layer.

Virtual Local Area Networks

Perhaps the most common method of providing separation at the link layer of the network is the use of *virtual local area networks* (VLANs). VLANs are enabled by network switches by applying a tag to each frame received at a port. This tag contains the VLAN identifier for that frame, based on the endpoint that sent it. The switch reads the tag to determine the port to which the frame should be forwarded. At the destination port, the tag is removed, and then the frame is sent to the destination host.

Properly configured, a VLAN enables various hosts to be part of the same network even if they are not physically connected to the same network equipment. Alternatively, a single switch could support multiple VLANs, greatly improving design and management of the network. Segmentation can also occur at the application level, preventing applications and services from interacting with others that may run on the same hardware. Keep in mind that segmenting a network at one layer doesn't carry over to the higher layers. In other words, simply implementing VLANs is not enough if you also desire to segment based on the application protocol.

Physical Segmentation

Though the use of VLANs for internal network segmentation is the norm nowadays, some organizations still use physical segmentation. For example, many networks are segmented from the rest of the Internet using a firewall. Internally, they may also use traditional switching instead of VLANs. In this model, every port on a given switch belongs to the same subnetwork. If you want to move a host to a different subnetwork, you must physically change the connection to a switch in the different subnet. If you are using wireless LANs, you could run into interesting problems if your access points are not all connected to the same switch. For this reason, physical network segmentation makes most sense when dealing with cyber-physical systems that don't (and shouldn't) move around, such as those in industrial controls systems. It also makes sense when you want to keep things a bit simpler. Today, many organizations use a mixture of both virtual and physical segmentation.

Jump Boxes

To facilitate outside connections to segmented parts of the network, administrators sometimes designate a specially configured machine called a *jump box*, or *jump server*. As the name suggests, these computers serve as jumping off points for external users to access protected parts of a network. The idea is to keep special users from logging into

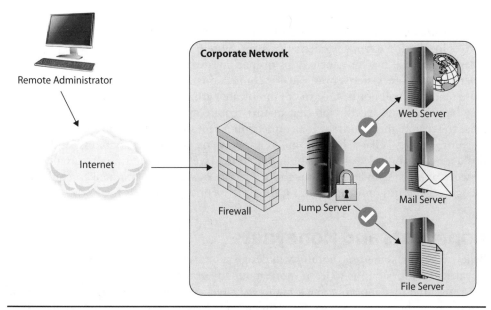

Figure 8-3 Network diagram of a simple jump box arrangement

a particularly important host using the same workstation they use for everything else. If that daily-use workstation were to become compromised, it could be used by an attacker to reach the sensitive nodes. If, on the other hand, these users are required to use a specially hardened jump box for these remote connections, it would be much more difficult for the attacker to reach the crown jewels.

A great benefit of jump boxes is that they serve as a chokepoint for outside users who want to gain access to a protected network. Accordingly, jump boxes often have high levels of activity logging enabled for auditing or forensic purposes. Figure 8-3 shows a very simple configuration of a jump box in a network environment. Notice the placement of the jump box in relation to the firewall device on the network. Although it may improve overall security to designate a sole point of access to the network, it's critical that the jump box is carefully monitored, because a compromise of this server may allow access to the rest of the network. This means disabling any services or applications that are not necessary, using strict ACLs, keeping up-to-date with software patches, and using multifactor authentication where possible.

 EXAM TIP You will likely see jump boxes on the exam in the context of privileged account users (for example, system admins) using them to remotely log into sensitive hosts.

System Isolation

Even if the systems reside within the same subnetwork or VLAN, some systems should be communicating only with certain other systems, and it becomes apparent that something is amiss if you see loads of traffic outside of the expect norms. One way to ensure that hosts in your network are talking only to the machines they're supposed to is to enforce *system isolation*. This can be achieved by implementing additional policies on network devices in addition to your segmentation plan. System isolation can begin with physically separating special machines or groups of machines with an *air gap*, which is a physical separation of these systems from outside connections. There are clearly tradeoffs in that these machines will not be able to communicate with the rest of the world. However, if they have only one specific job that doesn't require external connectivity, it may make sense to separate them entirely. If a connection is required, you may be able to use ACLs to enforce policy. Like a firewall, an ACL allows or denies certain access, and it does this depending on a set of rules applicable to the layer it is operating on, usually at the network or file-system level. Although the practice of system isolation takes a bit of fore-thought, the return on investment for the time spent to set it up is huge.

Honeypots and Honeynets

Believe it or not, sometimes admins will design systems to attract attackers. *Honeypots* are a favorite tool for admins to learn more about adversaries' goals by intentionally expos-ing a machine that appears to be a highly valuable, and sometimes unprotected, target. Although the honeypot may seem legitimate to the attacker, it is actually isolated from the normal network, and all its activity is being monitored and logged. This offers several benefits from a defensive point of view. By convincing an attacker to focus his efforts against the honeypot machine, an administrator can gain insight in the attacker's tactics, techniques, and procedures (TTPs). This can be used to predict behavior or aid in iden-tifying the attacker via historical data. Furthermore, honeypots may delay an attacker or force him to exhaust his resources in fruitless tasks.

Honeypots have been in use for several decades, but they have been difficult or costly to deploy because they often meant dedicating actual hardware to face attackers, thus reducing what could be used for production purposes. Furthermore, to engage an attacker for any significant amount of time, the honeypot needs to look like a real (and, ideally, valuable) network node, which means putting some thought into what software and data to put on it. This all takes lots of time and isn't practical for very large deploy-ments. Virtualization has come to the rescue and addresses many of the challenges associ-ated with administering these machines because the technology scales easily and rapidly.

With a honeynet, the idea is that if some is good, then more is better. A *honeynet* is an entire network designed to attract attackers. The benefits of its use are the same as with honeypots, but these networks are designed to look like real network environments, complete with real operating systems, applications, services, and associated network traf-fic. You can think of honeynets as a highly interactive set of honeypots, providing realistic feedback as the real network would.

For both honeypots and honeynets, the services are not actually used in production, so there shouldn't be any reason for legitimate interaction with the servers. You can therefore assume that any prolonged interaction with these services implies malicious intent. It follows that traffic from external hosts on the honeynet is the real deal and not as likely to be a false positive as in the real network. As with individual honeypots, all honeynet activity is monitored, recorded, and sometimes adjusted based on the desire of the administrators. Virtualization has also improved the performance of honeynets, allowing for varied network configurations on the same bare metal.

Asset Management

An asset is, by definition, anything of worth to an organization, including people, partners, equipment, facilities, reputation, and information. Though every asset needs to be protected, we restrict our discussion in this chapter to information technology assets, which include hardware, software, and the data that they store, process, and communicate. These assets, of course, exist in a context: They are acquired or created at a particular point in time through a specific process and (usually) for a purpose. They move through an organization, sometimes adding value and sometimes waiting to be useful. Eventually, they outlive their utility and must be disposed of appropriately, sometimes being replaced by a new asset. This is the asset lifecycle.

The process by which we manage an asset at every stage of its lifecycle is called *IT Asset Management* (ITAM). Effective ITAM provides a wealth of information that is critical to defending our information systems. For example, it lets us know the location, configuration, and user(s) of each computer. It enables us to know where each copy of a vulnerable software application exists so we can quickly patch it. Finally, it can tell us where every file containing sensitive data is stored so we can keep it from leaking out of our systems.

Asset Inventory

You cannot protect what you don't know you have. Though inventorying assets is not what most of us would consider glamorous work, it is nevertheless a critical aspect of managing vulnerabilities in your information systems. In fact, this aspect of security is so important that it is prominently featured at the top of the list of Center for Internet Security (CIS) critical security controls (CIS Controls). CIS Control 1 is the inventory and control of authorized and unauthorized hardware assets, and CIS Control 2 deals with the software running on those devices.

Developing and maintaining an accurate asset inventory can be remarkably difficult. It is not simply about knowing how many computers you have, but also includes what software is installed in them, how they are configured, and to whom they are assigned. You also need to track software licenses (even the ones that are not in use), cloud assets, and sensitive data, just to name a few. If you didn't inherit an accurate inventory from the last security professional at your organization, you're probably going to have to discover those assets through some combination of physical and logical inspections. Even after you know what you have, you'll need to track changes to these assets, such as software application updates or a computer reassignment.

Asset Tagging

Asset tagging is the practice of placing a uniquely identifiable label on an asset, ideally as part of its acquisition or creation process. For example, when you order a new computer, it should be tagged and added to the inventory immediately upon receipt, even before it is configured for use. If, in the next step, the asset will go to the IT department for imaging, the inventory will show that it is assigned to the staff member responsible for that. This person will configure the device and, you guessed it, update the inventory to show the baseline that was applied to it. When the asset is issued to its intended user, that person will be able to acknowledge receipt of it by checking the tag number. The asset tag enables anyone in the organization to check which particular piece of equipment they're looking at.

Though some organizations still use visual inspection for checking asset tags, automated asset management tools make this process a lot more efficient. For this, you need tags that are machine-readable, which come in three main forms: barcode, Quick Response (QR) code, and radio frequency identification (RFID). Some tags implement multiple forms, so a tag could have both a barcode and an RFID chip, for example. You will obviously also need a reader that is integrated with your inventory management system so you can scan a tag and see the data pertaining to the corresponding asset.

An advantage of using RFID tags is that you could place readers in key places to alert when an asset is being moved. This *geofencing* works similarly to anti-theft systems in retail stores. Because it tracks asset movement, geofencing can be used to discourage people from taking company assets out of the building or even from moving those assets from one room (or area) to another within the same building. For example, if a tagged smart-board on wheels is supposed to stay in the conference room, you could place a RFID reader by the door that would sound an alarm if someone tried to roll the board outside the room.

Change Management

You can see that a tremendous amount of effort goes into managing the recommended actions following a vulnerability scan. Although implementing every recommendation may seem like a good idea on the surface, we cannot do this all at once. Without a systematic approach to managing all the necessary security changes, we risk putting ourselves in a worse place than when we began. The purpose of establishing formal communication and change management procedures is to ensure that the right changes are made the first time, that services remain available, and that resources are used efficiently throughout the changes.

In the context of cybersecurity, *change management* is the process of identifying, analyzing, and implementing changes to information systems in a way that addresses both business and security requirements. This is not an IT or cybersecurity function but one that should involve the entire organization and requires executive leadership and support from all business areas. There are usually (at least) three key roles in this process:

- **Requestor** The person or unit that requests the change and needs to make the business case for it
- **Review board** A group of representatives from various units in the organization that is responsible for determining whether the change is worth making and, if so, what its priority is
- **Owner** The individual or unit that is responsible for implementing and documenting the change, which may or may not be the requestor

Routine change requests should be addressed by the board on a regularly scheduled basis. Many boards meet monthly but have procedures in place to handle emergency requests that can't wait until the next meeting. Each change is validated and analyzed for risk so that any required security controls (and their related costs) are considered in the final decision. If the change is approved, an owner is assigned to see it through completion.

Change Advisory Board

Many organizations use a change advisory board (CAB) to approve major changes to a company's policies and to assist with change management in the monitoring and assessment of changes. Members of the CAB often include any entity that could be adversely affected by the proposed changes, such as customers, managers, technical staff, and company leadership. When convening, members of the board ensure that all proposed changes address the issue and make sense from both business and technical perspectives.

Identity and Access Management

Before users are able to access resources, they must first prove they are who they claim to be and that they have been given the necessary rights or privileges to perform the actions they are requesting. Even after these steps are completed successfully and a user can access and use network resources, that user's activities must be tracked, and accountability must be enforced for whatever actions they take. *Identification* describes a method by which a subject (user, program, or process) claims to have a specific identity (username, account number, or e-mail address). *Authentication* is the process by which a system verifies the identity of the subject, usually by requiring a piece of information that only the claimed identity should have. This piece could be a password, passphrase, cryptographic key, personal identification number (PIN), anatomical attribute, or token. Together, the identification and authentication information (for example, username and password) make up the subject's *credentials*. Credentials are compared to information that has been previously stored for this subject. If the credentials match the stored information, the subject is authenticated. The system can then perform *authorization*, which is a check against some type of policy to verify that this user has indeed been authorized to access the requested resource and perform the requested actions.

Identity and access management (IAM) is a broad term that encompasses the use of different technologies and policies to identify, authenticate, and authorize users through automated means. It usually includes user account management, access control, credential management, single sign-on (SSO) functionality, rights and permissions management for user accounts, and the auditing and monitoring of all of these items. IAM enables organizations to create and manage digital identities' lifecycles (create, maintain, terminate) in a timely and automated fashion.

Privilege Management

Over time, as users change roles or move from one department to another, they often are assigned more and more access rights and permissions. This is commonly referred to as *authorization creep*. This can pose a large risk for a company, because too many users may have too much privileged access to company assets. In immature organizations, it may be considered easier for network administrators to grant users more access than less; then a user would not later ask for authorization changes that would require more work to be done on her profile as her role changed. It can also be difficult to know the exact access levels different individuals require. This is why privilege management is a critical security function that should be regularly performed and audited. Enforcing least privilege on user accounts should be an ongoing job, which means each user's permissions should be reviewed regularly to ensure the company is not putting itself at risk.

This is particularly true of administrator or other privileged accounts, because these allow users (legitimate or otherwise) to do things such as create new accounts (even other privileged ones), run processes remotely on other computers, and install or uninstall software. Because of the power privileged accounts have, they are frequently among the first targets for an attacker.

Here are some best practices for managing privileged accounts:

- Minimize the number of privileged accounts.
- Ensure that each administrator has a unique account (that is, no shared accounts).
- Elevate user privileges only when necessary, after which time the user should return to regular account privileges.
- Maintain an accurate, up-to-date account inventory.
- Monitor and log all privileged actions.
- Enforce multifactor authentication.

Sooner or later every account gets deprovisioned. For users, this is usually part of the termination procedures. For system accounts, this could happen because you got rid of a system or because some configuration change rendered an account unnecessary. If this is the case, you must ensure that this is included in your standard change management process. Whatever the type of account or the reason for getting rid of it, it is important that you document the change so you no longer track that account for reviewing purposes.

Multifactor Authentication

Authentication is still largely based on credentials consisting of a username and a password. If a password is the only factor used in authenticating the account, this approach is considered *single-factor authentication. Multifactor authentication* (MFA) just means that more than one authentication factor is used. *Two-factor authentication* (2FA) is perhaps the most common form of authentication and usually relies on a code that is valid for only a short time, which is provided to the user by another system. For example, when logging into your bank from a new computer, you may enter your username and password and then be required to provide a six-digit code. This code is provided by a separate application (such as Google Authenticator) and changes every 30 seconds. Some organizations will text a similar verification code to a user's mobile phone (though this approach is no longer considered secure) if, for example, you've forgotten your password and need to create a new one. For even more security, three-factor authentication (3FA), such as a smart card, PIN, and retinal scan, is used, particularly in government scenarios.

Single Sign-On

Employees typically need to access many different computers, communications services, web applications, and other resources throughout the course of a day to complete their tasks. This could require the employees to remember multiple credentials for these different resources. *Single sign-on* (SSO) enables users to authenticate only once and then be able to access all their authorized resources regardless of where they are.

Using SSO offers benefits both to the user and the administrator sides. Users need to remember only a single password or PIN, which reduces the fatigue associated with managing multiple passwords. Additionally, they'll save time by not having to reenter credentials for every service desired. For the administrator, using SSO means fewer calls from users who forget their passwords. Figure 8-4 shows the flow of an SSO request in Google using the Security Assertion Markup Language (SAML) standard, a widely used method of implementing SSO.

SAML provides access and authorization decisions using a system to exchange information between a user, the identity provider (IDP) such as Acme Corp., and the service provider (SP) such as Google. When a user requests access to a resource on the SP, the SP creates a request for identity verification for IDP. The IDP will provide feedback about the user, and the SP can make its decision on an access control based on its own internal rules and the positive or negative response from the IDP. If access is granted, a token is generated in lieu of the actual credentials and passed on to the SP.

Although SSO improves the user experience in accessing multiple systems, it does have a significant drawback in the potential increase in impact if the credentials are compromised. Using an SSO platform thus requires a greater focus on the protection of the user credentials. This is where including multiple factors and context-based solutions can provide strong protection against malicious activity. Furthermore, as SSO centralizes the authentication mechanism, that system becomes a critical asset and thus a target for attacks. Compromise of the SSO system, or loss of availability, means loss of access to the entire organization's suite of applications that rely on the SSO system.

Identity Federation

Federated identity is the concept of using a person's digital identity credentials to gain access to various services, often across organizations. Where SSO can be implemented either within a single organization or among multiple ones, federation applies only to the latter case. The user's identity credentials are provided by a broker known as the *federated identity manager*. When verifying her identity, the user needs only to authenticate with the identity manager; the application that's requesting the identity information also needs to trust the manager. Many popular platforms, such as Google, Amazon, and

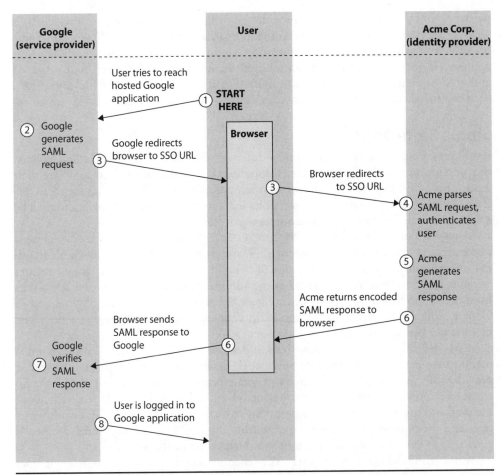

Figure 8-4 Single sign-on flow for a user-initiated request for identity verification

Twitter, take advantage of their large memberships to provide federated identity services for third-party websites, saving their users from having to create separate accounts for each site.

OpenID

OpenID is an open standard for user authentication by third parties. It is a lot like SAML, except that a user's credentials are maintained not by the user's company, but by a third party. Why is this useful? By relying on specialized identity providers (IDPs) such as Amazon, Google, or Steam, developers of Internet services (such as websites) don't need to develop their own authentication systems. Instead, they are free to use any IDP or group of IDPs that conforms to the OpenID standard. All that is required is that all parties use the same standard and that everyone trusts the IDP(s).

Figure 8-5
OpenID
process flow

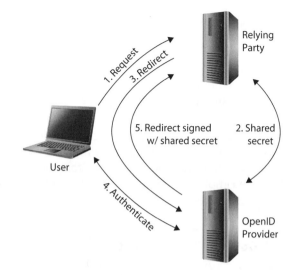

OpenID, currently in version 2.0, defines three roles:

- **End user** The user who wants to be authenticated in order to use a resource
- **Relying party** The server that owns the resource that the end user is trying to access
- **OpenID provider** The IDP (such as Google) on which the end user already has an account and which will authenticate the user to the relying party

You have probably encountered OpenID if you've tried to access a website and were presented with the option to log in using your Google identity credentials. (Oftentimes you see an option for Google and one for Facebook in the same window, but Facebook uses its own protocol, Facebook Connect.) In a typical use case, depicted in Figure 8-5, a user wants to visit a protected website. The user's agent (typically a web browser) requests the protected resource from the relying party. The relying party connects to the OpenID provider and creates a shared secret for this transaction. The server then redirects the user agent to the OpenID provider for authentication. Upon authentication by the provider, the user agent receives a redirect (containing an authentication token, also known as the user's OpenID) to the relying party. The authentication token contains a field signed with the shared secret, so the relying party is assured that the user is authenticated.

Role-Based Access Control

What if you have a tremendous number of resources that you don't want to manage on a case-by-case basis? *Role-based access control* (RBAC) enables you to grant permissions based on a user's role, or group. The focus for RBAC is at the role level, where the administrators define what the role can do. Users are able to do only what their role permissions allow them to do, so there is no need for an explicit denial to a resource for a given role. Ideally, there are a small number of roles in a system, regardless of the total number of users, so managing them becomes much easier with RBAC.

Let's say, for example, that we need a research and development analyst role. We develop this role not only to enable an individual to have access to all product and testing data, but also, and more importantly, to outline the tasks and operations that the role can carry out on this data. When the analyst role makes a request to access the new testing results on the file server, in the background the operating system reviews the role's access levels before allowing this operation to take place.

 NOTE Introducing roles also introduces the difference between rights being assigned *explicitly* and *implicitly*. If rights and permissions are assigned explicitly, they are assigned directly to a specific individual. If they are assigned implicitly, they are assigned to a role or group and the user inherits those attributes.

Attribute-Based Access Control

Attribute-based access control (ABAC) uses attributes of any part of a system to define allowable access. These attributes can belong to subjects, objects, actions, or contexts. Here are some possible attributes you could use to describe your ABAC policies:

- **Subjects**   Clearance, position title, department, years with the company, training certification on a specific platform, member of a project team, location
- **Objects**   Classification, files pertaining to a particular project, HR records, location, security system component
- **Actions**   Review, approve, comment, archive, configure, restart
- **Context**   Time of day, project status (open/closed), fiscal year, ongoing audit

As you can see, ABAC provides the most granularity of any of the access control models. It would be possible, for example, to define and enforce a policy that allows only directors to comment on (but not edit) files pertaining to a project that is currently being audited. This specificity is a two-edged sword, however, since it can lead to an excessive number of policies that could interact with each other in ways that are difficult to predict.

Mandatory Access Control

For environments that require additional levels of scrutiny for data access, such as those in military or intelligence organizations, the *mandatory access control* (MAC) model is sometimes used. As the name implies, MAC *requires* explicit authorization for a given user on a given object. The MAC model uses additional labels for multilevel security—Unclassified, Confidential, Secret, and Top Secret—that are applied to both the subject and object. When a user attempts to access a file, a comparison is made between the security labels on the file, called the *classification level*, and the security level of the subject, called the *clearance level*. Only users who have the appropriate security labels in their own profiles can access files at the equivalent level, and verifying this "need to know" is the main strength of MAC. Additionally, the administrator can restrict further propagation of the resource, even from the content creator.

Manual Review

Auditing and analysis of login events is critical for a successful incident investigation. All modern operating systems have a way to log successful and unsuccessful access attempts. Figure 8-6 shows the contents of an auth.log file indicating all enabled logging activity on a Linux server. Although it may be tempting to focus on the failed attempts, you should also pay attention to the successful logins, especially in relation to those failed attempts. The chance that the person logging in from 4.4.4.12 is an administrator who made a mistake the first couple of times is reasonable. However, when you combine this information with the knowledge that this device has just recently performed a suspicious network scan, you can assume that an innocent mistake is probably unlikely.

 EXAM TIP   When dealing with logs, consider the time zone difference for each device. Some may report in the time zone you operate in, some may use GMT, and others may be off altogether.

Event logs are similar to syslogs in the detail they provide about a system and connected components. Windows enables administrators to view all of a system's event logs with the Event Viewer utility. This makes it much easier to browse through the thousands of entries related to system activity, as shown in Figure 8-7. It's an essential tool for understanding the behavior of complex systems such as Windows—and particularly important for servers, which aren't designed to always provide feedback through the user interface.

In recent Windows operating systems, successful login events have an event ID of 4624, whereas login failure events have an ID of 4625, with error codes to specify the exact reason for the failure. In the Windows 10 Event Viewer, you can specify exactly which types of event you'd like to get more detail on using the Create Custom View command available from the Action pull-down menu. The resulting dialog is shown in Figure 8-8.

```
Apr  6 11:08:27 edda sshd[5761]: pam_unix(sshd:auth): authentication failure; logname=
uid=0 euid=0 tty=ssh ruser= rhost=4.4.4.12  user=root
Apr  6 11:08:29 edda sshd[5761]: Failed password for root from 4.4.4.12 port 52724 ssh2
Apr  6 11:08:49 edda sshd[5761]: message repeated 2 times: [ Failed password for root from
4.4.4.12 port 52724 ssh2]
Apr  6 11:08:49 edda sshd[5761]: Connection closed by 4.4.4.12 port 52724 [preauth]
Apr  6 11:08:49 edda sshd[5761]: PAM 2 more authentication failures; logname= uid=0 euid=0
tty=ssh ruser= rhost=4.4.4.12  user=root
Apr  6 11:10:48 edda sshd[5812]: Accepted password for root from 4.4.4.12 port 52793 ssh2
Apr  6 11:10:48 edda sshd[5812]: pam_unix(sshd:session): session opened for user root by
(uid=0)
Apr  6 11:10:48 edda systemd-logind[983]: New session 1079 of user root.
```

Figure 8-6   Snapshot of the auth.log entry in a Linux system

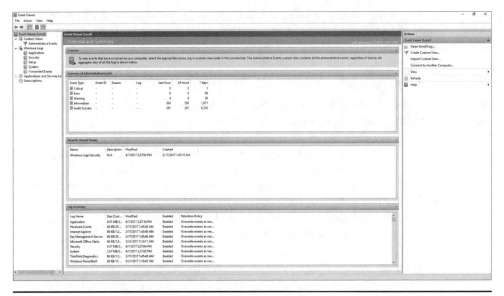

Figure 8-7 The Event Viewer main screen in Windows 10

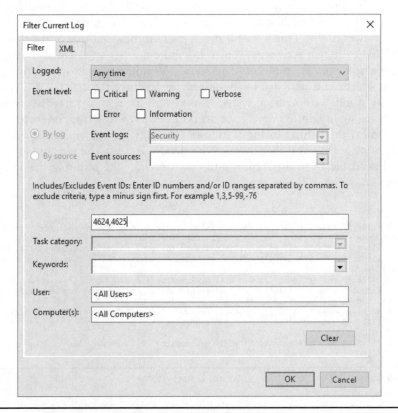

Figure 8-8 The Event Viewer dialog for filtering log information

Cloud Access Security Broker

Before the emergence of diverse cloud computing solutions, it was a lot simpler for us to control access to our information systems, monitor their performance, and log events in them. Doing this became even more difficult with the shift toward using mobile (and potentially personal) devices to access corporate resources. How do you keep track of what's happening when a dozen different cloud services are in use by your organization? A *cloud access security broker* (CASB) sits between each user and each cloud service, monitoring all activity, enforcing policies, and alerting you when something seems to be wrong.

We generally talk of four pillars to CASBs.

- **Visibility**   This is the ability to see who is doing what, where. Are your users connecting to unauthorized resources (such as cloud storage that is not controlled by your organization)? Are they constrained to services and data that they are authorized to use?

- **Threat protection**   This pillar detects (and, ideally, blocks) malware, insider threats, and compromised accounts. In this regard, features in this pillar look a lot like those of an intrusion detection or protection system but for your cloud services.

- **Compliance**   If you are in a regulated environment (and, let's face it, with data laws like the GDPR, most of us are), you want to ensure that any data in the cloud is monitored for compliance with appropriate regulations concerning who can access it, how, and when.

- **Data security**   Whether or not you are in a regulated environment, you likely have an abundance of data that needs protection in the cloud. This includes intellectual property, customer information, and future product releases.

Monitoring and Logging

It goes without saying that you need to keep an eye out for a lot of "stuff" in your environment. Otherwise, how would you know that one of your web servers crashed or that your DNS server has an extremely high CPU utilization? *Monitoring* is the practice of ensuring that your information systems remain available and are operating within expected performance envelopes. Though primarily an IT function, monitoring can provide early warnings to cybersecurity analysts when systems are under attack or have been compromised.

Real user monitoring (RUM) is a passive way to monitor the interactions of real users with applications or systems. It uses agents to capture metrics such as delay, jitter, and errors from the user's perspective. Although RUM captures the actual user experience fairly accurately, it tends to produce noisy data (such as incomplete transactions due to users changing their minds or losing mobile connectivity) and thus may require more back-end analysis. It also lacks the elements of predictability and regularity, which could mean that a problem won't be detected during low utilization periods.

An alternative approach to monitoring relies on synthetic interactions with the systems under observation. For example, you could write a simple script that periodically queries your DNS server and records the time it takes for it to respond. You could compare response times over time to determine what is normal and then generate an alert when the performance is suspiciously slow (or when there is no response at all). An advantage of this approach over RUM is that you can precisely control the manner and frequency of your observations. A disadvantage is that you are typically limited in terms of the complexity of the transactions. Whereas a real user can interact with your systems in a virtually infinite number of ways, the synthetic transactions will be significantly fewer and less complex.

What if you find something out of the norm while monitoring one of your systems? This is where you turn to logs. *Logging* is the practice of managing the log data produced by your systems. It involves determining what log data is generated by which systems, where it is sent (and how), and what happens to it once it is aggregated there. It starts with configuring your systems to record the data that will be useful to you in determining whether your systems are free of compromise. For example, you probably want to track login attempts, access to sensitive data, and use of privileged accounts, to name a few. Once you decide what to store, you will need a secure place to aggregate all this data so you can analyze it. Since adversaries will want to interfere with this process, you will have to take extra care to make it difficult to alter or delete the log data.

Turning these reams of raw data into actionable information is the goal of log analysis. In analysis, your first task is to *normalize* of the data, which involves translating the multitude of data formats you get from all over your environment into a common format. Normalization can be done by an agent on the sending system or by the receiving aggregator. Once the data is in the right format, your log analysis tool will typically do one or more of the following: pattern recognition, classification, and prediction.

Pattern recognition compares the log data to known good or bad patterns to find interesting events. This could be as simple as a large number of failed login attempts followed by a successful one, or it could involve detecting user activity at an odd time of day (for that user). *Classification* involves determining whether an observation belongs to a given class of events. For instance, you may want to differentiate peer-to-peer traffic from all other network activity to detect lateral movement by an adversary. Log analysis can also involve *predictive analytics*, which is the process of making a number of related observations and then predicting what the next observation ought to be. A simple example of this would be to observe the first two steps of a TCP three-way handshake (a SYN message followed by a SYN-ACK) and predict that the next interaction between these two hosts will be an ACK message. If you don't see that message, your prediction failed, and this may mean something is not right.

Encryption

Encryption is a method of transforming readable data, called *plaintext*, into a form that appears to be random and unreadable, which is called *ciphertext*. Plaintext is in a form that can be understood either by a person (such as a document) or by a computer (such as executable code). Once plaintext is transformed into ciphertext, neither human nor machine can use it until it is *decrypted*. Encryption enables the transmission of confidential information over insecure channels without unauthorized disclosure. The science behind encryption and decryption is called *cryptography*, and a system that encrypts and/or decrypts data is called a *cryptosystem*.

Although there can be several pieces to a cryptosystem, the two main pieces are the algorithms and the keys. Algorithms used in cryptography are complex mathematical formulas that dictate the rules of how the plaintext will be turned into ciphertext, and vice versa. A key is a string of random bits that will be used by the algorithm to add to the randomness of the encryption process. For two entities to be able to communicate using cryptography, they must use the same algorithm and, depending on the approach, the same or a complementary key. Let's dig into that last bit by exploring the two approaches to cryptography: symmetric and asymmetric.

Symmetric Cryptography

In a cryptosystem that uses *symmetric cryptography*, the sender and receiver use two instances of the same key for encryption and decryption, as shown in Figure 8-9. So the key has dual functionality, in that it can carry out both encryption and decryption processes. It turns out this approach is extremely fast and hard to break, even when using relatively small (such as 256-bit) keys. What's the catch? Well, if the secret key is compromised, all messages ever encrypted with that key can be decrypted and read by an intruder. Furthermore, if you want to separately share secret messages with multiple people, you'll need a different secret key for each, and you'll need to figure out a way to get that key to each person with whom you want to communicate securely. This is sometimes referred to as the key distribution problem.

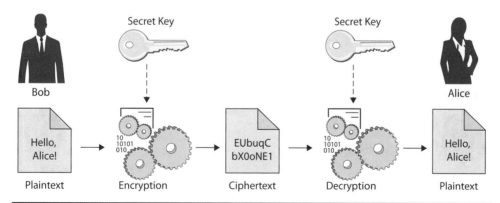

Figure 8-9 When using symmetric algorithms, the sender and receiver use the same key for encryption and decryption functions.

Asymmetric Cryptography

Another approach is to use different keys for encryption and decryption, or *asymmetric cryptography*. The key pairs are mathematically related to each other in a way that enables anything that is encrypted by one to be decrypted by the other. The relationship, however, is complex enough that you can't figure out what the other key is by analyzing the first one. In asymmetric cryptosystems, a user keeps one key (the private key) secret and publishes the complimentary key (the public key). So, if Bob wants to send Alice an encrypted message, he uses her public key and knows that only she can decrypt it. This is shown in Figure 8-10. What happens if someone attempts to use the same asymmetric key to encrypt and decrypt a message? They get gibberish.

So what happens if Alice sends a message encrypted with her private key? Then anyone with her public key can decrypt it. Although this doesn't make much sense in terms of confidentiality, it assures every recipient that it was really Alice who sent it and that it wasn't modified by anyone else in transit. This is the premise behind digital signatures.

Digital Signatures

Digital signatures are short sequences of data that prove that a larger data sequence (say, an e-mail message or a file) was created by a given person and has not been modified by anyone else after being signed. Suppose Alice wants to digitally sign an e-mail message. She first takes a hash of it and encrypts that hash with her private key. She then appends the resulting encrypted hash to her (possibly unencrypted) e-mail. If Bob receives the message and wants to ensure that Alice really sent it, he would decrypt the hash using Alice's public key and then compare the decrypted hash with a hash he computes on his own. If the two hashes match, Bob knows that Alice really sent the message and nobody else changed it along the way.

 NOTE In practice, digital signatures are handled by e-mail applications, so all the hashing and decryption are done automatically, not by you.

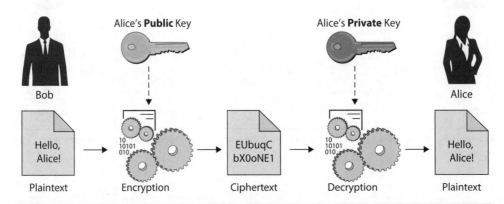

Figure 8-10 When using asymmetric algorithms, the sender encrypts a message with the recipient's public key, and the recipient decrypts it using a private key.

Symmetric vs. Asymmetric Cryptography

The biggest differences between symmetric and asymmetric cryptosystems boil down to key length and encryption/decryption time. Symmetric cryptography relies on keys that are random sequences of characters, and these can be fairly short (256 bits), but it would still require years for an attacker to conduct an effective brute-force attack against them. By contrast, asymmetric key pairs are not random with regard to each other (remember, they are related by a complex math formula), so they are easier to brute-force at a given key size compared to symmetric keys. For this reason, asymmetric keys have to be significantly longer to achieve the same level of security. The current NIST recommendation is that asymmetric key pairs be at least 2048 (and preferably 3072) bits in length.

The other big difference between these two approaches to cryptography is that symmetric cryptosystems (partly because they deal with smaller keys) are significantly faster than asymmetric ones. So, at this point, you may be wondering, why use asymmetric encryption at all if it is slower and requires bigger keys? The answer goes back to the key distribution problem we mentioned earlier. In practice, when transmitting information securely, we almost always use asymmetric cryptography to exchange secret keys between the parties involved, and then switch to symmetric cryptography for the actual transfer of information.

Certificate Management

When we discussed asymmetric key encryption, we talked about publishing one of the keys, but how exactly does that happen? If we allow anyone to push out a public key claiming to be Alice, then how do we know it really is her key and not an impostor's key? There are two ways in which the security community has tackled this problem. The first is through the use of a *web of trust*, which is a network of users who can vouch for each other and for other users wanting to join in. The advantage of this approach is that it is decentralized and free. It is a common approach for those who use Pretty Good Privacy (PGP) encryption.

Most businesses, however, prefer a more formal process for verifying identities associated with public keys, even if it costs more. What they need are trusted brokers who can vouch for the identity of a party using a particular key. These brokers are known as *certificate authorities* (CAs), and their job is to verify someone's identity and then digitally sign that public key, packaging it into a digital certificate or a public key certificate.

A *digital certificate* is a file that contains information about the certificate owner, the CA who issued it, the public key, its validity timeframe, and the CA's signature of the certificate itself. The de facto standard for digital certificates is X.509 and is defined by the Internet Engineering Task Force (IETF) in its RFC 5280.

When a CA issues a digital certificate, it has a specific validity period. When that time elapses, the certificate is no longer valid and the owner needs to get a new one. It is possible for the certificate to become invalid before it expires, however. For example, we stressed earlier the importance of keeping the private key protected, but what if it is compromised? In that case, the owner would have to invalidate the corresponding digital certificate and get a new one. Meanwhile, the public key in question may still be in use by others, so how do we notify the world to stop using it? A *certificate revocation list* (CRL), which is maintained by a revocation authority (RA) is the authoritative reference for certificates that are no longer trustworthy.

Digital certificate management involves the various actions undertaken by a certificate owner, its CA, and its RA to ensure that valid certificates are available and invalid ones are not used. From an organizational perspective, it is focused on acquiring certificates from a CA, deploying them to the appropriate systems, protecting the private keys, transitioning systems to a new certificate before the old ones expire, and reporting compromised (or otherwise no longer valid) certificates to the RA. It can also entail consulting a CRL to ensure that the continued validity of certificates belonging to other entities with whom the organization does business.

 CAUTION   Many organizations disable CRL checks because they can slow down essential business process. This decision introduces serious risks and should not be taken lightly.

Active Defense

Traditionally, defensive operations involve preparing for an attack by securing resources as best we can, waiting for attackers to make a move, and then trying to block them or, failing that, ejecting them from the defended environment. *Active defense* consists of adaptive measures aimed at increasing the amount of effort attackers need to exert to be successful, while reducing the effort for the defenders. Think of it as the difference between a boxer who stands still in the ring with fists raised and a good stance, and a boxer who uses footwork, slipping and bobbing to make it harder for an opponent to effectively land a punch.

One of the most well-known approaches to active defense is *moving target defense* (MTD). MTD makes it harder for an attacker to understand the layout of the defender's systems by frequently (or even constantly) changing the attack surface (such as IP addresses on critical resources). For example, suppose an attacker is conducting active reconnaissance, finds an interesting server at a certain IP address, and makes a note to come back and look for vulnerabilities when the scan is over. A few minutes later, he interrogates that address and finds it is no longer in use. Where did the server go? The defender was employing an MTD scheme in which the addresses are changed every few minutes. The attacker has to go back to square one and try to figure out where the server is, but, by the time he does this, the server will be moved again.

Another approach, which is sometimes bundled with MTD, is the use of honeypots or honeynets, as we discussed earlier in the chapter. Here, the idea is to make an attacker waste her time interacting with an environment that is simply a decoy. Because the honeypot is not a real system, the mere fact that someone is using it can alert the defenders to an attacker's presence and even enable them to observe the attacker and further confuse her.

Chapter Review

The typical enterprise IT environment is both complex and dynamic. Both of these features provide opportunities for attackers to compromise our information systems unless we understand our infrastructures and carefully manage changes within them. Two of the most important concepts we covered in this chapter are asset management, by which we keep track of all the components of our networks, and change management, which enables us to ensure that changes to these assets do not compromise our security. Another very important concept we covered is identity and access management, which involves the manner in which users are allowed to interact with our assets. It is critical that we carefully control these interactions to simultaneously enable organizational business functions while ensuring the protection of our assets.

Questions

1. Your boss is considering an initiative to move all services to the cloud as a cost-savings measure but is concerned about security. Your company has experienced two consecutive quarters of decreasing revenues. Some of your best talent got jittery and found jobs elsewhere, and you were forced to downsize. Faced with a financial downturn, your lifecycle IT refresh keeps getting pushed back. What do you recommend to your boss?

   A. Don't move to the cloud because it's really not any cheaper.

   B. Wait until after the lifecycle refresh and then move to the cloud.

   C. Move to the cloud; it will save money and could help lessen your team's burden.

   D. Don't move to the cloud, because you don't have the needed additional staff.

2. Which architecture would you choose if you were particularly interested in adaptive control of how data moves around the network?

   A. Software-defined

   B. Physical

   C. Serverless

   D. Virtual private network

3. Best practices for managing privileged accounts include all of the following *except* which?

   A. Maintain an accurate, up-to-date account inventory.

   B. Monitor and log all privileged actions.

   C. Ensure that you have only a single privileged account.

   D. Enforce multifactor authentication.

4. Which of the following is the best reason to implement active defense measures?

   A. Hacking back is the best way to stop attackers.

   B. They make it harder for the attacker and easier for the defender to achieve objectives.

   C. An organization lacks the resources to implement honeynets.

   D. Active defense measures are a better alternative to moving target defense (MTD).

Use the following scenario to answer Questions 5–8:

You just started a new job at a small online retailer. The company has never before had a cybersecurity analyst, and the IT staff has been taking care of security. You are surprised to learn that users are allowed to use Remote Desktop Protocol (RDP) connections from outside the network directly into their workstations. You tell your boss that you'd like to put in a change request to prevent RDP and are told there is no change management process. Your boss trusts your judgement and tells you to go ahead and make the change.

5. What would be the best way to proceed with the RDP change?

   A. Your boss approved it, so just go for it.

   B. Encourage everyone to use a virtual private network (VPN) for RDP.

   C. Take no further action until a change management process is established.

   D. Use this request as impetus to create a change management board.

6. You are given three months to implement a solution that allows selected users to access their desktops from their home computers. You start by blocking RDP at the firewall. Which of the following would *not* be your next step?

   A. Require users to use VPN.

   B. Require users to log in through a jump box.

   C. Require users to log in through a honeypot.

   D. Deploy a virtual desktop infrastructure for your users.

7. To improve user authentication, particularly when users remotely access their desktops, which of the following technologies would be most helpful?

   A. Asymmetric cryptography

   B. Symmetric cryptography

   C. Cloud access security broker

   D. Active defense

8. As you continue your remote desktop project, you realize that there are a lot of privileged accounts in your company. Which of the following actions should you take?

A. Minimize the number of accounts by having system administrators share one account and log all activity on it.

B. Deploy a single sign-on solution that requires multifactor authentication.

C. Allow everyone to have administrative access but deploy OpenID for federated authentication.

D. Minimize the number of accounts, ensure that there is no sharing of credentials, and require multifactor authentication.

Answers

1. **C.** A cloud migration makes sense particularly when your IT infrastructure is not new (that is, it will soon need a costly refresh) and/or your team is not up to the task of securing the on-premises systems.

2. **A.** A software-defined network architecture provides the most adaptability when it comes to managing traffic flows. You could also accomplish this in some VPCs, but that would not be as flexible and was not an option.

3. **C.** Though you certainly want to minimize the number of privileged accounts, having only one means one of two things are certain to happen: either you'll have multiple people sharing the same account, or you'll be locked out of making any changes to your environment if the account holder is not willing or able to do the job.

4. **B.** The main objective of active defense measures is to make it harder for the attacker and easier for the defender to achieve objectives. Two examples of this approach are moving target defense and honeynets.

5. **D.** The lack of a change management process is a serious shortcoming in your organization. Because your boss is allowing you to deal with the RDP risk, you could use this effort to jumpstart the change management process. B and C do nothing to mitigate the risk. Answer option A deals with the risk but doesn't fix the change management problem, so it's not the best answer.

6. **C.** A honeypot is a system deployed to attract attackers and should have no real business purpose. Each of the other three answers could reasonably be used to enable external access to sensitive internal resources.

7. **A.** Digital certificates, which employ asymmetric cryptography, can be used to ensure that both endpoints of a communications channel are who they claim to be and can thus help strengthen authentication. A CASB could help, but only for cloud computing assets.

8. **D.** Minimizing the number of privileged accounts, ensuring that there is no sharing of credentials, and requiring MFA are three of the best practices for privileged account management.

PART II

Software Assurance Best Practices

In this chapter you will learn:

- How to develop and implement a software development lifecycle
- General principles for secure software development
- Best practices for secure coding
- How to ensure the security of software

*Give me six hours to chop down a tree and I will
spend the first four sharpening the axe.*

—Abraham Lincoln

When you're developing software, most of the effort goes into either planning and design (in good teams) or debugging and fixes (in other teams). You are very unlikely to be working as a software developer if your principal role in your organization is cybersecurity analyst. You are, however, almost certainly going to be on the receiving end of the consequences for software that is developed in an insecure manner. Quite simply, a developer's skills, priorities, and incentives are very different from those of their security teammates. It is in everyone's best interest, then, to bridge the gap between these communities, which is why CompTIA included the objectives we cover here in the CySA+ exam.

Platforms and Software Architectures

In the last chapter, we discussed network architectures as the way in which we arrange and connect network devices to support our organizations. Here, we turn our attention to *software architectures*, which are the descriptions of the software components of an application and of the manner in which they interface with one another, with other applications, and with end users. The easiest way to get started is to consider the three fundamental tasks that any nontrivial application must do: interact with someone (whether it's a human or another application) to get or give data, do something useful with that data, and finally store (at least some) data.

Figure 9-1 shows a very high-level architecture. User interaction is handled by what is commonly called the presentation layer. This layer also interacts with the business logic

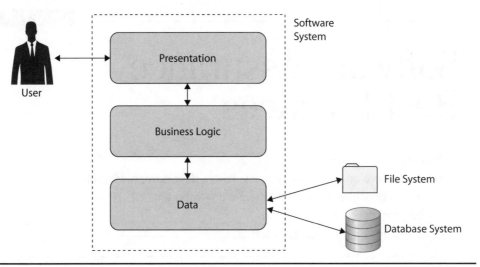

Figure 9-1 A high-level software architecture

of the application, which is where all of the real work gets done. Finally, if the application needs to store data persistently (and most business applications do), the data layer (sometimes called the data access layer) interacts with wherever it is that the data will be stored. This could be a file system or a database management system, for example.

All three tasks—giving or getting data, doing something useful with data, and storing data—can be performed within the same computing device or platform, called a *standalone application*. This type of architecture is becoming increasingly rare, but the venerable calculator and notepad applications that are installed by default with most operating systems are good examples. Office productivity suites with word processing and spreadsheet applications used to be good examples of this class of software. Nowadays, however, many of those applications have moved some or all of their tasks to the cloud.

 NOTE An application could have some cloud (or otherwise remote) functionality and still be considered standalone as long as that functionality is not essential for the application to work.

Distributed applications perform their core tasks on different, connected computing platforms. Multiplayer online games, network file systems, and cloud-based office productivity suites are obvious examples of distributed software. This is the most popular approach to architecting software, and its use is projected to continue to grow. Technically, this type of architecture consists of multiple related software systems that are typically called either distributed applications or distributed systems.

Even applications you may have thought of as standalone have at least some distributed functionality. For example, automatic software updates, which are prevalent in newer applications, make use of network connectivity to check for, download, and install patches and updates. While this is generally a good security feature, it could be subverted by a sophisticated adversary to compromise systems. This happened in 2019, for example, when ASUS, one of the world's largest computer manufacturers, had its trusted automatic software update tool compromised in what's been dubbed Operation ShadowHammer. The attackers used this tool to distribute and install backdoors on thousands of computers around the world, though they were particularly interested in only a few hundred specific target devices.

Client/Server

A client/server system, which is a kind of distributed application, requires that two (or more) separate applications interact with each other across a network connection in order for users to benefit from them. One application, the *client*, makes requests (and receives responses) from a *server* that is otherwise passively waiting for someone to call on it. Perhaps the most common example of a client/server application is your web browser, which is designed to connect to a web server. Sure, you could just use your browser to read local documents, but that's not really the way it's meant to be used. Most of us use our browsers to connect two tiers: a client and a server, which is why we might call it a *two-tier architecture*.

Generally, client/server systems are known as *n*-tier architectures, where *n* is a numerical variable that can assume any value; most of the time, only the development team would know the number of tiers in the architecture (which could change over time), even if it looks like just two to the user. Consider the example of browsing the Web, which is probably a two-tier architecture if you are reading a static web page on a small web server. If, on the other hand, you are browsing a typical commercial site, you will probably be going through many more tiers, like the four-tier system shown in Figure 9-2. In this example, a client (tier 1) connects to a web server (tier 2) that provides the static HTML, CSS, and some images. The dynamic content, however, is pulled by the web server from an application server (tier 3) that in turn gets the necessary data from a back-end database (tier 4).

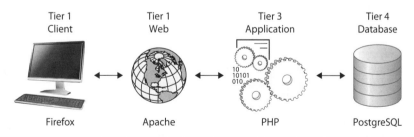

Tier 1 Client	Tier 1 Web	Tier 3 Application	Tier 4 Database
Firefox	Apache	PHP	PostgreSQL

Figure 9-2 A four-tier client/server system

Web Application

You know that a client/server system is a specific kind of distributed system. A *web application*, in turn, is a specific type of client/server system that is accessed through a web browser over a network connection, like the one shown in Figure 9-2. Examples of common web applications are web mail, online stores, and online banking.

Web applications are typically exposed to the most threats, because their very nature requires that they be accessible to anyone with an Internet connection. While some applications are deployed in a private network such as an intranet to reduce their exposure, many others are directly accessible from anywhere in the world. For this reason, we need to pay particular attention to ensuring that these applications are developed securely (we will cover secure coding later in this chapter), configured properly (we covered configuration management in the previous chapter), and monitored closely.

The OWASP Top Ten

The Open Web Application Security Project (OWASP) is an organization that deals specifically with web security issues. Along with a long list of tools, articles, and resources that developers can access to create secure software, OWASP also sponsors individual member meetings (chapters) throughout the world. The group provides development guidelines, testing procedures, and code review steps, but it is probably best known for the top ten web application security risk list that it maintains. The top security risks identified by this group as of the writing of this book are as follows:

1. Injection

2. Broken authentication

3. Sensitive data exposure

4. XML external entities (XXE)

5. Broken access control

6. Security misconfiguration

7. Cross-site scripting (XSS)

8. Insecure deserialization

9. Using components with known vulnerabilities

10. Insufficient logging and monitoring

This list represents the most common vulnerabilities that reside in web-based software and are exploited most often. You can find out more information pertaining to these vulnerabilities at https://owasp.org/www-project-top-ten/.

Mobile

Mobile devices such as smartphones and tablets run specialized operating systems and therefore require that applications be specifically developed for those systems. The vast majority of mobile devices in use today run either Apple iOS or Google Android systems. Applications (or apps) developed for one will not work on the other, so even if an app is available on both platforms, the actual software is likely to be different. This means that one version could have vulnerabilities that are not present in the other.

Most mobile apps used by businesses are actually distributed systems. Even if some of the data is stored on the local device, the app probably makes use of remote services to get, process, or store data, and these connections need to be protected. Another concern is that some data is almost certainly being stored locally, so it could possibly be accessed by another (malicious) app on the same device or by someone who gains physical access to the device. Additionally, authentication on the mobile device may not be as robust as on a traditional computer system. Many users rely on simple PINs to unlock their devices, for example, and these can be guessed or observed by an attacker who is nearby.

Embedded

An embedded system is a self-contained computer system (that is, it has its own processor, memory, and input/output devices) designed for a very specific purpose. An embedded device is part of (or embedded into) some other mechanical or electrical device or system. Embedded systems are typically cheap, rugged, and small, and they use very little power. They are usually built around microcontrollers, which are specialized devices that consist of a CPU, memory, and peripheral control interfaces. Microcontrollers have a very basic operating system, if they have one at all. A digital thermometer is an example of a very simple embedded system, and other examples of embedded systems include traffic lights and factory assembly line controllers. As you can see from these examples, embedded systems are frequently used to sense and/or act on a physical environment. For this reason, they are sometimes called cyber-physical systems.

The main challenge in securing embedded systems is that of ensuring the security of the software that drives them. Many vendors build their embedded systems around commercially available microprocessors, but they use their own proprietary code that is difficult, if not impossible, for a customer to audit. Depending on the risk tolerance of your organization, this may be acceptable as long as the embedded systems are standalone. The problem, however, is that these systems are increasingly shipping with some sort of network connectivity. For example, some organizations have discovered that some of their embedded devices have "phone home" features that are not documented. In some cases, this has resulted in potentially sensitive information being transmitted unencrypted to the manufacturer. If a full audit of the embedded device security is not possible, at a very minimum you should ensure that you see what data flows in and out of it across any network.

Another security issue presented by many embedded systems concerns the ability to update and patch them securely. Many embedded devices are deployed in environments where they have no Internet connectivity. Even if this is not the case and the devices can check for updates, establishing secure communications or verifying digitally signed code, both of which require processor-intensive cryptography, may not be possible on a cheap device.

System on a Chip

If you take the core elements of an embedded system (processor, memory, I/O) and put it all in a single integrated circuit, or chip, you would have a system on a chip (SoC). Sometimes, SoCs are used to build embedded devices, but you can see them just as frequently in smartphones, or in Internet of Things (IoT) devices. The beauty of a SoC is that, because all the components are miniaturized into the same chip, these devices tend to be very fast and consume less power than modular computers. This lack of modularity, by the way, is precisely the main drawback of SoCs: you can't just replace components because they're all stuck in the same chip.

Firmware

Whether you have a very simple SoC with no operating system running your household thermometer or a big server running your enterprise website, when you switch it on, some sort of software needs to be loaded into the CPU and executed so that your device can do something useful. *Firmware* is software that is stored in read-only, nonvolatile memory in a device and is executed when the device is powered on. Some devices' functionality is simple enough that they require only firmware to store their entire code base. More complex systems use firmware to load and perform power-on tests and then load and run the operating system from a secondary storage device. One way or another, virtually every computing device needs firmware to function.

So how do you update or patch firmware if it is read-only? If you have real read-only memory (ROM), there is no way (short of replacing the physical ROM chip) of doing this. However, most devices we come across use *electronically erasable programmable ROM* (EEPROM), which is a memory chip that can be wiped and overwritten. This enables device manufacturers to push out patches and updates to firmware as needed. Devices whose firmware is not updatable can present security risks if a vulnerability is ever found. At that point, your only choice is to isolate the vulnerable devices or replace them.

Service-Oriented Architecture

Up to this point, we have taken a platform-centric perspective on software architectures. In describing client/server, web application, mobile, and embedded systems software, we've focused on which platform executes the code. Another way to look at software architecture is to focus on the functionality or services being provided and not care where it is hosted, as long as it is available.

A *service-oriented architecture* (SOA) describes a system as a set of interconnected but self-contained components that communicate with each other and with their clients through standardized protocols. These protocols, called *application program interfaces* (APIs), establish a "language" that enables a component to make a request from another component and then interpret that second component's response. The requests that are defined by these APIs correspond to discrete business functions (such as estimate shipping costs to a postal code) that can be useful by themselves or can be assembled into more complex business processes. An SOA has three key characteristics: self-contained components, a standardized protocol (API) for requests/responses, and components that implement business functions.

SOAs are commonly built using web services standards that rely on HTTP as a standard communication protocol. Examples of these are the Simple Object Access Protocol (SOAP) and the representational state transfer (REST) and microservices architectures.

Simple Object Access Protocol

One of the first SOAs to become widely adopted is the Simple Object Access Protocol, a messaging protocol that uses XML over HTTP to enable clients to invoke processes on a remote host in a platform-agnostic way. SOAP enables you to have Linux and Windows machines working together as part of the same web service, for example. SOAP consists of three main components: Firstly, there is a message envelope that defines the messages that are allowed and how they are to be processed by the recipient. The second component of the protocol is a set of encoding rules used to define data types. Lastly, SOAP includes conventions on what remote procedures can be called and how to interpret their responses.

One of the key features of SOAP is that the message envelope allows the requestor to describe the actions that it expects from the various nodes that respond. This feature allows for things like routing tables that could specify the sequence and manner in which a series of SOAP nodes will take action on a given message. This can make it possible to finely control access as well as efficiently recover from failures along the way. This richness of features, however, comes at a cost: SOAP is not as simple as its name would imply. In fact, SOAP systems tend to be fairly complex and cumbersome, which is why many web service developers prefer more lightweight options like REST.

Extensible Markup Language

The term eXtensible Markup Language (XML) keeps coming up for good reasons. XML is a popular language to use if you want to mark up parts of a text document. If you've ever looked at raw Hypertext Markup Language (HTML) documents, you probably noticed the use of tags such as <title>CySA+</title> to mark up the beginning and end of a page's title. These tags enable both humans and machines to interpret text and process it (such as rendering it in a web browser) as the author intended it to. Similarly, XML enables the author of a text document to "explain" to a receiving computer what each part of the file means so that a receiving process can know what to do with it. Before XML, there was no standard way to do this, but nowadays there are a number of options including JavaScript Object Notation (JSON) and Yet Another Markup Language, or YAML Ain't Markup Language (YAML).

Representational State Transfer

REST architectures are among the most common in web services today. REST (or RESTful, as they're more commonly called) systems leverage the statelessness and standard operations of HTTP to focus on performance and reliability. Unlike SOAP, which is a standard protocol, REST is considered an architectural style. As such, it has no official definition. Instead, there is a pretty broad consensus on what it means to be RESTful, which includes the following features:

- The standard unit of information is a *resource*, which is uniquely identified through a uniform resource identifier (URI), such as an HTTP universal resource locator (URL).

- The system must implement a *client/server* architecture.

- The system must be *stateless*, which means that each request includes all the information needed to generate a response (that is, the server doesn't need to "remember" any past transactions involving the client).

- Responses must state whether or not they are *cacheable* by an intermediary such as a web proxy.

- The system must have a *uniform interface* that enables parts of it to be swapped or upgraded without breaking the whole thing.

Microservices

Another example of an SOA with which you should be acquainted is the *microservice*. Like REST, microservices are considered an architectural style rather than a standard, but there is broad consensus that they consist of small, decentralized, individually deployable services built around business capabilities. They also tend to be *loosely coupled*, which means there aren't a lot of dependencies between the individual services. As a result, they are quick to develop, test, and deploy and can be exchanged without breaking the larger system.

The decentralization of microservices can present a security challenge. How can you track adversarial behaviors through a system of microservices, where each service does one discrete task? The answer is *log aggregation*. Whereas microservices are decentralized, we want to log them in a centralized fashion so we can look for patterns that span multiple services and can point to malicious intent. Admittedly, you will need automation and perhaps data analytics or artificial intelligence to find these malicious events, but you won't have a chance at spotting them unless you aggregate the logs.

Security Assertions Markup Language

Technically, the Security Assertion Markup Language (SAML) is just that, a markup language standard. However, it is widely used to implement a specific kind of SOA for authentication and authorization services. We already saw an example of SAML in the last chapter when we were discussing single sign-on (SSO).

As a refresher, SAML defines two key roles: the Service Provider (SP) and the Identity Provider (IdP). A user who wants to access a service—say, a cloud-based sales application—requests access from the SP. The SP redirects the user to the IdP for authentication. The IdP authenticates the user and, assuming it is successful, redirects back to the SP. This last redirect includes a *SAML assertion*, which is a security statement (typically) from an IdP that SPs use to make access control decisions. SAML assertions can specify conditions (such as don't allow access before/after a certain time), user attributes (such as roles or group memberships), and the IdP's X.509 certificate if the assertion is digitally signed.

 NOTE Web browser redirects are essential to SAML working properly.

The Software Development Lifecycle

There are many approaches to building software, but they all follow some sort of predictable pattern called a *software development lifecycle* (SDLC). It starts with identifying an unmet need and ends with retiring the software, usually so that a new system can take its place. Whether you use formal or agile methodologies, you still have to identify and track the user or organizational needs; design, build, and test a solution; put that solution into a production environment; keep it running until it is no longer needed; and, finally, dispose of it without breaking anything else. In the sections that follow, we present the generic categories of effort within this lifecycle, though your organization may call these by other names. Along the way, we'll highlight how this all fits into the CySA+ exam.

 EXAM TIP You do not need to memorize the phases of the SDLC, but you do need to know how a cybersecurity analyst would contribute to the development effort at different points in it.

Requirements

All software development should start with the identification of the requirements that the finished product must satisfy. Even if those requirements are not explicitly listed in a formal document, they will exist somewhere before the first line of code is written. Generally speaking, there are two types of requirements: functional requirements that describe *what* the software must do, and nonfunctional requirements that describe *how* the software must do these things or what the software must be like. Left to their own devices, many software developers will focus their attention on the functionality and only begrudgingly (if at all) pay attention to the rest.

Functional Requirements

A *functional requirement* defines a function of a system in terms of inputs, processing, and outputs. For example, a software system may receive telemetry data from a temperature sensor, compare it to other data from that sensor, and display a graph showing how the values have changed for the day. This requirement is not encumbered with any specific constraints or limitations, which is the role of nonfunctional requirements.

Nonfunctional Requirements

A *nonfunctional requirement* defines a characteristic, constraint, or limitation of the system. Nonfunctional requirements are the main input to architectural designs for software systems. An example of a nonfunctional requirement, following the previous temperature scenario, would be that the system must be sensitive to temperature differences of one-tenth of a degree Fahrenheit and greater. Nonfunctional requirements are sometimes called *quality requirements*.

Security Requirements

The class of requirements in which we are most interested deals with security. A *security requirement* defines the behaviors and characteristics a system must possess to achieve and maintain an acceptable level of security by itself and in its interactions with other systems. Accordingly, this class includes both functional and nonfunctional aspects of the finished product.

Development

Once all the requirements have been identified, the development team starts developing or building the software system. The first step in this phase is to design an architecture that will address the nonfunctional requirements. Recall that nonfunctional requirements describe the characteristics of the system. On this architecture, the detailed code modules that address the features or functionality of the system are designed so that they satisfy the functional requirements. After the architecture and features are designed, software engineers start writing, integrating, and testing the code.

Testing is a critical part of developing secure code. Four types of testing are usually involved: The first is *unit testing*, which ensures that specific blocks (or units) of code behave as expected. This is frequently automated and involves a range of inputs that are reasonable, absurd, and at the boundary between these two extremes. The next level of testing is *integration testing* and involves ensuring that the outputs that one unit provides to another as inputs don't reveal any flaws. After integration testing, a product is usually ready for *system testing*, which is sort of like integration testing but for the entire system. Assuming everything is okay so far, we know that the product is built right (system verification). The final question to answer is whether we built the right product (system validation), and this is something only the intended user can determine. To take care of this final check, products go through a formal *acceptance testing*, at which point the customer certifies that the software meets the needs for which it was developed. At the end of the development phase, the system has passed all unit, integration, and system tests and is ready to be rolled out onto a production network.

Implementation

The implementation phase is the point at which frictions between the development and operations teams can start to become real problems, unless these two groups have been integrated beforehand. The challenges in this transitory phase include ensuring that the software will run properly on the target hardware systems, that it will integrate properly with other systems (for example, Active Directory), that it won't adversely affect the performance of any other system on the network, and that it doesn't compromise the security of the overall information system. If the organization used DevOps or DevSec-Ops (which we'll describe shortly), most of the thorny issues will have been identified and addressed by this stage, which means implementation becomes simply an issue of provisioning and final checks.

Operation and Maintenance

By most estimates, operation and maintenance (O&M) of software systems represents around 75 percent of the total cost of ownership (TCO). Somewhere between 20 and 35 percent of O&M costs are related to correcting vulnerabilities and other flaws that were not discovered during development. If you multiply these two figures together, you can see that typically organizations spend between 15 and 26 percent of the TCO for a software system fixing defects. This is the main driver for spending extra time in the design, secure development, and testing of the system before it goes into O&M. By this phase, the IT operations team has ownership of the software and is trying to keep it running in support of the business, while the software developers have usually moved on to the next project and see requests for fixes as distractions from their main efforts. This should highlight, once again, the need for secure software development before it ever touches a production network.

DevOps and DevSecOps

Historically, the software development and quality assurance teams would work together, but in isolation from the IT operations teams who would ultimately have to deal with the end product. Many problems have stemmed from poor collaboration between these two during the development process. It is not rare to have the IT team berating the developers because a feature push causes the former group to have to stay late, work on a weekend, or simply drop everything they were doing in order to "fix" something that the developers "broke." This friction makes a lot of sense when you consider that each team is incentivized by different outcomes. Developers want to push out finished code, usually under strict schedules. The IT staff, on the other hand, wants to keep the IT infrastructure operating effectively. A good way to solve this friction is to have both developers and operations staff (hence the term DevOps) work together throughout the software development process. *DevOps* is the practice of incorporating development, IT, and quality assurance (QA) staff into software development projects to align their incentives and enable frequent, efficient, and reliable releases of software products. Recently, the cybersecurity team is also being included in this multifunctional team, leading to the increasing use of the term *DevSecOps*, as shown in Figure 9-3.

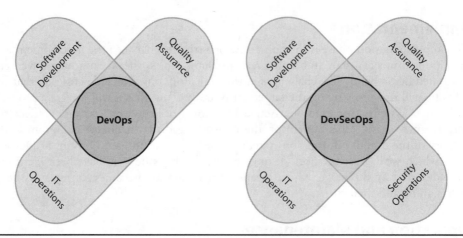

Figure 9-3 The functions involved in DevOps and in DevSecOps

Software Assessment Methods

So far, we've focused on the practices that would normally be performed by the software development or quality assurance team. As the project transitions from development to implementation, the IT operations and security teams typically perform additional security tests to ensure the confidentiality, integrity, and availability not only of the new software, but of the larger ecosystem once the new program is introduced. If an organization is using DevSecOps, some or most of these security tests could be performed as the software is being developed, because security personnel would be part of that phase as well. Otherwise, the development team gives the software to the security team for testing, and these individuals will almost certainly find flaws that will start a back-and-forth cycle that could delay final implementation.

User Acceptance Testing

Every software system is built to satisfy the needs of a set of users. Accordingly, the system is not deemed acceptable (or finished) until the users or their representatives declare that all the features have been implemented in ways that are acceptable to them. Depending on the development methodology used, user acceptance testing could happen before the end of the development phase or before the end of the implementation phase. Many organizations today use agile development methodologies that stress user involvement during the development process. This means that user acceptance testing may not be a formal event but rather a continuous engagement where representative users actively interact with the system.

Stress Testing

Another type of testing that also attempts to break software systems does so by creating conditions that the system would not reasonably be expected to encounter during normal conditions. *Stress testing* places extreme demands that are well beyond the planning thresholds of the software to determine how robust it is. The focus here is on attempting to compromise the availability of the system by creating a denial-of-service (DoS) condition.

The most common type of stress testing attempts to give the system *too much* of something (for example, simultaneous connections or data). During development, the team will build the software so that it handles a certain volume of activity or data. This volume may be specified as a nonfunctional requirement, or it may be arbitrarily determined by the team based on their experience. Typically, this value is determined by measuring or predicting the maximum load that the system is likely to be presented. To stress-test the system with regard to this value, the team would simply exceed it under different conditions and see what happens. The most common way to conduct these tests is by using scripts that generate thousands of simulated connections or by uploading exceptionally high volumes of data (either as many large files or fewer huge ones).

Not all stress tests are about overwhelming the software; it is also possible to underwhelm it. This type of stress testing provides the system with *too little* of something (for example, network bandwidth, CPU cycles, or memory). The idea here is to see how the system deals with a threat called *resource starvation,* in which an attacker intentionally causes the system to consume resources until none are left. A robust system would gracefully degrade its capabilities during an event like this but wouldn't fail altogether. Insufficient-resource tests are also useful to determine the absolute minimum configuration necessary for nominal system performance.

Security Regression Testing

Software is almost never written securely on the first attempt. Organizations with mature development processes will take steps like the ones we've discussed in this chapter to detect and fix software flaws and vulnerabilities before the system is put into production. Invariably, errors will be found, leading to fixes. The catch is that fixing a vulnerability may very well inadvertently break some other function of the system or even create a new set of vulnerabilities. *Security regression testing* is the formal process by which code that has been modified is tested to ensure that no features and security characteristics were compromised by the modifications. Obviously, regression testing is only as effective as the standardized suite of tests that were developed for it. If the tests provide insufficient coverage, regression testing may not reveal new flaws that may have been introduced during the corrective process.

Code Reviews

One of the best practices for quality assurance and secure coding is the *code review,* which is a systematic examination of the instructions included in a piece of software, performed by someone other than the author of the code. This approach is a hallmark of mature software development processes. In fact, in many organizations, developers are not allowed to push out their software modules until someone else has signed off on them after doing code reviews. Think of this as proofreading an important document before you send it to an important person. If you try to proofread it yourself, you will probably not catch all those embarrassing typos and grammatical errors as easily as if someone else were to check it.

Code reviews go way beyond checking for typos, though that is certainly one element of it. It all starts with a set of coding standards developed by the organization that wrote the software. Reviewers could be an internal team, an outsourced developer, or a commercial vendor. Obviously, code reviews of off-the-shelf commercial software are extremely rare unless the software is open source or you happen to be a major government agency. Still, each development shop will have a style guide or documented coding standards that cover everything from how to indent the code to when and how to use existing code libraries. Therefore, a preliminary step to the code review is to ensure that the author followed the team's standards. In addition to helping the maintainability of the software, this step gives the code reviewer a preview of the magnitude of work ahead; a sloppy coder will probably have a lot of other, harder-to-find defects in his code, and each of those defects is a potential security vulnerability.

Static Analysis Tools

We discussed static and dynamic analysis in Chapter 4 in the context of assessing software for vulnerabilities. That was an after-the-fact test to find defects in any software, whether it was developed in house or acquired from someone else. But what if we apply the same techniques as we develop the code instead of waiting until it is completed? Most mature software development organizations integrate static analysis tools into their software development process. When a developer finishes writing a software module and submits it to the code repository, a suite of tests that includes static analysis tools is run against it. If the module doesn't pass these tests, the submission is rejected and the developer has to fix whatever vulnerabilities were found. By integrating automated static analysis into the code submission process, we can greatly reduce the number of vulnerabilities in our software.

Dynamic Analysis Tools

Another way in which software developers reduce vulnerabilities is by integrating dynamic analysis tools into their development process. As we discussed in Chapter 4, dynamic program analysis is the examination of a program while it is being executed. There are many approaches to this. Some tools enable you to stop the program after each line of code is executed, or at user-defined checkpoints, much like a debugger. Others will map out the possible execution flows, determine which inputs would cause different branches to be followed, and then run the program with those inputs to see what happens. This last approach is usually infeasible for large programs because they may have too many possible flows to examine in a reasonable (or even an unreasonable) amount of time.

Code coverage is the measure of how much of a program is examined by a particular set of tests. While it may not be possible to examine every possible flow through a large application, we want to maximize the percentage of the program that is evaluated at a particular level of effort. Most programs don't need more than 75 percent code coverage to be considered well-tested. Above that point, the new bugs that may be uncovered come at a much higher cost than the developers can or want to pay. However, for mission- or safety-critical systems, especially ones with a smaller code base, 100 percent code coverage may be appropriate or even required.

EXAM TIP When taking the exam, keep in mind the differences between the two approaches: static program analysis does not require code execution; dynamic program analysis requires running it.

Formal Methods of Verifying Critical Software

When you need to be absolutely sure that a program will exactly do specific things and won't do any others, static and dynamic analysis may not be enough. *Formal methods* are mathematical approaches to specifying, developing, and verifying software. On the low end of the spectrum of these approaches, we have *formal specifications*, which are ways to describe the software requirements in very rigorous terms. Often, a formal specification looks like something out of an algebra textbook. This enables the developers to ensure consistency and completeness of the requirements, which cuts down on unexpected outputs. If you ever come across formal methods, this is what you will probably see.

Still, for some safety-critical systems (such as a medical infusion pump), it's not enough to formally specify them. In these cases, you want to be able to prove that they won't behave unexpectedly under any conditions. For this, the entire software development process is extremely rigorous. Beyond the formal specification, every property of the system is mathematically proven during the design and then verified after implementation. This level of rigor guarantees that the program will behave exactly as intended, but it comes with a huge cost. For this reason, only specific mission-critical components of larger software systems are normally developed using formal methods.

Secure Coding Best Practices

Perhaps the most important concept behind software development is that of quality. *Quality* can be defined as fitness for purpose—in other words, how good something is at whatever it is meant to do. A quality car will be good for transportation. We don't have to worry about it breaking down or failing to protect its occupants in a crash or being easy for a thief to steal. When we need to go somewhere, we simply get in the car and count on it safely taking us to wherever we need to go. Similarly, we don't have to worry about quality software crashing, corrupting our data under unforeseen circumstances, or being easy for someone to subvert. Sadly, many developers still think of functionality first (or only) when thinking about quality. When we look at things holistically, we should see that quality is the most important concept in developing secure software.

This, of course, is not a new problem. Secure software development has been a challenge for a few decades. Unsurprisingly, there is an established body of best practices to minimize the flaws and vulnerabilities in our code. You should be familiar with what some of the best-known advocates for secure coding recommend.

Input Validation

If there is one universal rule to developing secure software, it is this: don't *ever* trust any input entered by a user. This is not just an issue of protecting our systems against malicious attackers; it is equally applicable to innocent user errors. The best approach to validating inputs is to perform context-sensitive whitelisting. In other words, consider what is supposed to be happening within the software system at the specific points in which the input is elicited from the user, and then allow only the values that are appropriate. For example, if you are getting a credit card number from a user, you would allow only 16 consecutive numeric characters to be entered. Anything else would be disallowed.

Perhaps one of the most well-known examples of adversarial exploitation of improper user input validation is Structured Query Language (SQL) injection (SQLi). SQL is a language developed by IBM to query information in a database management system (DBMS). Because user credentials for web applications are commonly stored in a DBMS, many web apps will use SQL to authenticate their users. A typical *insecure* SQL query to accomplish this in PHP is shown here:

```
$result = mysql_query("SELECT * FROM userdb WHERE username='$form_username'
        AND password='$form_password'");
$num_rows = mysql_num_rows($result);
if($num_rows > 0){
    $authenticated = True;
else
    $authenticated = False;
```

Absent any validation of the user inputs, the user could provide the username **attacker' or 1=1 --** and **pawned** (or anything or nothing) for the password, which would result in the following query string:

```
"SELECT * FROM userdb WHERE username='attacker' or 1=1 --'
        AND password='pawned'"
```

If the DBMS for this web app is MySQL, that system will interpret anything after two dashes as a comment, which will be ignored. This means that the value in the password field is irrelevant, because it will never be evaluated by the database. The username can be anything (or empty), but because the logical condition 1=1 is always true, the query will return all the registered users. Because the number of users is greater than zero, the attacker will be authenticated.

Clearly, we need to validate inputs such as these, but should we do it on the client side or the server side? Client-side validation is often implemented through JavaScript and embedded within the code for the page containing the form. The advantage of this approach is that errors are caught at the point of entry, and the form is not submitted to the server until all values are validated. The disadvantage is that client-side validation is easily negated using commonly available and easy-to-use tools. The preferred approach is to do client-side validation to enhance the user experience of benign users, but double-check everything on the server side to ensure protection against malicious actors.

Output Encoding

Sometimes user inputs are displayed directly on a web page. If you've ever posted anything on social media or left a product review at a vendor's site, then you've provided input (your post or review) that a web application incorporated into an HTML document (the updated page). But what happens if your input includes HTML tags? This could be useful if you wanted to use boldface by stating you are very happy about your purchase. (The <b> tag in HTML denotes boldface.) It could just as easily be problematic if you included a <script> tag with some malicious JavaScript. The web browser will happily interpret benign and malicious HTML just the same.

Output encoding is a technique that converts user inputs into safe representations that a web browser cannot interpret as HTML. So, going back to our previous example, when you enter <b>very</b> in a purchase review, the HTML tags are "escaped" or rendered in a way that prevents the browser from interpreting them as valid HTML. This is the most important control to prevent cross-site scripting (XSS) attacks.

Session Management

As you may know, HTTP is a *stateless* protocol. This means that every web server is a bit of an amnesiac; it doesn't remember you from one request to another. A common way around this is to use cookies to sort of "remind" the server of who you are so that, even if you haven't authenticated, you can go from page to page on an online retail store and have the web application remember what you may have added to your shopping cart. Every time you request a resource (such as a web page or an image), your browser sends along a cookie so the web application can tell you apart from thousands of other visitors. The cookie is simply a text file that usually contains some sort of unique identifier for you. This identifier is typically known as the *session ID*.

A HTTP session is sequence of requests and responses associated with the same user, while session management is the process of securely handling these sessions. The word "securely" is critical because, absent security, sessions can easily be hijacked. Let's go back to the previous example of writing a session ID on a cookie to track a given user. That cookie is sent along with every request, so anyone who can sniff it off the network can impersonate the legitimate session user. All they would have to do is wait for you to send a request, intercept it, get the session ID from the cookie, and then send their own request to the server, pretending to be you.

Secure session management revolves around two basic principles: use HTTPS whenever a session is active, and ensure that session IDs are not easy to guess. HTTPS is essential whenever any sensitive information is exchanged, and this is particularly true of a session ID. Even if attackers can't break your secure connection to steal your ID, they could still try to guess it. Suppose a web application assigns session IDs sequentially. You send a request and see that your session ID is 1000. You clear your cache and cookies and interact with the application again, noticing that your new session ID is 1002. It would be reasonable to assume that there is another active session with an ID of 1001, so you rewrite your cookie to have that value and send another request to the web application. Voilà! You just took over someone else's session. Secure session management involves using IDs that are hard to guess (such as pseudo-random), long enough to prevent brute-force attacks, and generic so that they are not based on any identifiable information.

Authentication

Most software systems, particularly distributed ones, require user authentication. Sadly, this is an area that is often given insufficient attention during the development process. Many programmers seem to think that authentication is pretty easy, and a surprising number of them prefer to implement their own. The problem with weak authentication mechanisms in distributed software systems is that they can turn the platform into an easy foothold into the trusted network for the attacker. Additionally, because many people reuse passwords in multiple systems, obtaining credentials on one (poorly built or protected) system often leads to unauthorized access to others. If at all possible, we should avoid building our own authentication and rely instead on one of the approaches to SSO we've discussed before.

Regardless of whether you roll your own or use someone else's, here are some of the best practices for authentication in software systems:

- Use multifactor authentication (MFA) whenever possible.
- Enforce strong passwords with minimum lengths (to thwart brute-force attacks) as well as maximum lengths (to protect against long password DoS attacks).
- Never store or transmit passwords in plaintext anywhere (and use the appropriate cryptographic techniques to protect them).
- Implement failed-login account lockouts if at all possible (but beware of this making you vulnerable to account DoS attacks).
- Log all authentication attempts and ensure that the logs are periodically reviewed (or, better yet, generate appropriate alerts).

 NOTE We discussed authentication in general and SSO in particular in Chapter 8.

Data Protection

Passwords are not the only data that we need to protect in our software systems. Depending on their functions, these applications can contain personal data, financial information, or trade secrets. Generally speaking, data exists in one of three states: at rest, in motion, or in use. These states and their interrelations are shown in Figure 9-4.

Figure 9-4
The states of data

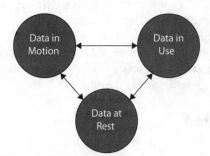

Data at Rest

Information in a software system spends most of its time waiting to be used. The term *data at rest* refers to data that resides in external or auxiliary storage devices, such as hard disk drives (HDDs), solid-state drives (SSDs), optical discs (CDs/DVDs), or even on magnetic tape. A challenge with protecting data in this state is that it is vulnerable not only to threat actors attempting to reach it over our systems and networks, but also to anyone who can gain physical access to the device. The best approach to protecting data at rest is to encrypt it. Every major operating system provides encryption means for individual files or entire volumes in a way that is almost completely transparent to the user. Similarly, every major database management system enables you to encrypt data deemed sensitive, such as passwords and credit card numbers.

Data in Motion

Data in motion is data that is moving between computing nodes over a data network such as the Internet. This is perhaps the riskiest time for our data—when it leaves the confines of our protected enclaves and ventures into that Wild West that is the Internet. The single best protection for our data while it is in motion (whether within or without our protected networks) is strong encryption such as that offered by Transport Layer Security (TLS) version 1.1 and later or IPSec, both of which support multiple cipher suites (though some of these are not as strong as others). By and large, TLS relies on digital certificates (we talked about those in the last chapter) to certify the identity of one or both endpoints. Another approach to protecting our data in motion is to use trusted channels between critical nodes. Virtual private networks (VPNs) are frequently used to provide secure connections between remote users and corporate resources. VPNs are also used to securely connect campuses or other nodes that are physically distant from each other. The trusted channels we thus create enable secure communications over shared or untrusted network infrastructure.

Data in Use

Data in use refers to data residing in primary storage devices, such as volatile memory (such as RAM), memory caches, or CPU registers. Typically, data remains in primary storage for relatively short periods of time while a process is using it. Note, however, that anything stored in volatile memory could persist there for extended periods (until power is shut down) in some cases. The point is that data in use is being touched by the CPU and will eventually go back to being data at rest or end up being deleted. Many people think this state is safe, but the Meltdown, Spectre, and BranchScope attacks that came to light in 2018 show how a clever attacker can exploit hardware features in most modern CPUs. Meltdown, which affects Intel and ARM microprocessors, works by exploiting the manner in which memory mapping occurs. Since cache memory is a lot faster than main memory, most modern CPUs include ways to keep frequently used data in the faster cache. Spectre and BranchScope, on the other hand, take advantage of a feature called speculative execution, which is meant to improve the performance of a process by guessing what future instructions will be, based on data available in the present. All three of these attacks go after data in use.

So, how do we protect our data in use? For now, it boils down to ensuring that our software is tested against these types of attacks. Obviously, this is a tricky proposition, since it is very difficult to identify and test for every possible software flaw. In the near future, whole-memory encryption will mitigate the risks described in this section, particularly when coupled with the storage of keys in CPU registers instead of in RAM. Until these changes are widely available, however, we must remain vigilant to the threats against our data while it is in use.

Parameterized Queries

A common threat against web applications is the SQL injection (SQLi) class of attacks. We already covered them in Chapter 7, but by way of review, SQLi enables an attacker to insert arbitrary code into a SQL query. This typically starts by inserting an escape character (such as a closing quote around a literal value), terminating the query that the application developer intended to execute based on the user input, and then inserting a new and malicious SQL command. The key to SQLi working is the insertion of user inputs directly into an SQL query. Think of it as cutting and pasting the input into the actual query that gets executed by the database.

A *parameterized query* (also known as a prepared statement) is a programming technique that treats user inputs as parameters to a function instead of substrings in a literal query. This means that the programmer can specify what values are expected (for example, a number, a date, a username) and validate that they conform to whatever limits are reasonable (such as a value range, a maximum length for a username) before integrating them into a query that will be executed. Properly implemented, parameterized queries are the best defense against SQLi attacks. They also highlight the importance of validating user inputs and program parameters.

Chapter Review

Although you may not spend much time (if at all) developing software, as a security analyst you will certainly have to deal with the consequences of any insecure coding practices. Your best bet is to be familiar with the processes and issues and be part of the solution. Simply attending the meetings, asking questions, and sharing your thoughts could be the keys to avoiding a catastrophic compromise that results from programmers who didn't realize the potential consequences of their (otherwise reasonable) decisions. Ideally, you are part of a cohesive DevSecOps team that consistently develops high-quality code. If this is not the case, maybe you can start your organization down this path.

As for the CySA+ exam, you should have an awareness of the major issues with developing secure code. You will be expected to know how a cybersecurity analyst can proactively or, if need be, reactively address vulnerabilities in custom software systems. Some of the key concepts here are the security tests you can run such as web app vulnerability scans, stress tests, and fuzzing. You may be asked about the SDLC, but you don't have to be an expert in it.

Questions

1. The practice of testing user input to reduce the impact of malformed user requests is referred to as what?

    A. Input validation

    B. Static code analysis

    C. Manual inspection

    D. Stress testing

2. Which phase in the software development lifecycle often highlights friction between developers and business units due to integration and performance issues?

    A. Implementation

    B. Design

    C. Planning

    D. Maintenance

3. To reduce the amount of data that must be examined and interpreted by a web application, what method can be used to catch errors before submission?

    A. Server-side validation

    B. Proxy validation

    C. Client-side validation

    D. Stress validation

4. What key process is often used to determine the usability and suitability of newly developed software before implementation across an organization?

    A. User acceptance testing

    B. Parameter validation

    C. Regression testing

    D. Data filtering

5. Which of the following is *not* a key characteristic of a Service Oriented Architecture (SOA)?

    A. Self-contained components

    B. Platform-centric architectures

    C. Standardized protocol for requests/responses

    D. Components that implement business functions

Use the following scenario and illustration to answer Questions 6–10:

Your accounting department's administrator has reached out to your team because one of the department's analysts has discovered a discrepancy in the accounting reports. Some of the department's paper documents do not match the stored versions, leading them to believe the database has been tampered with. This database is for internal access only, and you can assume that it hasn't been accessed from outside the corporate network. The administrator tells you that the database software was written several years ago by one individual and that they haven't been able to update the system since the initial rollout. You are also provided traffic capture data by the local admin to assist with the analysis.

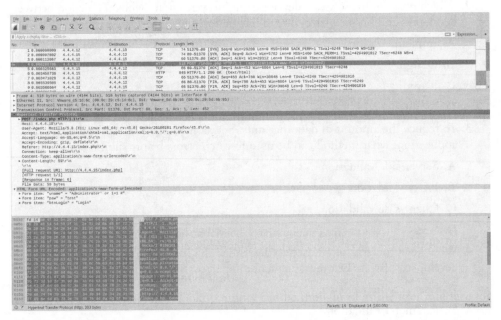

6. After hearing the description of how the software was developed by one person, what process do you know would have improved the software without needing to run it?

   A. Runtime analysis

   B. Just-in-time analysis

   C. Stress testing

   D. Code review

7. Based on the traffic-capture data, what is the most likely method used to gain unauthorized access to the web application?

   A. Regression

   B. Replay attack

   C. SQL injection

   D. Request forgery

8. What practice may have prevented this particular type attack from being successful?

    A. Network segmentation

    B. SSL

    C. Two-factor authentication

    D. Input validation

9. To prevent input from being interpreted as actual commands, what method should the developer have used?

    A. Regression testing

    B. Generic error messages

    C. Session tokens

    D. Parameterized queries

10. You have updated the server software and want to test it actively for new flaws. Which method is the *least* suitable for your requirement?

    A. User acceptance testing

    B. Static code analysis

    C. Stress testing

    D. Web app vulnerability scanning

Answers

    1. A. Input validation is an approach to protecting systems from abnormal user input by testing the data provided against appropriate values.

    2. A. Implementation is all about seeing how the software works in its production environment. Although problems are bound to surface, the most productive organizations have mature mechanisms for feedback for improvement.

    3. C. Client-side validation checks are performed on data in the user browser or application before the data is sent to the server. This practice is used alongside server-side validation to improve security and to reduce the load on the server.

    4. A. User acceptance testing is a method to determine whether a piece of software meets specifications and is suitable for the business processes.

    5. B. Service Oriented Architectures (SOAs) are, in many ways, the opposite of platform-centric ones. This is because SOA components can be replaced without impacting the services offered. One day you could be running Service A in the cloud and the next day on an embedded device, but as long as you provide an identical service, nobody would notice.

    6. D. Code review is the systematic examination of software by someone not involved in the initial development process. This ensures an unbiased perspective and promotes adherence to coding and security standards.

7. **C.** SQL injection is a technique of manipulating input to gain control of a web application's database server. Notice that the attacker submitted a uname value of "Administrator' or 1=1 #". The second part of the statement (1=1) always evaluates to true, so this is a way to trick a vulnerable system into giving up administrative access. It is effective and powerful, and often facilitates data manipulation or theft.

8. **D.** Input validation is the practice of constraining and sanitizing input data. This is an effective defense against all types of injection attacks by checking the type, length, format, and range of data against known good types.

9. **D.** Using parameterized queries is a developer practice for easily differentiating between code and user-provided input.

10. **B.** All methods except for static code analysis are considered active types of assessments.

## CHAPTER 10

Hardware Assurance Best Practices

In this chapter you will learn:

- How hardware can be used to create a root of trust
- How to use and update firmware securely
- How to protect data using hardware-based security
- Anti-tamper techniques for hardware

Every successful hardware has a software behind.

—Thiru Voonna

We began this second part of the book by discussing security solutions for infrastructure management. In the last chapter, we covered software assurance best practices. We now wrap up our discussion of software and systems security by addressing hardware assurance. It has been said that, if attackers can get their hands on your devices, they will eventually own anything in them. This may be true in some extreme cases of determined nation-state actors, but it is a risk that can largely be mitigated for the majority of threats to our environments. In this chapter, we turn our attention to the most important technologies and approaches to ensuring that our data remains protected even if the devices on which they are stored or processed become compromised. You don't need to be an expert in these issues to be a successful cybersecurity analyst, but you need to be aware of them so that you know when to apply the appropriate hardware solution to protect your systems. It all begins with building trust.

Hardware Root of Trust

We can go to great lengths to ensure that the software we develop is secure, to run it on operating systems that have been hardened by a myriad of security controls, and to monitor everything using advanced security tools. But if the physical devices on which these systems run are untrustworthy, then all our efforts are for naught. A *hardware root of trust* is a trusted execution environment, with built-in cryptographic functions for data protection, that is resistant to tampering. Each of these three elements is essential to establishing trust. A hardware root of trust is the foundation on which all security functions rest.

It can be integrated right into the motherboard or added as a module later. Virtually every modern computer, with the notable exception of inexpensive embedded and Internet of Things (IoT) devices, has a hardware root of trust, though many organizations don't use them—at least not to their full potential. There are fundamentally two ways to look at how to build this: on the motherboard or as an add-on module.

Trusted Platform Module

The *Trusted Platform Module* (TPM) is a system on a chip (SoC, discussed in Chapter 9) installed on the motherboard of modern computers that is dedicated to carrying out security functions involving the storage of cryptographic keys and digital certificates, symmetric and asymmetric encryption, and hashing. The TPM was devised by the Trusted Computing Group (TCG), an organization that promotes open standards to help strengthen computing platforms against security weaknesses and attacks.

The essence of the TPM lies in a protected and encapsulated microcontroller security chip that provides a safe haven for storing and processing critical security data such as keys, passwords, and digital certificates. The use of a dedicated and encoded hardware-based platform drastically improves the root of trust of the computing system, while allowing for a vastly superior implementation and integration of security features. The introduction of TPM has made it much harder to access information on computing devices without proper authorization and allows for effective detection of malicious configuration changes to a computing platform.

TPM Uses

The most common usage scenario of the TPM is to *bind* a hard disk drive, where the content of a given hard disk drive is affixed with a particular computing system. The content of the hard disk drive is encrypted, and the decryption key is stored away in the TPM chip. To ensure safe storage of the decryption key, it is further "wrapped" with another encryption key. Binding a hard disk drive makes its content basically inaccessible to other systems, and any attempt to retrieve the drive's content by attaching it to another system will be very difficult. However, in the event of the TPM chip's failure, the hard drive's content will be rendered useless, unless a backup of the key has been escrowed.

Another application of the TPM is *sealing* a system's state to a particular hardware and software configuration. Sealing a computing system through TPM is used to deter any attempts to tamper with a system's configurations. In practice, this is similar to how hashes are used to verify the integrity of files shared over the Internet (or any other untrusted medium). Sealing a system is fairly straightforward. The TPM generates hash values based on the system's configuration files and stores them in its memory. A sealed system will be activated only once the TPM verifies the integrity of the system's configuration by comparing it with the original "sealing" value.

The TPM is essentially a securely designed microcontroller with added modules to perform cryptographic functions. These modules allow for accelerated storage and processing of cryptographic keys, hash values, and pseudonumber sequences. The TPM's

Figure 10-1
Functional
components of a
trusted platform
module

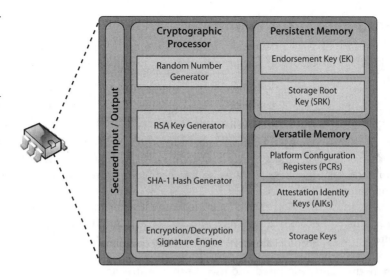

internal storage is based on random access memory (RAM), which retains its information when power is turned off and is therefore termed *nonvolatile RAM* (NVRAM). The TPM's internal memory is divided into two different segments: persistent (static) and versatile (dynamic) memory modules, as shown in Figure 10-1.

Persistent Memory   Two kinds of keys are present in the static memory:

- **Endorsement key (EK)**   A public/private key pair that is installed in the TPM at the time of manufacture and cannot be modified. The private key is always present inside the TPM, while the public key is used to verify the authenticity of the TPM itself. The EK, installed in the TPM, is unique to that TPM and its platform.

- **Storage root key (SRK)**   The master wrapping key used to secure the keys stored in the TPM.

Versatile Memory   Three kinds of keys (or values) are present in the versatile memory:

- **Platform Configuration Registers (PCRs)**   Used to store cryptographic hashes of data used for TPM's sealing functionality.

- **Attestation Identity Keys (AIKs)**   Used for the attestation of the TPM chip itself to service providers. The AIK is linked to the TPM's identity at the time of development, which in turn is linked to the TPM's EK. Therefore, the AIK ensures the integrity of the EK.

- **Storage keys**   Used to encrypt the storage media of the computer system.

Hardware Security Module

Whereas a TPM is a microchip installed on a motherboard, a *hardware security module* (HSM) is a removable expansion card or external device that can generate, store, and manage cryptographic keys. HSMs are commonly used to improve encryption/decryption performance by offloading these functions to a specialized module, thus freeing up the general-purpose microprocessor to take care of, well, general-purpose tasks. HSMs have become critical components for data confidentiality and integrity in digital business transactions. The US Federal Information Processing Standard (FIPS) 140-2 is perhaps the most widely recognized standard for evaluating the security of an HSM. This evaluation is important, because so much digital commerce nowadays relies on protections provided by HSM.

As with so many other cybersecurity technologies, the line between TPMs and HSMs gets blurred. TPMs are typically soldered onto a motherboard, but they can be added through a header. HSMs are almost always external devices, but you will occasionally see them as peripheral component interconnect (PCI) cards. In general, however, TPMs are permanently mounted and used for hardware-based assurance and key storage, while HSMs are removable (or altogether external) and are used for both hardware-accelerated cryptography and key storage.

eFuse

An *eFuse* is a single bit of nonvolatile memory that, once set to 1, can never be reverted to 0. The technology was invented by IBM in 2004 and is a form of one-time programmable (OTP) memory. It relies on a special compound that normally conducts electricity just fine, but if you apply a specific amount of power, its chemical composition changes and it becomes a resistor instead. Figure 10-2 illustrates these two states. Obviously, this requires special circuits that are able to apply that higher power safely only to the eFuse and then only when instructed to do so. Once an eFuse is programmed or blown, it cannot be reverted to its unprogrammed state.

Figure 10-2
Unprogrammed (left) and programmed (right) eFuses (source: IBM)

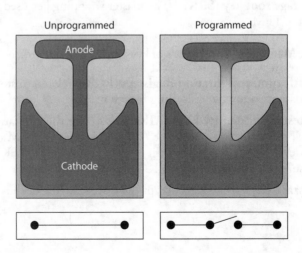

eFuses have two main security applications. The first is to disable access to certain functionality on a chip. It is very common for integrated circuits to have all sorts of test and development features during manufacturing that are not needed after the product is shipped. For example, you may want special circuitry that enables an engineer to bypass security controls to test specific features of a chip directly. This is great in the factory but problematic if an attacker is trying to hack a device. Putting an eFuse in front of these test circuits would offer an added layer of protection for the finished hardware.

Another security application of eFuses is to store data. For example, you could use them to keep track of the latest version of firmware you have loaded on a device. This would ensure that nobody could revert the firmware to an earlier (presumably less secure) version. If you have enough eFuses in the chip, you could store cryptographic keys on it, which could be used to verify the integrity of a firmware update by checking its digital signature before installing it. You could also use this approach to verify the integrity of an operating system before you load, which is what secure bootloaders do.

Firmware

Nothing of any significance happens on a computer unless it is done in a processor, and nothing happens in a processor unless it has a sequence of instructions to execute. So the question is, how do you get this jumpstarted when you power on a computer? The processor is typically hardwired to read a specific memory address when it is powered on. This memory address, which is typically a bunch of 0's, points to the start of section of nonvolatile memory containing instructions and data—in other words, the firmware. It's called firmware because it sits between the immutable hardware and the ever-changing software, so, in a sense, it is firm. Firmware is the set of instructions and data that take the device from a cold start to the point where it is doing its job (for very simple devices without an operating system) or executing the first instruction in its operating system code.

For many years after personal computers (PCs) emerged, the dominant approach to implementing firmware was the Basic Input/Output System (BIOS) developed by IBM in the 1980s. The very first PCs stored the BIOS firmware in read-only memory (ROM), which could not be overwritten or updated. Any data that needed updating (such as the BIOS password, the location, and types of disk drives) was stored in a battery-powered (but otherwise volatile) complementary metal-oxide-semiconductor (CMOS) RAM chip. If the battery died or was removed, all BIOS settings would revert to default values. The advent of electronically erasable programmable ROM (EEPROM) chips enabled the BIOS to be updated, but it did nothing to negate the need for battery-powered CMOS RAM. This changed when NVRAM became inexpensive and common and opened up a world of possibilities.

By this time, it started to become increasingly clear that BIOS had a large number of problematic features. For starters, it required booting from a local disk drive, which is problematic when you're trying to manage thousands of workstations in an enterprise environment. It is also built on a 16-bit architecture, which limits its memory address space to 1MB. Finally, it was only a de facto standard that was never really formalized. Even though BIOS is small, simple, and still in common use for embedded devices, something else was needed to power the newer generation of computers.

Unified Extensible Firmware Interface

The *Unified Extensible Firmware Interface* (UEFI) is a software interface standard that describes the way in which firmware executes its tasks. It is significantly more powerful and flexible than BIOS and has shipped in virtually every workstation and server built in the last ten years. UEFI solved all the problems that BIOS had and introduced some important new features. It enables the use of disk partitions larger than 2TB, and can be booted from disks, removable drives, the local network, and even the Internet using HTTP. Even before loading an OS, UEFI includes networking capabilities that allow for remote recovery and management. It is also modular and works on pretty much any CPU architecture.

The boot process, which has some important security implications, is summarized here:

1. **Security Phase (SEC)** Initializes a memory cache, establishes a software root of trust, and loads and executes any architecture-specific code needed
2. **Pre-EFI Initialization (PEI)** Loads and executes code to initialize (and potentially recover) key hardware such as main memory
3. **Driver Execution Environment (DXE)** Loads and executes the drivers that initialize the rest of the hardware
4. **Boot Device Select (BDS)** Initializes input and output devices and identifies and loads the boot options
5. **Transient System Load (TSL)** Invokes the UEFI user interface if available, or the system proceeds to boot the default operating system
6. **Runtime (RT)** Hands control to the OS loader

As you can see, most of what UEFI does is to figure out what code is needed, load it into memory, and execute it. By default, it trusts all the code that it uses, but there is an option to do better. *Secure boot* is a feature in UEFI that establishes the root of trust in the firmware. Here's how it works: UEFI firmware has a root certificate authority (CA) and a set of X.509 certificates belonging to trusted software vendors. Before running an executable (such as a loader or driver), the firmware checks the code's digital signature to ensure that it is trusted and has not been altered, and then it runs the executable if it checks out. UEFI also has a mechanism for blacklisting certificates and hashes and for updating all this data in a secure manner, as needed.

Measured Boot and Attestation

There may be situations when secure boot is not ideal. Keep in mind that you can only load and run code that is signed by the vendors whose certificates are in the firmware, which means no code developed in-house (unless you also modify the firmware) can be used. For certain mission-critical systems, it may be better to run code that is not verified than to keep the device from booting at all. Additionally, secure boot has no way of verifying configuration settings that are set locally and not at the factory.

Measured boot also starts with a firmware root of trust, but instead of verifying the digital signatures of code, it simply hashes them (a sort of "measurement") and stores the hash in a secure location. This creates an audit trail of code and data that was loaded. While this certainly has forensic value, it really doesn't help you detect problems or attacks in real time. *Attestation* is the process of securely sending the hashes to a management station. It thus makes sense to combine the two and use measured boot and attestation in situations where secure boot is impractical.

Of course, this raises the issue of how to securely compute, store, and transmit the hashes needed for secure boot and attestation. This challenge is typically addressed through the use of a TPM that is separate from the firmware. Whereas TPM is optional in secure boot (since the firmware can do the required tasks), it is almost always necessary for measured boot and attestation.

Trusted Firmware Updates

So far we've covered how secure and measured boots are supported by the firmware, but what happens when the firmware needs to be updated? One approach would be simply to overwrite the old firmware with the new one and hope it all works. Even if you independently verify the signatures before updating, you still run the chance that something goes wrong with either the firmware or the hardware and you end up with an unbootable device. A better approach may be to build the downloading, verification, and swapping functions right into the firmware and ensure that it has enough storage space for both versions as well as a small module that would swap from the old to the new when everything checks out. You can think of trusted firmware updates as an extension of secure boot in that the existing firmware is responsible for the security of the entire process.

Firstly, a trusted firmware update system needs enough storage for both the old and the new versions of the firmware simultaneously so you can restore to a known-good version if the download fails. It also needs a small bit of code (a bootloader) to tell the device which version to load up. Finally, and importantly, the system needs a cryptologic module with certificates and code to verify signatures. This last bit could exist in either the bootloader, the firmware images, or externally in a TPM.

Self-Encrypting Drive

Full-disk encryption (FDE) refers to approaches used to encrypt the entirety of data at rest on a disk drive. This can be accomplished in either software or hardware. A *self-encrypting drive* (SED) is a hardware-based approach to FDE in which a cryptographic module is integrated with the storage media into one package. Typically, this module is built right into the disk controller chip. Most SDEs are built in accordance with the TCG Opal 2.0 standard specification.

 NOTE Although SEDs can use onboard TPMs, that is not the norm.

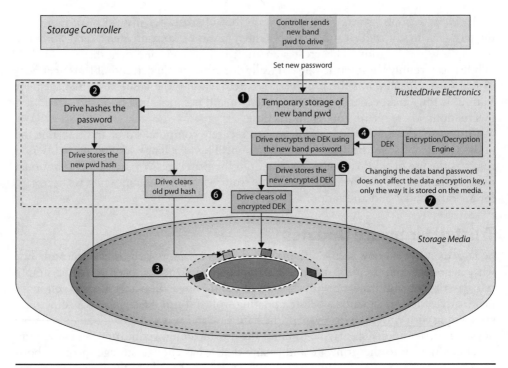

Figure 10-3 Changing the password on an SED

The data stored in an SED is encrypted using symmetric key encryption, and it's decrypted dynamically whenever the device is read. A write operation works the other way—that is, the plaintext data arrives at the drive and is encrypted automatically before being stored on disk. Because the SED has its own hardware-based encryption engine, it tends to be faster than software-based approaches.

Encryption typically uses the Advanced Encryption Standard (AES) and a 128- or 256-bit key. This data encryption key (DEK) is stored in nonvolatile memory within the cryptographic module and is itself encrypted with a password chosen by the user. Figure 10-3 shows that, if the user changes the password, the same DEK is encrypted with the new password, which means the whole disk doesn't have to be decrypted and then re-encrypted. If ever there is a need to securely wipe the contents of the SED, the cryptographic module is simply told to generate a new DEK. Since the drive contents were encrypted with the previous (now overwritten) key, that data is effectively wiped. As you can imagine, wiping an SED is almost instantaneous.

Bus Encryption

While the self-encrypting drive protects the data as it rests on the drive, it decrypts the data prior to transferring it to memory for use. This means that an attacker has three opportunities to access the plaintext data: on the external bus connecting the drive to the motherboard (which is sometimes an external cable), in memory, or on the bus between memory and the CPU. What if we moved the cryptographic module from the disk controller to the CPU? This would make it impossible for attackers to access the plaintext data outside the CPU itself, making their job that much harder.

Bus encryption means data and instructions are encrypted prior to being put on the internal bus, which means they are also encrypted everywhere else except when data is being processed. This approach requires a specialized chip, a *cryptoprocessor*, that combines traditional CPU features with a cryptographic module and specially protected memory for keys. If that sounds a lot like a TPM, it's because it usually is.

You won't see bus encryption in general-purpose computers, mostly because the cryptoprocessors are both more expensive and less capable (performance-wise) than regular CPUs. However, bus encryption is a common approach to protecting highly sensitive systems such as automated teller machines (ATMs), satellite television boxes, and military weapon systems. Bus encryption is also widely used for smart cards. All these examples are specialized systems that don't require a lot of processing power but do require a lot of protection from any attacker who gets his or her hands on them.

Secure Processing

By way of review, data can exist in one of three states: at rest, in motion, or in use. While we've seen how encryption can help us protect data in the first two states, it becomes a bit trickier when it is in use. The reason is that processors almost always need unencrypted code and data to work on.

There are three common ways to protect data while it's in use. The first is to create a specially protected part of the computer in which only trusted applications can run with little or no interaction with each other or those outside the trusted environment. Another approach is to build extensions into the processors that enable it to create miniature protected environments for each application (instead of putting them all together in one trusted environment). Finally, we can just write applications that temporarily lock the processor and/or other resources to ensure nobody interferes with them until they're done with a specific task. Let's take a look at these approaches in order.

Trusted Execution Environment

A *trusted execution environment* (TEE) is a software environment in which special applications and resources (such as files) have undergone rigorous checks to ensure that they are trustworthy and remain protected. Some TEEs, particularly those used in Apple products, are called *secure enclaves*, but the two terms are otherwise interchangeable. TEEs exist in parallel with untrusted *rich execution environments* (REEs) on the same platform, as shown in Figure 10-4. They are widely used in mobile devices and increasingly included in embedded and IoT ones as well, to ensure that certain critical applications and their data have guaranteed confidentiality, integrity, and availability. We're also starting to see them show up in other places, such as microservices and cloud services, where hardware resources are widely shared.

A TEE works by creating a trust boundary around itself and strictly controlling the way in which the untrusted REE interacts with the trusted applications. The TEE typically has its own hardware resources (such as a processor core, memory, or persistent

Figure 10-4
A typical TEE and
its related REE

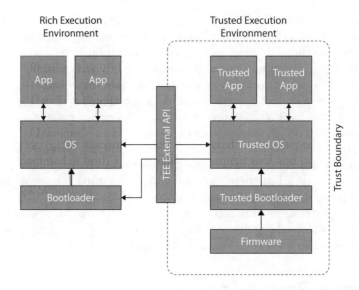

storage) that are unavailable to the REE. It also runs its own trusted OS that is separate from and independent of the one in the REE. The two environments interact through a restricted external application programming interface (API) that enables the rich OS to call a limited set of services provided by the REE.

NOTE The term "secure enclave" is most commonly associated with Apple products such as the iPhone, but it is otherwise equivalent to the term "trusted execution environment."

So, how do TPM, HSM, and TEE differ from each other? A TPM is usually an SoC soldered onto the motherboard to provide limited cryptographic functions. An HSM is a big TPM that plugs into a computer system to provide these functions at a much larger scale. A TEE can perform the functions of a TPM, but, unlike both the TPM and HSM, it is specifically designed to run trusted applications that may have nothing to do with cryptography.

Trusted execution starts with a secure boot, in which the firmware verifies the integrity of the trusted OS bootloader before executing it. In fact, every executable and driver in the TEE is verified to the hardware root of trust and restricted to its own assigned resources. Only specific applications that have undergone rigorous security assessments at the hands of trusted parties are deployed in the TEE by the device manufacturer. This enables trusted applications such as cryptography, identity, and payment systems to enjoy high levels of protection that would otherwise be impossible to attain.

EXAM TIP TEEs (and, by extension, secure enclaves) do not implement hardware roots of trust because they are implemented in software. However, TEEs typically rely on an underlying root of trust provided by a TPM on the device.

Processor Security Extensions

TEEs need hardware support, which all the major chip manufacturers provide in their chipsets. Security is baked into the chips of most modern microprocessors. These CPU packages become a security perimeter outside of which all data and code can exist in encrypted form. Before it can cross into the secure perimeter, everything can be decrypted and/or checked for integrity. Even once allowed inside, data and code are restricted by special controls that ensure what may be done with or to them. For all this to work, however, we need to enable the features through special instructions.

Processor security extensions are instructions that provide these security features in the CPU and can be used to support a TEE. They can, for example, enable programmers to designate special regions in memory as being encrypted and private for a given process. These regions are dynamically decrypted by the CPU while in use, which means any unauthorized process, including the OS or a hypervisor, is unable to access the plaintext stored in them. This feature is one of the building blocks of TEEs, which enables trusted applications to have their own protected memory.

Atomic Execution

Atomic execution is an approach to controlling the manner in which certain sections of a program run so that they cannot be interrupted between the start and end of the section. This prevents other processes from interfering with resources being used by the protected process. To enable this, the programmer designates a section of code as atomic by placing a lock around it. The compiler then leverages OS libraries that, in turn, invoke hardware protections during execution of that locked code segment. The catch is that if you do this too often, you would see some dramatic performance degradation in a modern multitithreaded OS. You want to use atomic execution as little as possible to protect critical resources and tasks.

Atomic execution protects against a class of attacks called time-of-check to time-of-use (TOCTOU). This type of attack exploits the dependency on the timing of events that take place in a multitasking OS. When running a program, an OS must carry out instruction 1, then instruction 2, then instruction 3, and so on. This is how it is written. If an attacker can get in between instructions 2 and 3 and manipulate something, she can control the result of these activities. Suppose instruction 1 verifies that a user is authorized to read an unimportant file that is passed as a link, say a help file. Instruction 2 then opens the file pointed to by the link, and instruction 3 closes it after it's been read by the user. If an attacker can interrupt this flow of execution after instruction 1, change the link to point to a sensitive document, and then allow instruction 2 to execute, the attacker will be able to read the sensitive file even though she isn't authorized to do so. By enforcing atomic execution of instructions 1 and 2 together, we would protect against TOCTOU attacks.

NOTE   This type of attack is also referred to as an asynchronous attack. Asynchronous describes a process in which the timing of each step may vary. The attacker gets in between these steps and modifies something.

Putting It All Together: Where Can Data Be Encrypted?

Data in a computer system can exist in three different places: in the processor, in memory, and in secondary storage such as a disk drive. Standard systems do not encrypt data in any of these three places by default. You can opt for using FED, such as a self-encrypting drive, to encrypt the data in secondary storage, but that leaves it exposed everywhere else in the system, including the external bus. The third option is to use bus encryption, which requires a cryptoprocessor that is (relatively) expensive and underpowered. You are unlikely to want this unless you *really* have to protect the data in situations where you assume the adversary will be able to hack your hardware. Finally, the most flexible (and common) balance of protection, performance, and cost is the use of TEEs that coexist with untrusted applications. Only the data within the TEE receives the full encryption treatment outside the CPU, leaving everything else to run on regular processor cores.

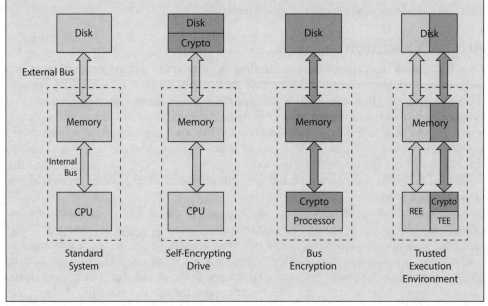

Trusted Foundry

In their novel *Ghost Fleet,* authors P.W. Singer and August Cole describe a string of battles that go terribly wrong for the United States. The cause, unbeknownst to the hapless Americans, is a sophisticated program that inserts undetectable backdoors into the computer chips that run everything from missiles to ships. Although the account is fictional, there have been multiple reports in the open media about counterfeit products introducing vulnerabilities into networks, including some in the military.

The threat is real. In 2004, the U.S. Department of Defense (DoD) instituted the Trusted Foundry Program. The goal is to ensure that mission-critical military and government systems can be developed and fielded using a supply chain that is hardened against external threats. A *trusted foundry* is an organization capable of developing prototype or production-grade microelectronics in a manner that ensures the integrity of their products. The trust is ensured by the National Security Agency through a special review process. At the time of this writing, 77 vendors were rated as trusted foundries and available to DoD customers.

Anti-Tamper Techniques

An important requirement in secure processing is that the hardware resources on which the processor relies is resistant to tampering. Anti-tamper techniques are developed to increase the cost (in terms of hours, money, and/or effort) for an attacker to gain access to or modify the programs and data stored on a hardware device. Determined attackers with virtually unlimited resources (such as some nation states) will eventually be able to compromise any device they can physically get a hold of. Still, we want to make doing so not worth their while.

A common class of physical attacks against integrated circuit chips, *microprobing* comes in at least two flavors: electronic and visual. In electronic microprobing, the attacker applies voltages to various conductors on the chip and observes the results. Once enough experiments have been conducted, it may be possible for the attacker to map the functions of the chip and extract useful information. One of the risks with this approach is that if an attacker applies the wrong amount of electricity in the right place, he could burn out a part of the chip and make the job a lot harder.

In a visual attack, the attacker carefully removes very thin slices of the chip, exposing its layers of individual components. This can be done with an abrasive rotary tool, with a laser cutter, or with chemicals. Using a microscope, it is possible to map out the circuitry and even read individual bits of memory. Obviously, this is a destructive process, but it can reveal a great deal of information given enough time.

There are a number of anti-tamper techniques to defeat microprobing and other physical attacks. Electronic microprobing can be thwarted through the generation of random signals. Integrated circuits normally have predictable clock signals and execute an instruction (or enable a state change) deterministically at particular points in time. But what if the chip were designed to use a random timing signal that only the manufacturer knew? Alternatively, what if each legitimate instruction were sandwiched between a random number of bogus ones? The result would be a chip that would be nearly impossible to understand, let alone manipulate by an adversary.

The attacker may decide to open the chip physically and visually examine its inner workings. One way to defeat this attack is to build a microscopic mesh between the chip and its plastic casing, and then monitor it for breaches. If the plastic casing is removed, whether chemically or mechanically, it is extremely likely to break one of the tiny conductors on the mesh. This event can trigger automatic zeroization of nonvolatile memory and perhaps even the firmware. These countermeasures, called active meshes, are commonly used in smart cards and in mobile subscriber identification modules (SIMs).

Chapter Review

Hardware roots of trust are essential to building secure systems, because without them, attackers can easily access and even modify data and code. Fortunately, most modern CPUs ship with a wealth of hardware features that can be used to provide this security. The catch is that many systems do not take full advantage of these features. It is a good idea to explore how your operating systems and applications leverage trusted platform modules and processor security extensions. If you're not taking advantage of these capabilities, it would be a good idea to look for opportunities to integrate them into your security architecture.

Questions

1. Which of the following implements a hardware root of trust?
   A. Trusted Platform Module (TPM)
   B. eFuse
   C. Unified Extensible Firmware Interface (UEFI)
   D. Trusted foundry

2. Which of the following is not a good use of eFuses?
   A. Keeping track of firmware updates
   B. Storing session keys
   C. Disabling test functions on a chip
   D. Storing digital certificates

3. Where is the data encrypted in a self-encrypting drive system?
   A. On the disk drive
   B. In memory
   C. On the bus
   D. All of the above

4. Where is the data encrypted in a bus encryption system?
   A. On the disk drive
   B. In memory
   C. On the bus
   D. All the above

5. What is the difference between a Trusted Platform Module (TPM) and a Hardware Security Module (HSM)?
   A. HSM is typically on the motherboard and TPM is an external device.
   B. Only an HSM can store multiple digital certificates.

    C. There is no difference, as both terms refer to the same type of devices.

    D. TPM is typically on the motherboard and HSM is an external device.

6. Which of the following is *not* a required feature in a TPM?

    A. Hashing

    B. Certificate revocation

    C. Certificate storage

    D. Encryption

7. Which of the following is true about changing the password on a self-encrypting drive?

    A. It requires re-encryption of stored data.

    B. The new password is encrypted with the existing data encryption key.

    C. It has no effect on the encrypted data.

    D. It causes a new data encryption key to be generated.

8. Which of these is true about processor security extensions?

    A. They are after-market additions by third parties.

    B. They must be disabled in order to establish trusted execution environments.

    C. They enable developers to encrypt memory associated with a process.

    D. Encryption is not normally one of their features.

Answers

1. A. Trusted Platform Modules (TPMs) and Hardware Security Modules (HSMs) are the two best examples of devices that implement a hardware root of trust.

2. B. Session keys are single-use symmetric keys used for one specific session, which means that they change all the time. Because eFuses are programmable only once, this use would quickly exhaust them.

3. A. Self-encrypting drives include a hardware module that decrypts the data prior to putting it on the external bus, so the data is protected only on the drive itself.

4. D. In systems that incorporate bus encryption, the data is decrypted only on the cryptoprocessor. This means that the data is encrypted everywhere else on the system.

5. D. In general, TPMs are permanently mounted on the motherboard and used for hardware-based assurance and key storage, while HSMs are removable or altogether external and are used for both hardware accelerated cryptography and key storage.

6. **B.** Certificate revocation is not a required feature in a TPM. TPMs must provide storage of cryptographic keys and digital certificates, symmetric and asymmetric encryption, and hashing.

7. **C.** When you change the password on a self-encrypting drive, the existing DEK is retained but is encrypted with the new password. This means the encrypted data on the disk remains unaltered.

8. **C.** Processor security extensions are instructions that provide security features in the CPU and can be used to support a TEE. They can, for example, enable programmers to designate special regions in memory as being encrypted and private for a given process.

PART III

Security Operations and Monitoring

## CHAPTER 11

Data Analysis in Security Monitoring Activities

In this chapter you will learn:

- Best practices for security data analytics using automated methods
- Common sources for system and event logs and methods of analyzing them for security operations.
- Advanced techniques for e-mail analysis
- Processes to help you continually improve your security operations

Experts often possess more data than judgment.

—Colin Powell

Modern corporate networks are incredibly diverse environments, with some generating gigabytes of data every day in just logging and event information. Scripting techniques and early monitoring utilities are quickly approaching the end of their usefulness, because the variety and volume of data now exceed what these techniques and utilities were designed to accommodate. Many organizations take an approach that it's not a matter of *if* a security breach will occur, but *when*. Discovering and preventing malicious behavior is just one of the many reasons organizations create a security operations center (SOC). Although security incidents will not always result in headline-grabbing events, organizations need to be able to introspect continuously into their network and make security-based decisions using accurate and timely data.

Security Data Analytics

Security monitoring requires that relevant data be identified before any sort of useful analysis can be performed. In security monitoring, data from various sources across the network must be collected, normalized, and visualized in a way that is useful for analysts to generate products for the purpose of making security decisions, auditing, and compliance. Managing information about your network environment requires a sound strategy and tactical tools for refining data into information, over to knowledge, and onto actionable wisdom. Figure 11-1 shows the relationship between data provided by tools at the tactical level and your goal of actionable intelligence. Data and information sources on

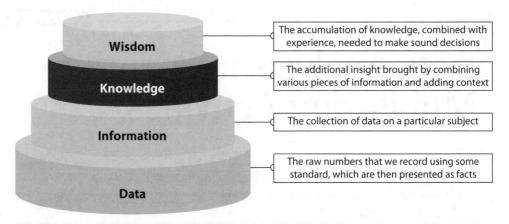

The accumulation of knowledge, combined with experience, needed to make sound decisions

The additional insight brought by combining various pieces of information and adding context

The collection of data on a particular subject

The raw numbers that we record using some standard, which are then presented as facts

Figure 11-1   How data, information, knowledge, and wisdom work together to create actionable intelligence

your network are at least as numerous as the devices on the network. Log data comes from network routers and switches, firewalls, vulnerability scanners, intrusion prevention and detection systems (IPS/IDS), unified threat management (UTM) systems, and mobile device management (MDM) providers. Additionally, each node may provide its own structured or unstructured data from the services it provides. Our goal in security data analytics is to see through the noise of all this network data to produce an accurate picture of the network activity, from which we make decisions in the best interest of our organizations.

Data Aggregation and Correlation

The process of collecting the correct data to inform business decisions can lead to frustration, particularly if the sources are heterogeneous. After all, data ought to be a benefit rather than an impediment to your security team. To understand why data organization is so critical to security operations, we must remember that no single source of data is going to provide all that is necessary to understand an incident. When detectives investigate a crime, for example, they take input from all manner of sources to compile the most complete picture possible. The video, eyewitness accounts, and forensics that they collect all play important parts in the analysis of the physical event. Before a detective can begin analyzing what happened, she must collect, tag, order, and display evidence in a way that it useful for analysis. A similar practice, called *data aggregation,* will enable your team to compare similar data types easily, regardless of their source. The first step in this process usually involves a *log manager* that collects and normalizes data from sources across the network. With the data consolidated and stored, it can then be displayed on a timeline for easy search and display.

Suppose you have gathered a ton of data in the aggregation process. At this point, you may have packet captures, NetFlow records, and log files collected from firewalls, IDSs, and system events. Where do you start? There are at least two schools of thought on this: you can start with the tool you have in front of you, or you can start with an observation.

When you start your analysis with a tool (or a set of tools), you may choose to follow a familiar workflow. The tools you use to capture the data often include at least some basic analysis tools. If nothing else, most of them offer filters that enable you to focus on items of interest. These features, however, will typically be helpful only in a pinch, when you don't have access to anything better. To perform real analysis work, you will need a comprehensive tool with which you can simultaneously look at all the data at your disposal. Broadly speaking, these tools fall into three categories:

- **Security information and event management (SIEM) systems**   These systems collect data from a variety of sensors, perform pattern matching and correlation of events, generate alerts, and provide dashboards that enable analysts to see the state of the network. One of the best-known commercial solutions is Splunk, shown in Figure 11-2, while on the open source side, the Elasticsearch-Logstash-Kibana (ELK) stack is very popular.

- **Big data analytics solutions**   These solutions are designed to deal with massive data sets that are typically beyond the range of SIEMs. The term *big data* refers to data sets so big in terms of volume (that is, the number of records), velocity (the rate at which new records are added), and variability (the number of different data formats) that traditional databases cannot handle them. Big data platforms are normally used to complement SIEMs, not replace them.

- **Locally developed analytics solutions**   These solutions are typically scripts developed in-house by security analysts. PowerShell and Python are popular languages used to develop these tools, which are typically built to perform very specific functions in addition to or in lieu of a SIEM.

Figure 11-2   A dashboard created using the Splunk Enterprise Security SIEM to display security data

Another approach to analysis is to start with an observation, regardless of the tool that enabled you to make it. Based on that observation, you make a hypothesis that would explain it. The next step is either to prove or to disprove that hypothesis through additional observations or experiments. If this sounds familiar, that is because we just described the scientific method, which, as you might imagine, has a lot to do with security analytics. This approach forces us to think beyond the tools at our disposal and ask questions whose answers may not be in the data we already have. If we limit ourselves and our questions to the tools and information in front of us, we will probably miss novel and potentially crippling attacks.

A Brief Detour on Statistics

Having a working knowledge of statistics is extremely beneficial to your ability to perform rigorous security analytics. Using statistics, we can formulate descriptive statements about data, such as "this is normal," and then state categorically that something is not "normal." Although we all have an intuitive sense of normality based on our experience, we are all subject to a variety of biases that will all too often lead us to the wrong conclusions. If you never learned (or have since forgotten) statistics, we highly recommend that you brush up on them. It is difficult to perform some sorts of analyses, such as anomaly and trend analyses, without knowledge of some basic statistics functions.

Data Analysis

Many security analytics tools provide built-in trend analysis functionality. Determining how the network changes over time is important as you assess whether countermeasures and compensating controls are effective. Many SIEMs can display source data in a *time series*, which is a method of plotting data points in chronological order. Indexing these points in a successive manner makes it much easier for you to detect anomalies, because you can roughly compare any single point to all other values. With a sufficient baseline, you'll find it easier to spot new events and unusual download activity.

Trend Analysis

Trend analysis is the study of patterns over time to determine how, when, and why they change. You can use a number of applications to perform trend analysis. Most commonly, trend analysis is applied to security by tracking evolving patterns of adversaries' behaviors. Every year, a number of well-known security firms will publish their trend analyses and make projections for the next year based on the patterns they discovered. This approach would, for example, prompt you to prioritize distributed-denial-of-service (DDoS) mitigations if these attacks are trending up and/or in the direction of your specific sector.

Trend analysis can be useful in the context of threat management as well as a way to determine the controls that are most appropriate within our architectures to mitigate those threats. The goal in any case, however, remains unchanged: to answer the question, "Given what we've been seeing in the past, what should we expect to see in the future?" When we talk about trend analysis, we are typically interested in *predictive analytics*.

Internal Trends

Internal trends can reveal emerging risk areas. For example, there may be a trend in your organization to store increasing amounts of data in cloud resources such as Dropbox. Although this may make perfect sense from a business perspective, it could entail new or increased risk exposure for confidentiality, availability, forensic investigations, or even regulatory compliance. By noting this trend, you will be better equipped to decide the point at which the risk warrants a policy change or the acquisition of a managed solution.

Temporal Trends

Temporal trends show patterns related to time. There are plenty of examples of an organization's systems being breached late on a Friday night in hopes that the incident will not be detected until three days later. Paradoxically, because fewer users will be on the network over the weekend, this should better enable alert defenders to detect the attack, since the background traffic would presumably be lower. Another temporal trend could be an uptick in events in the days leading up to the release of a quarterly statement, or an increase in phishing attempts around tax season. These trends can help us better prepare our technical and human assets for likely threats to come.

Spatial Trends

Trends can also exist in specific regions. Though we tend to think of cyberspace as being almost independent of the physical world, in truth every device exists in a very specific place (or series of places for mobile devices). It is a common practice, for instance, to give staff members a "burner" laptop when they travel to certain countries. This device is not allowed to connect to the corporate network, stores a limited set of files, and is digitally wiped immediately upon the user's return. This practice is the result of observing a trend of sophisticated compromises of devices traveling to particular countries. Another example would be the increasing connection of devices to free Wi-Fi networks at local coffee shops, which could lead to focused security awareness training and the mandated use of virtual private network (VPN) connections.

Historical Analysis

Whereas trend analysis tends to be forward-looking, historical analysis focuses on the past. It can help answer a number of questions, including "Have we seen this before?" and "What is normal behavior for this host?" This kind of analysis provides a reference point (or a line or a curve) against which we can compare other data points.

Historical data analysis is the practice of observing network behavior over a given period. The goal is to refine the network baseline by implementing changes based on observed trends. Through detailed examination of an attacker's past behavior, analysts can gain perspective on the techniques, tactics, and procedures (TTPs) of an attacker to inform decisions about defensive measures. The information obtained over the course of the process may prove useful in developing a viable defense plan, improving network efficiency, and actively thwarting adversarial behavior. Although it's useful to have a large body from which to build a predictive model, there is one inherent weakness to this method: the unpredictability of humans. Models are not a certainty, because it's impossible to predict the future. Using information gathered on past performance means a large assumption that the behavior will continue in a similar way moving forward. Security analysts therefore must consider present context when using historical data to forecast attacker behavior. A simple and obvious example is a threat actor who uses a certain technique to great success until a countermeasure is developed. Up until that point, the model was highly accurate, but with the hole now discovered and patched, the actor is likely to move on to something new, making your model less useful.

 EXAM TIP The difference between trend and historical analyses is small; most practitioners use the terms interchangeably. For purposes of the CySA+ exam, trend analysis helps predict future events, and historical analysis helps compare new observations to past ones.

Behavioral Analysis

Behavioral analysis is closely related to anomaly analysis in that it attempts to find anomalous behaviors. In fact, the two terms are oftentimes used interchangeably or in combination with each other, as in *network behavior anomaly analysis*. The difference, to the extent that there is one in practice, is quite small: behavioral analysis looks at multiple correlated data points to define anomalous behavior. For example, it may be normal behavior for a user to upload large files to an Amazon cloud platform during business hours, but it is abnormal for that user to upload large files to a Google cloud platform after hours. In behavioral analysis, data points relating to size, destination, and time are used together. In a strict interpretation of anomaly analysis, in which data points could be taken in isolation, you may have received two alerts: one for destination and one for time, which you would then have to correlate manually.

 EXAM TIP You should not see questions asking you to differentiate between anomaly analysis and behavioral analysis. You could, however, see questions in which you must recall that they are both examples of data correlation and analytics (as opposed to point-in-time data analysis). You should also remember that they both leverage baselines.

Heuristics

A *heuristic* is a "rule of thumb" or, more precisely, an approach based on experience rather than theory. There are problems in computing that are known to be provably unsolvable, and yet we are able to use heuristics to get results that are close enough to work for us. Heuristic analysis in cybersecurity is the application of heuristics to find threats in practical, if imperfect, ways. This type of analysis is commonly seen in malware detection. We know it is not possible to find malware with 100 percent accuracy, but we also know that the majority of malware samples exhibit certain characteristics or behaviors. These, then, become our heuristics for malware detection.

Next-generation firewalls (NGFs) are devices that, in addition to the usual firewall features, include capabilities such as malware detection. This detection is usually accomplished in NGFs through heuristic analysis of the inbound data. The first approach is to take a suspicious payload and open it in a specially instrumented virtual machine (VM) within or under the control of the NGF. The execution of the payload is then observed, looking for telltale malware actions such as replicating itself, adding user accounts, and scanning resources. Obviously, certain malware families might not attempt any of these actions, which is what makes this approach heuristic: it is practical, but not guaranteed.

Anomaly Analysis

Fundamentally, anomaly analysis attempts to answer the question, "Is this normal?" Obviously, we must have first established what normal means before we can answer the question. The process by which we learn the normal state or flows of a system is called *baselining*. Though we can create baselines for individual systems, it is sometimes more helpful to do so for collections of them. Anomaly analysis focuses on measuring the deviation from this baseline and determining whether that deviation is statistically significant. This last part is particularly important, because everything changes constantly. The purpose of anomaly analysis is to determine whether the change could be reasonably expected to be there in a normal situation, or whether it is worth investigating.

An example of an application of anomaly analysis would be a sudden increase in network traffic at a user's workstation. Without a baseline, we would not be able to determine whether or not the traffic spike is normal. Suppose that we have baselined that particular system and this amount of traffic is significantly higher than any other data point. The event could be classified as an outlier and deemed anomalous. But if we took a step back and looked at the baselines for clusters of workstations, we may find out that the event is consistent with workstations being used by a specific type of user (say, in the media team). Once in a while, one of them sends a large burst (say, to upload a finished film clip), but most of the time they are fairly quiet.

Endpoint Security

Focusing on security at the network level isn't sufficient to prepare for an attacker. While we aim to mitigate the great majority of threats at the network level, the tradeoff between usability and security emerges. We want network-based protection to be able to inspect traffic thoroughly, but not at the expense of network speed. Though keeping an eye on the network is important, it's impossible to see everything and respond quickly. Additionally, the target of malicious code is often the data that resides on the hosts. It doesn't make sense to strengthen the foundation of the network if the rest of it doesn't have a similar level of protection. It's therefore just as important to ensure that the hosts are fortified to withstand attacks and provide an easy way to give insight into what processes are running.

Malware

Recognizing malicious software and its associated behaviors is critical in protecting endpoints. There are several ways that security teams can detect and respond to malware, from looking for commonly used binaries to deconstructing the software itself. The latter usually requires in-depth understanding of the architecture of the processors on which the software is intended to run. Reversing binaries is significantly different for ARM processors compared to x86 processors, for example. The principles are the same, but the devil, as they say, is in the details.

Fingerprinting/Hashing

Sometimes we can save ourselves a lot of trouble by simply fingerprinting or hashing known-good or known-bad binary executable files. Just like fingerprints have an astronomically small probability of not being unique among humans, the result of running a file through a secure hashing function is extremely unlikely to be the same for any two files. The net result is that, when you compute the SHA-256 value of a known-good file like a Windows dynamic-link library (DLL), the probability of an adversary modifying that file in any way (even by changing a single bit) and having it produce the same hash value is remote. But we are getting ahead of ourselves here.

A hashing function is a one-way function that takes a variable-length sequence of data such as a file and produces a fixed-length result called a *hash value*, or message digest. For example, if you want to ensure that a given file does not get altered in an unauthorized fashion, you would calculate a hash value for the file and store it in a secure location. When you want to ensure the integrity of that file, you would perform the same hashing function and then compare the new result with the hash value you previously stored. If the two values are the same, you can be sure the file was not altered. If the two values are different, you would know the file was modified, either maliciously or otherwise, so you would then investigate the event.

We can also apply hashes to malware detection by comparing the hash of the suspicious file to a knowledge base of known-bad hashes. One of the indispensable tools in any analyst's toolkit is VirusTotal.com, a website owned and operated by Google that enables you to upload the hashes (or entire files) and see if anyone has already reported them as malicious or suspicious. Figure 11-3 shows the results of submitting a hash for a suspicious file that has been reported as malicious by 68 out of 73 respondents.

 NOTE Uploading binaries to VirusTotal will enable the entire worldwide community, potentially including the malware authors, to see that someone is suspicious about these files. There are many documented cases of threat actors modifying their code as soon as it shows up on VirusTotal.

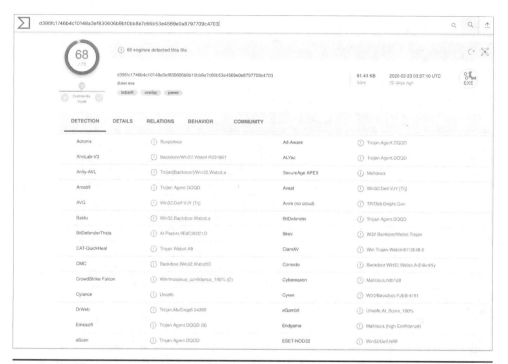

Figure 11-3 VirusTotal showing the given hash corresponds to a malicious file

Decomposition

We can tell you from personal experience that not every suspicious file is tracked by VirusTotal. Sometimes you have to dig into the code yourself to see what it does. In these situations, it is important to consider that computers and people understand completely different languages. The language of a computer, which is dictated by the architecture of its hardware, consists of patterns of 1's and 0's. People, on the other hand, use words that are put together according to syntactical rules. In order for people to tell computers what to do, which is what we call "programming," there must be some mechanism that translates the words that humans use into the binary digits that computers use. This is the job of the compiler and the assembler. As Figure 11-4 shows, a human programmer writes code in a high-level language like C, which is compiled to assembly language, which is in turn assembled into a binary executable.

Binary executables are specific to an operating system and processor family, which means that you cannot run a Linux program on a Windows machine. Windows programs are packaged in the Portable Executable (PE) format, in which every file starts with the 2-byte sequence 5A 4D (or 4D 5A, depending on which operating system you are using to inspect the file). By contrast, Linux executables are in the Executable and

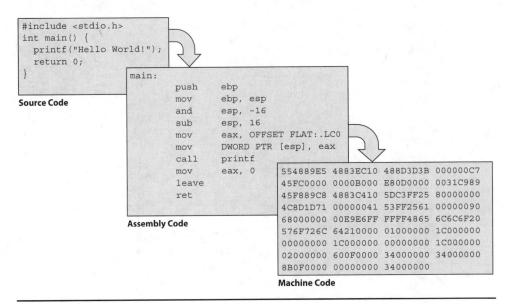

Figure 11-4   Source code being compiled and then assembled

Linkable Format (ELF), in which every file starts with the 4-byte sequence 7F 45 4C 46. These starting sequences, or "magic numbers," enable you to determine quickly which operating system is targeted by a given malware sample.

Generations of Computer Languages

In many aspects of our lives, successive generations are better than their predecessors and render the latter obsolete, but this is not so when describing computer languages. These generations, as listed next, coexist in modern computing systems and will likely continue to do so for the foreseeable future:

- **First generation**   When computers were first invented, the only way to program them was to do so in *machine language,* which is a sequence of operators and arguments represented as sequences of 1's and 0's, sometimes represented in hexadecimal. This language is very specific to a particular family of processors (for example, Intel x86) and very difficult for most programmers to understand.

- **Second generation**   Programming in machine language is absolutely no fun, and it is very prone to errors. It is not surprising that it didn't take us long to come up with a more human-friendly way to program. Assembly language represents machine language operators and arguments using easier to remember operation codes (opcodes) and symbolic variables. Because we still have to somehow get to machine language, we invented assemblers, which turn second-generation assembly language into machine code.

- **Third generation**   Assembly language was a huge step in the right direction, but it still required significant expertise about a particular computer architecture. Some very smart people decided it would be nice to be able to program in a more human-like language without having to know about the underlying architecture. This led to the invention of languages such as BASIC, Pascal, C/C++, Java, and Python. We still had to get down to machine language, but this time we got to leverage the assemblers we already created. The missing part is the compiler, which translates third-generation programming languages into assembly language, which is then turned by the assembler into machine code.

When we are analyzing malware, it is a rare thing to have access to the source code. Instead, all we usually get is a machine language binary file. To reverse-engineer this program, we need a disassembler, such as IDA Pro. The disassembler converts the machine language back into assembly language, which can then be analyzed by a reverse-engineer. Some decompilers also exist, but those are more "hit or miss," because there are many possible programs that would compile to a given assembly language file. This means that, on average, decompilers are not worth the effort.

Detect and Block

Two general types of malware detection for endpoint solutions appear in this category. The first, signature-based detection, compares hashes of files on the local machine to a list of known malicious files. Should there be a match, the endpoint software can quarantine the file and alert the administrator of its presence. Modern signature-based detection software is also capable of identifying families of malicious code.

But what happens when the malicious file is new to the environment and therefore doesn't have a signature? This may be where behavior-based malware detection can help. It monitors system processes for telltale signs of malware, which it then compares to known behaviors to generate a decision on the file. Behavior-based detection has become important because malware writers often use polymorphic (constantly changing) code, which makes it very difficult to detect using signature methods only.

There are limitations with both methods. False positives—files incorrectly identified as malware—can cause a range of problems. At best, they can be a nuisance, but if the detection software quarantines critical system files, the operating system may be rendered unusable. Scale this up several hundred or thousand endpoints, and it becomes catastrophic for productivity.

Fileless Malware

Unlike traditional malware, fileless, or in-memory, malware isn't written to disk but rather directly to memory. The challenge with this method is that it doesn't leave behind traces that can be used for rapid detecting. Additionally, since nothing is preserved to the file system, most forensics techniques used to reconstruct the series of event will not work.

Fileless malware will also use existing software, such a Windows PowerShell, and authorized protocols, like Domain Name System (DNS), to carry out malicious activities.

Sandbox

Creating malware takes a lot of effort, so writers will frequently test their code against the most popular detection software to make sure that it can't be seen before releasing it to the wild. Endpoint solutions have had to evolve their functionality to cover the blind spots of traditional detection by using a technique called *sandboxing*. Endpoint sandboxes can take the form of virtual machines that run on the host to provide a realistic but restricted operating system environment. As the file is executed, the sandbox is monitored for unusual behavior or system changes, and only when the file is verified as being safe can it be allowed to run on the host machine.

Historically, sandboxes were used by researchers to understand how malware was executing and evolving, but given the increase in local computing power and the advances in virtualization, sandboxes have become mainstream. Some malware writers have taken note of these trends and have started producing malware that can detect whether it's operating in a sandbox using built-in logic. In these cases, if the malware detects the sandbox, it will remain dormant to evade detection and will become active at some later point. This highlights the unending battle between malware creators and the professionals who defend our networks.

Cloud-Connected Protection

Like virtualization, the widespread use of cloud computing has allowed for significant advancements in malware detection. Many modern endpoint solutions use cloud computing to enhance protection by providing the foundation for rapid file reputation determination and behavioral analysis. Cloud-based security platforms use automatic sharing of threat details across the network to minimize the overall risk of infection from known and unknown threats. Were this to be done manually, it would take much more time for analysts to prepare, share, and update each zone separately.

Trust

The concept of inherent trust in networks presents many challenges for administrators. Humans make mistakes—they lose devices or unwittingly create entry points into the network for attackers. Thus, the practice of a *zero-trust environment* is emerging as the standard for enterprise networks, in which the network's design and administration consider that threats may come from external and internal sources.

User and Entity Behavior Analytics

One increasingly useful place for applied machine learning techniques for better security is through a practice called *user and entity behavior analytics* (UEBA). Unlike traditional techniques, which alert on suspicious activity based on statistical analysis and predefined rules, UEBA enables analysis to detect anomalous behavior quickly when there are deviations from normal patterns, without the need for predefined rules. To do this, a system would first collect information related to user behavior and trends on a network, usually from network and system logs. Over time, UEBA solutions allow for the creation of reliable baselines of user behavior patterns. From this point, UEBA can then continuously monitor future network behavior and alert upon deviations from the norm. UEBA effectiveness relies on machine learning techniques. As a result, the upfront costs for implementing such a solution may be quite high in terms of money and time to establish a baseline. It does, however, provide insight into suspicious behaviors on a network that traditional monitoring techniques cannot provide, because it does not need predefined rules.

Network

In a well-built information system, there is no shortage of data sources that may assist in providing insight into network activity. It is important that we consider both internal and external data when determining the threats we face. Each source will have its own strengths and limitations, which we can oftentimes balance out by carefully planning our monitoring activities. In the following sections, we consider some common sources of data that are available on almost any network that will help us identify security events faster and more accurately.

Domain Name System Analysis

The DNS architecture is a critical service, but it can add complexity for an analyst when it comes time to research a machine attempting to resolve a bad or suspicious domain. Many companies use an intermediate controller or other server to perform DNS resolution in a recursive manner, meaning that they perform DNS resolution on behalf of clients. The challenge is that this often means that the source IP address in an alert will generally be that of the intermediate server, rather than the originating client. This is problematic, because these servers are often meant to handle a significant number of DNS requests, so an analyst is likely to see the same set of IPs across most alerts of this type. Although it may be possible to tie the outbound traffic to the originating machine using a few pivots, the problem increases in complexity as the network grows.

Fortunately, many modern DNS servers allow for enhanced logging, which in turn can be integrated automatically into SIEM platforms. The tradeoff in performance due to the extra collection may be worth the increased visibility into the network. An alternative to enhanced logging is to re-architect the network so that originating client information is either piped directly to the SIEM or detected by a security device before the request reaches a resolver. This technique may add complexity to the network, but it can be ideal for smaller, geographically consolidated networks. It's important to note that storing these logs isn't free. DNS requests on a network are certainties, and enhanced logging on servers may require gigabytes of storage daily just for logs. More daunting than the storage requirements is the sheer volume of individual log entries. For an organization of thousands, this could mean logs on the order of tens or hundreds of millions.

From a detection and blocking point of view, one of the easiest ways to reduce the number of alerts is to compare the domains and IPs resolved against a blacklist and blocking them when there's a match. This is a comparatively low-cost method that leverages public or commercially vetted sources of known malicious infrastructures. A second technique is to look at the structure of the DNS requests and responses for anomalies. Malformed DNS traffic may be indicative of DNS manipulation, an attack on the DNS infrastructure, or advanced exfiltration techniques.

Domain Generation Algorithms

Domain generation algorithms (DGAs) are used to generate domains rapidly using seemingly random but predictable processes. Malware often depends on a fixed domain or IP addresses for command and control (C2) servers, and when domains are blocked, it is quite disruptive to the malware's operation. By cutting the link between the malware and the C2 servers, we can prevent infected machines from retrieving new commands, updates, and keys. Malware authors have discovered, however, that they may be able to prevent this countermeasure from being effective by writing their code in such a way that it can quickly switch to new domains rather than rely on a static target. This ensures that they can maintain connectivity even when domains are blocked, and that new domains are more difficult for us to guess for future blocking.

To ensure that the malware can anticipate what the new domains will be, it will need to have some degree of predictability, while remaining difficult to anticipate for the defender. After all, this is the only way this system can work in the case of lost connectivity. To achieve this, the DGA processes will use a *seed*, or an initial numerical or string input, and an interval for the switch to occur.

Figure 11-5 shows the Python code for a basic DGA. Line 12 of the code is the implementation of a *linear congruential generator* used to create pseudorandom numbers. This function, used alongside built-in functions, allows for the creation of domains as necessary. In this example, five randomized 20-character domains in the .com top-level domain (TLD) are created, as defined in line 19.

The output of this short bit of code with an input seed of 4 is listed next. Notice that the length of each of the domain names is 20 characters, and that they all terminate in a .com. Using a similar setup, and attacker can easily achieve the desired composition of C2 URLs.

```
zfpwdwpkoojpiuedzsxw.com
krhhxydhdsuxgtjbkmov.com
hzcgmckfoumfopqfvsjn.com
jdufqkienzhkzghckuoc.com
bbgvhhopcgnoynljlmmk.com
```

The adversary will also consider additional conditions when deploying DGAs, such as cost of domain registration, privacy of registration, and reliability of the infrastructure. Identifying the output pattern of the DGA or registrant contact information may be helpful in detecting DGA domains. Modern machine learning techniques have made it easier to classify these types of domains through various methods of quantifying randomness.

```
1    import argparse
2    from ctypes import c_int
3
4    class dgarand:
5
6        def __init__(self):
7            self.r = c_int()
8
9        def srand(self, seed):
10           self.r.value = seed
11
12       def rand(self):
13           self.r.value = 214013*self.r.value + 2531011
14           return (self.r.value >> 16) & 0x7FFF
15
16       def randint(self, lower, upper):
17           return lower + self.rand() % (upper - lower + 1)
18
19   def dga(r):
20       sld = ''.join([chr(r.randint(ord('a'), ord('z'))) for _ in range(20)])
21       return sld + '.com'
22
23
24   if __name__=="__main__":
25       parser = argparse.ArgumentParser()
26       parser.add_argument("seed", type=int)
27       args = parser.parse_args()
28       r = dgarand()
29       r.srand(args.seed)
30       for _ in range(5):
31           print(dga(r))
```

Figure 11-5   Python example of a DGA that outputs a list of pseudorandom domain names

Flow Analysis

Another approach to detecting anomalous behaviors on your networks is to look at where
the traffic is originating and terminating. If you are monitoring the communications of
your nodes in real-time and you suddenly see an odd endpoint, this could be an indicator of a compromise. Admittedly, however, you would end up with many false positives
every time someone decided to visit a new website. An approach to mitigating these false
alarms is to use automation (for example, scripts) to compare the anomalous endpoints
with the IP addresses of known or suspected malicious hosts.

 NOTE   The website VirusTotal.com is a helpful place to check quickly whether
a given URL has been reported as malicious and, if so, by whom.

Traffic analysis can also be done in the aggregate—you keep an eye on the volume of
traffic in a given portion of your system. A large increase in traffic coming to or from
a given host could indicate a compromise. Like our previous example on monitoring
unusual endpoints, this approach will lead to many false positives, absent some mechanism for pruning them. A useful open source tool we've used, Etherape, is shown in

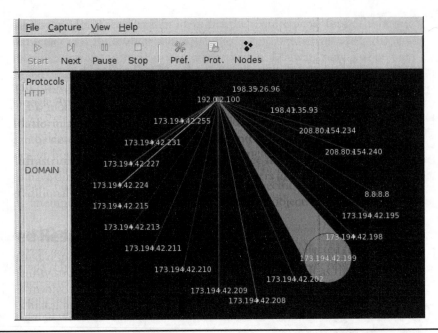

Figure 11-6 Etherape depicting a node transmitting a large amount of HTTP traffic

Figure 11-6. It graphically depicts all known endpoints, both internal and external to your organization, with circles around them to show how their size is proportional to the volume of traffic coming from them at any point in time. A host performing a port scan, for instance, would show up as a very large circle. Then again, so would a server that is streaming high-definition video. The takeaway on traffic analysis is that it is most useful as an early-warning technique that must be backed up or reinforced with additional analysis.

NetFlow Analysis

NetFlow is a system originally developed by Cisco in the late 1990s as a packet-switching technology. Although it didn't serve that role for long, it was repurposed to provide statistics on network traffic, which is why it is important to analysts today. It works by grouping into "flows" all packets that share the following characteristics:

- Arrival interface at the network device (for example, switch or router)
- Source and destination IP addresses
- Source and destination port numbers (or the value zero if not TCP or UDP)
- IP protocol
- IP type of service

When a packet arrives at a NetFlow-enabled network device and does not belong to any known flows, the device will create a new flow for it and start tracking any other related packets. After a preset amount of time elapses with no more packets in a flow, that flow is considered to be finished. The NetFlow-enabled device will aggregate statistics about the flow, such as duration, number of packets, and number of bytes, and then export the record. NetFlow collectors will then receive the data, clean it up a bit if necessary, and store it. The final component of the system is the analysis console, which enables analysts to examine the data and turn it into actionable information.

Notice that the flow data is available for analysis only *after* the flow has ended. This means that this type of analysis is better suited for forensic investigations than for real-time mitigation of attacks. Furthermore, NetFlow captures aggregate statistics and not detailed information about the packets. This type of analysis is helpful in the early stages of an investigation to point the analysts toward the specific packets that should be analyzed in detail (assuming the organization is also doing packet captures).

 NOTE CompTIA separates packet, traffic, and NetFlow as three distinct types of analysis for knowledge organization purposes. In reality, packet and NetFlow analyses are, by definition, types of traffic analysis. Depending on the purpose and approach, protocol and wireless analyses could also be considered types of traffic analysis. "Traffic analysis" is the umbrella term used with packet and NetFlow, and sometimes protocol and wireless analyses are considered subordinate types.

Packet Analysis

An analyst can glean a remarkable amount of information from packet capture data. In fact, given enough of it, an analyst can re-create a very precise timeline of events around any network security incident. In an ideal situation, strategically placed sensors throughout the network are doing full packet captures. The resulting data files contain a wealth of information but can consume enormous amounts of storage space. This can be a challenge, particularly for security teams with limited resources. Another challenge can be finding the useful data in a sea of packet captures.

Filters

Filters are commonly used in two ways: for capture and for display. The first use limits the amount of packets that are captured using some set criteria. For instance, an organization may choose to capture only packets whose source or destination address falls within a specific network range. An application of this could be a file server containing the organization's crown jewels. It may be that limiting packet captures just to those to or from that sensitive server mitigates risks, while minimizing undesirable impacts such as cost of storage or potential violations of privacy. The obvious problem with using capture filters is that packets that might be useful for an investigation may never have been captured.

The other approach is to capture everything but to use filters when looking at the data. Extending the previous example, analysts can choose to look only at the packets to or from the sensitive server, but if they discover that other packets may contain useful clues, they can simply change the display filter and gain visibility over those packets as well. It is almost always better to capture too much information than not enough.

 NOTE Performing full packet captures can have legal implications regarding privacy. Ensure that you consult your legal counsel before you start capturing.

TCP Streams

A noteworthy feature of the design of the Internet is that packets may take different routes to their destination and thus arrive at any given time and in any given order. Although this is normally taken care of by the Transport Control Protocol (TCP) at the transport layer, or via different mechanisms at the application layer for the connectionless User Datagram Protocol (UDP) traffic, such mechanisms are not available to an analyst when the packets are captured directly from the network. Packet analysis tools such as Wireshark offer the ability to reconstruct streams of TCP data. This is particularly useful to recover a malicious file that an employee may have inadvertently downloaded, or to see the full contents of web pages visited.

Encryption

One of the biggest problems with packet analysis is that it is of limited utility when dealing with data that has been encrypted. The analyst will have access to the headers, but the contents may be incomprehensible. Threat actors are known to use encryption to hide their deeds from prying eyes. A way to address this issue is the use of HTTPS (or "SSL") proxies, which are proxy servers that terminate Transport Layer Security (TLS) or Secure Sockets Layer (SSL) connections, effectively acting like a trusted man-in-the-middle that enables the organization to examine or capture the contents of the otherwise encrypted session. If an organization controls the configuration of all clients on its network, it is not difficult to add a certificate authority (CA) to its browsers so that the users will not notice anything odd when they connect to an encrypted site through a decrypting proxy.

 NOTE Using HTTPS proxies can have legal implications regarding privacy. Ensure that you consult your legal counsel before leveraging this capability.

Protocol Analysis

Whereas the focal point in packet analysis is the content of the packets under study, protocol analysis deals with the way in which the packets conform to the protocol they are supposed to be implementing. For instance, the Internet Control Message Protocol (ICMP) allows echo request and echo reply packets to have a payload as long as the total packet length is no greater than the network's maximum transmission unit (MTU). This feature was intended to support diagnostic messages, though in practice this is almost never seen. What we do see, however, are threat actors exploiting this protocol to establish ICMP tunnels in which two hosts create a clandestine communications channel using echo requests and replies. Conducting an analysis of ICMP would reveal these channels.

Another application of protocol analysis is in determining the security or, conversely, vulnerabilities of a given protocol. Suppose you purchase or develop an application for deployment in your organization's systems. How would you know the risks it would introduce unless you had a clear understanding of exactly how its protocols were expressed on your networks? Performing protocol analyses can be as simple as sniffing network traffic to ensure that all traffic is encrypted, or as complex as mathematical models and simulations to quantify the probabilities of unintended effects.

 EXAM TIP For the purposes of the CySA+ exam, you should focus on the sorts of protocol analyses that look at how well packets conform to established protocols.

Malware

One of the most common telltale signs of the presence of malware is anomalous network traffic activity. Security teams routinely monitor for unusual patterns and volumes, in addition to sensitive data leaving the network. While monitoring inbound and outbound activity on your corporate network will enable your team to identify potential attacks in progress, there is another aspect to network monitoring that may prepare the team for an imminent attack. By analyzing network traffic for requests to suspicious domains, security teams may be able to identify behavior associated with the preparation for malware delivery well before malware hits the network. Since modern malware often relies on command and control mechanisms to deliver specific instructions to victim machines, disrupting this communication can vastly improve the team's chances to avoid infection.

Log Review

Logs can be found everywhere; they are often generated by all manner of hardware and software across the network. The data that populate logs is a rich source of information that may enable us to piece together what may have happened during a security event. Understanding how to interpret this endless stream of information is critical to your determining what may have happened and how to prevent it from occurring in the future. When enabled, logging is frequently captured as a time series, meaning events are written with a timestamp associated with the action. Interpretation of these logs involves gathering, correlating, and analyzing that information in a central location such as a SIEM. Although it's tempting to believe that machines can do it all, at the end of the day, a security team's success will be defined by how well its human analysts can piece together the story of an incident. Automated security data analytics may take care of the bulk noise, but the real money is made by the analysts.

Packet Captures

Let's look back at our previous examples of network scanning to explore how an analyst may quickly piece together what happened during a suspected incident. Figure 11-7 gives a detailed list of specific interactions between two hosts on a network. We can see that, in under a second, the device located at IP address 4.4.4.12 sent numerous probes to two devices on various ports, indicative of a network scan.

In addition to source, destination, and port information, each exchange is assigned a unique identifier in this system. After the scan is complete a few minutes later, we can see that the device located at 4.4.4.12 establishes several connections over port 80 to a device with the 4.4.4.15 IP address, as shown in Figure 11-8. It's probably safe to assume that this is standard HTTP traffic, but it would be great if we were able to take a look. It's not unheard of for attackers to use well-known ports to hide their traffic.

Time	source_ip	source_port	destination_ip	destination_port
April 6th 2017, 17:57:33.151	4.4.4.12	47629	4.4.4.14	7002
April 6th 2017, 17:57:33.151	4.4.4.12	47629	4.4.4.15	1068
April 6th 2017, 17:57:33.151	4.4.4.12	47629	4.4.4.15	3784
April 6th 2017, 17:57:32.973	4.4.4.12	47629	4.4.4.14	1805
April 6th 2017, 17:57:32.973	4.4.4.12	47629	4.4.4.14	2875
April 6th 2017, 17:57:32.973	4.4.4.12	47629	4.4.4.15	50003
April 6th 2017, 17:57:32.973	4.4.4.12	47629	4.4.4.14	5800
April 6th 2017, 17:57:32.973	4.4.4.12	47629	4.4.4.15	1972
April 6th 2017, 17:57:32.972	4.4.4.12	47629	4.4.4.15	6881
April 6th 2017, 17:57:32.972	4.4.4.12	47629	4.4.4.15	50800
April 6th 2017, 17:57:32.972	4.4.4.12	47629	4.4.4.14	9535
April 6th 2017, 17:57:32.972	4.4.4.12	47629	4.4.4.15	541
April 6th 2017, 17:57:32.972	4.4.4.12	47629	4.4.4.15	2200
April 6th 2017, 17:57:32.972	4.4.4.12	47629	4.4.4.15	40911
April 6th 2017, 17:57:32.972	4.4.4.12	47629	4.4.4.15	1112
April 6th 2017, 17:57:32.972	4.4.4.12	47629	4.4.4.14	2260
April 6th 2017, 17:57:32.972	4.4.4.12	47629	4.4.4.15	6156

Figure 11-7 SIEM list view of all traffic originating from a single host during a network scan

Time	source_ip	source_port	destination_ip	destination_port
April 6th 2017, 18:04:52.299	4.4.4.12	40536	4.4.4.15	80
April 6th 2017, 18:03:05.237	4.4.4.12	40674	4.4.4.15	80
April 6th 2017, 18:03:00.234	4.4.4.12	40674	4.4.4.15	80
April 6th 2017, 18:02:17.218	4.4.4.12	40672	4.4.4.15	80
April 6th 2017, 18:02:12.213	4.4.4.12	40672	4.4.4.15	80
April 6th 2017, 18:02:06.207	4.4.4.12	40670	4.4.4.15	80
April 6th 2017, 18:02:01.204	4.4.4.12	40670	4.4.4.15	80
April 6th 2017, 18:01:56.200	4.4.4.12	40668	4.4.4.15	80
April 6th 2017, 18:01:51.197	4.4.4.12	40668	4.4.4.15	80
April 6th 2017, 17:59:58.104	4.4.4.12	4444	4.4.4.15	32772
April 6th 2017, 17:57:46.961	4.4.4.12	35494	4.4.4.15	631
April 6th 2017, 17:57:46.957	4.4.4.12	167	4.4.4.15	111
April 6th 2017, 17:57:46.957	4.4.4.12	492	4.4.4.15	111
April 6th 2017, 17:57:46.957	4.4.4.12	877	4.4.4.15	111
April 6th 2017, 17:57:46.957	4.4.4.12	916	4.4.4.15	111
April 6th 2017, 17:57:46.957	4.4.4.12	40338	4.4.4.15	443
April 6th 2017, 17:57:46.957	4.4.4.12	40348	4.4.4.15	443

Figure 11-8 Listing of HTTP exchange between 4.4.4.12 and 4.4.4.15 after scan completion

This view enables us to get more information about what happened during that time by linking directly to the packet capture of the exchange. The capture of the first exchange in that series shows a successful request of an HTML page. As we review the details in Figure 11-9, this appears to be the login page for an administrative portal.

Looking at the very next capture in Figure 11-10, we see evidence of a login bypass using SQL injection. The attacker entered **Administrator' or 1=1 #** as the username, indicated by the text in the uname field. When a user enters a username and password, a SQL query is created based on the input from the user. In this injection, the username is populated with a string that, when placed in the SQL query, forms an alternate SQL statement that the server will execute. This gets interpreted by the SQL server as follows:

```
SELECT * FROM users WHERE name='Administrator' or 1=1 #'
 and password='boguspassword'
```

```
4.4.4.12:40668_4.4.4.15:80-6-1121361501.pcap

Sensor Name: onion-eth1
Timestamp: 2017-04-06 18:01:51
Connection ID: CLI
Src IP: 4.4.4.12 (Unknown)
Dst IP: 4.4.4.15 (Unknown)
Src Port: 40668
Dst Port: 80
OS Fingerprint: 4.4.4.12:40668 - UNKNOWN [S20:64:1:60:M1460,S,T,N,W7::?:?] (up: 233 hrs)
OS Fingerprint: -> 4.4.4.15:80 (link: ethernet/modem)

SRC: GET / HTTP/1.1
SRC: Host: 4.4.4.15
SRC: User-Agent: Mozilla/5.0 (X11; Linux x86_64; rv:45.0) Gecko/20100101 Firefox/45.0
SRC: Accept: text/html,application/xhtml+xml,application/xml;q=0.9,*/*;q=0.8
SRC: Accept-Language: en-US,en;q=0.5
SRC: Accept-Encoding: gzip, deflate
SRC: Connection: keep-alive
SRC:
SRC:
DST: HTTP/1.1 200 OK
DST: Date: Thu, 06 Apr 2017 14:50:16 GMT
DST: Server: Apache/2.0.52 (CentOS)
DST: X-Powered-By: PHP/4.3.9
DST: Content-Length: 667
DST: Connection: close
DST: Content-Type: text/html; charset=UTF-8
DST:
DST: <html>
DST: <body>
DST: <form method="post" name="frmLogin" id="frmLogin" action="index.php">
DST: ..<table width="300" border="1" align="center" cellpadding="2" cellspacing="2">
DST: ..<tr>
DST: ...<td colspan='2' align='center'>
DST: ...<b>Remote System Administration Login</b>
DST: ...</td>
DST: ..</tr>
DST: ..<tr>
DST: ...<td width="150">Username</td>
DST: ...<td><input name="uname" type="text"></td>
DST: ..</tr>
DST: ..<tr>
DST: ...<td width="150">Password</td>
DST: ...<td>
DST: ...<input name="psw" type="password">
DST: ...</td>
DST: ..</tr>
DST: ..<tr>
DST: ...<td colspan="2" align="center">
DST: ...<input type="submit" name="btnLogin" value="Login">
DST: ...</td>
```

Figure 11-9   Packet capture details of first HTTP exchange between 4.4.4.12 and 4.4.4.15

Because the 1=1 portion will return true, the server doesn't bother to verify the real password and grants the user access. The note "Welcome to the Basic Administrative Web Console" in Figure 11-10 shows that the attacker has gained access.

Just because the attacker now has access to a protected area of the web server, it doesn't mean he has full access to the network. Nevertheless, this behavior is clearly malicious, and it's a lead that we should follow to the end. In the following subsections, we discuss how the approach to manual review we just presented using automation tools and packet captures can be extended to other sources of information.

 NOTE   Software-defined networking (SDN) addresses several challenges that make correlation difficult. Because the network is centrally controlled to optimize the performance, the SDN provider is also a perfect place to perform data collection. This reduces the need to perform collection, formatting, and normalizing tasks for each device. Rather, these tasks can be performed once across the entire network.

```
4.4.4.12:40670_4.4.4.15:80-6-159015080.pcap

Sensor Name: onion-eth1
Timestamp: 2017-04-06 18:02:06
Connection ID: CLI
Src IP: 4.4.4.12 (Unknown)
Dst IP: 4.4.4.15 (Unknown)
Src Port: 40670
Dst Port: 80
OS Fingerprint: 4.4.4.12:40670 - UNKNOWN [S20:64:1:60:M1460,S,T,N,W7::?:?] (up: 233 hrs)
OS Fingerprint: -> 4.4.4.15:80 (link: ethernet/modem)

SRC: POST /index.php HTTP/1.1
SRC: Host: 4.4.4.15
SRC: User-Agent: Mozilla/5.0 (X11; Linux x86_64; rv:45.0) Gecko/20100101 Firefox/45.0
SRC: Accept: text/html,application/xhtml+xml,application/xml;q=0.9,*/*;q=0.8
SRC: Accept-Language: en-US,en;q=0.5
SRC: Accept-Encoding: gzip, deflate
SRC: Referer: http://4.4.4.15/
SRC: Connection: keep-alive
SRC: Content-Type: application/x-www-form-urlencoded
SRC: Content-Length: 68
SRC:
SRC: uname=Administrator%27+or+1%3D1+%23&psw=boguspassword&btnLogin=Login
DST: HTTP/1.1 200 OK
DST: Date: Thu, 06 Apr 2017 14:50:26 GMT
DST: Server: Apache/2.0.52 (CentOS)
DST: X-Powered-By: PHP/4.3.9
DST: Content-Length: 586
DST: Connection: close
DST: Content-Type: text/html; charset=UTF-8
DST:
DST: <html>
DST: <body>
DST:
DST: <!-- Start of HTML when logged in as Administator -->
DST: .<form name="ping" action="pingit.php" method="post" target="_blank">
DST: ..<table width='600' border='1'>
DST: ..<tr valign='middle'>
DST: ...<td colspan='2' align='center'>
DST: ...<b>Welcome to the Basic Administrative Web Console<br></b>
DST: ...</td>
DST: ..</tr>
DST: ..<tr valign='middle'>
DST: ...<td align='center'>
DST: ....Ping a Machine on the Network:
DST: ...</td>
DST: ...<td align='center'>
DST: ....<input type="text" name="ip" size="30">
DST: ....<input type="submit" value="submit" name="submit">
```

Figure 11-10 Packet capture details of a second HTTP exchange between 4.4.4.12 and 4.4.4.15, showing evidence of a SQL injection

System Logs

Properly configured, end systems can log a much richer set of data than many network devices. This is partly because network traffic can be encrypted or obfuscated, but the actions on the end system are almost always easier to observe. Another reason why event logs matter tremendously is that there are many threat actors who will acquire domain credentials either before or at an early stage of their attack. Once they impersonate a legitimate user account, the data you capture from the network will be much less insightful to your analysis. It is a very common technique to use real credentials (stolen though they may be) to accomplish lateral movement throughout a targeted network.

System logs come in a variety of formats. The two formats with which you should be familiar as a CySA+ certified professional are the Windows Event Logs and the more generalized syslog. Both are intended to standardize the reporting of events that take place on a computing device. Though they are not the only formats in existence, most applications will generate events in the Windows format if they are running on a Microsoft operating system and in syslog format otherwise. Additionally, various products are available that will take input from one or the other (or any other format for that matter) and aggregate it into one event log.

Event Logs

Event logs provide detail about a system and connected components. Administrators can view all of a Windows system's event logs with the Event Viewer utility. This feature makes it much easier to browse through the thousands of entries related to system activity, as shown in Figure 11-11. It's an essential tool for understanding the behavior of complex systems like Windows—and particularly important for servers, which aren't designed to always provide feedback through the user interface.

In recent Windows operating systems, successful login events have an event ID of 4624, whereas login failure events are given an ID of 4625 with error codes to specify the exact reason for the failure. In the Windows 10 Event Viewer, you can specify exactly which types of event you'd like to get more details on using the Filter Current Log option in the side panel. The resulting dialog is shown in Figure 11-12.

NOTE The default location for the Linux operating system and applications logs is the /var/log directory. In the Windows environment, you can use the Event Viewer to view the event logs. For other network devices, the syslog location may vary.

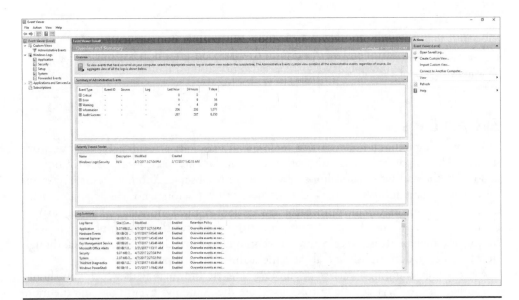

Figure 11-11 The Event Viewer main screen in Windows 10

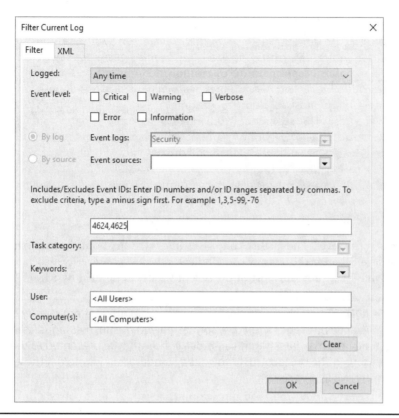

Figure 11-12   The Event Viewer prompt for filtering log information

Syslog

Syslog is a messaging protocol developed at the University of California, Berkeley, to standardize system event reporting. Syslog has become a standard reporting system used by operating systems and includes alerts related to security, applications, and the OS. The local syslog process in UNIX and Linux environments, syslogd, collects messages generated by the device and stores them locally on the file system. This includes embedded systems in routers, switches, and firewalls, which use variants and derivatives of the UNIX system. There is, however, no preinstalled syslog agent in the Windows environment. Syslog is a great way to consolidate logging data from a single machine, but the log files can also be sent to a centralized server for aggregation and analysis. Figure 11-13 shows the structure of a typical syslog hierarchy.

The syslog server will gather syslog data sent over UDP port 514 (or TCP port 514, if message delivery needs to be guaranteed). Analysis of aggregated syslog data is critical for security auditing, because the activities that an attacker will conduct on a system are bound to be reported by the syslog utility. These clues can be used to reconstruct the scene and perform remedial actions on the system. Each syslog message includes a facility

Figure 11-13
Typical hierarchy
for syslog
messaging

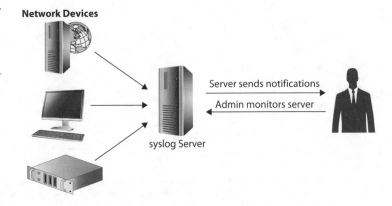

Network Devices

Server sends notifications

Admin monitors server

syslog Server

code and severity level. The facility code gives information about the originating source of the message, whereas the severity code indicates the level of severity associated with the message. Table 11-1 provides a list of the severity codes as defined by RFC 5424.

NOTE The syslog protocol for event messaging does not specify how exactly messages should be formatted. As a result, messages from different devices may have slight variations in how they're presented to the syslog server. The protocol just provides a standardized way to get the message from client to server.

Firewall Logs

It is widely accepted that firewalls alone cannot secure an information system, or even its perimeter. Still, they remain an important part of any security architecture. Fundamentally, a firewall is meant to restrict the flow of data between two or more network interfaces according to some set of rules. This means that it has to examine every packet that

Value	Severity	Keyword	Description
0	Emergency	emerg	System is unusable
1	Alert	alert	Action must be taken immediately
2	Critical	crit	Critical conditions
3	Error	err	Error conditions
4	Warning	warning	Warning conditions
5	Notice	notice	Normal but significant conditions
6	Informational	info	Informational messages
7	Debug	debug	Debug-level messages

Table 11-1 Syslog Severity Codes, Keywords, and Descriptions

arrives at any of the interfaces to ensure compliance with whatever policies are specified in the firewall's configuration files. Whether the packet is accepted, denied, or dropped, the firewall generates a log entry with details that can help us in preventing or recovering from incidents.

 NOTE It is important to pay attention to inbound as well as outbound traffic. Some incidents will be easier to detect in one versus the other.

The amount of information contained in firewall logs is configurable. Some fairly universal parameters logged by most firewalls are described in Table 11-2. By default, most firewalls provide ample logs for any incident response. But it can be frustrating and unhelpful to begin an investigation only to find out that a well-meaning firewall administrator pared down the amount of information that a device would store in its logs. There is no way to recover log data that was never captured.

Web Application Firewall Logs

Traditional firewalls were not designed to protect web applications and thus do not provide adequate defense against the wide-ranging types of attacks observed against Internet-connected services. Simply blocking traffic based on source and destination port numbers may not be sufficient to protect from attacks such as cross-site scripting (XSS) and SQL injection. Web application firewalls (WAFs) operate at the application layer to monitor and block potentially destructive traffic, usually over HTTP and HTTPS. Though HTTP traffic is stateless, WAFs are typically user, session, and application aware, meaning that they often operate as the first line of defense. Analyzing WAF logs can provide enormous insight into how attackers behave.

Field	Description
Timestamp	Date and time at which the packet was logged
Source Address	IP address of the source of the packet
Source Port	Port number at the source
Destination Address	IP address of the destination of the packet
Destination Port	Port number at the destination
Protocol	IP protocol of the packet (for example, TCP, UDP, or ICMP)
IN Interface	Firewall interface that received the packet
OUT Interface	Firewall interface on which the packet was forwarded (unless denied or dropped)
Rule Name	Firewall rule that was applied to the packet resulting in whatever action was taken
Action	Action taken by the firewall (for example, accept, deny, or drop)

Table 11-2 Typical Firewall Log Fields

Proxy Logs

In comparison to WAFs, which aim to protect applications and the servers they reside on, proxies protect clients. Most often proxies take requests from internal sources, such as those from clients on a local network, and determine how they should be handled based on predetermined rules. The proxy server is in an ideal position to log all web traffic originating from inside the network and can block and filter as necessary. Proxy logs can provide information to identify anomalous browsing behavior such as that associated with C2 connections.

Intrusion Detection/Prevention Systems

A step up from firewall logs in terms of valuable threat data are the logs and alerts of the intrusion detection system (IDS) and the intrusion prevention system (IPS). The difference between the two is that the former simply generates alerts when it detects suspected threats and the latter actively blocks them. What oftentimes makes these systems more helpful in finding threats is that their rule language is typically more powerful than that of a firewall. Whereas a firewall will allow, drop, or alert traffic on a fairly simple set of rules or heuristics, an IDS/IPS can look for very specific signatures or behaviors. Figure 11-14 shows a Snort alert triggered by a port scan originating in a neighboring node.

 EXAM TIP   A next-generation firewall (NGF) incorporates functionality from both traditional firewalls and IPSs. Unless otherwise stated in the exam, assume that an IDS/IPS is not an NGF.

IDSs can be measured along two dimensions: focus and approach. The focus can be on a host, which makes them host-based IDSs (HIDSs); otherwise, they are network-based IDSs (NIDSs). Their approach can be signature or anomaly based. Finally, they can be standalone or integrated into another platform, such as an NGF.

Snort

Snort is probably the best-known NIDS in the open source community. However, it is actually more than a NIDS, because it can operate as a packet analyzer or as a network intrusion prevention system (NIPS). Snort was originally developed by Martin Roesch in the late 1990s and has been under constant development ever since. It is most known for the richness of its rules language and the abundance of rules that exist for it.

Figure 11-14
Sample IDS alert

```
[**] [1:469:3] ICMP PING NMAP [**]
[Classification: Attempted Information Leak] [Priority: 2]
01/20-17:30:12.439889 192.168.192.7 -> 192.168.192.8
ICMP TTL:48 TOS:0x0 ID:63971 IpLen:20 DgmLen:28
Type:8 Code:0 ID:56127 Seq:45129 ECHO
[Xref -> http://www.whitehats.com/info/IDS162]

[**] [122:1:0] (portscan) TCP Portscan [**]
01/20-17:30:12.724540 192.168.192.7 -> 192.168.192.8
PROTO255 TTL:0 TOS:0x0 ID:0 IpLen:20 DgmLen:162 DF
```

Snort rules have two parts: the header and options. The header specifies the action Snort will take (for example, alert or drop) as well as the specific protocol, IP addresses, port numbers, and directionality (directional or bidirectional). The real power of the rules, however, is in the options. In this section of a rule, one can specify where exactly to look for signs of trouble as well as what message to display to the user or record in the logs.

The following rule, for example, shows how to detect a backdoor in the network:

```
alert tcp $EXTERNAL_NET any -> $HOME_NET 7597 (msg:"MALWARE-BACKDOOR QAZ Worm
Client Login access"; content:"qazwsx.hsq";)
```

In this case, we are looking for inbound TCP packets destined for port 7597 containing the text "qazwsx.hsq." If these are found, Snort will raise an alert that says "MALWARE-BACKDOOR QAZ Worm Client Login access." Note that many more options could be written into the rule, such as hashes of known malware.

 NOTE Threat Intelligence companies will often include Snort signatures for newly discovered threats as part of their subscription services.

Zeek

Zeek, formerly known as Bro, is both signature and anomaly based. Instead of only looking at individual packets and deciding whether or not they match a rule, it creates events that are inherently neither good nor bad; they simply alert you that something happened. An advantage in this approach is that Zeek will track sessions to ensure that they are behaving as you would expect, and it keeps track of their state. All this data is retained, which can help forensic investigations. These events are then compared to policies to see what actions, if any, are warranted, and it is here that Zeek's power really shines: the policies can do anything from sending an e-mail or a text message, to updating internal metrics, to disabling a user account.

Another powerful feature in Zeek is the ability to extract complete executables from network streams and send them to another system for malware analysis. This download feature is also helpful when you're performing forensic investigations in which, for example, you need to determine which files may have been exfiltrated by an attacker. Because all the events (which include embedded files) are stored, they are available for future analysis.

Suricata

Suricata can be thought of as a more powerful version of Snort, even though its architecture is quite different. It can use Snort signatures, but it can also do a lot more. Specifically, it is multithreaded (Snort isn't) and can even take advantage of hardware acceleration (that is, using the graphics accelerator to process packets). Like Zeek, it can also extract files from the packet flows for retention or analysis. Like both Zeek and Snort, Suricata can be used as an IPS.

```
Apr  6 11:08:27 edda sshd[5761]: pam_unix(sshd:auth): authentication failure; logname=
uid=0 euid=0 tty=ssh ruser= rhost=4.4.4.12  user=root
Apr  6 11:08:29 edda sshd[5761]: Failed password for root from 4.4.4.12 port 52724 ssh2
Apr  6 11:08:49 edda sshd[5761]: message repeated 2 times: [ Failed password for root from
4.4.4.12 port 52724 ssh2]
Apr  6 11:08:49 edda sshd[5761]: Connection closed by 4.4.4.12 port 52724 [preauth]
Apr  6 11:08:49 edda sshd[5761]: PAM 2 more authentication failures; logname= uid=0 euid=0
tty=ssh ruser= rhost=4.4.4.12  user=root
Apr  6 11:10:48 edda sshd[5812]: Accepted password for root from 4.4.4.12 port 52793 ssh2
Apr  6 11:10:48 edda sshd[5812]: pam_unix(sshd:session): session opened for user root by
(uid=0)
Apr  6 11:10:48 edda systemd-logind[983]: New session 1079 of user root.
```

Figure 11-15   Snapshot of the auth.log entry in a Linux system

Authentication Logs

Auditing and analysis of login events is critical for a successful incident investigation. All modern operating systems have a way to log successful and unsuccessful attempts. Figure 11-15 shows the contents of the auth.log file indicating all enabled logging activity on a Linux server. Although it might be tempting to focus on the failed attempts, you should also pay attention to the successful logins, especially in relation to those failed attempts. There's a reasonable chance that the person logging in from 4.4.4.12 is an administrator who made a mistake the first couple of times. However, when you combine this information with the knowledge that this device has just recently performed a suspicious network scan, the likelihood that this is an innocent mistake goes way down.

 EXAM TIP   When dealing with logs, consider the time zone difference for each device. Some may report in the time zone you operate in, some may use GMT, whereas others may be off altogether. This is why using a centralized time source on your network is critical. Be sure to note any time zone differences that may exist when analyzing logs from different sources, as may be the case when answering simulation questions on the exam.

Impact Analysis

A successful attack can affect an organization in many ways. The results of an attack are referred to as *impact*. Impact will vary depending on the details of the event but can involve many tangible and intangible costs to an organization, its employees, and customers over the short and long terms. We'll cover a few key concepts of impact analysis that are important to security operations here. To help categorize the types of impact to an organization, we can break them down into phases that align with the response and recovery process. In a triage phase, for example, analysts will work quickly to determine what's impacted and will then begin to formulate a plan to communicate with stakeholders about what's occurred. Efforts in this phase are usually focused on stopping an adversary's ongoing activity and assessing immediate damage. As an analyst, your goals here are often to react to what's happened, assess the *immediate impact*, prevent further damage, and take steps that will better inform near-term decisions from leaders.

The *localized impact*, the immediate effects that the incident has on availability, confidentiality, or integrity of an asset, is often a priority, but there may be broader organizational impacts that are not always apparent. Sometimes the true nature of an impact will not reveal itself until further investigation is performed after the initial discovery of a security event. After initial triage, security teams will move beyond reacting and on to determining the next steps to reduce disruptions to long-term business processes and stakeholder relationships. This phase, characterized by actions taken perhaps weeks and months after then initial event, tends to focus heavily on auditing and determining the *total impact* to the organization. In some cases, there may be regulatory requirements related to disclosure, or even law enforcement proceedings.

The final phase of impact analysis is all about repairing the damage discovered throughout the previous two phases and installing measures to prevent or reduce the likelihood of a similar event in the future. This could mean pitching the case for investment in a new detection technology or refining security awareness training for employees—both important goals in making the company more resilient as you move forward.

Availability Analysis

Sometimes our focus rests on protecting the confidentiality and integrity of our systems at the expense of preventing threats to their availability. Availability analysis is focused on determining the likelihood that our systems will be available to authorized users in a variety of scenarios. Perhaps the most common of these is the mitigation of DDoS attacks, which is in part accomplished by acquiring the services of an anti-DDoS company such as Akamai. These services can be expensive, so an availability analysis could help make the business case for them by determining at which point the local controls would not be able to keep up with a DDoS attack and how much money the organization would lose per unit of time that it was unavailable to its customers.

Another application of availability analysis is in determining the consequences of the loss of a given asset or set of assets. For example, what would be the effects on the business processes of the loss of a web server, or the data server storing the accounting data, or the CEO's computer? Obviously, you cannot realistically analyze every asset in your system, but there are key resources whose unavailability could cripple an organization. Performing an availability analysis of those resources can shed light on how to mitigate the risk of their loss.

Resource-Monitoring Tools

There are several tools that will enable you to monitor the performance and uptime of your systems. At a very high level, these tools track the availability of your network devices, including workstations, servers, switches, routers, and indeed anything that can run an agent or send data to a plug-in. For each of those devices, tools can monitor specific metrics such as processor or disk utilization. When things go wrong, the tools will log the event and then either send a notification via e-mail or text or take a specific action by running an event handler that can correct the problem. Many of them are remarkably easy to set up and often scalable enough to handle complex environments and procedures, giving you visibility on anything affecting availability.

PART III

Security Information and Event Management Review

SIEM systems are at least as much analytical as they are collective tools. All tools in this category perform four basic functions: collect, store, analyze, and report. Most of the collected information comes from various systems' logs, which are exported and sent to the SIEM system. The SIEM will then typically normalize the format of the data from these disparate sources so that they can be compared with each other. It then stores everything in a system that is optimized for quick retrieval, which is needed to analyze vast amounts of information. Whether the analysis is the result of a simple user query or the end product of sophisticated processes of correlation, SIEMs also have to produce a variety of reports for different purposes that range from internal lessons learned to regulatory compliance.

Figure 11-16 shows a SIEM dashboard that displays the security events collected over a fixed period of time. This particular SIEM is based on the ELK stack, a popular solution for security analysts who need large-volume data collection, a log parsing engine, and search functions. From the total number of raw logs (more than 3000 in this case), the ELK stack generates a customizable interface with sorted data and provides color-coded charts for each type.

From these charts, you can see the most commonly used protocols and most talkative clients at a glance. Unusual activity is also very easy to identify. Take a look at the "Top Destination Ports" chart shown in Figure 11-17. Given a timeframe of only a few minutes, is there any good reason why one client attempts to contact another over so many ports? Without diving deeply into the raw data, you can see that there is almost certainly scanning activity occurring here.

Many SIEM solutions offer the ability to craft *correlation* rules to derive more meaningful information from observed patterns across different sources. For example, if you observe traffic to UDP or TCP port 53 that is not directed to an approved DNS server,

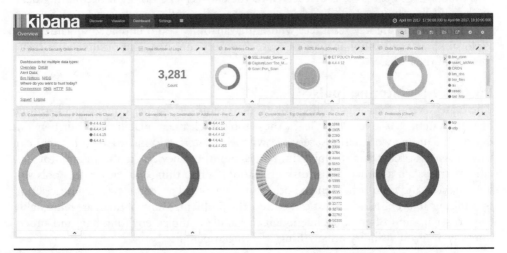

Figure 11-16   SIEM dashboard showing aggregated event data from various network sources

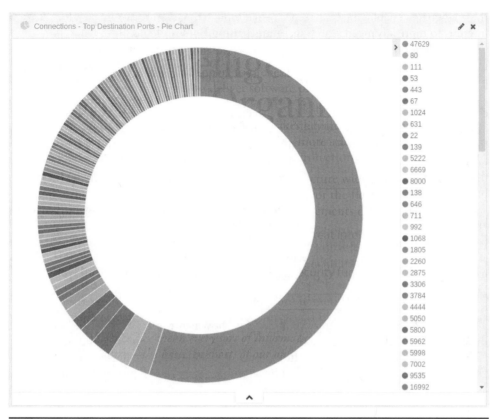

Figure 11-17 SIEM dashboard chart showing all destination ports for the traffic data collected

this could be evidence of a rogue DNS server present in your network. You are taking observations from two or more sources to inform a decision about which activities to investigate further. If these activities are connected to known-bad IP addresses, for example, you will have more confidence in making a determination about blocking or redirecting that traffic.

Query Writing

The ability to search data is a core function of any data aggregation platform. Although the syntax for querying may be specific to a platform, it is often easy for you to understand and migrate from one to the other. After all, the whole point of collecting and normalizing this data is to be able to derive meaning from it at a later point through manual or automated retrieval. The platform's features and functions are often heavily driven from searches. Search languages such as Splunk Search Processing Language (SPL), Kibana Query Language (KQL), and Apache Lucene are similar in that they all use a high structure and scalable approach, much like the commonly used SQL. Each of these languages enables analysts to perform simple string searches or queries for terms of interest, to more advanced search techniques using Boolean logic.

Depending on the platform used, you may be able to search and then perform automated actions such as alert delivery via scripting. The most commonly supported types include shell, batch, Perl, and Python scripts. Among your most important considerations when creating automation scripts are using the appropriate working directories, configuring the environment correctly, and ensuring that arguments are passed correctly. Splunk provides representational state transfer (REST) endpoints that enable analysts to initiate searches and retrieve results automatically. Splunk also provides supporting for advanced Python scripting via the open source Enterprise SDK for Python. Using the SDK, an analyst could programmatically interact with the Splunk engine using the provided wrapper functionality to leverage the Splunk REST API.

It may also be possible to script at the system level and interact directly with platforms via their APIs, cURL, or other mechanisms. The only limiting factor in this case would be the permissions granted to the account attempting to perform the actions. ELK, for example, enables queries to be performed easily via a built-in console or via cURL. A basic search can be performed on a specific index using this format:

```
GET /index_name/_search?q=field:search_term
```

The results of this search using the console is shown in the Figure 11-18. Notice that the output is displayed alongside the search box for ease of testing and troubleshooting.

Passing data using built-system functionalities such as piping and redirection can be used to test functionality quickly or for low-volume processing. Piping is a useful function

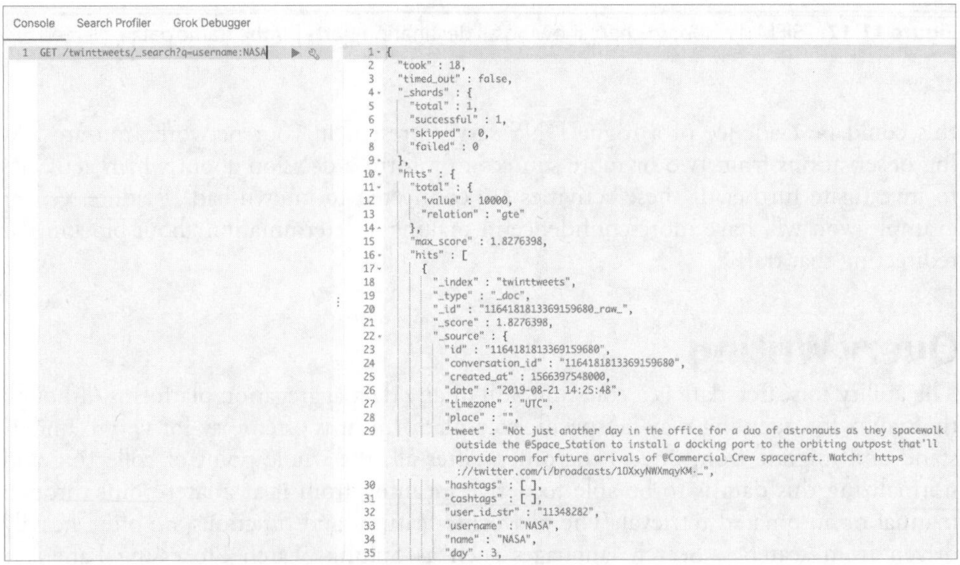

Figure 11-18   Basic search performed from the built-in Elasticsearch console

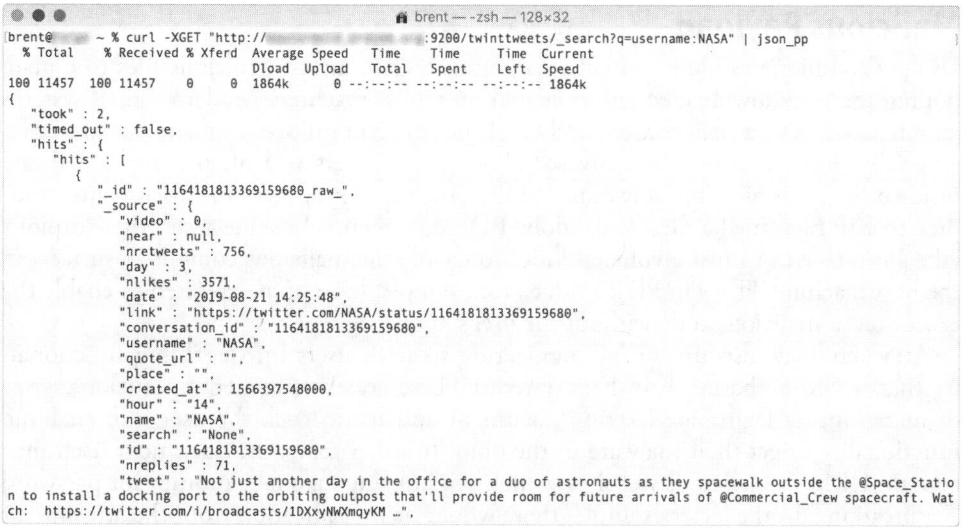

Figure 11-19 Basic search performed via cURL in the command line

in that it enables the standard output (stdout) of a command to be connected to standard in (stdin) of another command. The following example shows how we can perform the same operation we performed from the Elasticsearch console via the command line. We can then pipe the output of the cURL operation directly to the json_pp utility, which is used to "pretty-print" various formats, adding stylistic and spacing changes to make the output easier to read by an analyst.

```
curl -XGET "http://serveraddress:9200/index_name/_search?q= field:search_
term" | json_pp
```

Figure 11-19 shows an excerpt of the same search we performed in the Elasticsearch console, but using cURL here instead. Note the neat structure of the response. The key/value pairs of the JSON can be easily identified for follow-on processing.

E-mail Analysis

Depending on the size of your organization, you may see thousands or even millions of e-mails traverse your networks. Within that massive set of traffic are malicious messages that threat actors use to target employees of your organization. Hoping to gain access to internal systems or sensitive personal and financial information, these threat actors craft their messages to be as realistic as possible. Phishing remains a top attack vector for threat actors of all sophistication levels. Furthermore, many modern phishing e-mails are made using techniques that make the messages indistinguishable from legitimate e-mail. A scalable and automated e-mail analysis process, therefore, is required to provide the most protection against increasingly convincing phishing attempts.

Malicious Payload

Using a technique as old as e-mail itself, attackers can attach malicious files to e-mail, hoping they are downloaded and executed on a host machine. Modern e-mail systems automatically scan e-mail content and block a significant number of malicious messages, especially those that are obviously so. However, attackers will often conceal malware inside other types of commonly e-mailed files, such as word processing documents, compressed ZIP files, media files, and Adobe PDF documents. In some cases, these exploits take advantage of software vulnerabilities to enable the malicious content to surface on the host machine. Flaws in PDF readers, for example, have been exploited to enable the execution of malicious code without the user's knowledge.

Attackers may also use social engineering to trick users into enabling functionality that would be harmful to their systems. These attackers embed a malicious script or macro into a legitimate looking document and try to trick the user into enabling functionality to get their malware in the door. In other cases, the document itself may contain URLs to malicious websites, which might bring up a fake prompt for password "verification." In the latter example, there would be no indication from the attachment that anything unusual was happening, since there may not be malicious content in the document itself.

DomainKeys Identified Mail

The DomainKeys Identified Mail (DKIM) standard was introduced as a way for e-mail senders to provide a method for recipients to verify messages. It specifically offers three services: identity verification, identification of an identity as known or unknown, and determination of an identity as trusted or untrusted. DKIM uses a pair of keys, one private and one public, to verify messages. The organization's public key is published to DNS records, which will later be queried for and used by recipients. When sending a message using DKIM, the sender includes a special signature header in all outgoing messages. The DKIM header will include a hash of the e-mail header, a hash of some portion of the body, and information about the function used to compute the hash, as shown here:

```
DKIM-Signature: v=1; a=rsa-sha256; d=example.com; s=test;
c=relaxed/relaxed; q=dns/txt; t=1126524832; x=1149015927;
h=from:to:subject:date:keywords:keywords;
bh=kWwVkljr4/RXuFhWzCWO8PPyulPPHzyVhGYICEk1NWg=;
b=dzdVyOfAKCdLXdJOc9G2q8LoXSlEniSbav+yuU4zGeeruD00lszZVoG4ZHRNiYzR
```

Upon receiving a message, the destination server will look up the previously published public key and use this key to verify the message. With this process, DKIM can effectively protect against spam and spoofing, and it can also alert recipients to the possibility of message tampering. Importantly, DKIM is not intended to give insight into the intent of the sender, protect against tampering after verification, or prescribe any actions for the recipient to take in the event in a verification failure.

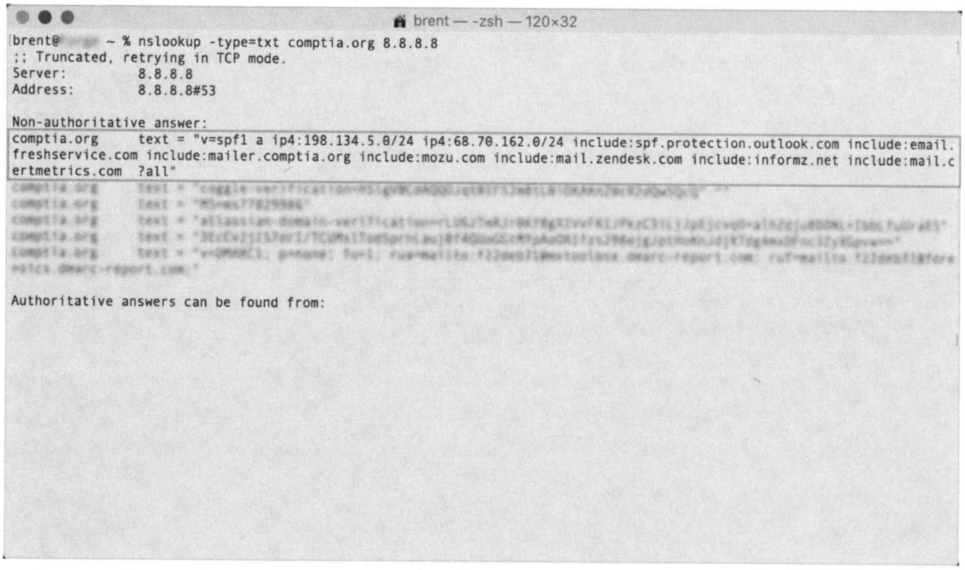

Figure 11-20 TXT records from DNS lookup of comptia.org highlighting SPF entry

Sender Policy Framework

The Simple Mail Transfer Protocol (SMTP) enables users to send e-mails to recipients and explicitly specify the source without any built-in checks. This means that anyone can send a message claiming to be from anyone. This default lack of verification has enabled spammers to send messages claiming to be from legitimate sources for decades. The Sender Policy Framework (SPF) enables domain owners to prevent such e-mail spoofing using their domains by leveraging DNS functionality. An SPF TXT record lists the authorized mail servers associated with a domain. Before a message is fully received by a recipient server, that server will verify the sender's SPF information in DNS records. Once this is verified, the entirety of the message can be downloaded. If a message is sent from a server that's not in that TXT record, the recipient's server can categorize that e-mail as suspicious and mark it for further analysis. SPF TXT information can be manually queried for, as shown in Figure 11-20, which shows the Google public DNS server, 8.8.8.8, reporting several TXT records for the domain comptia.org, including SPF information.

Domain-Based Message Authentication, Reporting, and Conformance

Domain-based Message Authentication, Reporting, and Conformance (DMARC) is an e-mail authentication protocol designed to give e-mail domain owners the ability to prevent spoofing and reduce spam that appears to originate from their domain. Like SPF, an entry is created in the domain owner's DNS record. DMARC can be used to

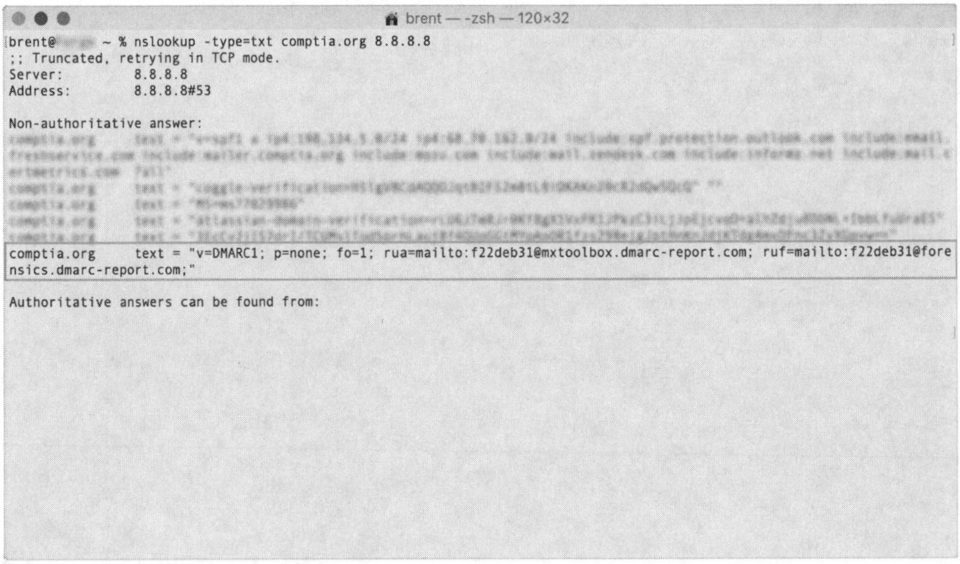

Figure 11-21 TXT records from DNS lookup of comptia.org highlighting DMARC entry

tell receiving servers how to handle messages that appear to be spoofed using a legitimate domain. DMARC uses SPF and DKIM to verify that messages are authentic, so it's important that both SPF and DKIM are correctly configured for the DMARC policy to work properly. Once the DMARC DNS entry is published, any e-mail server receiving a message that appears to be from the domain can check against DNS records, authenticated via SPF or DKIM. The results are passed to the DMARC module along with message author's domain. Messages that fail SPF, DKIM, or domain tests may invoke the organization's DMARC policy. DMARC also makes it possible to record the results of these checks into a daily report, which can be sent to domain owners, usually on a daily basis. This allows for DMARC policies to be improved or other changes in infrastructure to be made. As with SPF, DMARC TXT information can be manually queried for using any number of DNS utilities. Figure 11-21 shows the DMARC entry from a DNS lookup.

Header

An e-mail header is the portion of a message that contains details about the sender, the route taken, and the recipient. Analysts can use this information to detect spoofed or suspicious e-mails that have made it past filters. These headers are usually hidden from view, but mail clients or web services will allow the message header to be viewed with no more than a few clicks. Figure 11-22 is an excerpt of an e-mail header, featuring several details about the message's journey from sender to recipient inbox. Note the SPF and DKIM verdicts are captured in the header information along with various server addresses.

```
Delivered-To: *******@gmail.com
Received: by 10.80.176.68 with SMTP id i62csp9339161edd;
        Thu, 28 Dec 2017 13:53:30 -0800 (PST)
X-Received: by 10.200.50.206 with SMTP id a14mr43214491qtb.59.1514498010046;
        Thu, 28 Dec 2017 13:53:30 -0800 (PST)
ARC-Seal: i=1; a=rsa-sha256; t=1514498010; cv=none;
        d=google.com; s=arc-20160816;
        b=a70ETRryKuMDlF6QOgYwid5R45Raxn4p9VhYYyxt+uEaZxl8hSSWCQ0DamwY7KSuwW
         RBKXtQhxN     YyYrolYz       e8jhgwb       R2HXDxSxT       FKEt3ajA
         
Retu   Path:
<32WdFWggTD5YDE-H4FBO022EKDJI.6EE6B4.2ECA4DD4J7C0IKJ0PK6C08B.2EC@scoutcamp.bounces.google.
com>
Received: from mail-sor-f69.google.com (mail-sor-f69.google.com. [209.85.220.69])
        by mx.google.com with SMTPS id l46sor25230779qtc.113.2017.12.28.13.53.29
        for <*******@gmail.com>
        (Google Transport Security);
        Thu, 28 Dec 2017 13:53:30 -0800 (PST)
Received-SPF: pass (google.com: domain of
32wdfwggtd5yde-h4fbo022ekdji.6ee6b4.2eca4dd4j7c0ikj0pk6c08b.2ec@scoutcamp.bounces.google.
com designates 209.85.220.69 as permitted sender) client-ip=209.85.220.69;
Authentication-Results: mx.google.com;
        dkim=pass header.i=@google.com header.s=20161025 header.b=RZ0mFn+4;
        spf=pass (google.com: domain of
32wdfwggtd5yde-h4fbo022ekdji.6ee6b4.2eca4dd4j7c0ikj0pk6c08b.2ec@scoutcamp.bounces.google.
com designates 209.85.220.69 as permitted sender)
smtp.mailfrom=32WdFWggTD5YDE-H4FBO022EKDJI.6EE6B4.2ECA4DD4J7C0IKJ0PK6C08B.2EC@scoutcamp.
bounces.google.com;
        dmarc=pass (p=REJECT sp=REJECT dis=NONE) header.from=accounts.google.com
DKIM-Signature: v=1; a=rsa-sha256; c=relaxed/relaxed;
        d=google.com; s=20161025;
        h=mime-version:date:reply-to:feedback-id:message-id:subject:from:to;
        bh=XYYcYIdqi7ODvBTOHbIrzfOLsG8K+kZQDQ2ohki1Qao=;
        b=RZ0mFn+49r+Y/csjnRzyT6x405/wNDFbg8NIn+iMqdqhqXAta+yfv4ALV660R2r4bx
         7D909058kwsrIL2l3iwvjeargmPkLWMAaFIEkipinwm42NX7BPrSFUM5C4Vgl+quLDBA
         /AHMF/d5+N2oaR9OsCQYwpdaDwsXX8EX/4VUdofzfhwzdjpo4cDYHwMlHEXUHxMqYTo8
         oRUItOIdgiYpSsuNBUClK9TdsLIy6e1eLrazCrM0yuEJrFv3MWWNJdbd08NUPQWT4C0yr
         nFGL6c19Cfaw+AnUuwVCN30zHHcGIevM1nwh+C8rLt8lnKsP+LUIaePRXGxHlJHz+BzG
         MbbA==
X-Google-DKIM-Signature: v=1; a=rsa-sha256; c=relaxed/relaxed;
        d=1e100.net; s=20161025;
        h     message       mime-v       date: r      o:feedb
        
A      KCQjUw       uKfcO       e3lRIu       3V9Iu       t4+ny       Hd80=
MIME-Version: 1.0
X-Received: by 10.200.27.122 with SMTP id p55mr23522587qtk.53.1514498009758; Thu, 28 Dec
2017 13:53:29 -0800 (PST)
Date: Thu, 28 Dec 2017 21:53:27 +0000 (UTC)
Reply-To: Google <no-reply@accounts.google.com>
X-Google-Id: 196013
Feedback-ID: 91-anexp#rmd-standalone:account-notifier
X-Account-Notification-Type: 91-anexp#rmd-standalone
X-Notifications:
GAMMA:<c252089a83cb90e8.1514498009303.100215992.10009979.en.859a425a228ae17a@google.com>
Message-ID:
<c252089a83cb90e8.1514498009303.100215992.10009979.en.859a425a228ae17a@google.com>
Subject: Security alert
From: Google <no-reply@accounts.google.com>
To: *******@gmail.com
Content-Type: multipart/alternative; boundary="94eb2c0b29f6e27cf605616d8b72"
```

Figure 11-22   Sample e-mail header

Phishing

Many attackers know that the best route into a network is through a careless or untrained employee. In a social engineering campaign, an attacker uses deception, often influenced by the profile they've built about the target, to manipulate the target into performing an act that may not be in their best interest. These attacks come in many forms—from advanced phishing e-mails that seem to originate from a legitimate source, to phone calls requesting personal information. Phishing attacks continue to be a challenge for network defenders because they are becoming increasingly convincing, fooling recipients into divulging sensitive information with regularity. Despite the most advanced technical countermeasures, the human element remains the most vulnerable part of the network. Handling phishing is no trivial task. Attacker techniques are evolving, and although several best practices have been established for basic protection, the very nature of e-mail is such that there aren't always visible indicators that you can use to ensure that a message is genuine, especially at scale.

Forwarding

Some larger organizations have dedicated inboxes to review suspicious e-mails that have somehow passed through e-mail filters. Encouraging users who've received these kinds of messages to forward them to a special inbox may help analysts determine what changes need to be made to e-mail rules to prevent these e-mails from getting through in the future. It's also useful to analyze reports in aggregate to assist in identifying large-scale or targeted attacks. This technique is useful not only in helping security teams identify false negatives, or bad e-mails that have marked as good, but false positives as well. It can just be frustrating to have legitimate e-mails continuously sent to a junk folder.

 TIP Users provide the most useful information to a security team by forwarding an e-mail in its entirety, with headers and body intact, rather than just copying and pasting the text within it the e-mail. It's also often possible for users to attach multiple e-mails in a forwarded message.

Digital Signatures and Encryption

A digital signature provides sender verification, message integrity, and nonrepudiation, or the assurance that a sender cannot deny having sent a message. This kind of signature requires the presence of public and private cryptographic keys: When crafting a message, the sender signs the message locally with the sender's private key. Upon receipt, the recipient verifies it on his device by using the sender's public key. Digital messages are used today in a similar way that sealing wax was used centuries ago to seal documents. Wax sealant was used to verify that documents were unopened, and when sealed with a custom signet ring, the sealant could also be used to verify the sender's identity. Because these rings were difficult to replicate, recipients could have reasonable certainty that the message was legitimate if they recognized the seal.

Secure/Multipurpose Internet Mail Extensions (S/MIME) and Pretty Good Privacy (PGP) are the most common protocols used on the Internet for authentication and privacy when sending e-mails. PGP offers cryptographic privacy and authentication for many kinds of transmissions, but it is widely used for signing, encrypting, and decrypting electronic data to protect e-mail content. S/MIME is included in many browsers and e-mail clients and also supports encryption, key exchanges, and message signing. S/MIME and PGP can both provide authentication, message integrity, and nonrepudiation. In practice, S/MIME is often used in commercial setting, while PGP tends to be used by individuals.

 CAUTION   It's critical for your users to understand that a digital signature is not the same as the signature block routinely used in outgoing messages. Signature block content is simply additional text that is automatically or manually inserted with messages to enable users to share contact information. It offers no security advantage, whereas a digital signature provides verification of the sender's authenticity and message integrity.

Embedded Links

Some security devices perform real-time analysis of inbound messages for the presence of URLs and domains and modify the messages so that links are either disabled or redirected to a valid domain. Some Software as a Service (SaaS) e-mail platforms, such as Microsoft Office 365 and Google Gmail, offer functionality to identify links behind standard and short URLs, scan linked images for malicious content, and intercept user clicks to untrusted domains.

Impersonation

It might be easy to think that almost every phishing e-mail contains malicious URLs, a malware attachment, or some other malicious technical content. This is not always the case, however, as we look at how effective impersonation attacks have been over the past few years. Impersonation attacks are highly targeted efforts designed to trick victims into performing actions such as wiring money to attacker accounts. Often these victims are directly connected to key decision-makers, such as members of a finance team and executive assistants. By pretending to be a CEO, for example, an attacker may use tailored language to convince her targets to perform the requested task without thinking twice. Detecting and blocking these types of attacks cannot be done with technical tools alone. Key staff must be aware of current attacker trends and take the required training to resist them.

Chapter Review

Acquiring data through continuous network monitoring is only the beginning. The real work is in analyzing that data and turning it into information. In this chapter, we covered a multitude of approaches and tools that you can use to perform basic security operations and monitoring, along with some security analytics. There is no one-size-fits-all answer for any of this, so it is critical that, as cybersecurity analysts, we have enough familiarity with our options to choose the right one for a particular job. The topics brought up in this chapter are by no means exhaustive, but they should serve you well for the CySA+ exam and, perhaps more importantly, as a starting point for further lifelong exploration.

Questions

1. What method describes efforts to identify malicious activity by looking for suspicious characteristics that can be found in unknown software, for which signatures may not be available?

   A. Predictive analysis

   B. Trend analysis

   C. Point-in-time analysis

   D. Heuristic analysis

2. What class of technique includes a means for attackers to rapidly generate domain names using seemingly random ways to circumvent blacklisting?

   A. Domain Name System analysis

   B. Domain generation algorithm

   C. Web application firewall

   D. Pseudorandom number generator

3. Your team is looking to aggregate logs from network devices, security appliances, and endpoints across your network so that they can be quickly queried and viewed in a visual manner. What is the most appropriate tool for this task?

   A. Event Viewer

   B. SIEM

   C. Activity Monitor

   D. IDS

4. Which of the following provides aggregate statistics for network traffic, suitable for forensic investigations?

   A. NetFlow

   B. Syslog

   C. UEBA

   D. Logstash

Use the following scenario to answer Questions 5–8:

A colleague from your organization's finance team reaches out by phone to get your advice on a suspicious e-mail he just received. The e-mail is marked as urgent and appears to be sent from the company's chief financial officer herself. In the body of the message are instructions to wire $10 million to an account the analyst has never seen before. Both the CEO and CFO are currently traveling internationally and they are unavailable. Your colleague is worried that his inaction may delay an important project from moving forward, as is stressed in the e-mail, but given the sum of the request, he is concerned about the possibility of the message being a scam.

5. You want to collect details about the message to include any information about e-mail servers involved in its delivery. What must you ask your colleague to include when forwarding the e-mail to you?

   A. Syslogs

   B. Payload

   C. DNS forwarders

   D. Header

6. Although the e-mail address used appears to belong to the company's CFO, the mail server used to send the message is not associated with your organization. What technology would you recommend being put in place moving forward to prevent occurrences like this?

   A. DKIM

   B. IDS

   C. SPF

   D. WAF

7. In addition to the previous improvement, you want to leverage the organization's public keys to ensure that e-mails bearing the company domain are indeed sent and authorized by the organization. What technology would best fit this requirement?

   A. SPF

   B. SIEM

   C. DKIM

   D. DMARC

8. A week later, you read an industry report that outlines a surging trend in e-mail messages similar to the one received by your colleague. What is this type of attack most commonly referred to as?

   A. Impersonation

   B. Malicious payload

   C. Embedded links

   D. Trojan

PART III

Answers

1. **D.** Heuristic analysis is a method of detecting malicious activity by observing software's behavior and is often done with the assistance of virtualization and decomposition technologies.

2. **B.** Domain generation algorithms (DGAs) are sets of instructions designed to generate domain names in rapid, seemingly random fashion to ensure that command and control instructions can be exchanged as defenders block and take down malicious domains.

3. **B.** Security information and event management (SIEM) systems provide a way to aggregate endpoint, network, and system log data while also offering tools for analytics and visualization.

4. **A.** NetFlow is a network protocol developed by Cisco for monitoring network traffic and collecting information such as source and destination IP address, IP protocol, and source and destination ports.

5. **D.** The header is a portion of the e-mail message that contains information about the sender, recipient, route, and authentication details.

6. **C.** The Sender Policy Framework (SPF) enables domain owners to prevent e-mail spoofing by specifying the authorized mail servers in its DNS records.

7. **C.** The DomainKeys Identified Mail (DKIM) standard was introduced as a way for e-mail senders to provide a method for recipients to verify messages using a pair of private and public keys to sign and verify messages.

8. **A.** Impersonation attacks are highly targeted efforts designed to trick victims into performing actions such as wiring money to attacker accounts.

Implement Configuration Changes to Existing Controls to Improve Security

In this chapter you will learn:

- How to manage access with user and group policies
- How to control execution of applications with whitelisting and blacklisting
- Various methods of network and application filtering
- How to read and write common intrusion detection rules

The world is full of obvious things which nobody by any chance ever observes.

—Sir Arthur Conan Doyle

Information systems are composed of increasingly complex components designed to meet a variety of business needs. Maintaining a high level of security requires careful thought about how the network is architected, configured, and deployed. Understanding that the environment is in a constant state of change requires that an agile process to security configuration management is in place. In this chapter, we'll discuss some best practices for developing controls to maintain a high level of awareness of your system's activities while keeping security incidents at a minimum.

Permissions

Permissions are the access privileges that administrators and resource owners grant to users to various resources. The operating system manages and enforces these privileges to allow and deny users access to specific objects. In computing, a *user* is an entity that can be authenticated against a system. The user can represent a person, or it can be created specifically to perform a task, as is the case with service accounts. It's a best practice to avoid managing users on an individual basis and to focus instead on managing a collection, or *group*, of users. Administrative permissions should be assigned judiciously, and users should have only the minimum number of permissions possible. This makes it much easier for administrators to understand the capabilities of users.

NOTE A best practice for both software developers and systems administrators is to stick to the principle of *least privilege*. This policy ensures that a user or function has the lowest level of access required to accomplish necessary tasks. This means that processes should never run as root and that users should avoid logging in as administrators. Failing to stick to this practice creates a huge vulnerability for systems, because attackers will often target applications running as root or accounts with privileged access.

Users

Depending on the version of Windows you're running, you'll be able to view and adjust user properties using the Local Group Policy Editor, Local Security Policy Editor, or Default Domain Policy Group Policy object (GPO). Figure 12-1 shows the Windows 10 Local Users and Groups (lusrmgr.msc) utility related to user and group management. This lists all of the current users on this particular system. From here, an administrator can modify user properties to restrict certain types of account activity.

Let's take a look at the Guest account, an account that enables unauthenticated network users to log on without a password. (It's advisable that you disable the Guest account whenever possible to force all network users to authenticate before being able to

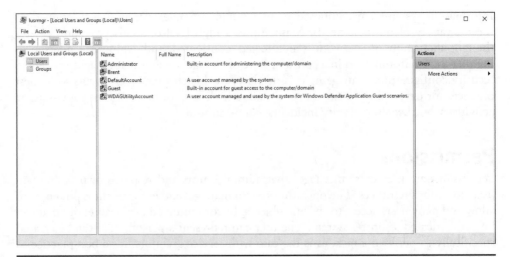

Figure 12-1 List of users in the Windows 10 Local Users and Groups utility

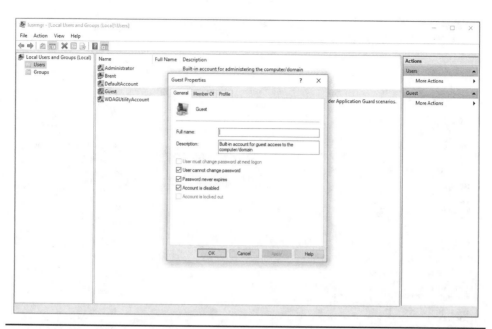

Figure 12-2 Guest Properties in the Windows 10 Local Users and Groups utility

access resources.) Figure 12-2 shows the properties of the Guest account and includes a few options that can be changed directly from the lusrmgr.msc utility.

In Red Hat versions of Linux, you can get a listing of all users on the system using the **lslogins** command. This command provides information about the user identifier (UID), the username, and any recent login activity. For detailed information about the currently logged in user, you can use the same **lslogins** command followed by the username. With this information, shown in Figure 12-3, you may be able to recover useful information about how the account was, or may be, used or if there are changes that can be made to harden the account.

Groups

Groups can contain multiple user accounts, each of which, by default, inherits the privileges assigned to the group. Inversely, users can belong to many groups and given access to files and directories by their owners. All major operating systems have a similar concept of groups, which offer many benefits for management. First is the ability to assign and remove users' privileges quickly based on their status in the organization. For example, you may give all members of your human resources team access to certain employee

PART III

```
● ● ●                          brent — brent@rhel:~ — ssh rhel — 120×42
[brent@rhel ~]$ lslogins
  UID USER              PROC PWD-LOCK PWD-DENY  LAST-LOGIN GECOS
    0 root               167                               root
    1 bin                  0                               bin
    2 daemon               0                               daemon
    3 adm                  0                               adm
    4 lp                   0                               lp
    5 sync                 0                               sync
    6 shutdown             0                    Apr13/09:22 shutdown
    7 halt                 0                               halt
    8 mail                 0                               mail
   11 operator             0                               operator
   12 games                0                               games
   14 ftp                  0                               FTP User
   59 tss                  0                               Account used by the trousers package to sandbox the tcsd daemon
   74 sshd                 0                               Privilege-separated SSH
   81 dbus                 0                               System message bus
   89 postfix              0
   99 nobody               0                               Nobody
  192 systemd-network      0                               systemd Network Management
  998 chrony               0
  999 polkitd              0                               User for polkitd
 1000 brent                2                    11:50:20 brent
[brent@rhel ~]$ lslogins brent
Username:                     brent
UID:                          1000
Gecos field:                  brent
Home directory:               /home/brent
Shell:                        /bin/bash
No login:                     no
Primary group:                brent
GID:                          1000
Supplementary groups:         wheel
Supplementary group IDs:      10
Last login:                   11:50:20
Last terminal:                pts/0
Last hostname:                10.10.0.100
Hushed:                       no
Running processes:            2

Last logs:
```

Figure 12-3 Output of **lslogins** user information command in Red Hat Enterprise Linux

directories; if your organization configures groups to reflect its company structure, it would be easy to assign the requisite access to a new HR team member by simply adding the user to the group.

As with users, you can view group-related properties in Windows 10 using the lusrmgr.msc utility. Figure 12-4, for example, shows all groups on the system along with descriptions of their purpose. Note that many of these groups are not generated by a user or administrator, but are included with the operating system by default. These include special service groups required to run services.

In the Red Hat environment, you can get a similar listing of available groups, as shown in Figure 12-5. The command **getent group** (get entries), will list all of the groups of the system, while the **groups** command will show the groups that the currently logged on user belongs to.

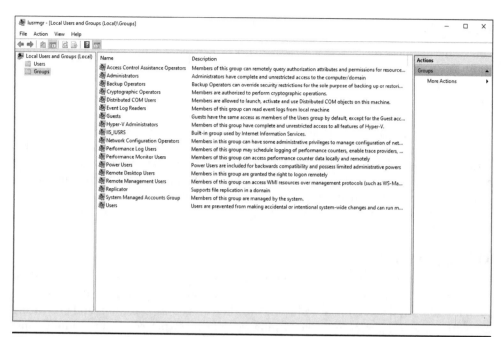

Figure 12-4 Listing of groups in the Windows 10 Local Users and Groups utility

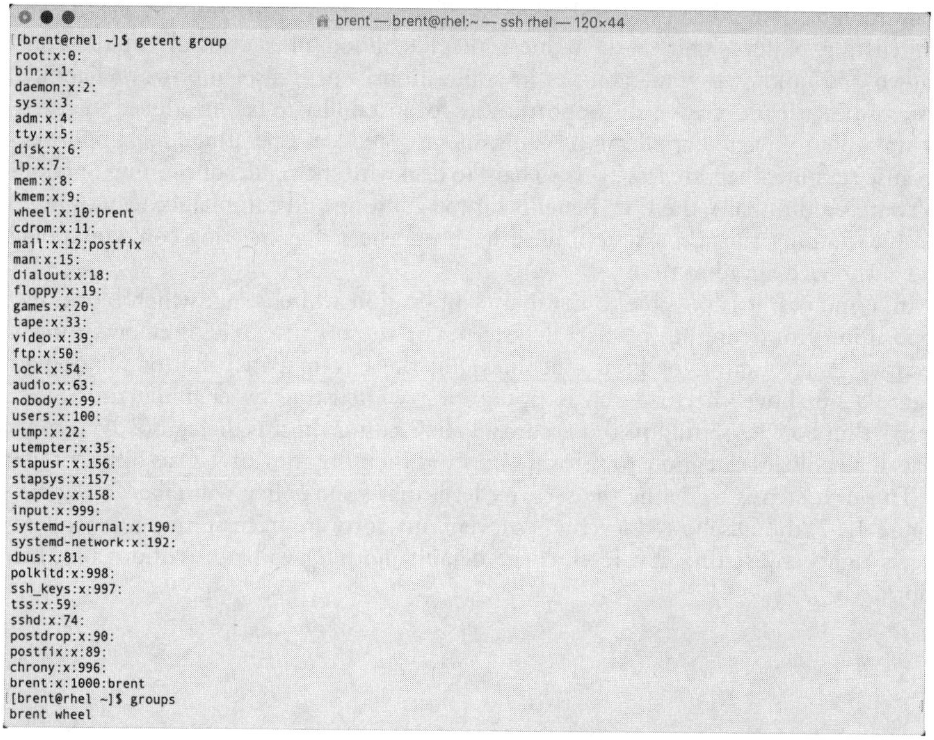

Figure 12-5 Listing of groups in a Red Hat Linux operating system

Blacklisting

For many decades, the vast majority networks operated on the concept of inherent trust—trust that the devices on the network belonged there and should have unfettered access to resources across the environment. As reality set in, defenders moved quickly to find ways to identify qualities that it could use to make decisions about access. With blacklisting, administrators and security teams can simply exclude known threats from the system. Although this may appear straightforward, the effectiveness of blacklisting relies heavily on the condition that the list of known threats is current and accurate. This turns out to be a major Achilles' heel to security operations, because potentially abusive activity will always be permitted until it's explicitly identified as bad. For this reason, blacklisting is not really effective against new actors and attack patterns never seen before, such as zero-day threats.

Whitelisting

For organizations looking to take a more proactive approach to malware mitigation, whitelisting is among the most effective methods of preventing unfamiliar applications from installing and executing. By allowing only known good applications and preventing everything else from executing, whitelisting can effectively protect organizations against the introduction of known malware, variants of old malware, and even unknown malware. While the primary goal of application whitelisting is often to reduce the chances of malware infection, it brings several other benefits to a security program. A well-designed whitelisting policy significantly reduces the likelihood of successful exploitation of known and unknown vulnerabilities in applications, but it also improves reliability of systems because it decreases the opportunities for instability to be introduced to a system by way of an unvetted applications. This, in turn, reduces operating and support costs because staff members are less likely to have to deal with the results of running unfamiliar software. Additionally, there are benefits for the auditing and compliance team, since, by default, nothing runs on a system until it's been assessed as meeting company policies and authorized ahead of time.

In Windows, it is possible to configure application whitelisting, which blocks every application from running by default, except for those you explicitly allow, using the Security Policy Editor for local machines and the Group Policy Editor for domains. Figure 12-6 shows the first step in application whitelisting by enabling the Software Restriction Policies setting in the Security Policy Editor. In this dialog box, you can see that these policies can apply to software files and their libraries, as well as different users.

The next step is to define the security level that your policy will take. As shown in Figure 12-7, the Disallowed level will prevent any software from running, regardless of access rights. By setting this level as the default, nothing will run without first being approved.

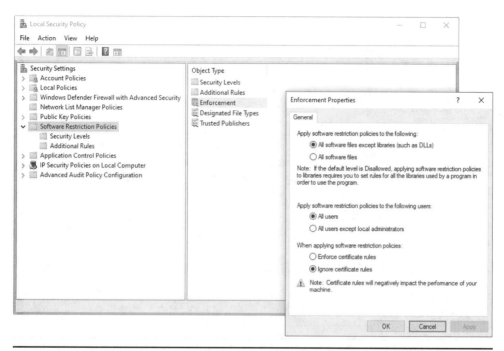

Figure 12-6   Enforcement properties of the Software Restriction Policies in Windows 10

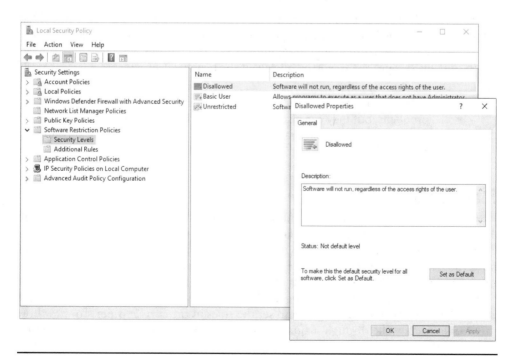

Figure 12-7   Security levels of the Software Restriction Policies in Windows 10

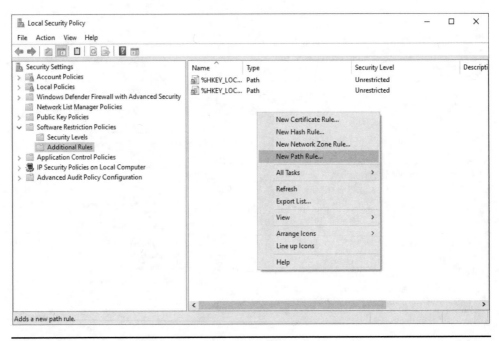

Figure 12-8   Creation of whitelisting rules within Software Restriction Policies in Windows 10

To specify the applications that can be run, we can add the application paths. New rules must be specified for each new application that the user or administration wants to add using the functionality shown in Figure 12-8. Although this will take some time to set up initially, this process is worthwhile given the benefits it provides.

NOTE   The terms "blacklist" and "whitelist" apply to the execution of programs, but they can also be used in describing the restrictions around access to a resource on the network, IP traffic, Internet domains, and even receipt of e-mail.

Firewalls

Firewalls use explicit rules that control how they handle traffic, with many set to deny all incoming connections and allow all outgoing connections by default. Anyone attempting to reach your machine will thus be unable to connect, while any outbound attempts from the same machine would be able to reach the outside world. Sometimes, firewalls come with no default, in which case these baseline rules must be added.

Setting up firewalls depends on the hardware or software used, but the concepts are always the same: enable the firewall and set the rules in the correct order. To show how this works, we'll walk through a quick setup using ufw, or Uncomplicated Firewall.

```
○ ● ●                    🏠 brent — brent@budgie-dev: ~ — ssh budgie — 96×32
[brent@budgie-dev:~$ sudo ufw default deny incoming                                    ]
Default incoming policy changed to 'deny'
(be sure to update your rules accordingly)
[brent@budgie-dev:~$ sudo ufw default allow outgoing                                   ]
Default outgoing policy changed to 'allow'
(be sure to update your rules accordingly)
[brent@budgie-dev:~$ sudo ufw allow ssh                                                ]
Rules updated
Rules updated (v6)
[brent@budgie-dev:~$ sudo ufw enable                                                   ]
[Command may disrupt existing ssh connections. Proceed with operation (y|n)? y         ]
Firewall is active and enabled on system startup
[brent@budgie-dev:~$ sudo ufw allow 80                                                 ]
Rule added
Rule added (v6)
[brent@budgie-dev:~$ sudo ufw allow https                                              ]
Rule added
Rule added (v6)
[brent@budgie-dev:~$ sudo ufw status verbose                                           ]
Status: active
Logging: on (low)
Default: deny (incoming), allow (outgoing), disabled (routed)
New profiles: skip

To                      Action      From
--                      ------      ----
22/tcp                  ALLOW IN    Anywhere
80                      ALLOW IN    Anywhere
443/tcp                 ALLOW IN    Anywhere
22/tcp (v6)             ALLOW IN    Anywhere (v6)
80 (v6)                 ALLOW IN    Anywhere (v6)
443/tcp (v6)            ALLOW IN    Anywhere (v6)
```

Figure 12-9   Basic setup of Uncomplicated Firewall (ufw) from the command line

This command-line software serves as an interface to iptables, an IP packet filter utility that interfaces directly with the Linux kernel. As shown in Figure 12-9, we begin by explicitly denying all incoming traffic that does not already have an established connection, and allowing outgoing traffic. The software provides feedback that the default policy was changed. Adding basic rules using ufw is as simple as calling the utility and providing the rule (allow, deny, or reject), the direction (incoming, outgoing, or routed), and the policy details. Rules can be applied to specific IP addresses, subnets, ports, protocols, or any combination of these.

For this exercise, we'll allow the SSH service and enable the firewall. The feedback lets us know that the rules have been implemented and the firewall is active. We can continue to add simple rules and see the full list of rules by asking for the firewall's status with **sudo ufw status verbose**. Checking the iptables configuration, we can see that the rules recently added via ufw also reflect in iptables, as indicated in Figure 12-10, with the command **sudo iptables –S**. In the case of iptables and many other firewalls, the order of the rules is important. Each rule in a list is interpreted in order as a packet comes in for inspection and verdict. In iptables, rules are processed in sequential order as presented in

```
●●●                    brent — brent@budgie-dev: ~ — ssh budgie — 108×32
-A ufw-before-input -p icmp -m icmp --icmp-type 3 -j ACCEPT
-A ufw-before-input -p icmp -m icmp --icmp-type 11 -j ACCEPT
-A ufw-before-input -p icmp -m icmp --icmp-type 12 -j ACCEPT
-A ufw-before-input -p icmp -m icmp --icmp-type 8 -j ACCEPT
-A ufw-before-input -p udp -m udp --sport 67 --dport 68 -j ACCEPT
-A ufw-before-input -j ufw-not-local
-A ufw-before-input -d 224.0.0.251/32 -p udp -m udp --dport 5353 -j ACCEPT
-A ufw-before-input -d 239.255.255.250/32 -p udp -m udp --dport 1900 -j ACCEPT
-A ufw-before-input -j ufw-user-input
-A ufw-before-output -o lo -j ACCEPT
-A ufw-before-output -m conntrack --ctstate RELATED,ESTABLISHED -j ACCEPT
-A ufw-before-output -j ufw-user-output
-A ufw-logging-allow -m limit --limit 3/min --limit-burst 10 -j LOG --log-prefix "[UFW ALLOW] "
-A ufw-logging-deny -m conntrack --ctstate INVALID -m limit --limit 3/min --limit-burst 10 -j RETURN
-A ufw-logging-deny -m limit --limit 3/min --limit-burst 10 -j LOG --log-prefix "[UFW BLOCK] "
-A ufw-not-local -m addrtype --dst-type LOCAL -j RETURN
-A ufw-not-local -m addrtype --dst-type MULTICAST -j RETURN
-A ufw-not-local -m addrtype --dst-type BROADCAST -j RETURN
-A ufw-not-local -m limit --limit 3/min --limit-burst 10 -j ufw-logging-deny
-A ufw-not-local -j DROP
-A ufw-skip-to-policy-forward -j DROP
-A ufw-skip-to-policy-input -j DROP
-A ufw-skip-to-policy-output -j ACCEPT
-A ufw-track-output -p tcp -m conntrack --ctstate NEW -j ACCEPT
-A ufw-track-output -p udp -m conntrack --ctstate NEW -j ACCEPT
-A ufw-user-input -p tcp -m tcp --dport 22 -j ACCEPT
-A ufw-user-input -p tcp -m tcp --dport 80 -j ACCEPT
-A ufw-user-input -p udp -m udp --dport 80 -j ACCEPT
-A ufw-user-input -p tcp -m tcp --dport 443 -j ACCEPT
-A ufw-user-limit -m limit --limit 3/min -j LOG --log-prefix "[UFW LIMIT BLOCK] "
-A ufw-user-limit -j REJECT --reject-with icmp-port-unreachable
-A ufw-user-limit-accept -j ACCEPT
```

Figure 12-10   Output of iptables listing the active firewall rules on the system

the rules file, with an action occurring if there is a match for a rule. In the case of a match, no other rules will be processed for that packet.

In this walkthrough, you can see that ufw automatically populates the correct port for the SSH and HTTPS services. It does this by quickly referencing the /etc/services file on our Linux machine to see the ports associated with those services. This file serves as an excellent reference for well-known ports, or those ports that are normally reserved for specific services. A snapshot of the /etc/service file is shown in Figure 12-11.

TIP   When building firewall rules, there are two approaches: either deny everything by default and explicitly allow selected traffic, or allow everything first and deny selected traffic. Blocking specific traffic, rather than blocking all traffic by default, means that your organization will have to know exactly how bad traffic arrives for this approach to be most effective. Not only is it hard to develop and maintain a complete list of ports that need to be blocked, but it takes a significant amount or work in an environment that is subject to change frequently. It's far more efficient and safer to deny all traffic by default and allow only the specific communication that you need to pass through.

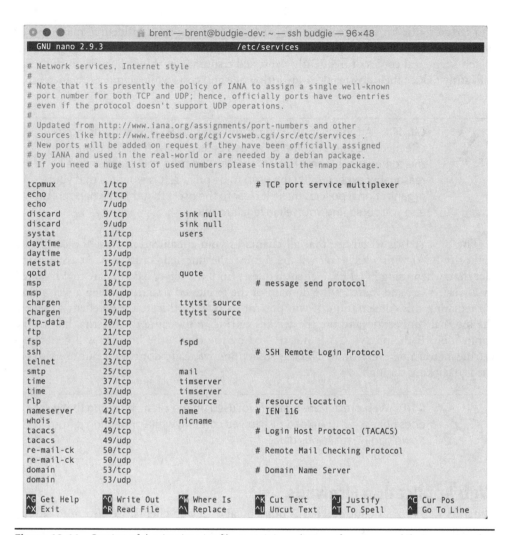

Figure 12-11 Portion of the /etc/service file containing a listing of services and their associated port numbers

Web Proxies

A *web proxy* is a system that intercepts and then forwards web traffic between clients and servers. Such proxies are commonly used to carry out content filtering to ensure that Internet use conforms to the organization's acceptable-use policy. They can block unacceptable web traffic, provide logs with detailed information pertaining to the sites specific users visited, monitor bandwidth usage statistics, block restricted website usage, and screen traffic for specific keywords (such as adult content, confidential markings, or Social Security numbers). The proxy servers can be configured to act as caching servers, which keep local copies of frequently requested resources, enabling organizations to significantly reduce their upstream bandwidth usage and costs while significantly increasing performance.

 CAUTION Although most web proxies support HTTPS traffic, doing so effectively requires additional steps. For starters, you will be examining the contents of a "conversation" that is encrypted and can therefore be reasonably expected (by the user) to be private. It is essential that your organizational policies make it clear to the users that this can happen, or else you could find yourself in legal trouble.

The next step is to ensure that all clients in your organization trust the certificate authority (CA) with which you will be signing the internal certificates. At issue is the fact that, when using HTTPS, a client requests from the server a certificate that is issued by a trusted CA and matches the domain of the requested resource. When a web proxy is mediating this conversation, it will present its own certificate to the client to secure the internal connection and use the server's certificate to secure the external half of the connection. If the proxy's CA is not trusted by the client, the browser will generate a certificate warning every time, which is something we really don't want our users to get used to clicking through.

 NOTE Web proxies are typically focused on the client, ensuring that it does not access or upload disallowed content, while protecting it from downloading malicious data.

Web Application Firewalls

Whereas web proxies control what is done by and to web clients, we also need similar protections for our servers so that we have assurance that external clients won't be able to attack them easily. A *web application firewall* (WAF) is a system that mediates external traffic to a protected server. The WAF is configured for the specific web apps (or classes of web apps) that it is intended to protect. In other words, the WAF "speaks the language" of the web app so as to identify unusual or disallowed requests to it. It is able to determine which URLs, directories, and parameters are acceptable and which are suspicious. This is something that traditional firewalls cannot do.

Operating System Firewalls

Every major operating system comes with its own version of a firewall that can, and should, be used in addition to network-based firewalls. Windows, macOS, and Linux all have built-in firewall options as well as support for third-party offerings. Windows Defender, Microsoft's unified security solution for its operating systems, includes a

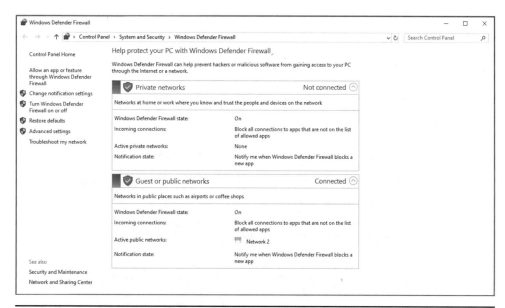

Figure 12-12   Windows Defender Firewall status screen in Windows 10

host-based utility called Windows Defender Firewall, shown in Figure 12-12, which was introduced as a component to the Windows XP operating system (and known at the time as the Internet Connection Firewall). Since then, its functionality has been improved along with support for more granular control over traffic rules. As an administrator of an Active Directory domain, you can manage Windows Defender Firewall settings via Group Policies. This will enable deployment of consistent rules across the network to hosts and Windows servers, with the capability to prevent those having local administrator permissions from disabling the feature.

In the macOS environment, the built-in application firewall can be accessed from the Security & Privacy section of the System Preferences. As Figure 12-13 shows, the configuration screen of the firewall enables users to adjust filtering settings that apply to the local machine. Note that macOS includes an option to block all incoming connections, and while this may be useful for certain situations, doing so will result in services such as messaging, file sharing, and media streaming to fail. Another useful feature of the application firewall is the ability to whitelist applications automatically if they are signed by a valid CA. Rather than prompting the user to authorize them every time, these programs are automatically added to the list of allowed apps. Applications included with macOS, for example, are signed by Apple and are allowed to receive incoming connections when this setting is enabled. As an application firewall, this functionality works in parallel with any IP-based firewall solutions that exist on the host or the network.

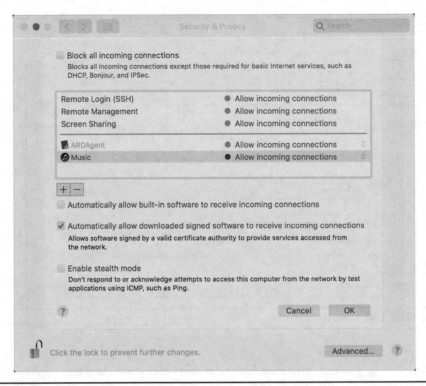

Figure 12-13 The macOS 10.15 firewall options

Intrusion Prevention System Rules

A key function of protecting the network is the recognition of suspicious behavior across the network or on the hosts. The purpose of intrusion detection systems (IDSs) is to identify suspicious behavior. Available as a software product or a hardware appliance, an IDS works by regularly reviewing network events and analyzing them for signs of intrusion. IDSs aren't just firewalls, which allow or deny traffic based on prescribed rulesets. Rather, IDSs work either by matching to previously identified malicious activity or by analyzing network traffic for indicators. Some IDSs have additional functionality to act on recognition of malicious activity and to stop traffic or quarantine hosts. These devices are called intrusion prevention systems (IPSs). If the IDS/IPS is set up to work across a network, it can also be referred to as a network IDS (NIDS), whereas a system that operates on an individual host is called a host IDS (HIDS).

Among the most commonly deployed IPS technologies are Zeek (formerly known as Bro), Suricata, and Snort. Though Zeek and Snort were not originally designed to be used strictly as IPSs, their powerful scripting languages certainly allow for it. Suricata, a free and open source tool, was designed to inspect network traffic using a similar set of rules and signature language. Suricata stands out in that it can handle multiple gigabits of traffic, it provides an intuitive user interface, and it's capable of sending alert messages through a number of methods.

Snort Rule Building

Snort's scripting language, though not as powerful as the others, is still flexible and sufficient to stop any network threat for which you can develop a signature. Snort does not automatically log everything it sees on the network, which may be attractive if you have limited means to store large amounts of event data.

Snort rules are highly customizable and enable creators to communicate relevant information in both and human- and machine-readable formats. The basic structure of a Snort rule is shown here:

```
<snort action> <protocol> <src IP> <src PORT> <direction> <dst IP>
<dst port> (msg:"Description of what's being detected and why"; <optional
classtype> ;<optional snort ID (sid)>; <optional revision (rev)
number>;)
```

The key to good rule-making is to make them as focused as possible for each entry. The following rule, for example, will flood an alert queue because it's way too broad and will fire on every web connection:

```
alert tcp any any -> any 80 (msg:"A web connection was made!";
flow: stateless; rev:1;)
```

Instead, we should try to scope the rule down to affect a specific type of traffic. The next rule alerts on any occurrence of DNS requests to a .buzz domain. According to Spamhaus, an international organization that tracks spam activity, the .buzz domain was the number one abused top-level domain, with nearly two-thirds of those domains being classified as malicious:

```
alert udp $HOME_NET any -> any 53 (msg:"Suspicious DNS requests for a .buzz
domain"; flow:to_server; byte_test:1,!&,0xF8,2; content:"|02|buzz|00|";
reference:url,www.spamhaus.org/statistics/tlds; sid:12345; rev:1;)
```

Updating Snort rules is straightforward in smaller networks but can be a challenge with many sensors. Fortunately, there are several tools available to use in lieu of manual updates. PulledPork and OinkMaster are two solutions that automatically install the latest ruleset based on the currently installed version of Snort. Written in Perl, these scripts can be initiated by a cron job to download, parse, and modify rules. Both include a useful feature to maintain the state of any rule throughout updates. In other words, there's no need to comment out a troublesome or noisy rule with every update; the scripts will detect which rules are currently disabled and keep the same status with subsequent upgrades.

Zeek Logs

At its core, Zeek does two things: it captures all sorts of events (labeling them neither good nor bad) and then runs scripts that analyze the events looking for anomalies that may indicate a security incident. These scripts can take actions ranging from sending a warning message to changing configurations on systems in order to thwart a threat. In addition to built-in anomaly detection, Zeek also provides a signature language for doing

Log File	Description	Field Descriptions
conn.log	TCP/UDP/ICMP connections	Conn::Info
dhcp.log	DHCP leases	DHCP::Info
dns.log	DNS activity	DNS::Info
ftp.log	FTP activity	FTP::Info
http.log	HTTP requests and replies	HTTP::Info
irc.log	IRC commands and responses	IRC::Info
kerberos.log	Kerberos	KRB::Info
modbus.log	Modbus commands and responses	Modbus::Info
mysql.log	MySQL	MySQL::Info
ntlm.log	NT LAN Manager (NTLM)	NTLM::Info
ntp.log	Network Time Protocol	NTP::Info
radius.log	RADIUS authentication attempts	RADIUS::Info
rdp.log	RDP	RDP::Info
sip.log	SIP	SIP::Info
smb_cmd.log	SMB commands	SMB::CmdInfo
smtp.log	SMTP transactions	SMTP::Info
snmp.log	SNMP messages	SNMP::Info
socks.log	SOCKS proxy requests	SOCKS::Info
ssh.log	SSH connections	SSH::Info
ssl.log	SSL/TLS handshake info	SSL::Info
syslog.log	Syslog messages	Syslog::Info

Table 12-1 Network Protocol Log File Automatically Generated by Zeek

low-level pattern matching similar to that done with Snort. What's noteworthy is that despite the support for signatures, it's not Zeek's preferred detection method. Table 12-1 includes a sampling of the network protocols logged by Zeek.

Suricata Rule-Building

Suricata was built to be a high-performance IDS, IPS, and network monitoring tool. As a result, it has a full feature set of inspection, detection, and network capture capabilities. Suricata is highly scalable in its ability to run many processing threads as well as its use of Lua scripting to perform advanced pattern matching.

Suricata rules consist of three components that work together to deal with traffic extremely quickly. The first component is the *action* that the software will take upon a match of the rule. These actions are pass, drop, reject, and alert. For the *pass* action, the scanning engine will complete scanning of the packet and proceed to the next packet, skipping all rules that may follow. For *drop*, the engine silently discards the packet without any notification to the sender. In contrast, *reject* actively denies the packet and responds with a reset or ICMP-error packet. Finally, a packet may generate an *alert* action, which will notify the administrator of the event. The second component of the rule is the *header*, which defines the protocol, addresses, ports, and direction of the rule. Suricata lower level protocols such as TCP, UDP, ICMP and IP, as well as application layer protocols such as HTTP, FTP, SMTP, and DNS, may be enabled or disabled in the Suricata configuration file to save resources. Finally, the *rule options* component is used to add definition to the rule.

TIP   Writing custom rules is a great way to defend your organization, because they will be the most relevant to your network. It's very important to maintain the correct syntax, noting where commas, semicolons, and parentheses are expected. Failure to include those could mean that the rule may not work at all.

Host-Based Intrusion Prevention Systems

As the name implies, a host-based IPS (HIPS) inspects and responds only to the traffic moving in and out of a host's network interfaces. Unlike its network-based brethren, the protection is afforded to one device only. Although this may seem like a limitation, it actually enhances the overall security posture by enabling the HIPS to become finely tuned to the traffic and pattern at one specific host. This is doubly true of an HIPS, which doesn't just rely on signatures but also incorporates behavioral or heuristic approaches.

The only reason why you *wouldn't* have HIPSs deployed across your organization is if your organization doesn't have the resources (for example, time, money, or personnel) to install and maintain them properly. Apart from the direct costs for the licenses, an anomaly-based HIPS can require a significant amount of human supervision during the training period (that is, the period of time it takes the system to learn what "right" looks like). During this period, you may also have to deal with loss of productivity as the HIPS incorrectly classifies benign traffic as malicious. Some of these costs can be mitigated by ensuring that your solution includes centralized management and monitoring capabilities.

Data Loss Prevention

Digital communication increases the speed and ease of interaction, particularly across great distances. Combine this with the decreasing cost of storage, and the security risks are apparent. Even in the days before high-capacity, removable storage and high-speed connections, unauthorized removal of data from organizations was the bane of auditors and security professionals. What's more, the increasing trends related to data spillage are not restricted to actors who are knowingly malicious. In fact, many employees don't realize that it may be inappropriate to store the company's sensitive materials on a personal device or in unapproved cloud-based storage. At the core of the challenge is a tension between IT and the security team's responsibility to ensure that data is protected when stored, accessed, and transmitted, while simultaneously allowing the unencumbered flow of data to support business operations.

Without a complete digital *and* physical lockdown of data, total prevention of data loss is all but impossible. However, with a well-defined approach to dealing with data in use, data in transit, and data at rest, security teams may be able to reduce the risk of spillage to a manageable and acceptable level. Data loss prevention (DLP) technology is a major component of the measures to reduce the occurrences of intentionally and inadvertently leaked corporate data. Many DLP solutions work in a similar fashion to IDSs by inspecting the type of traffic moving across the network, attempting to classify it, and making a go or no-go decision based on the aggregate of signals. After all, if you can detect a certain type of traffic coming in, you ought to be able to do the same for outbound traffic. Early iterations of DLP were most effective with predictable patterns such as Social Security and credit card numbers. These techniques are less useful for more complex patterns, however, or in the presence of complicated encoding or encryption methods, in which case deep inspection of packets becomes much more difficult.

Some Software as a Service (SaaS) platforms feature DLP solutions that can be enabled to help your organization comply with business standards and industry regulations. Microsoft Office 365 and Google G Suite are two notable services that make DLP solutions available to be applied to media such as e-mails, word processing documents, presentation documents, spreadsheets, and compressed files. Many of these solutions require the configuration of rules that take into account data content, such as the exact type of data you're looking to monitor for, as well as the context, such as who is involved the data exchange. With SaaS solutions, your organization may have to expose at least part of its data with the service provider, which may introduce unacceptable exposure in itself. Cloud service providers are in a unique position because they have to comply with privacy laws and data protection legislation that applies to data that is not their own. It's in their best interest to use DLP, in addition to many other technologies, to protect data in use, in transit, and at rest. It may be incumbent on you, as your organization's security lead, to read, understand, and communicate a cloud provider's terms of service and policies regarding data protection.

Endpoint Detection and Response

Endpoint detection and response (EDR) solutions take the monitoring and detection aspects of traditional antimalware technologies and add investigation, response, and mitigation functionalities. EDR is an evolution in security tools designed to protect the endpoints in a network. Given that these solutions operate at the edges of the network with the requirement to work on a diverse set of systems and configurations, EDRs solutions are often built to be highly scalable, lightweight, and cross-platform. EDRs include an array of tools that provide the following capabilities:

- **Monitor**   Log and aggregate endpoint activity to facilitate trend analysis.
- **Detect**   Find threats with the continuous analysis of monitored data.
- **Respond**   Address malicious activity on the network by stopping activity or removing the offending asset.

Though an EDR solution may allow for faster remediation of a security issue, its use results in some disadvantages. EDR products collect a lot of data on an endpoint—everything from operating system logs, application activity, process activity, and user interactions to file changes. Storing this information for a large organization will require significant storage and compute resources. Accordingly, the data alone doesn't do much for security if there are no analysts to consume the data and make adjustments to ensure that the response actions are appropriate and effective. Running an EDR without the necessary support staff is a recipe for disaster. Understanding the goals of the security plan will help you determine if prevention is a factor in your EDR decision versus strictly remediation. Depending on the use cases, EDR can replace existing technologies or serve to augment them. It's important to be able to determine which EDR solution will be right for you.

Network Access Control

In an effort to enforce security standards for endpoints beyond baselining and group policies, engineers developed the concept of Network Access Control (NAC). NAC provides deeper visibility into endpoints and enables you to perform policy enforcement checks before the device is allowed to connect to the network. NAC ties in features such as role-based access control (RBAC), verification of endpoint malware protection, and version checks to address a wide swath of security requirements. Some NAC solutions offer transparent remediation for noncompliant devices. In principle, this solution reduces the need for user intervention while streamlining security and management operations for administrators.

There are, however, a few concerns about NAC that affect both user privacy and network performance. Some of NAC's features, particularly version checking and remediation, can require enormous resources. Imagine several hundred noncompliant nodes joining the network simultaneously and all getting their versions of Adobe Flash, Java, and Internet Explorer updated at once. This often means the network takes a hit—plus, the users may not be able to use their machines while the updates are applied. Furthermore, NAC normally requires some type of agent to verify the status of the endpoint's software and system configuration. This collected data can have major implications on user data privacy should it be misused.

As with all other solutions, we must take into consideration the implications for both the network and the user when developing policy for deployment of NAC solutions. The IEEE 802.1X standard was the de facto NAC standard for many years. While it supported some restrictions based on network policy, its utility diminished as networks became more complex. Furthermore, 802.1X solutions often delivered a binary decision on network access: either permit or deny. The increasing number of networks transitioning to support "bring your own device" (BYOD) required more flexibility in NAC solutions. Modern NAC solutions support several frameworks, each with its own restrictions, to ensure endpoint compliance when attempting to join the enterprise network. Administrators may choose from a variety of responses that modern NAC solutions provide in the case of a violation. Based on the severity of the incident, they may completely block the device's access, quarantine it, generate an alert, or attempt to remediate the endpoint. We'll look at some of the most commonly used solutions next.

Time-Based Solution

Does your remote network access need to be active at 3 A.M.? If not, then a time-based network solution could be right for you. Time-based solutions can provide network access for fixed intervals or recurring timeframes, and they can enforce time limits for guest access. Some more advanced devices can even assign different time policies for different groups.

Rule-Based Solution

NAC solutions will query the host to verify operating system version, security software version, the presence of prohibited data or applications, or any other criteria as defined by the list of rules. These rules may also include hardware configurations, such as the presence of unauthorized storage devices. Often, they can share this information back into the network to inform changes to other devices. Additionally, many NAC solutions are capable of operating in a passive mode, running only as a monitor functionality and reporting violations when they occur.

Role-Based Solution

In smaller networks, limiting the interaction between nodes manually is a manageable exercise, but as the network size increases, this becomes exponentially more difficult, in part because of the variety of endpoint configurations that may exist. Using RBAC, NAC solutions can assist in limiting the interaction between nodes to prevent unauthorized data disclosure, either accidental or intentional. As discussed, RBAC provides users with a set of authorized actions necessary to fulfill their roles in the organization. NAC may reference the existing RBAC policies using whatever directory services are in use across the network and enforce them accordingly. This helps with data loss prevention (DLP), because the process of locating sensitive information across various parts of the network becomes much faster. NAC, therefore, can serve as a DLP solution, even when data has left the network, because it can either verify the presence of host-based DLP tools or conduct the verification itself using RBAC integration.

Location-Based Solution

Along with the surge in BYOD in networks, an increasing number of employees are working away from the network infrastructure, relying on virtual private network (VPN) software or cloud services to gain access to company resources. NAC can consider device location when making its decision on access, providing the two main benefits of identity verification and more accurate asset tracking.

Sinkholing

Sinkholing is a technique used to mitigate malicious and abusive traffic by routing it to an internal server or dropping it altogether. Like DNS spoofing, sinkholing provides a response to a DNS query that does not resolve to the actual IP address. The difference, however, is that DNS sinkholes target the addresses for known malicious domains, such as those associated with a botnet, and return an IP address that does not resolve correctly or that is defined by the administrator.

Here's a scenario of how this helps with securing a network: Suppose you receive a notice that a website, www.malware.evil, is serving malicious content and you confirm that several machines in your network are attempting to contact that server. You have no way to determine which computers are infected until they attempt to resolve that hostname, but if they are able to resolve it, they may be able to connect to the malicious site and download additional tools that will make remediation harder. If you create a DNS sinkhole to resolve www.malware.evil to your special server at address 10.10.10.50, you can easily check your logs to determine which machines are infected. Any host that attempts to contact 10.10.10.50 is likely to be infected, because there would otherwise be no reason to connect there. All the while, the attackers will not be able to further the compromise because they won't receive feedback from the affected hosts.

Malware Signatures

Signature-based detection remains a common form of malware mitigation, and when combined with other detection techniques, it can have a major impact on defenses. For decades, both network- and host-based security mechanisms have worked by examining data for specific patterns. In the case of network traffic, security appliances would be on the lookout for indicators of compromise from packet data, while host-based detection technologies might rely on finding certain strings of bytes in a file that are associated with previously observed malware.

YARA rules are useful for identifying and classifying malware based on a well-defined rule-based approach. These rules were designed to be easy to understand and consist of two parts: the strings definition and the condition. The strings definition portion specifies the patterns that will be searched for in the file. Each string is described with identifier consisting of a $ character followed by a name. Text, hexadecimal, and regular expression (regex) characters can be used to describe the pattern. The condition section defines the logic under which the rule is rule, using Boolean expressions. Here's an example of a simple rule that will be true if either the specified text or hexadecimal patterns match:

```
rule MyGreatRule
{
    strings:
        $text_string1 = "malware"
        $hex_string1 = { A1 12 B2 34 C3 46 }
    condition:
        $text_string1 or $hex_string1
}
```

The Yara Rules Project (https://github.com/Yara-Rules/) is a public repository where different YARA signatures are compiled, classified, and updated. Another useful resource is the Exploit Database (http://www.exploit-db.com). Although it's primarily used as a collection of software vulnerabilities and their associated exploits, the database is also an excellent resource for security professionals looking to understand how software may be compromised and planning to begin writing their own methods of detection.

 EXAM TIP For the CySA+ exam, it will be important for you to understand the pros and cons of using signature-based methods and the techniques attackers have adopted to evade these technologies.

Sandboxing

Modern malware has become increasingly sophisticated in several ways that make it particularly difficult for signature-based systems to detect and stop attacks. In addition to the constant threat of zero-days, security analysts should be familiar with defenses against malware with polymorphic capabilities. *Polymorphic* malware is able to change certain characteristics about itself by modifying its own code to make it impossible for signature-based detection solutions to discover it.

In Chapter 4, we touched on the concept of dynamic analysis as a way to assess the security of software based on the behaviors the executable performs. Many of the same techniques used for analyzing known software for bugs can be used to find out what unknown software might be trying to do. Many security products offer *sandboxes*, or specially instrumented virtual environments, to perform basic dynamic analysis of unknown software. Often used at scale, sandboxes provide a safe environment to observe and even "detonate" malware without fear of damaging production systems. Most sandboxes work in a similar fashion: they rapidly spin up preconfigured operating systems, execute the software, observe what's happening, and report on what was seen. With sandboxes, analysts can monitor network traffic, examine processes, and compare both to normal baseline behaviors to discover activity deviations more easily. From here, they can take steps to determine whether the software in question is legitimate or malicious by checking memory artifacts, log files, and system settings for suspicious changes.

As a security analyst, you should be aware that sandboxes do have a few drawbacks, which you should consider when formulating and executing your defense plan. First, sandboxes can be costly to run, and if your team opts for a custom solution, it may take significant time to set it up and instrument it correctly. With regard to customization, it's often difficult for malware to be executed in automated sandboxes with additional options enabled, as one might observe in actual usage of the software. Additionally, even with the most comprehensive setup, sandboxes may not record all events. Finally, modern malware is very smart—it will often detect when it is running in a virtual machine and will behave differently if such an environment is detected. Sometimes the malware will check for legitimate data to be present on a system before executing. While sandboxes may not be able to report definitively that a piece of software is malware, it does provide insight into what the software is trying to do so that analysts like you can take the appropriate steps.

Port Security

We're always amazed at how many consumer products designed to be connected to a network leave unnecessary services running, sometimes with default credentials. If you do not require a service, it should be disabled. Running unnecessary services not only means more avenues of approach for an adversary, but it also means more work to administer. Services that aren't used are essentially wasted energy, because the computers will still run the software, waiting for a connection that is not likely to happen by a legitimate user.

 EXAM TIP For the exam, you'll need to remember a few things about ports. The UDP and TCP ports between 0 and 1023 are known as the *well-known ports* because they are used for commonly used services. Some notable well-known ports are 20 (FTP), 22 (SSH), 25 (SMTP), and 80 (HTTP). Ports 1024 to 49151 are *registered* ports, and ports above 49151 are *ephemeral* or *dynamic* ports.

Chapter Review

Traditional network-based firewalls were historically the workhorses of a security program, continuously checking for undesirable traffic as it came into the network and filtering out those packets. Over time, however, the model of simply deploying an appliance has proven insufficient to handle the scale and complexity of modern network environments. Although modern firewalls still check, control, and block incoming or outgoing network traffic, additional options are now available that operate at different layers of the network, offering advanced introspection capabilities and integrating seamlessly with other network security devices. A comprehensive approach that uses the principle of least privilege at it relates to access, enhanced logging, and smart whitelisting rules can help keep adversaries out of your network.

Questions

1. Which of the following is a Linux command-line utility available in some distributions and used to configure the system's firewall?

   A. YARA

   B. ufw

   C. Snort

   D. Zeek

2. What kind of detection method works by examining data for specific patterns?

   A. Signature-based

   B. Role-based

   C. Rule-based

   D. Behavior-based

3. A colleague has sent you a large list of malicious domains that you want to prevent users from visiting. Additionally, you want to present a warning to those users that the site they are attempting to visit is known to be malicious and give them options to participate in security training. What is the name of technique that you'll use to initially prevent them from accessing those sites?

    A. DNS relay

    B. DNS masquerading

    C. DNS sinkholing

    D. DNS spoofing

Use the following scenario to answer Questions 4–8:

Your security team has been asked to develop configurations for a new, high-security research and development environment within your network. Access to this network is restricted to those with the appropriate levels of clearance, because it contains detailed plans for the next five years' worth of products.

4. Which approach do you choose to control the kinds of programs that are executed in this environment?

    A. Whitelisting

    B. Blacklisting

    C. Firewalling

    D. Sinkholing

5. What solution may assist with protecting against development plans and schematics leaving the network as a result of unauthorized access to your system?

    A. Firewall

    B. DLP

    C. IPS

    D. IDS

6. Citing multiple benefits, a colleague suggests the use of a web proxy. Which one of the following statements about web proxy servers is correct?

    A. The system is unable to restrict content.

    B. The system can block malicious traffic.

    C. The system reduces speeds across the network.

    D. The system cannot increase available bandwidth.

7. Your team needs to be able to subscribe to feeds of intrusion detection rules or create their own rules. Which of the following technologies will *not* satisfy this requirement?

   A. Windows Defender

   B. Snort

   C. Zeek

   D. Suricata

8. Because of the nature of the network, the team often comes across unknown files. What technique may be useful to evaluate files for malicious or suspicious behaviors?

   A. Signatures

   B. Firewall

   C. Sandboxing

   D. IDS

Answers

1. **B.** The ufw utility is a popular Linux firewall utility created to simplify configuration of the iptables.

2. **A.** Signature-based methods can be used to quickly identify specific versions of malware packages that have already been seen.

3. **C.** You can use DNS sinkholing to send traffic visitors who attempt to visit a known bad domain to an alternate address, which acts as the sinkhole for that traffic.

4. **A.** Whitelisting is a proactive approach to malware mitigation; whitelisting allows only known-good applications to run and prevents everything else from executing.

5. **B.** A DLP control is used to identify and block attempts at exfiltration of sensitive information out of the network.

6. **B.** Web proxies are versatile and capable of acting as filters for malicious traffic and unwanted network content. Additionally, they may be able to increase the speed of loading web pages by creating local copies of frequently accessed sites.

7. **A.** Although Windows Defender is an extremely effective security tool, it is an antimalware technology component of the Windows operating system. Snort, Zeek, and Suricata are all intrusion detection and prevention systems.

8. **C.** Sandboxes are capable of analyzing files for malicious or questionable behaviors.

The Importance of Proactive Threat Hunting

In this chapter you will learn:

- The key benefits of creating a threat-hunting capability
- The process for threat hunting
- Tactics for threat hunting
- How to integrate threat hunting results into other security operations

A hunt based only on trophies taken falls far short of what the ultimate goal should be.

—Fred Bear

The majority of security concepts we cover in this book are reactive. Whether they involve tuning endpoint detection tools to identify potential incidents or blocking traffic based on how it is attempting to traverse the network, these techniques rely on your being able to observe behavior and make a call on whether it is good or bad—as it is happening or shortly thereafter. Threat hunting, on the other hand, is a proactive and iterative approach to defense, rooted in a mindset that the attacker is already in your system. Whether true or not, this approach means making fewer potentially damaging assumptions about your organization's security posture. Furthermore, just because a breach isn't visible doesn't mean it hasn't already occurred—it just means that your traditional methods of detection have failed. Often categorized as a type of active defense, threat hunting requires analysts to see beyond alerts and dig deep to find malicious actors in the network that may have slipped past defenses. Primarily driven by a human, threat hunting can benefit tremendously from technologies such as machine learning (ML) and user and entity behavior analytics (UEBA). The practice can never be fully automated, however, because hunting often requires analysts to step into the minds of attackers and see things from their point of view. This kind of assumed perspective helps us as defenders in determining the most likely techniques used by an attacker, and, combined with intelligence about motivations, can help clarify what exactly we need to look for in our network to uncover them. The best threat-hunting efforts combine the data and capabilities of advanced security operations with the strong, individual analytical skills of those on the hunting team.

Though different from other key defense processes such as vulnerability management, incident response, digital forensics, and detection, threat hunting requires familiarity with these processes and the skills associated with them. As a result, many good threat-hunting teams tend to be quite diverse, comprising members with rich experience in all of these domains. Hunting also requires that analysts have enhanced awareness about their network, including an accurate catalog of sensitive assets as well as an understanding of normal traffic patterns. Like threat intelligence analysts, so much of what differentiates threat hunters is their ability to add context to raw data to tell a story. Unlike threat intelligence analysts, however, threat hunters look for what already might be present and unseen instead of what might occur in the future. Ideally, this practice helps us find adversaries hiding in the network well before attackers have the chance to fully step through a kill chain and fulfill their goals.

Threat hunting is similar to hunting wildlife in a couple of key ways. For success in either, the hunter must develop the means to identify and distinguish the patterns of the target, track that target activity throughout the environment, and eventually take a successful shot. Any successful hunting campaign, therefore, requires identification of the target and comprehensive preparation on the part of the hunter. Before embarking on a threat-hunting effort, you'll also need to have a comprehensive plan in place. Because so many time- and resource-consuming activities are associated with threat hunting, knowing what you're going to do and preparing for overall success is key to maximizing the chances of a fruitful effort. Like traditional hunters, analysts will put tremendous effort into preparing their tools, clarifying plans for how they will test their assumptions, and thinking about what to do after the hunt is complete.

At the very least, you should address the following questions before moving forward:

- What is the purpose of the hunt?
- Where will it be conducted?
- What resources do I need to conduct the hunt?
- Who are the key stakeholders?
- What is the desired outcome of the hunt?

You can visualize the entire threat-hunting process in several ways, but for the sake of this exam, we'll use a four-stage cycle that's quite similar to the *intelligence cycle* described in Chapter 1. Many organizations already conduct hunting for threats in a manner similar to what we describe, whether they realize it or not. Ensuring that the process is repeatable and scalable is easier with a framework such as the one shown in Figure 13-1.

Figure 13-1
A four-phase
threat-hunting
process

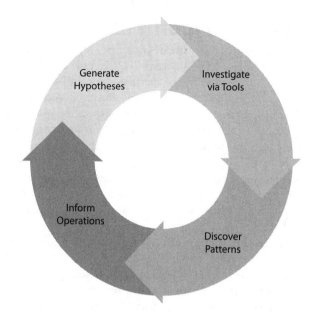

Generate Hypotheses

Investigate via Tools

Discover Patterns

Inform Operations

Assume Breach

For decades, security professionals operated with a primary focus on perimeter defense and incident response. Most believed that, with the right investment in outward-facing detection and prevention technologies, attackers would be thwarted most of the time. And in those rare occasions that something got through, incident responders would be able to hone in on and evict attackers quickly from the network. This mindset, as the community has come to learn, is flawed in many ways. Attackers, it turns out, can be patient, clever, and, in some cases, very well resourced. In many cases, attackers had more knowledge of technical details and flaws in a target network than the owners did. Years of increasingly successful and stealthy breaches proved to defenders that not only were attackers capable of launching a successful attack, but the defenders themselves weren't nearly as situationally aware as they had once believed.

Enter the concept of *assumption of breach*, which is a simple acceptance of the possibly that an attacker is already in the network and working toward an end goal. This comes with an inherent acknowledgement that, as we've covered throughout the book so far, no system is entirely free of vulnerabilities. Accordingly, security professionals cannot accurately claim that there haven't been intruders in their networks in a given timeframe.

Establishing a Hypothesis

With threat-hunting goals identified, your first step in the hunting effort is to develop a hypothesis (or hypotheses) for the hunt. Each hypothesis will be a simple statement with your ideas about the threats in the environment and how you foresee identifying and removing them. Hypotheses can be seen as educated guesses that need two key components to be valid: The first is some observable aspect that goes beyond an analyst's hunch. There must be some phenomena occurring, or not occurring, on the network that can be captured in a consistent way. This ensures that the event can be analyzed and compared in a consistent way. This leads to the second component of every good hypothesis—it must be testable. Good hunters will know what technologies to leverage to get the signal they need to begin analysis and testing of hypotheses. The initial signal for a hypothesis is often derived from one or more of the sources listed in Table 13-1.

Along with the two previously discussed components, keep in mind that, at some point, you may have to communicate your process, findings, or results to stakeholders. Your hypotheses, therefore, should be clear and concise. Attempting to move forward with an improperly formed hypothesis not only adds unnecessary confusion, but it may also result in dubious results. As with poor requirements in the intelligence cycle, poorly defined hypotheses can lead to substantial wastes of resources and time.

Type	Description	Example
Analytics-driven	Hypothesis informed by leveraging advanced technologies such as ML and UEBA	Your UEBA systems are showing significant spikes in outbound traffic from several workstations associated with members of your company's research and development arm, all of whom are currently out of the office at a technical conference. You suggest that all recent e-mails be checked for phishing messages and their endpoints scanned for unusual processes.
Situation-driven	Hypotheses developed with understanding of the organization's environment, network topology, and key assets, and how they change over time	Your organization is in the final steps of acquiring a small startup with an inexperienced security team. This startup has been the victim of several attackers' campaigns over the last few years. Attackers will likely leverage their existing access to the startup's network and the new connections to your organization to facilitate attacks against your company. You suggest that enhanced monitoring be configured at those connection points.

Table 13-1 Types of Threat-Hunting Hypotheses Sources

Type	Description	Example
Intelligence-driven	Hypotheses derived through use of indicators of compromise (IOCs); adversary tactics, techniques, and procedures (TTPs); and reports from vendors or internal threat intelligence teams	An external vendor recently sent a report describing the TTPs of a new organized crime outfit operating out of Eastern Europe. This group uses *bulletproof hosting* servers in the same region to send tailored phishing messages. You suggest that messages originating from that region undergo additional scrutiny or filtering based on the IOCs contained in the report.
Experience-driven	Hypotheses informed by lessons learned from previous hunts and knowledge gained from experiences in security operations	Having previously worked at a major ISP, you have firsthand experience with how the Border Gateway Protocol (BGP) is meant to operate. You have also observed suspicious behaviors purportedly conducted by a well-resourced, nation-level threat actor. You suggest that your current team develop and maintain an intelligence-sharing agreement with your old team and with other ISP security teams to gain insight into recent BGP abuses.

Table 13-1 Types of Threat-Hunting Hypotheses Sources *(continued)*

Profiling Threat Actors and Activities

In Chapter 2, we covered the MITRE ATT&CK framework, an important tool resource for understanding adversary tactics and techniques in the context of developing threat intelligence. As profiling attackers and their associated activities is a critical step in threat hunting, ATT&CK is incredibly useful. As a reminder, MITRE ATT&CK is meant to serve as a globally accessible knowledge base of adversary tactics and techniques based on real-world observations. It's the "real-world" feedback about the framework that adds so much relevancy to threat-hunting efforts. Another major benefit of the ATT&CK framework is that it enables threat hunters to discover potential activity based on the stage of the attack lifecycle. Tactics and techniques are grouped within a matrix that can be used for confirming hunting hypotheses. The ATT&CK matrix includes the following categories:

- **Initial access**   The techniques used by the adversary to obtain a foothold within a network
- **Execution**   The techniques that enable adversaries to run their code on a target system
- **Persistence**   The techniques that enable an adversary to maintain long-term access to a target system

- **Privilege escalation**   The techniques used by an adversary to gain higher level privilege, such as administrator or root
- **Defense evasion**   The techniques used by attackers to circumvent security mechanisms or obfuscate their behavior to avoid detection
- **Credential access**   The techniques developed to capture legitimate user credentials
- **Discovery**   The techniques used by adversaries to obtain information about systems and networks, often to assist in targeting and exploit development
- **Lateral movement**   The techniques that enable an attacker to move from one system to another within a network
- **Collection**   The techniques used by an adversary to aggregate information about target systems
- **Command and control**   The techniques leveraged by an attacker to enable communication between victim machines and those under their control
- **Exfiltration**   The techniques used to get data out of a compromised network and into an environment controlled by the attacker
- **Impact**   The techniques used by an attacker to impact legitimate users' access to a system

Threat hunting isn't spared from one of the biggest problems in defending modern systems: the volume and complexity of attacks. MITRE ATT&CK and other frameworks help hunters figure out what to focus on by providing a foundation that brings some order to the chaos. It provides a common language when describing attacker behavior that can be applied to long-term detection and remediation efforts, but it also means that new members are able to wrap their head about these activities and the impact they have more quickly.

 EXAM TIP   The CySA+ exam does not focus exclusively on MITRE ATT&CK in the context of threat hunting, but this framework is used broadly across the domains covered in the test. ATT&CK provides an excellent model for thinking about all adversary behavior, and it will be useful for you to be able to describe adversary behavior and how you, as an analyst, might develop your hunting plan using its language.

Threat-Hunting Tactics

Threat hunting, like nearly every other aspect of defense, benefits from automation and specialized tooling. Threat hunters can use a huge range of software tools to sift through the vast amount of data to help them make sense of what they are seeing and improve the hunting process. Getting starting on threat-hunting efforts does not often require additional investment in technology, because threat hunters often rely on existing security tools to achieve their key tasks. Firewalls, endpoint protection software, and intrusion detection systems (IDSs), for example, can be used to help reveal indicators of compromise. Additionally, hunters can use security information and event management (SIEM) solutions to aggregate vast amounts of log and traffic data to enable statistical analyses and visualization tools to present trends and highlight anomalies in useful ways. A security team doesn't have to spend significant amounts of money to get a threat-hunting effort started. In fact, one of the simplest threat-hunting tools is the spreadsheet, which can often be an effective tool for storing and retrieving observed data. Taking cues from the popularity and usefulness of tabular formats, the MITRE ATT&CK Navigator tool (https://mitre-attack.github.io/attack-navigator/enterprise/) presents its rich content in a manner similar to spreadsheet software to help defenders develop and execute their plans.

High-Impact TTPs

A common method of conducting a hunt is to go through every phase of an attack model and determine the areas or phases the team is most concerned with. The exact techniques your team will use to hunt malicious activity will depend largely on what you're trying to defend against. Each phase will involve several associated TTPs that an adversary may normally use. To hunt the adversary—and defend yourself—you must understand your enemy's tendencies. You can use the previously discussed techniques to uncover some common, but high-impact TTPs as you begin your hunting efforts. We'll take some time to explore TTPs aligned with the ATT&CK categories; although this is by no means an exhaustive list, it will provide an excellent starting point for the hunt.

Initial Access and Discovery

The TTPs associated with initial access and discovery have a lot to do with how the adversary is targeting your systems and how they decide to go one way instead of another. Accordingly, behaviors associated with enumeration are the focus here. The attacker works to determine details about a local host, the larger networks, and their configuration. As a hunter, you're looking for artifacts that indicate that an attacker is trying to get the network layout, lists of users and groups, and lists of privileges allotted to particular users. Querying to look for the use of tools such as nmap or the Windows net.exe utility is a good start.

Persistence

Uncovering TTPs related to persistence is all about how attackers maintain access even after the host is rebooted. Quite often, attackers will use built-in functionality, such as Windows scheduled tasks execution and registry or macOS login items, to ensure that their malicious programs or scripts can maintain a long-term presence on the system. Tasks can be also scheduled remotely, provided that an attacker has the correct credentials via remote procedure calls. Another example of using built-in resources, adversaries will sometimes run malicious code by using a legitimate process to load and execute the code. In this case, an attacker would leverage the trusted relationship between the system and the process to hide the malicious activity. This method, often a dynamic-link library (DLL) injection in the Windows environment, can lead to attacks taking hold of a process' memory and permissions. The most common way malware persists on macOS is via a LaunchAgent, a component of the OS service management service. Each macOS user can have LaunchAgents with configuration files stored in his or her own Library folder. These files specify the code that should be run every time that user logs in, or they may contain their own commands to execute directly.

Lateral Movement and Privilege Escalation

With some degree of confidence that they have established a foothold in a network, attackers will move to explore the network via lateral movement—attempting to access services beyond the level of the accounts they're currently in control of via some kind of privilege escalation. A common TTP associated with attackers looking to target credentials is a Pass-the-Hash (PtH) attack. This method involves capturing password hash values that can be used to authenticate against other systems at a later point. Importantly, PtH techniques do not require the actual password or even an attempt to crack the credentials. Another particularly effective means to traverse quickly across systems is a built-in functionality that's heavily used across many industries: Remote Desktop Protocol (RDP). As with the scheduled tasks and login items persistence techniques, attackers will use tools that they know are unlikely to be blocked because of their popularity. If RDP is enabled on a system and an attacker is in possession of those account credentials, he or she can access and exploit the target system in a manner that becomes very difficult to detect in most environments.

Command and Control

During the course of an attack, attackers may at some point need to have their malware reach out to get new instructions from them. In many cases, attackers will seek to hide in plain sight by blending in their command and control (C2) with routine network traffic and piping their instructions using standard ports and protocols that are unlikely to be blocked. Hoping that their communications are lost or ignored in the sheer volume of traffic, attackers rely on some assumptions made by the security team about where legitimate traffic resides. This reliance does have the possibility of backfiring, however, as may be the case if a detection team invests resources into deep inspection of traffic across heavily used ports and protocols. If this is known to be the case, an attack can bypass heavily monitored ports by sending data through uncommon ports to enable attackers to operate stealthily as they work toward their end goals.

Exfiltration

Many of the stealthier data exfiltration techniques rely on concepts similar to those used for sending and receiving C2 information. Domain Name System (DNS) tunneling, for example, is a method of sending data via encoded DNS queries. Because DNS is a critical and foundational protocol of the Internet, it's often untouched by network defenders. As long as the malicious encoding is consistent and the traffic conforms to DNS standards, it's incredibly hard to detect. Hunters will need to keep an eye out for abnormal DNS queries or unusually high query volumes.

Although detecting abnormal domains may be straightforward for a human, it's not as simple for machines to read a string of characters and understand its significance linguistically or culturally. This is where we can leverage mathematical principles to assist us with identifying suspicious DNS queries. In the context of information, *entropy* is the measure of randomness in data. *High entropy domains* are those that have comparatively large amounts of randomness, and in terms of readability, these domains are likely to appear as gibberish to a human. As we covered in Chapter 11, malware may use *domain generation algorithms* (DGAs) for its C2 exchanges to try to avoid detection. Because it is very difficult to block domains one at a time, this technique can be an effective way to facilitate that communication. For DGAs, the entropy associated with the generated domain is usually higher than that of standard domains. You can perform your own analysis of by first collecting legitimate DNS logs and determining the frequencies for characters that appear in those domains names to establish a threshold. When you use entropy alongside other hunting techniques, it increases your confidence in the harm of a particular signal. For example, if you have uncovered an unsigned application on a host acting strangely, you can then look for network connections that appear to have a relatively high degree of entropy to reduce the likelihood of a false positive or false negative.

PART III

Content Delivery Networks

Although computer-generated domain names are often used by many malware strains to define their C2 channels, DGAs have legitimate uses in content delivery networks (CDNs). These cloud-based services have become extremely popular in recent years. Many websites use CDNs to deliver dynamic content such as video and images to customers across the globe. The challenge is that many CDN domains are generated using DGAs, though they usually include some clue as to who the CDN provider is. Keeping track of each CDN's unique characteristics and naming schemes will improve your hunt queries.

Searching

Armed with collection and aggregation capabilities, and a hypothesis that you hope to prove, you can begin the hunt by simply searching for data that answers your questions. Querying, as we've discussed throughout the book, is a core functionality of any security data aggregator that can yield quick results, assuming that the questions are well defined and that queries are well written. Searching rapidly scopes down the working data set and prevents analyst overload. Always keep in mind that searching too broadly for general entities may produce far too many results to be useful to an analyst, or it may present the analyst with extra work. On the other hand, searching too specifically can result in the network producing too few results in the hunt.

Clustering

Two type of grouping techniques are particularly useful for threat hunting. Both borrowing from statistical analysis, these techniques are effective when you're looking for shared features in very large data sets. *Clustering analysis* is a technique often used in statistics to identify groups of data points based on certain criteria, such as occurrence.

Good clustering methods will result in groups that have high interclass similarity, meaning that they are indeed quite similar to one another based on the predefined criteria, as well as low interclass similarity, meaning that they can be sufficiently differentiated from other clusters.

Grouping

Like clustering, *grouping* is a means to categorize similar data points by taking a set of unique features and determining the artifacts that fit the criteria. Grouping requires that these defining features be described in advances, where clustering does not require this to be the case. Often, one of the features used in a clustering criterion is time, as would be the case in looking for a particular type of command executed in a certain timeframe.

Stacking

Stacking, or stack counting, is a basic technique for identifying outliers in data. It involves counting the number of occurrences of a particular value, sorting them, and investigating the extreme outliers. Stacking is less useful with very large data sets because the outliers in these collections can themselves be quite large. Stacking is most effective with data sets that produce a finite number of results, when the criteria are carefully designed, and with the use of automation or statistics technology. This is a case in which spreadsheet software can be very helpful in your hunting.

Delivering Results

Threat hunting is not a purely academic endeavor; there is the potential for immediate and massive operational impact as a result of your team's hunting efforts. In most cases, an analyst is conducting a hunt in a business environment, which is subject to the external influences of the organization, environmental restrictions, and business needs. Because of this, analysts must work to exhaust every possible avenue toward achieving well-defined criteria for success or completion of the hunt. Time is money, and hunting can be a costly undertaking. As important as knowing why you're starting the hunt in the first place is knowing when to stop. The criteria for ending the hunt should therefore be included when the scope of this hunt is defined.

Although hunts do not necessarily fall under the traditional incident response process, they will require similar levels of investment in personnel and technology. Unlike the incident response process, though, it's important that you accept that hunts may not yield results, and it may be difficult to determine when to stop. After all, it's expected that at any point during the hunt, an analyst may uncover a lead that will require more investigation. It's often impossible to tell at the onset what the exact time required for a particular hunt may be. Part science and part art, the process benefits from the freedom for an analyst to explore all possibilities and think "outside the box," but the process must also be sound and responsible. Good hunters will often know when they've hit the point of diminishing returns and need to refocus. If your security organization is just beginning its journey of building a threat-hunting team, it may be worthwhile to prescribe periodic reviews to make sure that your hunts are producing results.

Documenting the Process

At the beginning of this chapter, we described the importance of developing your hypothesis when starting out on a hunt. Hypotheses are best guesses, and sometimes even the most educated guesses are incorrect. Threat hunters routinely get to a point at which they've exhausted all possibilities and resources and will just have to move on. Hunts are often short, lasting only a few days or weeks, and highly scoped, so the act of accepting a fruitless outcome will be necessary for a good threat-hunting team to be prepared to tackle the next challenge. Even if a hunt results in no necessary action, it can nevertheless be very beneficial. Regardless of the outcome, documentation of the hypothesis, why it was or was not correct, any inhibitors to the process, and what you can do better next time will be useful for future efforts. Documentation is a crucial part of the hunt process, and if you forget to document every step you've taken, every lead you've tracked down, and everything you've learned, you'll simply have to reinvent the wheel the next time.

Reducing the Attack Surface Area and Bundling Critical Assets

Every year, SANS Institute conducts a Threat Hunting Survey to identify trends related to threat-hunting programs from a variety of organizations. From the 575 responses in 2019, for example, SANS discovered that although hunting efforts were usually conducted by more mature organizations or those that have been targeted in the past, nearly three-quarters of the organizations noticed marked increases in their security programs. The top two areas of significant improvement, according to the study, were in quality and coverage of detection and the reduction of attack surface exposure. In both areas, more than 90 percent of the respondents reported at least some improvement, with over a third reporting significant improvement. Incident responders were able not only to identify the source of malicious activity more quickly, but they were able to remediate the situation in a timely manner. Similarly, threat hunting enabled defenders to shore up protections around their important assets. It's noteworthy that although threat hunting is a separate effort from the traditional vulnerability management process, it still can produce impactful results with respect to mitigation efforts such as device hardening and network architecture. The survey showed that threat hunting, despite being focused on uncovering malicious actor behaviors not previously detected, was still able to inform decisions without the need for an active threat to be present.

We've learned from covering vulnerability management concepts in Chapter 3 that it's virtually impossible to defend something if you're not aware of its existence on your network. An area of improvement for any organization is its asset management. Network diagrams and asset catalogs are constantly in a state of evolution, and the natural lag between updates to that information and when the security team recognizes these changes creates a period of increased vulnerability. Threat hunting turns out to be an extremely effective way to gather knowledge about your network environment and its users.

Thinking as an attacker, you may often find it faster and more reliable to take advantage of known exploits, should a vulnerability present itself, rather than risk detection or failure with a new technique. This is in part why vulnerability scanners are such a useful part of the attacker's arsenal. Figure 13-2 shows the output from Nessus, a popular

Figure 13-2 Nessus scan results organized by hosts

vulnerability scanner, listing the found vulnerabilities by hosts. Nessus, like many other vulnerability scanners, provides the final list of vulnerabilities, along with their type, count, and other details about exploitability in the scan results that can assist defenders in hardening their hosts (see Figure 13-3). Attackers can use the information from the various scan results to determine priority of targets as they work toward their goals.

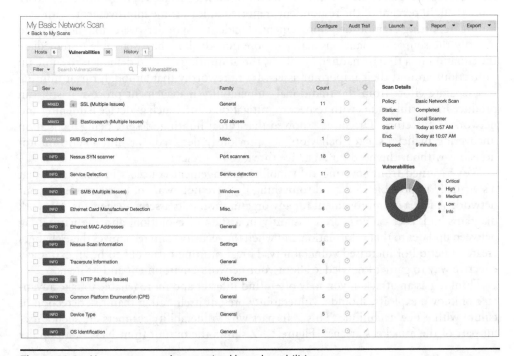

Figure 13-3 Nessus scan results organized by vulnerabilities

Attack Vectors

In studying how attackers make their ways onto networks, hunters find the same handful of attack vectors time and time again. At a macro level, attackers gain access to systems using one or a combination of the following: malware, vulnerability exploitation, social engineering, or insiders. Beginning with malware, attackers have found lots of success with code designed to evade detection by antivirus software and security appliances. As these malware programs can often be aware of their surroundings, modern strains can even choose to become active only when they detect the correct environmental conditions, all while appearing benign in virtualized environments. Furthermore, much of the malware in use today is *polymorphic*, taking on a different look as it moves to and from infected systems. This common quality is rendering many legacy detection techniques ineffective.

Vulnerability exploitation, whether by way of a zero-day exploit or using well-known techniques, continues to prove useful for attackers. Zero-days, especially, give attackers a massive advantage in terms of time. Knowing that activity associated with a zero-day is far less likely to be detected using traditional methods, an attacker can often dwell in a system for months while defenders are none the wiser. Exploits of all kinds are particularly lucrative for attackers, because they don't often require victim interaction. An attacker can scan a network, discover vulnerable conditions, launch the attack, and take advantage of the access. Another frequently observed technique is brute-force attacks for credentials.

For as long as attackers have been attempting to access networks, social engineering has been an integral part of their plan. By tricking their victims into divulging sensitive information or making decisions that weaken a part of the system, intruders use various methods such as pretexting, or communicating a false motive, to mislead and confuse. This is often very effective when used to deliver malware, because the privileged position gained as a result of a successful social-engineering effort means fewer security hurdles to overcome for delivery of the malicious code.

If attackers cannot get an unwitting actor to help them gain access, they will sometimes resort to enticing willing insiders to do their bidding. Motivated by any number of factors, insider actors continue to pose a massive threat that attackers are all too eager to take advantage of. What's worse is that it's incredibly difficult to detect insider vectors, because they "belong" on the network. Threat hunters will find that identifying insider behaviors related to attacker activities is among the most challenging categories of hunts.

Integrated Intelligence

Threat intelligence teams have a special relationship with threat hunters. Both are laser-focused on TTPs and want to know how attackers have behaved in the past and how they are likely to act in a given situation. Threat intelligence feeds inform threat hunters of the latest attacker TTPs, connected infrastructure, victimology, and attacker tendencies. Hunters, in pursuing these leads, can provide feedback to the intelligence team as to the effectiveness, accuracy, and relevance of the provided data. Additionally, they may also be able to feed back into the intelligence process so that the threat intelligence team can update their own products to reflect reality. In closing out a hunt, the team will want to communicate relevant malicious activity and new intelligence back to the operations and threat intelligence team. In a sense, threat hunters are extensions of threat-intelligence efforts, scouting for new information while verifying assumptions and assessments from the intelligence team.

Improving Detection Capabilities

A hunt, whether successful or not, will yield products that are directly relevant to detection efforts. At the very least, hunters will have a better understanding of TTPs and how they may be used against the environment. In the best scenario, the team can use examples of how TTPs were leveraged successfully to inform modifications to infrastructure, improvements to detection logic, and changes in policy. Technical modifications can result from discoveries about the network, servers, and endpoints during a hunt. One common change is the addition of security devices and sensors to areas to which the team lacks visibility. Additionally, a team may discover that a sensor is in place, but not correctly configured to catch the malicious behavior. Improvements to misconfigured sensors, network blind spots, and data collection not only helps the hunting team should they have to embark on a similar venture in the future, but prepares the detection team to catch potentially malicious events and actors more effectively.

Following are a few benefits that a threat-hunting team's output will provide to a detection team:

- Updated firewall and intrusion detection/prevention system (IDS/IPS) rules
- Updated alert logic for SIEM platforms
- Updated alert logic for endpoint detection and response (EDR) platforms
- Improvements to sensor placement across the network
- Improvements to asset visibility

Although not strictly within the purview of the detection team, the following benefits across the security team will affect the detection's team's efficiency and reduce excess noise on the network:

- Changes in your organization's development process
- Changes to security training across the organization
- Changes to quality assurance and quality control processes

 TIP The threat-hunting process often yields a plethora of new ideas for techniques to try and places to look. Be sure to record these ideas, as they will be great starting points for the next hunt.

Chapter Review

Today's threat actors are often well organized and highly capable, making it increasingly difficult to detect if and when they may have made it past your defensive measures and how they are currently operating on your network. Threat hunting is an analyst-centric process that enables organizations to look for malicious activity that may have slipped past detection and prevention mechanisms. It is a proactive process that relies on a skilled analyst to focus on attacker behavior and the evidence that they leave behind when they're moving about the network. Threat hunting is driven by a modern philosophy being adopted by more and more information security processionals: assume compromise and act accordingly. Accepting the very real possibility that no network is completely impenetrable and that attackers may have made it past defenses is a prudent mindset that demonstrates and mature and realistic view of security.

Conducting a hunt requires the smart application of tools and techniques to extract and make sense of highly detailed information across the environment. Part science and part art, the process is repeatable and scalable, while allowing for hunter intuition to guide the pursuit of signs of intrusion, no matter how insignificant they may seem. Bringing your analytical capacity and your understanding of the details will be more impactful than any single piece of technology. With the assistance of tools, you can bring threat-hunting techniques to bear to find evil and enable attacks to be discovered earlier in the attack cycle.

Questions

1. Which is not an example of a built-in resource in the Windows environment that attackers often leverage for persistence?

   A. LaunchAgents

   B. Windows registry modification

   C. DLL injection

   D. Scheduled tasks

2. What term is used for a key assumption that differentiates the threat hunter mindset from traditional security processes?

   A. Walled garden

   B. Defense in depth

   C. Least privilege

   D. Assume breach

3. Which increasingly popular framework provides an effective means to profile attacker behavior?

   A. PHI

   B. NIST

   C. MITRE ATT&CK

   D. CERT

4. What is the primary difference between threat intelligence and threat hunting as they relate to security operations?

   A. Threat intelligence aims to add context to observed activity on the network, while threat hunting does not.

   B. Threat hunting requires an understanding of adversary TTPs, while threat intelligence does not.

   C. Threat intelligence requires skills and experience in all facets of security operations, while threat hunting is focused on forensics.

   D. Threat hunting focuses on what may have already occurred, while threat intelligence can be used to estimate an adversary's future behaviors.

5. Which of the following is *not* a best practice when closing out threat-hunting efforts?

   A. Document areas of improvement.

   B. Document shortcomings in sensor placement, tooling, and techniques.

   C. Provide feedback to the intelligence team as to the effectiveness, accuracy, and relevance of their data.

   D. Continue the hunt until all attackers and their TTPs are identified.

6. Which of the following hypothesis types is developed to address the inherently dynamic nature of a network and its changing qualities?

   A. Analytics-driven

   B. Situation-driven

   C. Intelligence-driven

   D. Experience-driven

7. High-entropy domains are often associated with what category of TTPs?

   A. Command and control

   B. Exfiltration

   C. B and C

   D. None of the above

8. Which of the following threat-hunting tactics consists of querying data for specific results?

   A. Searching

   B. Grouping

   C. Clustering

   D. Stacking

Answers

1. A. LaunchAgent creation is the most commonly used method in the macOS environment, but it does not exist in Windows. In a Windows environment, registry modification, dynamic-link library (DLL) injection, and scheduled task execution are all methods that attackers use to maintain persistence.

2. D. Assume breach is a strategy based on the assumption that an attacker is already in your network, or that your defenses may be insufficient to detect all attacker behavior. The "assume breach" mindset promotes ongoing testing and refinement of detection and response techniques.

3. C. MITRE ATT&CK is used to help hunters figure out what to focus on by providing common language when describing attacker behavior.

4. D. Threat hunters are primarily focused on what may have already occurred on the network and will often look for what already may be present and unseen, instead of what might occur in the future.

5. D. Although analysts must work to exhaust every resource in proving their hypothesis, sometimes a hunt will come up dry. It's important to communicate the criteria for ending the hunt during the hunt preparation.

6. B. Situation-driven hypotheses are developed with an understanding of the organization's environment, network topology, and key assets, and how they change over time.

7. C. High-entropy domains have large amounts of randomness and often appear as gibberish to a human. They are often used to enable communications related to C2 and exfiltration.

8. A. Searching is often the simplest way to begin hunting. It involves using any number of data platforms, or aggregators, to looking for artifacts of interest.

Automation Concepts and Technologies

In this chapter you will learn:

- The role of automation technologies in modern security operations
- Best practices for employing orchestration technologies
- Best practices for building automation workflows and playbooks
- Tips for automating data enrichment at scale

No one can whistle a symphony. It takes an orchestra to play it.

—H. E. Luccock

Security processes can be tedious and repetitive. Whether you're testing software functionality for vulnerabilities or responding to frequently occurring incidents, you'll find a certain level of automation to be helpful, particularly for activities that do not require a great amount of analyst intervention. The goal is not to move the entire security process to full automation, but rather to create the conditions that enable the analyst to focus brainpower on complex activities. Vulnerability management analysts can automate repeated security testing cases, for example, and instead focus their energies on exploring the in-depth weaknesses in critical services and validating complex logic flaws. Even the best analysts make mistakes, and although they work their best to secure systems, they do introduce limitations and occasional inconsistencies that may lead to catastrophic results. As systems get more and more complex, the number of steps required to perform a remediation may scale accordingly, meaning that the sheer number of actions needed to secure a system correctly grows, as do the demands on an analyst's attention.

To assist with this, there are several automation concepts that we'll cover. Automation, in the context of security operations, refers to the application of technologies that leverage standards and protocols to perform specific, common security functions for a fixed period or indefinitely. One of the primary driving forces in high efficiency productions systems such as lean manufacturing and the famously effective Toyota Production System is the principle of *jidoka*. Loosely translated as "automation with a human touch," *jidoka* enables machines to identify flaws in their own operations or enables operators in a manufacturing environment to quickly flag suspicious mechanical behaviors. This principle has allowed for operators to be freed from continuously controlling machines, leaving the machines free to operate autonomously until they reach an error condition, at which point a human has the opportunity to exercise judgement to diagnose and remediate the issue. Most importantly, at least for efficiency, is that *jidoka* frees the human worker to concentrate on other tasks while the machines run. We can apply *jidoka* concepts to security to take advantage of standardization, consistent processes, and automation to reduce the cognitive burden on an analyst. Using specifications on how to interface with various systems across the network, along with using automated alerting and report delivery mechanisms, we can see massive increases in efficiency while experiencing fewer mistakes and employee turnover as a result of burnout. Additionally, the use of specifications and standards means that troubleshooting becomes more effective, more accurate, and less costly, while allowing for easy auditing of systems.

Workflow Orchestration

Automation proves itself as a crucial tool for security teams to address threats in a timely manner, while also providing some relief for analysts in avoiding repetitive tasks. However, to be truly effective, automation is often confined to systems using the same protocol. Although many analysts, including the authors, find success using scripting languages such as Python to interact with various system interfaces, there are challenges with managing these interactions at scale and across homogenous environments. Keeping tabs with how actions are initiated and under what conditions, and what reporting should be provided, gets especially tricky as new systems and technologies are added to the picture. Orchestration is the step beyond automation that aims to provide an instrumentation and management layer for systems automation. Good security orchestration solutions not only connect various tools seamlessly, but they do so in a manner that enables the tools to talk to one another and provide feedback to the orchestration system without requiring major adjustments to the tools themselves. This is often achieved by making use of the tools' application programming interfaces (APIs).

Security Orchestration, Automation, and Response Platforms

As teams begin to integrate and automate their suites of security tools, they invariably come up against the limitations of specialized scripting and in-house expertise. Getting disparate security systems to work nicely together to respond to suspicious activity is not a trivial task, so a market for security orchestration, automation, and response (SOAR) tools has emerged to meet this need. Described by IT research firm Gartner as "technologies that enable organizations to collect security threats data and alerts from different sources, where incident analysis and triage can be performed leveraging a combination of human and machine power to help define, prioritize and drive standardized incident response activities according to a standard workflow," SOAR tools enable teams to automate frequent tasks within a particular technology as well as coordinate actions from

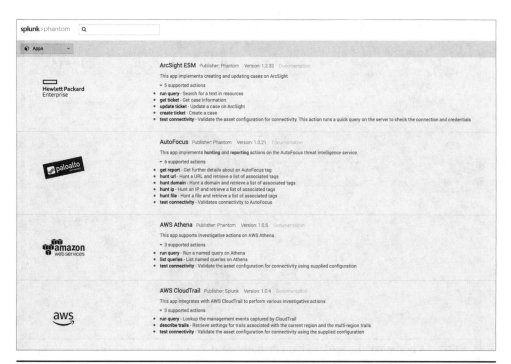

Figure 14-1 A portion of Splunk Phantom's supported apps

different systems using repeatable workflows. Like SIEM platforms, SOAR solutions often include comprehensive dashboard and reporting capabilities; in many cases, they work alongside security information and event management (SIEM) solutions to help security teams maximize analyst productivity.

Recently acquired by Splunk and renamed Splunk Phantom, the orchestration platform formerly known as Phantom Cyber is one of the most comprehensive SOAR platforms in the industry. Coupled with a customizable dashboard functionality, the Phantom platform aims to address two of the major challenges facing security teams: back-end integration and front-end presentation. Offered in both commercial and community versions, Phantom provides out-of-the-box integration with hundreds of data sources, instances, and devices by way of apps. Figure 14-1 shows a snapshot of a few popular apps and the supported actions related to those technologies. It's also possible to create custom app integrations with third-party services and APIs.

As far as dashboard options, Phantom provides a multitude of metrics to help you gain a better understanding of both system and analyst performance, as shown in Figure 14-2. Analysts can get a sense of what kinds of events are in the pipeline, what playbooks are used most often, and the return on investment (ROI) associated with these events and actions. ROI, measured in both time and dollars saved, can be particularly powerful for a team in communicating the value of investment in orchestration platforms and training.

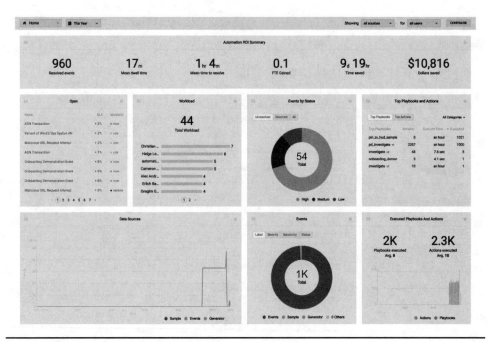

Figure 14-2 A portion of the Splunk Phantom dashboard

Orchestration Playbooks

Playbooks are workflows that help you visualize and execute processes across your security footprint in accordance with your orchestration rules. Also referred to as runbooks, these steps can be fully automated, or they can require human intervention at any point along the process. Much like a recipe, the playbooks prescribe the steps to be taken in a formulaic, often highly scalable, manner. In viewing their security process like an algorithm, security teams can eliminate much of the need for a manual intervention, save for the most critical decision points.

There are a few key components that all playbooks must have to be complete. The first is the *initiating condition*, or the rules that must be triggered to begin the steps within the rest of the playbook. Often, this is the presence of artifacts that meet whatever your organization defines as a security incident. Sometimes, the condition may be preventative in nature and be initiated on some schedule. This initial condition may set off a series of actions across many security devices, each of which would normally take a bit of human interaction. This step alone significantly reduces the investment in analyst resources, an investment that may not always lead to anything actionable.

Next are the *process steps* that the playbook will invoke by interacting with the various technologies across the organization. In many cases, security orchestration solutions have software libraries that enable the platform to interact seamlessly with technology. For some of the more popular orchestration solutions, these libraries are written in coordination with the service provider to ensure maximum compatibility. For the analyst, this means minimal configuration on the front end, while in other cases, libraries may need to be initially set up manually within the orchestration software.

Finally, an *end state* must be defined for the playbook. Whether this comes in the form of an action to remediate or an e-mailed report, ensuring that a well-defined outcome is reached is important for the long-term viability of the playbook. It may well be the case that the outcome for one playbook is the initiating activity for another. This is the concept of *chaining*, in which multiple playbook actions originate from a single initiating condition. This final stage of a playbook's operation will often include some kind of reporting and auditing functionality. While these products may not be directly applicable to the remediation effort itself, they are incredibly valuable for improving overall orchestration performance as well as meeting any regulatory requirements that your organization may have.

Orchestration is not only a tremendous time-saver, but it also facilitates a diversity of actions, such as technical and nontechnical, to be included in an overall security strategy. For example, a process such as an e-mail phishing investigation will benefit from a number of orchestration-enabled tasks. Enriching e-mail addresses and suspicious links by pulling data from threat intelligence data sets, and then forwarding those domains to a blacklisting service should they have a negative verdict, are tasks well suited for automation. Furthermore, trend reports can be generated over time and sent automatically to a training and communications team for use in improving user training. These kinds of tasks replace what may normally be a few hours of cutting, pasting, and formatting. When these tasks are chained within an orchestration tool, an initial signal about a suspicious e-mail can begin several processes without the need of any manual intervention from the security team, as shown in Figure 14-3.

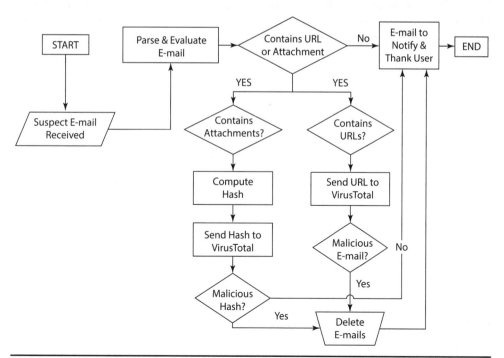

Figure 14-3 Example playbook for automated processing of suspected phishing e-mail

Many modern SOAR platforms include the ability to design playbooks both textually and visually. Using the previous phishing workflow, you can see an example of the visual breakout of a playbook related to phishing in Figure 14-4. Named phishing_investigate_ and_respond, this playbook extracts various artifacts from suspected phishing messages and uses them as input in querying enrichment sources, or data sources, that can give additional context to the artifacts Any attachments, for example, are submitted to file reputation providers for a verdict, while URLs, domains, and IP addresses are passed to sources that can provide details on those aspects.

Depending on how the playbooks are configured, they may call upon one or more apps to interact with data sources or devices. Playbook design becomes crucially important as event volume increases to ensure that the processes remain resilient while not overly consuming outside resources, such as API calls.

In some situations, your organization may require that a human take action because of the potential for adverse impact in the case of a false positive. Imagine for a moment that your organization has a highly synchronized identity management system, and a number of playbooks are related to account takeover actions. If one of these is written to perform a fully automated account reset, to include session logouts, password and token resets, and two-factor authentication (2FA) reset, it could cause catastrophic effects if initiated on a low-quality signal, such as the presence of suspicious language in an e-mail. A full account reset, therefore, is an action for which a fully automated action may be the best; it makes sense to check and double-check that the action is warranted, however. Another action that may benefit from human interaction is blocking of domains. Depending on your organization's network access policies, you may want to fully domain block or prompt an analyst for review. Domains, as you may recall, often have associated reputations. This reputation can change over time, or it may be the case that a domain

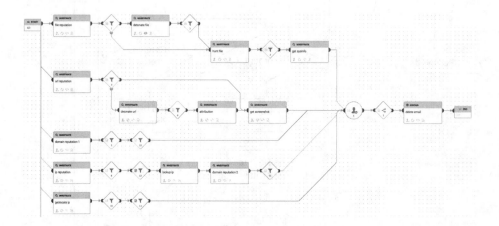

Figure 14-4 Splunk Phantom visual playbook editor

Figure 14-5   Splunk Phantom visual playbook editor showing a human prompt for action

has earned a malicious reputation as a result of a temporary loss of control but has since recovered. Considering these possibilities, it may be prudent to require analyst input into the blocking process, as shown in Figure 14-5. Phantom's editor allows for prompts to team members for approval.

Data Enrichment

A major part of any security analyst's day is to track down and investigate suspicious activity. Investigation into the source and meaning behind observables is a core skill of any detection analyst, incident responder, threat hunter, or threat intelligence analyst. Enrichments are the actions that lead to additional insights into data, and they involve a set of tasks that can be a draw on resources. Often repetitive, enrichment is a perfect candidate for automation. Any way that we can make it easier for analysts to initiate an enrichment request means more time for them to focus on really challenging tasks. Depending on how automation is set up, enrichment may even occur as the initial is delivered, meaning that the analyst sees the original signal coupled with useful context to make a far more informed decision. This practice leads not only to far lower response times, but it also reduces the chance for human error, since there are fewer areas for a human to interact directly in the process.

Scripting

The command line has traditionally been the fastest way for analysts to interface directly with a platform's core operations. Security team members and systems administrators have been building custom scripts to perform management tasks for as long as the command line interface (CLI) has been around. In its simplest form, this has come in the form of a cron job, a task that leverages the time-based scheduling utility cron in Linux/Unix environments. This tool enables users to run commands or scripts at specified intervals. As one of the most useful OS tools, cron is often used by systems administrators to perform maintenance-related tasks, but it can be employed by any knowledgeable user to schedule recurring tasks. The cron service runs in the background and constantly checks the contents of the cron table, or crontab, located at /etc/crontab, or in the /etc/cron.*/ directories for upcoming jobs. The crontab is a configuration file that lists the commands to be run, along with their schedule.

Administrators can maintain a system-wide crontab or enable individual users to maintain their own individual crontab files using the following syntax:

```
# <minute (0 - 59)>
#  <hour (0 - 23)>
#   <day of the month (1 - 31)>
#    <month (1 - 12)>
#     <day of the week (0 - 6)>
#
#
#
# * * * * * <command to execute>
```

Another common method of automating across the network is using the Secure Shell (SSH) protocol to issue commands via the CLI. It's possible to initiate SSH connections with scripting to perform any number of actions, but this method is prone to inconsistency and error. Copying files is also possible, using a secure connection with the **scp** command. Short for *secure copy*, this command enables copying to and from remote servers using SSH connections. The following command is an example of how **scp** can be used to connect to a remote server and retrieve a file named log.txt from remote server server2.example.com using user1 credentials:

```
scp user1@server2.example.com:/user1/logs/log.txt /Desktop/all_logs/
```

Python Scripting

Python has emerged as one of the most popular languages for security analysts in part because of its scripting potential. As a fast, multiplatform language, it has experienced rapid adoption and expansion with a vast collection of libraries, modules, and documentation specifically focused on computer security. It's also a language that's designed to be somewhat human-readable and to perform with the fewest lines of code possible. Getting into Python scripting is easy and well worth the investment for any security analyst. Let's consider a simple case of trying to extract IP addresses from a large log file provided by a security device. The contents of this fictional log file are displayed here:

```
suspicious_ips.log
26533    10.1.0.219    97
78082    10.7.0.7    10
48285    10.3.0.167    68
10080    10.4.0.168    19
26105    10.2.0.244    67
94692    10.8.0.8    20
22809    10.9.0.98    88
73733    10.2.0.103    49
75349    10.5.0.134    95
90112    10.4.0.255    70
19323    10.10.0.86    19
17138    10.3.0.139    43
38027    10.4.0.26    14
51923    10.0.0.248    59
```

```
54575    10.0.0.188    76
37940    10.5.0.40     76
53647    10.1.0.105    30
24139    10.5.0.199    78
37041    10.9.0.131    100
37527    10.4.0.244    85
```

With a few lines and the commonly used regular expression library re, we can quickly create a Python script capable of identifying just the IP addresses. The simple script we've written is shown next. It takes a file input, identified here as suspicious_traffic.log, and uses the regular expression Python library to search quickly through the contents for strings that resemble the structure of an IP address:

```
import re
file = open("suspicious_traffic.log")
for line in file:
        ipaddr = re.findall( r'[0-9]+(?:\.[0-9]+){3}', line )
        print ipaddr
```

Using this script, we can generate a list of just the IP addresses, as shown next. We can ship this list directly to another service, such as IP reputation, in an automated fashion, or add it to a blacklist.

```
['10.1.0.219']
['10.7.0.7']
['10.3.0.167']
['10.4.0.168']
['10.2.0.244']
['10.8.0.8']
['10.9.0.98']
['10.2.0.103']
['10.5.0.134']
['10.4.0.255']
['10.10.0.86']
['10.3.0.139']
['10.4.0.26']
['10.0.0.248']
['10.0.0.188']
['10.5.0.40']
['10.1.0.105']
['10.5.0.199']
['10.9.0.131']
['10.4.0.244']
```

PowerShell Scripting

PowerShell is an automation framework developed by Microsoft that includes both a command shell and support for advanced scripting. Normally used by systems administrators to maintain and configure systems, PowerShell can also be used by security teams to obtain a great deal of information from endpoints and servers that is useful for tasks such as incident response, detection, and vulnerability management.

Application Programming Interface Integration

APIs have ushered in several major improvements in how computers talk to each other at scale. APIs are becoming the primary mechanism for both standard user interactions and systems administration for a number of reasons. First, APIs simplify how systems integrate with each other, because they provide a standard language to communicate while maintaining strong security and control over the system owner's data. APIs are also efficient, since providing API access often means that the content and the mechanism to distribute the content can be created once using accepted standards and formats and published to those with access. This means that the data is able to be accessed by a broader audience that just those with knowledge about the supporting technologies behind the data. Thus, APIs are often used to distribute data effectively to a variety of consumers without the need to provide special instructions and without negatively impacting the requestor's workflow. Finally, APIs provide a way for machines to communicate at high volume in a scalable and repeatable fashion, which is ideal for automation.

As API usage has increased across the Web, two popular ways to exchange information emerged to dominate Internet-based machine-to-machine communications. The first is the Simple Object Access Protocol (SOAP). APIs designed with SOAP use Extensible Markup Language (XML) as the message format transmitting through HTTP or SMTP. The second is Representational State Transfer (REST). We'll take a deeper look into REST here, because it is the style you'll most likely interface with on a regular basis and it offers flexibility and a greater variety of data formats.

Representational State Transfer

Representational State Transfer is a term coined by computer scientist Dr. Roy Fielding, one of the principal authors of the HTTP specification. In his 2000 doctoral dissertation, Dr. Fielding described his designs of software architecture that provided interoperability between computers across networks. Web services that use the REST convention are referred to as *RESTful* APIs. Though many think of REST as a protocol, it is, in fact, an architectural style and thus has no rigid rules. REST interactions are characterized by six principles, however:

- **Client/server** The REST architecture style follows a model in which a client queries a server for particular resources, communicated over HTTP.

- **Stateless** No client information is stored on the server, and each client must contain all of the information necessary for the server to interpret the request.

- **Cacheable** In some cases, it may not be necessary for the client to query the server for a request that has already occurred. If marked as such, a client may store a server's response for reuse later on.

- **Uniform interface** Simplicity is a guiding principle for the architectural style and is realized with several constraints:
    - **Identification of resources** Requests identify resources (most often using URLs), which are separate from what is returned from the servers (represented as HTML, XML, or JSON)

- **Manipulation of resources through representations** A client should have enough information to modify a representation of a resource.

- **Self-descriptive messages** Responses should contain enough information about how to interpret them.

- **Hypermedia as the engine of application state** There is no need for the REST client to have special knowledge about how to interact with the server, because hypermedia, most often HTTP, is the means of information exchange.

- **Layered system** A client will not be able to tell if it's connected directly to the end server, an intermediary along the way, proxies, or load balancers. This means that security can be strongly applied based on system restrictions and that the server may respond in whatever manner it deems most efficient.

- **Code on demand (optional)** REST enables client functionality to be extended by enabling servers to respond to applets or scripts that can be executed client-side.

RESTful APIs use a portion of the HTTP response message to provide feedback to a requestor about the results of the response. Status codes fall under one of the five categories listed here:

Category	Description
1xx: Informational	Communicates transfer protocol–level information
2xx: Success	Indicates that the client's request was accepted successfully
3xx: Redirection	Indicates that the client must take some additional action to complete their request
4xx: Client Error	Indicates that the client takes responsibility for the error status codes
5xx: Server Error	Indicates that the server takes responsibility for the error status codes

Furthermore, you may be able to recover more detailed error messages depending on what API service is being used. VirusTotal, for example, provides the following detailed breakout of error codes that can be used to troubleshoot problematic interactions:

Error code	HTTP code	Description
AlreadyExistsError	409	The resource already exists.
AuthenticationRequiredError	401	The operation requires an authenticated user. Verify that you have provided your API key.
BadRequestError	400	The API request is invalid or malformed. The message usually provides details about why the request is not valid.
ForbiddenError	403	You are not allowed to perform the requested operation.
InvalidArgumentError	400	Some of the provided arguments are incorrect.

Error code	HTTP code	Description
NotFoundError	404	The requested resource was not found.
QuotaExceededError	429	You have exceeded one of your quotas (minute, daily, or monthly). Daily quotas are reset every day at 00:00 UTC.
TooManyRequestsError	429	There are too many requests.
UserNotActiveError	401	The user account is not active. Make sure you've properly activated your account by following the link sent to your e-mail.
WrongCredentialsError	401	The provided API key is incorrect.
TransientError	503	This is a transient server error; a retry might work.

 TIP The CySA+ does not require you to have in-depth knowledge of HTTP response and error codes, but understanding how to use them in troubleshooting automation is important. Knowing where to adjust your code based on the HTTP responses is a critical skill for API automation.

Automating API Calls

Many tools can assist you with crafting an API call and then converting it to code for use in other systems or as a subordinate function in custom software and scripting. Postman, Insomnia, and Swagger Codegen are three popular API clients that can assist with generating API calls for automation. We'll use the VirusTotal API and Insomnia to show some of the most common capabilities of these API clients. According to the VirusTotal documentation, the API endpoint associated with domain enrichment requires a **GET** HTTP method. In many cases, you can copy the commands provided via the API documentation or use the code that follows in your API client:

```
curl --request GET \
  --url https://www.virustotal.com/api/v3/domains/{domain} \
  --header 'x-apikey: <your API key>'
```

This **curl** request automatically gets parsed in the client. After the appropriate credentials are added via an API key, OAuth credentials, login/password pair, bearer token, or other means, the request is ready to send. Figure 14-6 shows a complete request sent to the VirusTotal domain enrichment endpoint. You'll notice the HTTP 200 response, indicating that the exchange was successful.

The full contents of the HTTP response are under the Preview tab in this client, as shown in Figure 14-7. In this case, 92 verdicts are returned for the domain google, with 83 being "harmless" and 9 being "undetected."

```
* Preparing request to https://www.virustotal.com/api/v3/domains/google.com
* Using libcurl/7.69.1 OpenSSL/1.1.1g zlib/1.2.11 brotli/1.0.7 libidn2/2.1.1 libssh2/1.9.0 nghttp2/1.40.0
* Current time is 2020-06-09T00:23:48.991Z
* Disable timeout
* Enable automatic URL encoding
* Enable SSL validation
* Enable cookie sending with jar of 0 cookies
* 17 bytes stray data read before trying h2 connection
* Found bundle for host www.virustotal.com: 0x7fea1b5fa6b0 [can multiplex]
* Re-using existing connection! (#0) with host www.virustotal.com
* Connected to www.virustotal.com (74.125.34.46) port 443 (#0)
* Using Stream ID: 7 (easy handle 0x7fea2810ba00)

> GET /api/v3/domains/google.com HTTP/2
> Host: www.virustotal.com
> user-agent: insomnia/2020.2.1
> x-apikey: 
> accept: */*

< HTTP/2 200
< cache-control: no-cache
< content-type: application/json; charset=utf-8
< x-cloud-trace-context: 73b8c77b7392d832b0a17829136c7098
< date: Tue, 09 Jun 2020 00:23:49 GMT
< server: Google Frontend
< content-length: 35387
```

Figure 14-6   Request sent to the VirusTotal domain enrichment API endpoint via API client

```
Preview ▼      Header      Cookie      Timeline

1    {
2      "data": {
3        "attributes": {
4          "categories": {},
5          "creation_date": 874296000,
6          "last_analysis_results": {   },
560        "last_analysis_stats": {
561          "harmless": 83,
562          "malicious": 0,
563          "suspicious": 0,
564          "timeout": 0,
565          "undetected": 9
566        },
567        "last_dns_records": [   ],
697        "last_dns_records_date": 1591571255,
698        "last_https_certificate": {   },
834        "last_https_certificate_date": 1591571256,
835        "last_modification_date": 1591662201,
836        "last_update_date": 1568043544,
837        "popularity_ranks": {   },
859        "registrar": "MarkMonitor Inc.",
860        "reputation": 135,
861        "tags": [],
862        "total_votes": {
863          "harmless": 45,
864          "malicious": 12
865        },
866        "whois": "Admin Country: US\nAdmin Organization: Google LLC\nAdmin State/Province: CA\nCreation Date: 1997-09-
     15T00:00:00-0700\nCreation Date: 1997-09-15T04:00:00Z\nDNSSEC: unsigned\nDomain Name: GOOGLE.COM\nDomain Name:
     google.com\nDomain Status: clientDeleteProhibited (https://www.icann.org/epp#clientDeleteProhibited)\nDomain Status:
     clientDeleteProhibited https://icann.org/epp#clientDeleteProhibited\nDomain Status: clientTransferProhibited
     (https://www.icann.org/epp#clientTransferProhibited)\nDomain Status: clientTransferProhibited
     https://icann.org/epp#clientTransferProhibited\nDomain Status: clientUpdateProhibited
     (https://www.icann.org/epp#clientUpdateProhibited)\nDomain Status: clientUpdateProhibited
     https://icann.org/epp#clientUpdateProhibited\nDomain Status: serverDeleteProhibited
     (https://www.icann.org/epp#serverDeleteProhibited)\nDomain Status: serverDeleteProhibited\nDomain Status: serverTransferProhibited
     (https://www.icann.org/epp#serverTransferProhibited)\nDomain Status: serverTransferProhibited
     https://icann.org/epp#serverTransferProhibited\nDomain Status: serverUpdateProhibited
     (https://www.icann.org/epp#serverUpdateProhibited)\nDomain Status: serverUpdateProhibited
     https://icann.org/epp#serverUpdateProhibited\nName Server: NS1.GOOGLE.COM\nName Server: NS2.GOOGLE.COM\nName Server:
```

Figure 14-7   Preview of the API response in the Insomnia API client

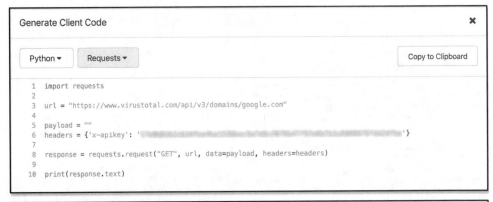

Figure 14-8   Two examples of API client code–generation features

Once you're satisfied with the exchange between client and server, you can export this request to many popular programming languages or scripting formats. Insomnia comes with built-in ability to generate client code in several languages, some with multiple library options. Figure 14-8 shows the client code for both Python using the requests library and the PowerShell Invoke–WebRequest cmdlet (pronounced "command-let"). This can be used as-is or added to existing code to extend functionality.

Automated Malware Signature Creation

YARA, the popular tool used by defenders to identify and classify malware samples, provides a simple but powerful way to perform rule-based detection of malicious files. As we covered in Chapter 12, each YARA description, or rule, is constructed with two parts: the strings definition and the condition. The YARA structure makes it easy to define malware using well-understood Boolean expressions, particularly at scale. Dr. Florian Roth wrote a simple yet powerful YARA rule generation software called yarGen, which facilitates the creation of YARA rules for malware by searching for the strings found in malware files, while ignoring those that also appear in benign files. The utility ships with

Figure 14-9 Usage of the yarGen YARA rule generation utility

a database that helps you get started by simply calling the script and specifying the directory of interest. In this case, we've populated a directory with EICAR (European Institute for Computer Antivirus Research) test files, a set of files used to test antimalware software without your having to use the real thing. Working from the yarGen directory, we call the program with the command **python yarGen.py -m <location of the EICAR directory>**. The resulting operation is shown in Figure 14-9.

After a few moments, yarGen generates a file called yargen_rules.yar, which contains YARA rules generated for any malicious filed found in that directory (see Figure 14-10). These generated rules contain meta, strings, and condition fields and can be directly imported by any software that can use YARA rules for immediate deployment.

Outside of YARA rules, a notable open source project called BASS is designed to generate antivirus signatures automatically from existing samples. Maintained by Talos, Cisco's threat intelligence and research group, BASS's creation was developed as a way to generate more efficient pattern-based signatures. Given the sheer scale of sample volume the Talos research team sees—in excess of 1.5 million unique samples per day—they were in need of a way to avoid maintaining massive databases of hash-based signatures while keeping a low memory profile and remaining performant. The BASS solution, which is Python-based and implemented using a series of Docker containers, is a framework for generating pattern-based signatures, keeping all the benefits of hash-based solutions without requiring a massive increase in resources as it scales.

```
⊙ ● ●                    brent — brent@budgie-dev: ~/Documents/yarGen — ssh budgie — 120×56
/*
    YARA Rule Set
    Author: yarGen Rule Generator
    Date: 2020-06-08
    Identifier: eicar
    Reference: https://github.com/Neo23x0/yarGen
*/

/* Rule Set ----------------------------------------------------------------- */

rule eicarcom2 {
    meta:
        description = "eicar - file eicarcom2.zip"
        author = "yarGen Rule Generator"
        reference = "https://github.com/Neo23x0/yarGen"
        date = "2020-06-08"
        hash1 = "e1105070ba828007508566e28a2b8d4c65d192e9eaf3b7868382b7cae747b397"
    strings:
        $s1 = "eicar.comX5O!P%@AP[4\\PZX54(P^)7CC)7}$EICAR-STANDARD-ANTIVIRUS-TEST-FILE!$H+H*PK" fullword ascii
        $s2 = "eicar.comPK" fullword ascii
        $s3 = "eicar_com.zipPK" fullword ascii
    condition:
        uint16(0) == 0x4b50 and filesize < 1KB and
        all of them
}

rule eicar_com {
    meta:
        description = "eicar - file eicar.com.txt"
        author = "yarGen Rule Generator"
        reference = "https://github.com/Neo23x0/yarGen"
        date = "2020-06-08"
        hash1 = "275a021bbfb6489e54d471899f7db9d1663fc695ec2fe2a2c4538aabf651fd0f"
    strings:
        $s1 = "X5O!P%@AP[4\\PZX54(P^)7CC)7}$EICAR-STANDARD-ANTIVIRUS-TEST-FILE!$H+H*" fullword ascii
    condition:
        uint16(0) == 0x3558 and filesize < 1KB and
        all of them
}

rule eicar_com_2 {
    meta:
        description = "eicar - file eicar_com.zip"
        author = "yarGen Rule Generator"
        reference = "https://github.com/Neo23x0/yarGen"
        date = "2020-06-08"
        hash1 = "2546dcffc5ad854d4ddc64fbf056871cd5a00f2471cb7a5bfd4ac23b6e9eedad"
    strings:
        $s1 = "eicar.comX5O!P%@AP[4\\PZX54(P^)7CC)7}$EICAR-STANDARD-ANTIVIRUS-TEST-FILE!$H+H*PK" fullword ascii
        $s2 = "eicar.comPK" fullword ascii
    condition:
        uint16(0) == 0x4b50 and filesize < 1KB and
        all of them
}
/* Super Rules ------------------------------------------------------------- */
```

Figure 14-10   Contents of the yarGen-generated YARA rules

Threat Feed Combination

In Chapter 2, we covered various types of threat data that can be used to support incident response, vulnerability management, risk management, reverse engineering, and
detection efforts. The sheer number of observables, indicators, and context continuously
pulled from internal and external sources often requires an abstraction layer to normalize and manage so much data. Since it's often not possible to affect the format in which
feeds arrive, analysts spend a lot of time formatting and tuning feeds to make them most

relevant to the operational environment. As you may recall, understanding what an indicator may mean to your network, given the right context, cannot always be automated, but a good deal of the processing and normalization that occurs before analysis can be handled by automation. This includes pulling in threat feeds via APIs and extracting unstructured threat data from public sources, and then parsing that raw data to uncover information relevant to your network. Getting the data stored in a manner that be easily referenced by security teams is not a trivial task, but with proper automation, it will take a great deal of strain off of analysts.

Putting new threat intelligence into operations can often be a manual and time-consuming process. These activities, which include searching for new indicators across various systems, are especially well-suited for automation. Once the threat data is analyzed, analysts can share threat intelligence products automatically to response teams and lead enforcement of new rules across the entire network. Improvements such as changes to firewall rules can many times be managed by the team's SIEM and SOAR platforms.

Machine Learning

Machine learning (ML) is the field of computer science that, when applied to security operations, can uncover previously unseen patterns and assist in decision-making without being specially configured for it. A particularly exciting aspect of ML techniques is that they improve automatically through experience. As more data is provided to the model, the more accurately it will be able to detect patterns automatically. When applied to security, ML usually falls into one of a few major applications. First are algorithms to look at past network data to predict future activity. These techniques are currently being used to process high volumes of data and identify patterns to make predictions about most likely changes in network traffic and adversary behaviors. With mathematical techniques at its core, ML techniques are often effective at mining information to discover patterns and assigning information into categories.

Another area where ML techniques thrive is in malware and botnet detection. As security techniques move away from signature-based techniques to those driven by behavior pattern recognition, ML is an especially useful tool. Even as attackers use polymorphism to avoid detection, ML remains effective, as it looks for behavior that deviates from the intended operation of the software. Spotting unusual activity, whether it's related to how a host is communicating, how data is being moved, or how a process is behaving, is something that fraud and detection teams are focused on improving on a daily basis.

Despite all of the promise, there is no silver bullet ML algorithm that is effective against every type of security challenge. Models that are developed for specific purposes routinely outperform general algorithms, but they are still not nearly as effective as the human decision-making process. With respect to automation, however, ML brings additional techniques to reduce the amount of noise that an analyst is initially faced with when embarking on the hunt for a potential security event.

Use of Automation Protocols and Standards

The use of automation to realize industry best practices and standards has been in practice for years in the commercial and government spaces. As an early developer of hardening standards, the Defense Information Systems Agency (DISA) has made significant investments in the promotion of these standards through products such as the Security Technical Implementation Guides (STIGs). STIGs, as we covered in Chapter 3, are meant to drive security strategies as well as prescribe technical measures for improving the security of networks and endpoints. Although STIGs describe how to minimize exposure and improve network resilience through proper configuration, patch management, and network design, they don't necessarily define methods to ensure that this can be done automatically and at scale. This is where automation standards come into play.

Security Content Automation Protocol

We introduced the Security Content Automation Protocol (SCAP) in Chapter 3 as a framework that uses specific standards for the assessment and reporting of vulnerabilities of the technologies in an organization. The current technical specification of SCAP, version 1.3, is covered by NIST SP 800-126 Revision 3. SCAP 1.3 comprises twelve component specifications in five categories:

- **Languages** The collection of standard vocabularies and conventions for expressing security policy, technical check techniques, and assessment results
  - Extensible Configuration Checklist Description Format (XCCDF)
  - Open Vulnerability and Assessment Language (OVAL)
  - Open Checklist Interactive Language (OCIL)
- **Reporting formats** The necessary constructs to express collected information in standardized formats
  - Asset Reporting Format (ARF)
  - Asset Identification (AID)
- **Identification schemes** The means to identify key concepts such as software products, vulnerabilities, and configuration items using standardized identifier formats, and to associate individual identifiers with additional data pertaining to the subject of the identifier
  - Common Platform Enumeration (CPE)
  - Software Identification (SWID) Tags
  - Common Configuration Enumeration (CCE)
  - Common Vulnerabilities and Exposures (CVE)

- **Measurement and scoring systems** Evaluation of specific characteristics of a security weakness and any scoring that reflects their relative severity
  - Common Vulnerability Scoring System (CVSS)
  - Common Configuration Scoring System (CCSS)
- **Integrity** An SCAP integrity specification that helps to preserve the integrity of SCAP content and results
  - Trust Model for Security Automation Data (TMSAD)

OpenSCAP

A great public resource is the OpenSCAP project, a community-driven effort that provides a wide variety of hardening guides and configuration baselines. The policies are presented in a manner accessible by organizations of all types and sizes. Even better, the content provided is customizable to address the specific needs or your environment.

To demonstrate a vulnerability scan, we'll use the oscap command-line tool provided by the OpenSCAP Base on a Red Hat 7 test system. The first step in evaluating a system is to download the definitions file, which will be used to scan the system for known vulnerabilities caused by missing patches. The Center for Internet Security (CIS) maintains an OVAL database (https://oval.cisecurity.org/repository) organized by platform. Alternatively, you can go directly to operating systems developers such as Red Hat and Canonical to retrieve the definitions file. For our demonstration, we're using both the Extensible Configuration Checklist Description Format (XCCDF) definitions provided by Red Hat and begin the scan using the command that follows. We'll specify the output to be a human-readable report using the **--report** option and a machine-readable report using the **--results** option. Both files will include details about what was tested, the test results, along with some recommendations for mitigation if applicable. Where available, Common Vulnerabilities and Exposures (CVE) identifiers are linked with the National Vulnerability Database (NVD) where additional information such as CVE description, Common Vulnerability Scoring System (CVSS) score, and CVSS vector are stored. Red Hat provides a complete overview of oscap usage on its site: https://access .redhat.com/documentation/en-us/red_hat_enterprise_linux/6/html/security_guide/ sect-using_oscap.

```
oscap xccdf eval --results results.xml --report report.html
com.redhat.rhsa-all.xccdf.xml
```

The utility will take a few moments to run and will provide the requested output on screen as well as written to the XML and HTML files created using the **--report** option. Figure 14-11 shows the on-screen output, while Figure 14-12 shows the HTML report created.

```
● ● ●                    🖥 brent — brent@rhel:~/scap — ssh rhel — 108×36
[brent@rhel scap]$ oscap xccdf eval --results results.xml --report report.html com.redhat.rhsa-all.xccdf.xml
Title    RHBA-2007:0304: Updated kernel packages available for Red Hat Enterprise Linux 4 Update 5 (None)
Rule     oval-com.redhat.rhba-def-20070304
Ident    RHBA-2007:0304
Ident    CVE-2005-2873
Ident    CVE-2005-3257
Ident    CVE-2006-0557
Ident    CVE-2006-1863
Ident    CVE-2007-1592
Ident    CVE-2007-3379
Result   pass

Title    RHBA-2007:0331: conga bug fix update (None)
Rule     oval-com.redhat.rhba-def-20070331
Ident    RHBA-2007:0331
Ident    CVE-2007-0240
Ident    CVE-2007-1462
Result   pass

Title    RHBA-2007:0565: tcp_wrappers bug fix update (None)
Rule     oval-com.redhat.rhba-def-20070565
Ident    RHBA-2007:0565
Ident    CVE-2009-0786
Result   pass

Title    RHBA-2008:0314: Updated kernel packages for Red Hat Enterprise Linux 5.2 (None)
Rule     oval-com.redhat.rhba-def-20080314
Ident    RHBA-2008:0314
Ident    CVE-2007-5906
Ident    CVE-2008-2365
Result   pass

Title    RHBA-2009:0070: util-linux bug-fix update (None)
Rule     oval-com.redhat.rhba-def-20090070
Ident    RHBA-2009:0070
Ident    CVE-2008-1926
```

Figure 14-11   Execution and immediate output of the oscap utility run on a Red Hat 7 system

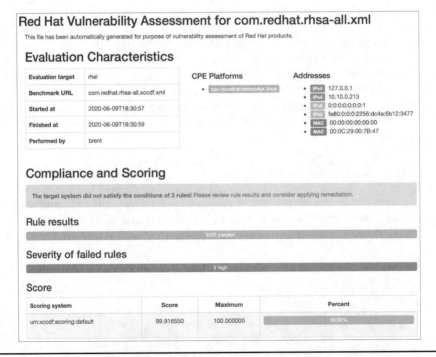

Red Hat Vulnerability Assessment for com.redhat.rhsa-all.xml

This file has been automatically generated for purpose of vulnerability assessment of Red Hat products.

Evaluation Characteristics

Evaluation target	rhel
Benchmark URL	com.redhat.rhsa-all.xccdf.xml
Started at	2020-06-09T18:30:57
Finished at	2020-06-09T18:30:59
Performed by	brent

CPE Platforms
- cpe:/o:redhat:enterprise_linux

Addresses
- IPv4 127.0.0.1
- IPv4 10.10.0.213
- IPv6 0:0:0:0:0:0:0:1
- IPv6 fe80:0:0:0:2256:dc4a:6b12:3477
- MAC   00:00:00:00:00:00
- MAC   00:0C:29:00:7B:47

Compliance and Scoring

The target system did not satisfy the conditions of 3 rules! Please review rule results and consider applying remediation.

Rule results

3592 passed

Severity of failed rules

3 high

Score

Scoring system	Score	Maximum	Percent
urn:xccdf:scoring:default	99.916550	100.000000	99.92%

Figure 14-12   Excerpt of the HTML report summary after execution of the oscap utility

Rule Overview

☐ pass	☑ fail	☐ notchecked	Search through XCCDF rules	Search
☐ fixed	☐ error	☐ notapplicable	Group rules by: Default	
☐ informational	☐ unknown			

Title	Severity	Result
▼ Red Hat Vulnerability Assessment for com.redhat.rhsa-all.xml (3x fail)		
RHSA-2020:2082: kernel security and bug fix update (Important)	high	fail
RHSA-2020:2337: git security update (Important)	high	fail
RHSA-2020:2344: bind security update (Important)	high	fail

Show all result details

Red Hat and Red Hat Enterprise Linux are either registered trademarks or trademarks of Red Hat, Inc. in the United States and other countries. All other names are registered trademarks or trademarks of their respective companies.

Figure 14-13 Excerpt of the HTML report after execution of the oscap utility, highlighting the failed rules

As part of the report, oscap provide details on the categories of rules along with their results. This test located three failures, as indicated in Figure 14-13. At minimum, oscap provides the name of the rule, a description of the vulnerability, and the severity.

Results from this test can be used to initiate a manual review of affected systems, or with the use of scripting or a SOAR platform, they can be queued for automated actions such as the quarantine of affected systems.

Software Engineering

Many of the techniques we've discussed so far, while incredibly effective, have a difficult time keeping pace with the rate of modern software engineering. With companies frequently looking to push changes to their internal and customer-facing solutions, there just isn't time to test these new versions with the same completeness and rigor every time. The solution is to integrate security directly into the software development process, driven by smart automation technologies. Using DevOps or DevSecOps, various security tasks for software development such as static code analysis and fuzzing can be automated. Additionally, with the help of cloud technologies, consistent development environments can be guaranteed as they are provisioned on the fly. Security's ideal place has always been as an integral part of the software development lifecycle, and not simply as an afterthought. This is the core principle of DevSecOps: it's about built-in security that is important at the beginning of an app's lifecycle and remains so throughout its deployment and up through its eventual deprecation.

Continuous Integration

Continuous integration is the practice of merging the various changes in code made by contributors back to the main branch, or effort, of a code base as early and often as possible. Changes can be quickly validated and pushed to production because versioning and management of merging is done smartly by platforms such as git. Along with the automated building and running of various quality tests against these builds, software security tests can be performed. Continuous integration ensures that the software remains at a high level of usability and security, thanks to automation, much of which is hidden from the developer as applications are checked and validated.

Continuous Delivery

Continuous delivery activities follow continuous integration and consist of tasks related to getting the software out of the development system and to the end user. This usually means testing so that the software is released in a timely manner after final validation occurs. Security can be automated in this stage as well, occurring alongside the release cycle that your software follows. In practice, this may mean that software is continuously tested for correctness and that findings are automatically included as input into the next development cycle.

Continuous Deployment

Continuous deployment follows continuous delivery and ensures that changes made previously in the cycle are incorporated and released to end users automatically, save for failed tests. Although continuous deployment is a great way to accelerate the feedback loop with your users, it doesn't fit well with traditional software assessment techniques, since software can potentially be in front of end users just minutes after it is compiled. Security will serve as a very visible bottleneck only if it is not integrated seamlessly. Automated security activities that usually occur at deployment include runtime security and compliance checks, disabling of unnecessary services and ports, removal of development tools, enablement of security mechanisms, and enforcement of audit and logging policies.

Chapter Review

At its core, security operations and management are centered on securing systems by shoring up defenses throughout the entire lifecycle of the organization's processes. As we've covered throughout the book, it takes a well-planned approach to identify and deploy technical, managerial, and operational steps that are repeatable and scalable to adjust with the organization's evolution. Alongside the challenge of pivoting to accommodate ever-changing operational environments and threat models is the need to be able to reliably perform the same processes time and time again. Detection, for example, suffers tremendously if monitoring rules are not applied consistently and quickly enough to identify suspicious behavior confidently over time.

Questions

1. Which of the following is *not* an example of an identification scheme in SCAP 1.3?

   A. Common Platform Enumeration (CPE)

   B. Common Data Enumeration (CDE)

   C. Software Identification (SWID) Tags

   D. Common Vulnerabilities and Exposures (CVE)

2. What describes the first event of a playbook process that triggers the rest of the steps?

    A. Reporting

    B. Process steps

    C. End-state

    D. Initiating condition

3. In the REST architecture style, which of the following is *not* one of the commonly returned data types in server responses?

    A. XML

    B. SCAP

    C. HTML

    D. JSON

4. Which HTTP status code indicates some kind of client error?

    A. 100

    B. 202

    C. 301

    D. 403

5. Which of the following utilities, found in most versions of Linux, is useful for scheduling recurring tasks?

    A. cron

    B. whois

    C. scp

    D. oscap

6. Which of the following hypothesis types is developed to address the inherently dynamic nature of a network and its changing qualities?

    A. The playbook may result in additional work for the analyst.

    B. The playbook may cause a significant impact to the end user in the case of a false positive.

    C. The playbook actions may change over time and require human tuning.

    D. Orchestration systems require that a human be present during the full duration of its operation.

7. DevSecOps is used an approach for integration security in which stages of software engineering cycles?

    A. Continuous integration

    B. Continuous delivery

    C. Continuous deployment

    D. All of the above

8. What is the name of the framework that supports automated configuration and vulnerability checking based on specific standards?

   A. STIG

   B. SCAP

   C. XCCDF

   D. BASS

Answers

1. **B.** Common Data Enumeration (CDE) is not part of SCAP 1.3, identification schemes, which include Common Platform Enumeration (CPE), Software Identification (SWID) Tags, Common Configuration Enumeration (CCE), and Common Vulnerabilities and Exposures (CVE).

2. **D.** The initiating condition can be defined as the rules that must be completed to trigger the remaining steps in the playbook.

3. **B.** SCAP is not a data type. Although REST does not mandate a data type, the most commonly returned response data types are XML, HTML, and JSON.

4. **D.** HTTP status codes (RFC 7231) in the 400 range indicate a client error. In this case, 403 (Forbidden) indicates that the server understood the request but is refusing to fulfill it because of insufficient privileges.

5. **A.** cron is a standard Linux/Unix utility that is used to schedule commands or scripts for automatic execution at specific intervals.

6. **B.** In some situations, such as a full account reset, your organization may require that a human take action because of the potential for adverse impacts in the case of a false positive.

7. **D.** DevSecOps is about integrating automated security functionality at all stages in the lifecycle of application development.

8. **B.** The Security Content Automation Protocol (SCAP) is a framework that uses specific standards for the assessment and reporting of vulnerabilities of the technologies in an organization.

PART IV

Incident Response

The Importance of the Incident Response Process

In this chapter you will learn:
- The importance of establishing communications processes
- Which factors contribute to data criticality
- How unauthorized disclosure may create serious adverse effects

I am prepared for the worst, but hope for the best.

—Benjamin Disraeli

While it's hard to predict when and how an incident will occur, it's certain that it will happen at some point. Good security teams will have taken measures to minimize the likelihood of an incident by hardening the network and addressing vulnerabilities, but great teams will go a step further and outline the exact steps necessary to address an incident when the time comes. The activities necessary to accomplish the goals of detecting and recovering from an incident are collectively referred to as the *incident response process*. For it to be effective, the overall process should be well documented, tailored to your organization, and well understood by all stakeholders across the organization, from your response team to public relations to the C-suite. Each incident starts with the detection of some potentially malicious security event—an event which may generate an alert that may require further investigation, analysis, and remediation. A well-designed IR process makes it more likely that your organization moves efficiently from initial detection to resumption of normal operations, while also keeping stakeholder apprised of progress.

Establishing a Communication Process

A variety of team members and stakeholders are involved in incident responses, each with different capabilities and priorities. Maintaining effective communications with them all can help you make the experience of responding to an incident far more manageable. Ineffective communication processes with internal and/or external stakeholders can endanger your entire organization, even if you have a textbook-perfect response to an incident.

We cannot be exhaustive in our treatment of how best to communicate during incident responses in this chapter, but we hope to convey the importance of interpersonal and interorganizational communications in these few pages. Even the best-handled technical incident response can be overshadowed very quickly by an organization's inability to communicate effectively—internally and/or externally.

Internal Communications

One of the key parts of any incident response plan is the process by which trusted internal parties are kept abreast of and consulted about an incident and how it will be dealt with. It is not uncommon, at least for the more significant incidents, to designate a war room where key decision-makers and stakeholders can meet for periodic updates and feedback regarding the incident response. In between these meetings, the room can serve as a clearinghouse for information about the response activities, where at least one knowledgeable member of the incident response (IR) team will be stationed for the duration of the response activity. The war room can be a physical space, but a virtual one may work as well, depending on your organization.

In addition to hosting regular meetings (formal or otherwise) in the war room, it may be necessary to establish a secure communications channel with which to keep key personnel up-to-date on the progress of the response. This could involve group texts, e-mails, or a chat room—but it must include all the key personnel who may have a role or stake in the issue. When it comes to internal communications with stakeholders after an incident, there is no such thing as too much information.

External Communications

Communications outside of the organization, on the other hand, must be carefully controlled. Sensible reports have a way of getting turned into misleading and potentially damaging sound bites. For this reason, a trained professional should be assigned the role of handling external communications. Some of these communications, after all, may be restricted by regulatory or statutory requirements.

The first and most important sector for external communications comprises government entities, which could include the US Securities Exchange Commission (SEC), the Federal Bureau of Investigation (FBI), or some other government entity. If your organization is required to communicate with these entities in the course of an incident response, your legal team must assist in crafting any and all messages. If you manage to ignore these communication requirements, your organization will pay a price. This is not to say that government stakeholders are adversarial, but that when the process is regulated by laws or regulations, the stakes are much higher.

Next on the list of importance are customers. Though your organization may be affected by regulatory requirements with regard to compromised customer data, our focus here is on keeping the public informed so that it perceives transparency and trustworthiness from the organization. This is particularly important when the situation is interesting enough to make headlines or "go viral" on social media. Just as lawyers are critical to government communications, the organization's media relations (or equivalent) team will carry the day when it comes to communicating with the masses. The goal here is to assuage fears and concerns as well as control the narrative to keep it factually correct. To this end, press releases and social media posts should be templated even before an event occurs to make it easier to push out information quickly.

Another group with which we may have to communicate deliberately and effectively includes the key partners, such as business collaborators, select shareholders, and investors. The goal of this communications thrust is to convey the impact of the incident on the business's bottom line. If the event could drive down the price of the company's stock, the conversation has to include ways in which the company will mitigate such losses. If the event could spread to the systems of partner organizations, the focus should be on how to mitigate that risk. In any event, the business leaders should carry on these conversations, albeit with substantial support from the senior response team leaders.

Response Coordination with Relevant Entities

In the midst of an incident response, it is all too easy to get so focused on the technical challenges that we forget about the human element, which is arguably at least as important. Here, we focus our discussion on the various roles involved and the manner in which we must ensure that these roles are communicating effectively with one another and with those outside the organization.

The key roles required in incident responses can be determined beforehand based on established escalation thresholds. The in-house technical team will always be involved, of course, but when and how are others brought in? This depends on the analysis that your organization performed as part of developing the IR plan. Figure 15-1 shows a typical escalation model used by many organizations.

The technical team is unlikely to involve management in routine responses, such as a response to an e-mail with a malicious link or attachment that somehow gets to a user's inbox, but is not clicked. You still have to respond to this and will probably notify others (such as the user, supervisor, and threat intelligence team) of the attempt, but management will not be in the loop at every step of the response. The situation is different, however, when the incident or response has a direct impact on the organization, such as when you have to reboot a production server to eradicate malware installed on it. Management needs to be closely involved in decision-making in this more serious scenario.

At some point, the skills and abilities of the in-house team will probably be insufficient to deal effectively with an incident; this is when the response is escalated and you bring in contractors to augment your team or even take over aspects of the response. Obviously, this is an expensive move, so you want to consider this option carefully, and management will almost certainly be involved in that decision. Finally, some incidents

Figure 15-1
Typical role
escalation model

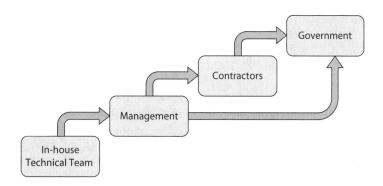

require government involvement. Typically, though not always, this comes in the form of notifying and perhaps bringing in a law enforcement agency such as the FBI or Secret Service. This may happen with or without your organization calling in external contractors, but it will always involve senior leadership. Whatever the process, it's important to publish the specific escalation paths for your analysts and maintain a call list, or contact information for those required to be notified or involved in case of an incident. Let's take a look at some of the issues involved with each of the roles involved.

 EXAM TIP For the purposes of the CySA+ exam, *IR stakeholders* are individuals and teams who are part of your organization and who have a role in helping with some aspects of an incident response.

The term *stakeholder* is broad and could include a very large set of people. Each stakeholder has a critical role to play in some (maybe even most), but not all, responses. These supporting stakeholders normally will not be accustomed to executing response operations, as the direct players are. You and the IR team must make extra efforts to ensure that each stakeholder knows what to do and how to do it when bad things happen.

Legal Counsel

Whenever an incident response escalates to the point of involving government agencies such as law enforcement, you will almost certainly be coordinating with your organization's legal counsel. Apart from requirements to report criminal or state-sponsored attacks on your systems, your organization may be affected by regulatory considerations such as those discussed in Chapter 5. For instance, if you work in an organization covered by the Health Insurance Portability and Accountability Act (HIPAA) and you are responding to an incident that compromised the protected health information (PHI) of 500 or more people, your organization will have some very specific reporting requirements that will have to be reviewed by your legal and/or compliance team(s).

The law is a remarkably complicated field, so even actions that may seem innocuous to many of us can have some onerous legal implications. Though some lawyers are very knowledgeable in complex technological and cybersecurity issues, most have only a cursory familiarity with them. In our experience, starting a dialogue early with the legal team and then maintaining a regular, ongoing conversation are critical to staying out of career-ending trouble.

Human Resources

The likeliest involvement of human resources (HR) staff in a response occurs when the team determines that a member of the organization probably had a role in the incident. The role need not be malicious, mind you, because it could involve a failure to comply with policies (for example, connecting a thumb drive into a computer when that is not allowed) or repeated failures to apply security awareness training (for example, clicking a link in an e-mail even after a few rounds of remedial security training). Malicious, careless, or otherwise, the actions of our teammates can and do lead to serious incidents. Disciplinary action in those cases requires HR involvement.

In other situations, as well, you may need to involve HR in the response, such as when overtime is required to deal with the response, or when key people need to be called in from time off or vacation. The safest bet is to involve HR in your IR planning process and especially in your drills, and let them tell you what, if any, involvement they should have in various scenarios.

Public Relations

Managing communications with your customers and investors is critical to recovering from an incident successfully. What, when, and how you divulge information is of strategic importance, so you're better off leaving it to the professionals who, most likely, reside in your organization's marketing or public relations (PR) department. If your organization has a dedicated strategic communications, media, or public affairs team, it should also be involved in the response process.

As with every other aspect of IR, planning and practice are the keys to success. When it comes to the PR team, this fact may be more applicable than it is with some other teams. These individuals, who are probably only vaguely aware of the intricate technical details of a compromise and incident response, will be the public face of the incident to the broad community. Their main goal is to mitigate any damage to the customers' and investors' trust in the organization. To do this, they need to have just the right amount of technical information, must be able to present it in a manner that is approachable to broad audiences, and must present information that can be dissected into effective sound bites (or "tweets"). For this, the PR team will rely heavily on members of the technical team who are able to translate "techno-speak" into something the average person can understand.

 EXAM TIP When you see references on the exam to the public relations team, think of this as whatever part of the organization communicates directly with the general public. Don't overthink the question if your organization calls this team something else.

Internal Staff

The composition of the technical team that responds to an incident will usually depend on the incident itself. Some responses will involve a single analyst, while others may involve dozens of technical personnel from many different departments. Clearly, there is no one-size-fits-all team; you need to pull in the right people to deal with each problem. The part that should be prescribed ahead of time is the manner in which we assemble the team and, most importantly, who is calling the shots during the various stages of incident response. If you don't build this into your plan, and then periodically test it, you will likely lose precious hours (or even days) in the "food fight" that will likely ensue during a major incident.

A best practice is to leverage your risk management plan to identify likely threats to your systems. Then, for every threat possibility (or at least the major ones), you can "wargame" a response in a handful of ways. At each major decision point in the process, you should ask, who decides? Whatever your answer, the next question should be, does that person have the required authority? If the person is lacking authority, you have to determine whether someone else should make the decision or whether that authority should be delegated in writing to the decider. These are important determinations: you don't want to be in the midst of an incident response and have to sit on your hands for a few hours while the decision is vetted up and down the corporate chain.

Tales from the Trenches: Pulling the Plug

We recently ran a large, multisector cyber-exercise for a major US city. At one point, the red team attackers compromised the domain controller for a financial institution and created a new domain admin account with which they were expanding their footprint. An astute analyst in the security operations center (SOC) detected this and sent a change request to have the host taken offline. The SOC team watched the box and could tell the attacker was not active, but it was a tense wait. At the first sign of a remote login to that account, and having yet received no response from the change approval authority, the SOC director pulled the plug himself. When the exercise referees challenged him on his "unauthorized" move, he was able to produce a response plan that explicitly delegated the authority to take systems offline if they appeared to have been compromised and posed an immediate risk to the security of the network. He was able to stop an attack quickly because his organization had anticipated this scenario and granted the appropriate authority to the technical staff. The red team was not happy.

Contractors and External Parties

No matter how skilled or well-resourced an internal technical team is, you may have to bring in "hired guns" at some point. Very few organizations, for example, are capable of responding to incidents involving nation-state offensive operators. Calling in the cavalry, however, requires a significant degree of prior coordination and communication. Apart from the obvious service contract with the IR firm, you have to plan and test exactly how they would come into your facility, what they would have access to, who would be watching and supporting them, and what (if any) parts of your system are off limits to them. These IR companies are very experienced in doing this sort of thing and can usually provide a step-by-step guide as well as templates for nondisclosure agreements (NDAs) and contracts. What they cannot do for you is train your staff (technical or otherwise) on how to deal with them once they descend upon your networks. This is where rehearsals and tests come in handy—in communicating to every stakeholder in your organization what a contractor response would look like and what each role would be.

It is possible to go too far in embedding IR contractors, however. Some organizations outsource all IR as a perceived cost-saving measure. Their rationale is that they pay only for what they need, because qualified personnel are hard to find, slow to develop in-house, and expensive. The truth of the matter, however, is that this approach is fundamentally flawed in at least two ways: First, incident response is inextricably linked with critical business processes whose nuances are difficult for third parties to grasp. This is why you will always need at least one qualified, hands-on incident responder who is part of the organization and can at least translate technical actions into business impacts. Second, IR can be at least as much about interpersonal communications and trust as it is about technical controls. External parties will have a much more difficult time dealing with the many individuals involved. One way or another, you are better off having some internal IR capability and augmenting it to a lesser or greater degree with external contractors.

Law Enforcement

A number of incidents will require you to involve a law enforcement agency (LEA). Sometimes, the laws that establish these requirements also have very specific timelines, lest you incur civil or even criminal penalties. In other cases, there may not be a requirement to involve an LEA, but it may be a very good idea to do so. If you (or your team) don't know which incidents fall into the two categories of required and recommended reporting, you may want to put that pretty high on your priority conversations list with your leadership and legal counsel.

When an LEA is involved, they will bring their own perspective on the response process. Whereas you are focused on mitigation and recovery, and management is keen on business continuity, the LEA will be driven by the need to preserve evidence (which should be, but is not always, an element of your IR plan anyway). These three sets of goals can be at odds with each other, particularly if you don't have a thorough, realistic, and rehearsed plan in place. If your first meeting with representatives from an LEA occurs during an actual incident response, you will likely struggle with it more than you would if you rehearse this part of the plan before an incident occurs.

Senior Leadership

Incident response almost always has some level of direct impact (sometimes catastrophic) on an organization's business processes. For this reason, the IR team should include key senior leaders from every affected business unit. Their involvement is more than about providing support; it will help shape the response process to minimize disruptions, address regulatory issues, and provide an interface into the affected personnel in their units as well as to higher level leaders within the organization. Effective incident response efforts almost always require the direct and active involvement of management as part of a multidisciplinary response team.

Integrating these business leaders into the team is not a trivial effort. Even if they are as knowledgeable and passionate about cybersecurity as you are (which is exceptionally rare in the wild), their priorities will oftentimes be at odds with yours. Consider a compromise of a server that is responsible for interfacing with your internal accounting systems and your external payment processing gateway. You know that every second you keep that compromised box on the network, you risk further compromises or massive exfiltration of customer data. Still, every second the box is off the network will cause the company significantly in terms of lost sales and revenue. If you approach the appropriate business managers for the first time when you are facing this serious situation, things will not go well for anybody. If, on the other hand, there is a process in place with which they're all familiar and supportive, the outcome will be better, faster, and less risky.

It's not always required for senior leaders to get involved in incidents. In fact, they are unlikely to be involved in any but the most serious of incidents, but you still need their buy-in and support to ensure that you get the appropriate resources from other business areas. Keeping leaders informed of situations in which you may need their support is a balancing act—you don't want to take too much of their time (or bring them into an active role), but they need to have enough awareness of the incident that a short call to them for help will make things happen.

Another way in which members of management are stakeholders for IR is not so much in what they do, but in what they don't do. Consider an incident that takes priority over some routine upgrades you were supposed to do for one of your business units. If that unit's leadership is not aware of what IR is in general, or of the importance of the ongoing response in particular, it could create unnecessary distractions at a time when you can least afford them. Effective communications with leadership can build trust and provide you a buffer in times of need.

Factors Contributing to Data Criticality

Although we take measures to protect all kinds of data on our networks, there are some types of data that need special consideration with regard to storage and transmission. Unauthorized disclosure of the following types of data may have serious, adverse effects on the associated business, government, or individual.

Personally Identifiable Information

Personally identifiable information (PII) is data that can be used to identify an individual. This information can be unique, such as a Social Security number or biometric profile, or it may be used with other data to trace back to an individual, as is the case with name and date of birth. This information is often used by criminals to conduct identity theft, fraud, or any other crime that targets an individual. Depending on the regulatory environment in which your organization operates, you may have to meet additional requirements with respect to the handling of PII, in addition to following federal and state laws. The US Privacy Act of 1974, for example, established strict rules regarding the collection, storage, use, and sharing of PII when it is provided to federal entities. In the US Department of Defense (DoD), documents that contain PII are required to have appropriate markings or a data cover sheet, which is shown in Figure 15-2.

PII is sometimes referred to as *sensitive personal information*. Often used in a government or military context, this type of data may include biometric data, genetic information, sexual orientation, and membership in professional or union groups. This data requires protection because of the risk of personal harm that could result from its disclosure, alteration, or destruction.

Privacy Act Data Cover Sheet

To be used on
all documents
containing personal
information

DOCUMENTS ENCLOSED ARE SUBJECT TO THE PRIVACY ACT OF 1974

Contents shall not be disclosed, discussed, or shared with individuals unless they have a direct need-to-know in the performance of their official duties. Deliver this/these document(s) directly to the intended recipient. **DO NOT** drop off with a third-party.

The enclosed document(s) may contain personal or privileged information and should be treated as "For Official Use Only." Unauthorized disclosure of this information may result in CIVIL and CRIMINAL penalties. If you are not the intended recipient or believe that you have received this document(s) in error, do not copy, disseminate or otherwise use the information and contact the owner/creator or your Privacy Act officer regarding the document(s).

Privacy Act Data Cover Sheet

DD FORM 2923, SEP 2010

Figure 15-2   Department of Defense Form 2923, Privacy Act Data Cover Sheet

PART IV

Personal Health Information

The Health Insurance Portability and Accountability Act of 1996 (HIPAA) is a law that establishes standards to protect individuals' personal health information (PHI). PHI is any data that relates to an individual's past, present, or future physical or mental health conditions. Usually, this information is handled by a healthcare provider, employer, public health authority, or school. HIPAA requires appropriate safeguards to protect the privacy of PHI, and it regulates what can be shared and with whom with and without patient authorization. HIPAA prescribes specific reporting requirements for violations, with significant penalties for HIPAA violations and the unauthorized disclosure of PHI, including fines and jail sentences for criminally liable parties.

High-Value Assets

High value assets (HVAs) include information or systems that are critical to the function of an organization. Loss of access to this data would have catastrophic impact to the organization's ability to operate because of the presence of sensitive controls, instructions, or operating data housed in the system. Given its importance, this data is often targeted by adversaries seeking to disrupt an organization via data destruction or ransomware attacks.

Payment Card Information

Consumer privacy and the protection of financial data has been an increasingly visible topic in recent years. Mandates by the European Union's General Data Protection Regulation (GDPR), for example, have introduced a sweeping number of protections for the handling of personal data, which includes financial information. Although similar modern legislation is being developed in the United States, some existing rules apply to US financial services companies and the way they handle this information. The Gramm-Leach-Bliley Act (GLBA) of 1999, for example, covers all US-regulated financial services corporations. The GLBA applies to banks, securities firms, and most entities handling financial data and requires that these entities take measures to protect their customers' data from threats to its confidentiality and integrity. The Federal Trade Commission's Financial Privacy Rule governs the collection of customers' personal financial information and identifies requirements regarding privacy disclosure on a recurring basis. Penalties for noncompliance include fines for institutions of up to $100,000 for each violation. Individuals found in violation may face fines of up to $10,000 for each violation and imprisonment for up to five years.

In previous chapters, we discussed the importance of having technical controls in place to remain compliant with standards such as the Payment Card Industry Data Security Standard (PCI DSS). PCI DSS was created to reduce credit card fraud and protect cardholder information. As a global standard for protecting stored, processed, or transmitted data, it prescribes general guidelines based on industry best practices, as shown in Figure 15-3.

Goals	PCI DSS Requirements
Build and Maintain a Secure Network and Systems	1. Install and maintain a firewall configuration to protect cardholder data 2. Do not use vendor-supplied defaults for system passwords and other security parameters
Protect Cardholder Data	3. Protect stored cardholder data 4. Encrypt transmission of cardholder data across open, public networks
Maintain a Vulnerability Management Program	5. Protect all systems against malware and regularly update anti-virus software or programs 6. Develop and maintain secure system and applications
Implement Strong Access Control Measures	7. Restrict access to cardholder data by business need to know 8. Identify and authenticate access to system components 9. Restrict physical access to cardholder data
Regularly Monitor and Test Networks	10. Track and monitor all access to network resources and cardholder data 11. Regularly test security systems and processes
Maintain an Information Security Policy	12. Maintain a policy that addresses information security for all personnel

Figure 15-3 PCI DSS goals and requirements for merchants and other entities involved in payment card processing

PCI DSS does not specifically identify what technologies should be used to achieve the associated goals. Rather, it offers broad requirements that may have multiple options for compliance. Though used globally, PCI DSS is not federally mandated in the United States. Although some US states enact their own laws to prevent unauthorized disclosure and abuse of payment card information, PCI DSS remains the de facto standard for payment card information protection.

Intellectual Property

Intellectual property, the lifeblood of a business, includes the business's collective knowledge about how to create some unique thing, or the actual unique creations that enable an organization to distinguish itself from its competition. As with tangible property owners, there are laws that govern the rights of an intellectual property owner. As a security professional, you should be aware of the intellectual property that resides on the network so that you can implement appropriate measures to prevent its unauthorized disclosure. Even with the best technical measures in place to protect their intellectual property, companies are still vulnerable to its exposure. Your IR policy must incorporate the latest legal guidance to ensure that employees understand the importance of protecting this information as well as the consequences of unauthorized disclosure for both the company and themselves.

Intellectual Property Types

Intellectual property encompasses four general categories: patents, copyrights, trademarks, and trade secrets. When an inventor develops a new and useful process or thing and patents it, the patent provides the holder the exclusive privilege to make, use, market, and sell that process or thing. Copyright is the tangible manifestation of an original creative expression, whether it's a book, a musical piece, a painting, or even an architectural design. Copyright protection extends to works that are published and unpublished. It's important to note that under the "fair use" doctrine, any criticism, commentary, or teaching based on the copyrighted work may be used to justify a violation of the copyright. A trademark, which includes brand name, is the unique sign, design, or expression that a business uses to differentiate it or its products from other businesses or products. Unlike trademarks, which companies want the public to be aware of, trade secrets are specially protected intellectual property that details how a company produces something it considers of value. Trade secrets are different from the other forms of intellectual property in that their details are usually not disclosed to any registration party or otherwise.

NOTE   Intellectual property laws vary by country. The protections granted by patents issued by the US Patent Office, for example, are enforceable only in the United States and US territories. It's important to understand the local laws that govern intellectual property where you operate so that you know what is protected under these laws.

Operation Aurora

In 2010, Google disclosed that it was the victim of a sophisticated attack that appeared to have two distinct goals: monitor the e-mail communications of human rights activists and gain control of sensitive source code. The campaign, determined to be part of a larger campaign called "Operation Aurora," was linked to several other breaches and intellectual property thefts at high-profile companies such as Adobe, Juniper Networks, Northrop Grumman, and Morgan Stanley. The attackers used a zero-day exploit for Internet Explorer, encrypted tunneling, and clever obfuscation methods, indicative of an advanced persistent threat (APT) actor. The exploit enabled malware to load onto corporate computers, at which point intellectual property could be funneled out of the network. Google and others claimed that the operation was perpetrated by the Chinese government. As a direct result, Google shut down its search engine service in China, despite the country having more than twice as many Internet users as the United States.

Corporate Confidential Information

Information about the internal operations of a company, or *corporate confidential information,* may include correspondence about upcoming changes to the company hierarchy, details about a marketing campaign, or any other information that may not be suitable for public consumption. Corporate confidential information is often referred to as proprietary information. You will often see the markings on corporate documents indicating that dissemination of the information should be tightly controlled.

Accounting Data

Financial data about a company requires special handling and protection in a similar way to other sensitive data, even if it may not reveal PHI or PII. In a way, financial data gives insight into the health of an organization and should be treated similarly to PHI. Corporate policies and legal guidance may dictate the general procedures for handling and storing the information, but you should take extra steps to protect accounting data from those within the company who do not have a "need to know."

Mergers and Acquisitions

Data about upcoming company mergers and acquisitions is another type of sensitive corporate information whose misuse is most often associated with fraud and conspiracy. If information about an upcoming acquisition were to be prematurely disclosed because of a malicious actor, it could have grave consequences on the finances of both companies. Companies about to be acquired may be vulnerable to manipulation or loss of their competitive advantage. If an employee trades a public company's stock because of privileged knowledge about a company's finances or a pending acquisition that is not yet public knowledge, he has committed a serious crime called *insider trading,* which can incur both civil and criminal penalties. The US Securities and Exchange Commission's Fair Disclosure regulation mandates that if special knowledge about a company is disclosed to one shareholder, it must be disclosed to the public.

Chapter Review

Regardless of the size of your organization, it will eventually be a target of an attacker. The methods used by attackers to conduct their campaigns and the nature of the attacks are constantly evolving. Individual actors, activist groups, and nation-states have far easier access to malicious software than ever before. The question becomes less *if* you will have an incident and more *when*. Ignoring this fact will place you and your organization in a precarious position.

Proper IR planning will indicate to your customers, stakeholders, and key leadership that you take security seriously and will instill confidence in the company. Should an incident occur, your preparation will enable you to identify the scope of damage quickly, because you will have identified the data that requires special handling and protection, including PII, PHI, intellectual property, corporate confidential information, and financial information about your organization.

PART IV

Certain types of security incidents are affected by a variety of regulatory requirements that must be addressed. Between laws that require organizations to notify customers that their information was put at risk and stipulations of the rules of your organization's regulatory environment, preparation and response isn't just an effort for your security team. You must assist organizational leadership in communicating the goals of the security policy and the importance of the employees' roles in supporting it. Aside from the benefit of having a smoother recovery, having a comprehensive incident-handling process regarding special data may protect you, your company, and company leaders from civil or criminal procedures should your organization be brought to court for failing to protect sensitive data. Once you've gotten buy-in from organization leadership for your IR plan, you need to continue to refine and improve it as threats evolve. If an unfortunate situation occurs and you need to put the plan into play, you will have the confidence and support to meet the challenge successfully.

Questions

1. Of the following types of data, which is likeliest to require notification of US government entities if it is compromised, lest your organization incur fines or jail sentences?

   A. Personally identifiable information (PII)

   B. Accounting data

   C. Personal health information (PHI)

   D. Intellectual property

2. Which of the following is *not* one of the four categories of protected intellectual property?

   A. Patents

   B. Trade secrets

   C. Items covered by the fair use doctrine

   D. Copyrights

3. What makes disclosure of information about mergers and acquisitions so important?

   A. Its disclosure may violate Securities and Exchange Commission's regulations and trigger notification requirements.

   B. The information is regulated by the PCI DSS.

   C. If the information is disclosed to the public, it could lead to charges of insider trading.

   D. The information can give an unfair advantage to the company being acquired.

4. When decisions are made to make changes that will incur significant costs or to involve law enforcement organizations in an incident, which of the following parties will be notified?

   A. Senior leaders

   B. Contractors

   C. Public relations staff

   D. Technical staff

5. When would you consult your legal department as a part of an incident response?

   A. In the case of a loss of more than 1 terabyte of data

   B. Immediately after the discovery of the incident

   C. In cases of compromise of sensitive information such as PHI

   D. When business processes are at risk because of a failed recovery operation

6. What term refers to members of your organization who have a role in helping with some aspects of some incident responses?

   A. Public relations

   B. Insiders

   C. Shareholders

   D. Stakeholders

7. What is the most appropriate course of action regarding communication with organizational leadership in the event of an incident?

   A. Provide details only if you're unable to restore services.

   B. Provide details until after law enforcement is notified.

   C. Forward the full technical details on the affected server(s).

   D. Provide updates on your progress and estimated time of service restoration.

8. What incident response tool should be in place to ensure that the appropriate staff is alerted when it is determined that they may be needed?

   A. Indicators of compromise list

   B. Call list

   C. Triage playbook

   D. Developers' e-mails

Answers

1. **C.** The protection of personal health information (PHI) is strictly regulated in the United States, and its disclosure could result in civil or criminal penalties. None of the other types of information listed is normally afforded this level of sensitivity.

2. **C.** Items covered by the fair use doctrine are not one of the four categories of intellectual property, though they are protected by copyright. Intellectual property comprises four categories: patents, copyrights, trademarks, and trade secrets.

3. **A.** The Securities and Exchange Commission's Fair Disclosure regulation mandates that if information about a company merger or acquisition is disclosed to one shareholder, it must be disclosed to the public as well; otherwise, the shareholder has an unfair advantage and could be charged with insider trading. This type of information must be carefully controlled.

4. **A.** Decisions to employ changes that will incur significant costs or to involve external law enforcement agencies will likely require the involvement of organizational senior leadership. Senior leaders will provide guidance with regard to company priorities, will assist in addressing regulatory issues, and will provide the support necessary to move through the IR process.

5. **C.** Regulatory reporting requirements must be met when dealing with compromises of sensitive data such as protected health information. Because compromises of PHI can lead to civil penalties or even criminal charges, it is important that you consult legal counsel.

6. **D.** Stakeholders are individuals and teams who are part of your organization and have a role in helping with some aspects of some incident response.

7. **D.** Organizational leadership should be given enough information to enable them to provide guidance and support. Management needs to be closely involved in critical decision-making points.

8. **B.** A call list provides escalation guidance and contact information for personnel who should be contacted in the event of an incident.

Appropriate Incident Response Procedures

In this chapter you will learn:

- The major steps of the incident response cycle
- How to prepare for security incidents
- Incident detection and analysis techniques
- How to contain, or reduce the spread, of an incident
- How to eradicate and recover from an incident
- What to do post incident

Predicting rain doesn't count. Building arks does.

—Warren Buffett

Although we commonly use the terms interchangeably, there are subtle differences between an *event,* which is any occurrence that can be observed, verified, and documented, and an *incident,* which is one or more related negative events that compromise an organization's security posture. *Incident response* is the process of negating the effects of an incident on an information system.

In Chapter 1, we covered the benefits of using the intelligence cycle in enabling us to understand and scale key intelligence tasks by breaking them down into distinct phases. As with generating and sharing intelligence, incident response benefits from using a framework to help us more easily understand an adversary's activities and respond to them using repeatable, scalable methods. The incident response cycle comprises the major steps required to prepare for an intrusion, identify the activities associated with it, develop and deploy the analytical techniques to understand it, and execute the plans to return the system to an operational state. In this chapter we'll cover these phases, which are shown in Figure 16-1, and we'll discuss techniques associated with each step and how these defensive efforts can make our responses more effective.

There are many incident response models, but all share some basic characteristics. They all require us to take some preparatory actions before anything bad happens, to identify and analyze an event to determine the appropriate counteractions, to correct the problem(s), and finally to keep the incident from happening again. Clearly, efforts to prevent future occurrences tie back to our preparatory actions, to create a cycle.

387

Figure 16-1
The incident
response cycle

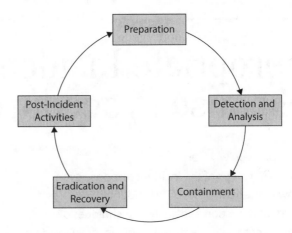

Preparation

As highlighted several times throughout the book, preparing for security incidents requires a sound methodology that aims to reduce the amount of uncertainty as much as possible. By prescribing technical and operational best practices, we may effectively protect critical business from compromise and sensitive data from exposure by identifying key assets and response priorities. Though we do as much as we can to be proactive in stopping threats, we must face the reality that at some point, something will get through, so we must be prepared to recognize and remediate as quickly and completely as possible. In thinking about how a security team might approach preparing for the unknown, we need to keep a few things in mind. Because preparation will likely affect a wide range of teams outside of the security team itself, it's important that we get buy-in from all levels of the organization as soon as possible.

Preparation will involve many technical and nontechnical steps. For the purpose of the CySA+ exam, we'll cover three elements of preparation: training, testing, and documentation. Although many of the technical steps associated with each element will not involve everyone in the organization, it will require the cooperation of teams directly impacted by changes to the network and its operations. For example, installing new security devices, deploying detection signatures, and ensuring that patches are applied may not need direct involvement from end users, but the organization's network architecture, systems administrators, and IT support teams must be aware so that they can make the appropriate modifications on their end to enforce and validate these changes. For all users in the organization, the processes and documentation techniques used must be as user-friendly as possible, particularly for a nontechnical audience.

Training

While laying the groundwork for effective response, you'll need to ensure that the first few steps performed by everyone involved are the right ones. After all, the more quickly and accurately a team can identify an issue, the better off you all will be as you work through the entire incident response (IR) process. Seeing the signs becomes will be easier as you and your team gain experience, but providing good training on the fundamentals can save the organization time and money in the long run. Training should include technical training for the IR team, college courses, or various professional education classes. Several universities well known for their computer science and security programs offer distance learning, which makes it easier for students to access training opportunities.

In many cases, the nontechnical staff make up a majority of the organization, and invariably these users will be the ones who are most exposed to the signs of a potential security incident. Devising a defense against continuous and persistent threats such as phishing is imperative. Key to a collective defense is training users in how to identify such threats and how to handle situations in which they may have been tricked into providing sensitive information. To be clear, training nontechnical staff members who are targets of these types of malicious activities should include training in how to communicate what they observe. This will significantly raise the organization's level of preparation.

Testing

In IR, as with so many tasks in life, practice makes perfect. And getting to a state of proficiency takes time and discipline. Practice sessions are useful in gauging the team's ability to respond appropriately, identifying areas for improvement, and increasing team members' confidence in themselves and in the IR process. Without practice that simulates live conditions, incident responders may not be able to articulate their observations in a realistic manner. Through realistic testing, your team members can validate assumptions and dispel unhelpful preconceptions about the incidents they may face.

There are many parallels between the development and execution of IR and military operations plans. For the last few decades, the US military has participated in a large-scale multinational series of exercises called Cobra Gold. Held annually since 1982, Cobra Gold began as a way for the United States and Thailand to strengthen ties between their militaries. While the focus has changed from year to year, the goals of having a venue for the participating nations to conduct military, humanitarian, and disaster relief exercises have remained. One of the authors had the pleasure of preparing an expeditionary communications team for multiple deployments to Thailand in support of Cobra Gold. As part of the preparation, these military units conducted various types of exercises used to evaluate its readiness, or ability to do its job. Year after year, the units were evaluated using four major types of exercises during the event. The four, listed next, can also be used to evaluate your organization's IR plan, procedures, and capabilities. As with military war gaming, the goals of IR exercises are to test strategies, vet procedures, and clarify the effects of deploying countermeasures in a manner that doesn't put any resources directly in front of the adversary.

- **Walkthrough** Walkthroughs offer the most basic kind of testing for team members. They often accompany training because they can be performed in a classroom setting, with little or no additional resources required. Walkthroughs are designed to familiarize participants with response steps, crisis communications plans, and their roles and responsibilities as defined in the plans.

PART IV

- **Tabletop exercises** Tabletop exercise are live sessions in which members of the security team come together to discuss their roles and responses to hypothetical situations. What separates tabletop exercises from simple walkthroughs is that participants are often guided through a scenario, complete with changes in the environment and simulated actor behaviors. The goal is to highlight how various teams execute their parts of the plan, what improvements may be needed, and what it's like to operate in uncertainty through discussion.

- **Functional exercises** Functional exercises enable teams to test their understanding of the IR plan, identifying and executing specific technical tasks related to the response effort. In this kind of exercise, a team may be presented with a scenario and be expected to identify the correct tool to use, how to use the tool, and how to create report results. The focus of functional exercises is often to assess an individual's mastery of a particular technique.

- **Full-scale exercises** A full-scale exercise is an experience modeled as closely as possible to a real event. Often performed in real-time, full-scale exercises test everything from the participants' detailed understanding of the organization's IR process to specific individual and team tasks. This type of exercise is useful in measuring performance compared to program objectives.

Keep in mind when formulating the testing plan that the tests and the testing process should be easy to understand for everyone involved. Understanding that plans are designed with little space for improvisation, the exercises should be designed so that all participants can clearly understand the roles they play and the tasks they are expected to perform. It's also desirable for key leaders to be involved in testing by providing initial requirements and vision, participating in a tabletop event, observing a full-scale exercise, or simply providing feedback at the conclusion of the assessments. Although the results from training aren't always immediately apparent, judging the effectiveness of a response strategy by using a well-designed testing plan can tell you if your team is on the right path, and it provides clear steps forward if adjustments are necessary.

Documentation

All good processes need to be recorded so that they can be referenced, shared, and sometimes improved upon. Documentation is arguably the most important nontechnical tool at your team's disposal. Even for the most seasoned team, it pays to have well-documented steps at the ready. Documentation extends beyond just recording the steps of your organization's IR process; any information about systems that may be useful to responders throughout the process, such as network configurations, system settings, and system points of contact, should also be included.

Detection and Analysis

Detecting and analyzing events is the first step in putting practice into play. Often referred to as identification, this step in the process may use a number of automated detection techniques to increase team efficiency. An automated detection and analysis process is much more scalable and reliable that manual processes because it can be used consistently to highlight behavior patterns of interest that a human analyst may miss.

Characteristics of Severity Level Classification

It's important to have a clear reference point to know the true scope of impact during a suspected incident—simply noting that the network seems slow will not be enough to make a good determination regarding what to do next. As responders, it's important that you lay out a clear set of criteria to determine how to classify a security event. This classification process, sometimes known as the *scope of impact,* is the formal determination of whether an event is enough of a deviation from normal operations to be called an incident and the degree to which services have been affected. Keep in mind that some actions you perform in the course of your duties as a systems administrator may trigger security devices and appear to be an attack. Documenting these types of legitimate anomalies will reduce the number of false positives and enable you to have more confidence in your alerting system. In the case of a legitimate attack, you must collect as much data as possible from sources throughout your network, such as log files from network devices. Gathering as much information as possible in this step will help you decide on the next steps for your incident responders.

Once an event is confirmed to be legitimate, you should quickly communicate with your team to identify who needs to be contacted outside of your security group and key leadership. Whom you contact may be dictated by your local policy—and in some cases, laws or regulations. Opening communication channels early will also ensure that you get the appropriate support for any major changes to the organization's resources. We covered the communications process in detail in Chapter 7.

For some organizations, the mere mention of a successful breach can be damaging, regardless of what was compromised. In 2011, for example, security company RSA was the victim of a major breach. RSA's SecureID, a line of two-factor authentication token-based products, was used by more than 40 million businesses at the time for the purposes of securing their own network. These tokens, which were the cornerstone of RSA's authentication service, were revealed to have been compromised after a thorough investigation of the incident, prompting the company to replace all 40 million tokens. Almost as damaging as the financial cost of replacing the tokens was the damage done to RSA's reputation after the story made headlines throughout the world. How could a security company be the victim of a hack? In serious cases like this, it's critical that only those playing a role in the IR and decision-makers be informed of the breach. This will help reduce confusion across the organization as a clear path forward is determined. In addition, you do not want an attacker to be tipped off that he has been discovered. As an incident responder, you must be prepared to inform leadership of your technical assessment so that they can make informed decisions in the best interest of the organization.

Downtime

Networks exist to provide resources to those who need them, when they need them. Without a network and services that are available when they need to be, nothing can be accomplished. Every other metric in determining network performance such as stability, throughput, scalability, and storage all require the network to be up. The decision on

PART IV

Figure 16-2
Maximum
tolerable
downtime

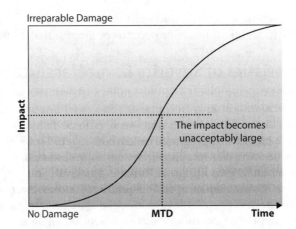

whether to take a network completely offline to handle a breach is not a small one by any measure. Understanding that a complete shutdown of the network may not be possible, you should isolate infected systems to prevent additional damage. The priority here is to prevent additional losses and minimize impacts on the organization. This is not dissimilar to triage in a hospital emergency room: your team must work quickly to perform triage on your network to determine the extent of the damage and prevent additional harm, all while keeping the organization running.

The key is to determine which of the organization's critical systems are needed for survival and estimate the outage time that can be tolerated by the company as a result of an incident. The outage time that can be endured by an organization is referred to as the *maximum tolerable downtime* (MTD), which is illustrated in Figure 16-2.

Following are some sample MTD estimates for several systems arranged by criticality. These estimates will vary from organization to organization and from business unit to business unit:

- **Nonessential** 30 days
- **Normal** 7 days
- **Important** 72 hours
- **Urgent** 24 hours
- **Critical** Minutes to hours

Each business function and asset should be placed in one of these categories, depending on how long the organization can survive without it. These estimates will help you determine how to prioritize response team efforts to restore these assets. The shorter the MTD, the higher the priority of the function in question. Thus, for example, the systems classified as Urgent should be addressed before those classified as Normal.

Recovery Time

Time is money, and the faster you can restore your network to a safe operating condition, the better it is for the organization's bottom line. Although there may be serious financial implications for every second a network asset is offline, you should not sacrifice speed for completeness. You should keep lines of communication open with organization management to determine acceptable limits to downtime. Having a sense of what the key performance indicators (KPIs) are for detection and remediation will help clear up confusion, manage expectations, and potentially enable you to demonstrate your team's preparedness should you exceed these limits. Setting recovery time limits may also be useful in the long run for the reputation of the team and may be useful in securing additional budgets for training and tools.

The recovery time objective (RTO) is oftentimes used, particularly in the context of disaster recovery, to denote the earliest time within which a business process must be restored after an incident to avoid unacceptable consequences associated with a break in business processes. The RTO value is smaller than the MTD value, because the MTD value represents the time after which an inability to recover significant operations will mean severe and perhaps irreparable damage to the organization's reputation or bottom line. The RTO assumes that there is a period of acceptable downtime. This means that an organization can be out of production for a certain period of time (RTO) and can still get back on its feet. But if it cannot get production up and running within the MTD window, the organization may be sinking too fast to recover properly.

Data Integrity

Taking down a network isn't always the goal of a network intrusion. For malicious actors, tampering with data may be enough to disrupt operations, and it may provide them with the outcome they were looking for. Financial transaction records, personal data, and professional correspondence are types of data that are especially susceptible to data tampering. There are cases when attacks on data are obvious, such as those involving ransomware. In these situations, malware will encrypt data files on a system so the users cannot access them without submitting payment for the decryption keys. However, it may not always be apparent that an attack on data integrity has taken place. It may be that you are able to discover the unauthorized insertion, modification, or deletion of data only after a detailed inspection. This illustrates why it's critical to back up data and system configurations, and to keep them sufficiently segregated from the network so that they are not themselves affected by the attack. An easily deployable backup solution will allow for very rapid restoration of services. The authors will caution, however, that much like Schrödinger's cat, the condition of any backup is unknown until a restore is attempted. In other words, having a backup alone isn't enough; it must be verified over time to ensure that it's free from corruption and malware.

Ransomware

Organized crime groups frequently set up malicious sites that serve malware convincingly disguised as games or other files. The malware contained in these files is often installed silently without user knowledge and encrypts a portion of the host's system, requiring payment for the decryption keys. For these groups, this is a source of significant and reliable income, because so many users and organizations have

(continued)

PART IV

poor backup habits. Ironically, these groups rarely renege on an exchange, because it would be very damaging to their business model. After all, if you knew that you'd never see your data again, what would be the point of submitting payment?

Economic Impacts

It's difficult to predict the second- and third-order effects of network intrusions. Even if some costs are straightforward to calculate, the complete economic impact of a network breach is difficult to quantify. A fine levied against an organization that had not adequately secured its workers' personal information is an immediate and obvious cost, but how does one accurately calculate the future losses due to identity theft, or the damage to the reputation of the organization resulting from the lack of confidence? It's critical to include questions like these in your discussion with stakeholders when determining courses of action for dealing with an incident.

Another consideration in calculating the economic scope of an incident is the value of the assets involved. The value placed on information is relative to the parties involved, what work was required to develop it, how much it costs to maintain, what damage would result if it were lost or destroyed, what enemies would pay for it, and what liability penalties could be endured. If an organization does not know the value of the information and the other assets it is trying to protect, it does not know how much money and time it should spend on protecting or restoring them. If the calculated value of a company's trade secret is $x, then the total cost of protecting or restoring it should be some value less than $x.

The previous examples refer to assessing the value of information and protecting it, but this logic also applies to an organization's facilities, systems, and resources. The value of facilities must be assessed, along with all printers, workstations, servers, peripheral devices, supplies, and employees. You do not know how much is in danger of being lost if you don't know what you have and what it is worth in the first place.

System Process Criticality

As part of your preparation, you must determine what processes are considered essential for the business's operation. These processes are associated with tasks that must be accomplished with a certain level of consistency for the business to remain competitive. Each business's list of critical processes will be different, but it's important to identify those early so that they can be the first to be brought back up during a recovery. The critical process lists aren't restricted to technical assets only; they should include the essential staff required to get these critical systems back online and keep them operational. It's important to educate members across the organization as to what these core processes are, how their work directly supports the goals of the processes, and how they benefit from successful operations. This is effective in getting the appropriate level of buy-in required for successfully responding to incidents and recovering from any resulting damage.

EXAM TIP Criticality and probability are the primary components of risk analysis. Whereas *probability* describes the chance of a future event occurring, *criticality* is the impact of that future event. Criticality is often expressed by degree, such as high, moderate, or low. Low criticality indicates little impact to business operations, moderate indicates impaired or degraded performance, and high indicates a significant impairment of business functions. For the CySA+ exam, it's important to be able to describe incident response priorities based on factors such as criticality.

Data Correlation

Security information and event management (SIEM) systems do a fantastic job of collecting and presenting massive amounts of data from disparate sources in a single view. They assist the analyst in determining whether an event is indeed malicious and what behaviors may be connected with the event. With a smart policy for how logs are captured and sent to a SIEM, we can spend more time investigating what actually happened versus trying to figure out if the data we needed even recorded. SIEMs are critical when it comes to categorizing attacks. Because a primary goal of IR is to enable the organization to figure out what happened and get back to normal operations, it helps to be able to say with a fair degree of certainty that something we observe on the network is tied to some stage of an attack. If we're able to determine with confidence that an attack is early in its attack cycle, we can prioritize actions to remove it from the network. Compare this with finding evidence of exfiltration after the fact, and you can see why time is so critical in tying into and referencing across massive data sets.

Reverse Engineering

Reverse engineering (RE) is the detailed examination of a product to learn what it does and how it works. In the context of IR, RE relates exclusively to malware. The idea is to analyze the binary code to find, for example, the IP addresses or host/domain names it uses for command and control (C2) nodes, the techniques it employs to achieve permanence in an infected host, or the unique characteristics that could be used as a signature for the malware.

Generally speaking, there are two approaches to reverse engineering malware. The first doesn't really care about what the binary *is*, but rather with what the binary *does*. This approach, sometimes called *dynamic code analysis,* requires a sandbox in which to execute the malware. This sandbox creates an environment that looks like a real operating system to the malware and provides such things as access to a file system, network interface, memory, and anything else the malware asks for. Each request is carefully documented to establish a timeline of behavior that enables us to understand what it does. The main advantage of dynamic malware analysis is that it tends to be significantly faster and requires less expertise than the alternative (described next). It can be particularly helpful for code that has been heavily obfuscated by its authors. The biggest disadvantage is that it doesn't reveal all that the malware does, but rather all that it did during its execution in the sandbox. Some malware will actually check to see if it is being run in a sandbox before doing anything interesting. Additionally, some malware doesn't immediately do anything nefarious, waiting instead for a certain condition to be met (for example, a time bomb that activates only at a particular date and time).

The alternative to dynamic code analysis is, unsurprisingly, *static code analysis*. In this approach to malware RE, a highly skilled analyst will either disassemble or decompile the binary code to translate its 1's and 0's into either assembly language or whichever higher level language it was created in. This enables a reverse-engineer to see all possible functions of the malware, not just the ones that it exhibited during a limited run in a sandbox. It is then possible, for example, to see all the domains the malware would reach out to given the right conditions, as well as the various ways in which it would permanently insert itself into its host. This last insight enables the IR team to look for evidence that any of the other persistence mechanisms exist in other hosts that were not considered infected up to that point.

NOTE In Chapter 4, we covered the function of compilers and disassemblers and the relationship between machine language and assembly language. While the goals for incident response may not be to understand the exact mechanics of malware, reverse engineering may provide insight into other indicators of compromise and allow the IR analyst to more completely describe the impact of the incident.

Containment

Once you know that a threat agent has compromised the security of your information system, your first order of business is to keep things from getting worse. Containment comprises a set of actions that attempts to deny the threat agent the ability or means to cause further damage. The goal is to prevent or reduce the spread of this incident while you strive to eradicate it. This is akin to confining highly contagious patients in an isolation room of a hospital until they can be cured to keep others from becoming infected. A proper containment process buys the IR team time for a proper investigation and determination of the incident's root cause. The containment should be based on the category of the attack (that is, whether it was internal or external), the assets affected by the incident, and the criticality of those assets. Containment approaches can be proactive or reactive. Which is the best approach depends on the environment and the category of the attack. In some cases, the best action may be to disconnect the affected system from the network. However, this reactive approach could cause a denial of service or limit functionality of critical systems.

NOTE Remember that preserving evidence is an important part of containment. You never know when a seemingly routine response will end up in court.

Segmentation

A well-designed security architecture will segment your information systems by some set of criteria such as function (for example, finance or HR) or sensitivity (for example, unclassified or secret). *Segmentation* divides a network into subnetworks (or segments) so that hosts in different segments are not able to communicate directly with each other. This can be done by either physically wiring separate networks or by logically assigning devices to separate virtual local area networks (VLANs). In either case, traffic between network segments must go through some sort of gateway device, which is oftentimes a router with the appropriate access control lists (ACLs). For example, the accounting division may have its own VLAN that prevents users in the research and development (R&D) division from directly accessing the financial data servers. If certain R&D users had legitimate needs for such access, they would have to be added to the gateway device's ACL, which could place restrictions based on source/destination addresses, time of day, or even specific applications and data to be accessed.

The advantages of network segmentation during IR should be pretty obvious: compromises can be constrained to the network segment in which they started. To be clear, it is still possible to go from one segment to another, as was the case with the R&D users example. Some VLANs may also have vulnerabilities that could enable an attacker to jump from one to another without going through the gateway. Still, segmentation provides an important layer of defense that can help contain an incident. Without it, the resulting "flat" network will make it more difficult to contain an incident.

Isolation

Although it is certainly helpful to segment the network as part of its architectural design, we already saw that this can still allow an attacker to move easily between hosts on the same subnet. As part of your preparations for IR, it is helpful to establish an isolation VLAN, much like hospitals prepare isolation rooms before any contagious patients actually need them. The IR team would then have the ability to move any compromised or suspicious hosts quickly to this VLAN until they can be further analyzed. The isolation VLAN would have no connectivity to the rest of the network, which would prevent the spread of any malware. This isolation would also prevent compromised hosts from communicating with external hosts such as C2 nodes. About the only downside to using isolation VLANs is that some advanced malware can detect this situation and then take steps to eradicate itself from the infected hosts. Although this may sound wonderful from an IR perspective, it does hinder your ability to understand what happened and how the compromise was executed so that you can keep it from happening in the future.

While a host is in isolation, the response team is safely able to observe its behaviors to gain information about the nature of the incident. By monitoring its network traffic, you can discover external hosts (for example, C2 nodes and tool repositories) that may be part of the compromise. This enables you to contact other organizations and get their help in shutting down whatever infrastructure the attackers are using. You can also monitor the compromised host's running processes and file system to see where the malware resides and what it is trying to do on the live system. All this helps you better understand the incident and come up with the best way to eradicate it. It also enables you to create indicators of compromise (IOCs) that you can then share with others such as the Computer Emergency Readiness Team (CERT) or an Information Sharing and Analysis Center (ISAC).

Removal

At some point in the response process, you may have to remove compromised hosts from the network altogether. This can happen after isolation or immediately upon noticing the compromise, depending on the situation. Isolation is ideal if you have the means to study the behaviors and gain actionable intelligence, or if you're overwhelmed by a large number of potentially compromised hosts that need to be triaged. Still, one way or another, some of the compromised hosts will come off the network permanently.

When you remove a host from the network, you need to decide whether you will keep it powered on, shut it down and preserve it, or simply rebuild it. Ideally, the criteria for making this decision are already spelled out in the IR plan. Here are some of the factors to consider in this situation:

- **Threat intelligence value** A compromised computer can be a treasure trove of information about the tactics, techniques, procedures (TTPs), and tools used by an adversary—particularly a sophisticated or unique one. If you have a threat intelligence capability in your organization and can gain new or valuable information from a compromised host, you may want to keep it running until its analysis is completed.

- **Crime scene evidence** Almost every intentional compromise of a computer system is a criminal act in many countries, including the United States. Even if you don't plan to pursue a criminal or civil case against the perpetrators, it is possible that future IR activities change your mind and would benefit from the evidentiary value of a removed host. If you have the resources, it may be worth your effort to make forensic images of the primary storage (for example, RAM) before you shut it down and of secondary storage (for example, the file system) before or after you power it off.

- **Ability to restore** It is not a happy moment for anybody in our line of work when we discover that, though we did everything by the book, we removed and disposed of a compromised computer that contained critical business information that was not replicated or backed up anywhere else. If we took and retained a forensic image of the drive, we could mitigate this risk, but otherwise, someone is going to have a bad day. This is yet another reason why you should, to the extent that your resources allow, keep as much of a removed host as possible.

The removal process should be well documented in the IR plan so that the right issues are considered by the right people at the right time. We address chain of custody and related issues in Chapter 18, but for now, suffice it so say that what you do with a removed computer can come back and haunt you if you don't do it properly.

Eradication and Recovery

Once the incident is contained, you turn your attention to the eradication process, in which you return all systems to a known-good state. It is important that you gather evidence before you recover systems, because in many cases you won't know that you need legally admissible evidence until days, weeks, or even months after an incident. It pays, then, to treat each incident as if it will eventually end up in a court of justice.

Once all relevant evidence is captured, you can fix all that was broken. The aim is to restore full, trustworthy functionality to the organization. For hosts that were compromised, the best practice is simply to reinstall the system from a gold master image and then restore data from the most recent backup that occurred prior to the attack.

 NOTE   An attacked or infected system should never be trusted, because you do not necessarily know all the changes that have taken place and the true extent of the damage. Some malicious code could still be hiding somewhere. Systems should be rebuilt to ensure they are trustworthy again.

Recovery in an IR is focused on ensuring that you have identified the corresponding attack vectors and implemented effective countermeasures against them. This stage presumes that you have analyzed the incident and verified the manner in which it was conducted. This analysis can be a separate post-mortem activity or can take place in parallel with the response.

Vulnerability Mitigation

Recovering from an incident means not only getting things back to where they were, but also not allowing your environment to be disrupted in the same way again. Mitigation involves eradicating the cause of any event or incident once it's been accurately identified. By definition, every incident occurs because a threat actor exploits a vulnerability and compromises the security of an information system. It stands to reason, then, that after recovering from an incident, you would want to scan your systems for other instances of that same (or a related) vulnerability.

Although it is true that you will never be able to protect against every vulnerability, it is also true that you have a responsibility to mitigate those that have been successfully exploited, whether or not you thought they posed a high risk before the incident. The reason is that you now know that the probability of a threat actor exploiting it is 100 percent, because it already happened. And if it happened once, it is likely to happen again absent a change in your controls. The inescapable conclusion is that, after an incident, you need to implement a control that will prevent a recurrence of the exploitation and develop a plug-in for your favorite scanner that will test all systems for any residual vulnerabilities.

For the best results, this process will require coordination with your vulnerability management process, since vulnerability mitigation activities will likely change your environment. Your team will need a method to determine, first, if the vulnerability was sufficiently addressed and, second, if any of these changes introduce new conditions that

PART IV

may make the system more vulnerable to attack. Your vulnerability management team will often be a great ally to advise IR efforts related to vulnerability mitigation, since they will likely have the most experience and resources related to identifying, prioritizing, and remediating software flaws.

Sanitization

According to NIST Special Publication 800-88 Revision 1, *Guidelines for Media Sanitization, sanitization* refers to the process by which access to data on a given medium is made infeasible for a given level of effort. These levels of effort, in the context of IR, can be cursory and sophisticated. What we call cursory sanitization can be accomplished by simply reformatting a drive. It may be sufficient against run-of-the-mill attackers who look for large groups of easy victims and don't put too much effort into digging their hooks deeply into any one victim. On the other hand, there are sophisticated attackers who may have deliberately targeted your organization and will go to great lengths to persist in your systems or, if repelled, compromise them again. This class of threat actor requires more advanced approaches to sanitization.

The challenge, of course, is that you don't always know which kind of attacker is responsible for the incident. For this reason, simply reformatting a drive is a risky approach. Instead, we recommend one of the following techniques, listed in increasing level of effectiveness at ensuring the adversary is definitely denied access to data on the medium:

- **Overwriting** Overwriting data entails replacing the 1's and 0's that represent it on storage media with random or fixed patterns of 1's and 0's to render the original data unrecoverable. This should be done at least once (for example, overwriting the medium with 1's, 0's, or a pattern of these), but it may have to be done more than that to ensure that all data is destroyed.

- **Encryption** Many mobile devices take this quick and secure approach to render data unusable. The data stored on the medium is encrypted using a strong key. To render the data unrecoverable, the system securely deletes the encryption key, which is many times faster than deleting the encrypted data. Recovering the data in this scenario is typically computationally infeasible.

- **Degaussing** This is the process of removing or reducing the magnetic field patterns on conventional disk drives or tapes. In essence, a powerful magnetic force is applied to the media, which results in the wiping of the data and sometimes the destruction of the motors that drive the platters. Note that degaussing typically renders the drive unusable.

- **Physical destruction** Perhaps the best way to combat data remanence is simply to destroy the physical media. The two most commonly used approaches to destroying media are to shred them or expose them to caustic or corrosive chemicals. Another approach is incineration.

Reconstruction

Once a compromised host's media is sanitized, your next step is to rebuild the host to its pristine state. The best approach to doing this is to ensure you have created known-good, hardened images of the various standard configurations for hosts on your network. These images are sometimes called *gold masters* and facilitate the process of rebuilding a compromised host. This reconstruction is significantly more difficult if you manually reinstall the operating system, configure it so it is hardened, and then install the various applications and/or services that were in the original host. We don't know anybody who, having gone through this dreadful process once, doesn't invest the time to build and maintain gold images thereafter.

Another aspect of reconstruction is the restoration of data to the host. Again, there is one best practice here, which is to ensure you have up-to-date backups of the system data files. This is also key for quickly and inexpensively dealing with ransomware incidents. Sadly, in too many organizations, backups are the responsibility of individual users. If your organization does not enforce centrally managed backups of all systems, your only other hope is to ensure that data is maintained in a managed data store such as a file server.

Secure Disposal

When you're disposing of media or devices as a result of an IR, any of the four techniques covered earlier (overwriting, encryption, degaussing, or physical destruction) may work, depending on the device. Overwriting is usually feasible only with regard to hard disk drives and may not be available on some solid-state drives. Encryption-based purging can be found in multiple workstation, server, and mobile operating systems, but not in all. Degaussing works on magnetic media only, but some of the most advanced magnetic drives use stronger fields to store data and may render older degaussers inadequate. Note that we have not mentioned network devices such as switches and routers, which typically don't offer any of these alternatives. In the end, the only way to dispose of these devices securely is by physically destroying them using an accredited process or service provider. This physical destruction involves the shredding, pulverizing, disintegration, or incineration of the device. Although this may seem extreme, it is sometimes the only secure alternative left.

Patching

Many of the most damaging incidents are the result of an unpatched software flaw. This vulnerability can exist for a variety of reasons, including failure to update a known vulnerability or the existence of a heretofore unknown vulnerability, also known as a zero-day. As part of the IR, the team must determine which cause is the case. The first would indicate an internal failure to keep patches updated, whereas the second would all but require notification to the vendor of the product that was exploited so a patch can be developed.

Many organizations rely on endpoint protection that is not centrally managed, particularly in a "bring-your-own-device" (BYOD) environment. As a result, if a user or device fails to download and install an available patch, it causes a vulnerability, which can become an incident. If this is the case in your organization and you are unable to change the policy to require centralized patching, you should also assume that some number of endpoints will fail to be patched, so you should develop compensatory controls elsewhere in your security architecture. For example, by implementing network access control (NAC), you can test any device attempting to connect to the network for patching, updates, antimalware, and any other policies you want to enforce. If the endpoint fails any of the checks, it is placed in a quarantine network that may allow Internet access (particularly for downloading patches) but keeps the device from joining the organizational network and potentially spreading malware.

If, on the other hand, your organization uses centralized patches and updates, the vulnerability was known, and still it was successfully exploited, this points to a failure within whatever system or processes you are using for patching. Part of the response would be to identify the failure, correct it, and then validate that the fix is effective at preventing a repeated incident in the future.

Restoration of Permissions

For many organizations, recovery is an effort to restore business operations quickly and efficiently, and it also addresses the issues that contributed to that incident. Although recovery operations will ideally leave you in a better place than before the incident, just getting access back at the same level is a win for many security teams. This often means that a reliable backup solution is in place and that backups have been tested and verified as part of the overall IR process. The rebuilding phase is not the time to discover that a backup you relied on was never verified.

Validating Permissions

There are two principal reasons for validating permissions before you wrap up your IR activities. The first is that inappropriately elevated permissions may have been a cause of the incident in the first place. It is not uncommon for organizations to allow excessive privileges for their users. One of the most common reasons we've heard is that if the users don't have administrative privileges on their devices, they won't be able to install whatever applications they'd like to try out in the name of improving their efficiency. Of course, we know better, but this may be an organizational culture issue that is beyond your power to change. Still, documenting the incidents (and their severity) that are the direct result of excessive privileges may, over time, move the needle in the direction of common sense.

Not all permissions issues can be blamed on end users. Time and again, we've seen system or domain admins who do all their work (including surfing the Web) using their admin account. Furthermore, most of us have heard of (or had to deal with) the discovery that a system admin who left the organization months or even years ago still has a valid account. The aftermath of an IR provides a great opportunity to double-check on issues like these.

Finally, it is very common for interactive attackers to create or hijack administrative accounts so that they can do their nefarious deeds undetected. Although it may be odd to see an anonymous user in Russia accessing sensitive resources on your network, you probably wouldn't get too suspicious if you saw one of your fellow admin staff members moving those files around. If there is any evidence that the incident leveraged an administrative account, it would be a good idea to delete that account and, if necessary, issue a new one to the victimized administrator. While you're at it, you may want to validate that all other accounts are needed and protected.

Restoration of Services and Verification of Logging

There are two key aspects of restoring services that the security team will be directly involved in. The first is developing and executing out the processes for network service validation and testing to certify all systems as operational and provide the requisite level of service. The second, and less obvious aspect, is to ensure that, along with certifying that any vulnerable services are taken offline, the team has the necessary visibility and logging to detect any future issues.

Post-Incident Activities

No effective business process would be complete without some sort of introspection or opportunity to learn from and adapt to our experiences. This is the role of the corrective actions phase of an incident response. It is here that we apply the lessons learned and information gained from the process to improve our posture in the future.

Lessons-Learned Report

In our time in the Army, it was virtually unheard of to conduct any sort of operation (training or real world), or run any event of any size, without having a "hotwash" (a quick huddle immediately after the event to discuss the good, the bad, and the ugly) and/or an after-action review (AAR) to document issues and recommendations formally. It has been very heartening to see the same diligence in most nongovernmental organizations in the aftermath of incidents. Although there is no single best way to capture lessons learned, we'll present one that has served us well in a variety of situations and sectors.

The general approach is that every participant in the operation is encouraged or required to provide his or her observations in the following format:

- **Issue**   A brief (usually a single sentence) label for an important (from the participant's perspective) issue that arose during the operation

- **Discussion**   A (usually a paragraph long) description of what was observed and why it is important to remember or learn from it for the future

- **Recommendation**   A recommendation that usually starts with a "sustain" or "improve" label if the contributor felt the team's response was effective or ineffective (respectively)

Every participant's input is collected and organized before the AAR. Usually all inputs are discussed during the review session, but occasionally the facilitator will choose to disregard some if he or she thinks they are repetitive (of others' inputs) or would be detrimental to the session. As the issues are discussed, they are refined and updated with other team members' inputs. At the conclusion of the AAR, the group (or the person in charge) decides which issues deserve to be captured as lessons learned, and those find their way into a final report. Depending on your organization, these lessons-learned reports may be sent to management, kept locally, and/or sent to a higher echelon clearinghouse.

Change Control Process

During the lessons learned or AAR process, the team will discuss and document important recommendations for changes. Although these changes may make perfect sense to the IR team, you must be careful about assuming that they should automatically be made. Every organization should have some sort of change control process. Oftentimes, this mechanism takes the form of a change control board (CCB), which consists of representatives of the various business units as well as other relevant stakeholders. Whether or not there is a board, the process is designed to ensure that no significant changes are made to any critical systems without careful consideration by all who may be affected.

Going back to an earlier example about an incident triggered by a BYOD policy in which every user could control software patching on their own devices, it is possible that the incident response team will determine that this is an unacceptable state of affairs and recommend that all devices on the network be centrally managed. This decision makes perfect sense from an information security perspective, but it would probably face some challenges in the legal and human resources departments. The change control process is the appropriate way to consider all perspectives and arrive at sensible and effective changes to the systems.

Updates to Response Plan

Regardless of whether the change control process implements any of the recommendations from the IR team, the response plan should be reviewed and, if appropriate, updated. Whereas the change control process implements organization-wide changes, the response team has much more control over the response plan. Absent sweeping changes, some compensation can happen at the IR team level.

As shown earlier in Figure 16-1, incident management is a process. In the aftermath of an event, we take actions that enable us to prepare better for future incidents, which starts the process all over again. Any changes to this lifecycle should be considered from the perspectives of the stakeholders with which we started this chapter. This will ensure that the IR team is creating changes that make sense in the broader organizational context. To get stakeholders' perspectives, establishing and maintaining positive communications is paramount.

Summary Report

The post-incident report can be a very short one-pager or a lengthy treatise; it all depends on the severity and impact of the incident. Whatever the case, you must consider who will read the report and what interests and concerns will shape the manner in which they interpret it. Before you even begin to write it, you should consider one question: What is the purpose of this report? If the goal is to ensure that the IR team remembers some of the technical details of the response that worked (or didn't), then you may want to write it in a way that persuades future responders to consider these lessons. This report would be very different from a report intended to persuade senior management to modify a popular BYOD policy to enhance security even if some are unhappy as a result. In the first case, the report would likely be technologically focused, whereas in the latter case, it would focus on the business's bottom line.

Indicator of Compromise Generation

We already mentioned the creation of IOCs as part of isolation efforts in the containment phase of the response. Now you can leverage those IOCs by incorporating them into your network monitoring plan. Most organizations would add these indicators to rules in their intrusion detection or prevention system (IDS/IPS). You can also cast a wider net by providing the IOCs to business partners or even competitors in your sector. This is where organizations such as the US-CERT and the ISACs can be helpful in keeping large groups of organizations protected against known attacks.

Monitoring

So you have successfully responded to the incident, implemented new controls, and ran updated vulnerability scans to ensure that everything is on the up and up. These are all important preventive measures, but you still need to ensure that you improve your ability to react to a return by the same (or a similar) actor. Armed with all the information on the adversary's TTPs, you now need to update your monitoring plan to help you better detect similar attacks.

Chapter Review

This chapter sets the stage for the rest of our discussion on incident responses. It started and ended with a focus on the interpersonal element of IR. Even before we discussed the technical process itself, we discussed the preparatory steps that you as a security analyst, along with your greater security team, must take. Many an incident has turned into an RGE (resume-generating event) for highly skilled responders who did not understand the importance of practice, execution, and communication.

The technical part, by comparison, is a lot more straightforward. The incident recovery and post-incident response process consists of five discrete phases: preparation, detection and analysis, containment, eradication and recovery, and post-incident activities. Your effectiveness in this process is largely dictated by the amount of preparation you and your teammates put into it. If you have a good grasp on the risks facing your organization, develop a sensible plan, and rehearse it with all the key players periodically, you will likely do very well when your adversaries breach your defenses. In the next few chapters, we get into the details of the key areas of technical response.

Questions

1. The process of dissecting a sample of malicious software to determine its purpose is referred to as what?

    A. Segmentation

    B. Frequency analysis

    C. Traffic analysis

    D. Reverse engineering

2. During the IR process, when is a good time to perform a vulnerability scan to determine the effectiveness of corrective actions?

    A. Change control process

    B. Reverse engineering

    C. Removal

    D. Eradication and recovery

3. What is the key goal of the containment stage of an IR process?

    A. To limit further damage from occurring

    B. To get services back up and running

    C. To communicate goals and objectives of the IR plan

    D. To prevent data follow-on actions by the adversary exfiltration

Use the following scenario to answer Questions 4–8:

You receive an alert about a compromised device on your network. Users are reporting that they are receiving strange messages in their inboxes and having problems sending e-mails. Your technical team reports unusual network traffic from the mail server. The team has analyzed the associated logs and confirmed that a mail server has been infected with malware.

4. You immediately remove the server from the network and route all traffic to a backup server. What stage are you currently operating in?

    A. Preparation

    B. Containment

    C. Eradication

    D. Validation

5. Now that the device is no longer on the production network, you want to restore services. Before you rebuild the original server to a known-good condition, you want to preserve the current condition of the server for later inspection. What is the first step you want to take?

   A. Format the hard drive.

   B. Reinstall the latest operating systems and patches.

   C. Make a forensic image of all connected media.

   D. Update the antivirus definitions on the server and save all configurations.

6. Your team has identified the strain of malware that took advantage of a bug in your mail server version to gain elevated privileges. Because you cannot be sure what else was affected on that server, what is your best course of action?

   A. Immediately update the mail server software.

   B. Reimage the server's hard drive.

   C. Write additional firewall rules to allow only e-mail–related traffic to reach the server.

   D. Submit a request for next-generation antivirus for the mail server.

7. Your team believes it has eradicated the malware from the primary server. You attempt to bring affected systems back into the production environment in a responsible manner. Which of the following tasks will *not* be a part of this phase?

   A. Applying the latest patches to server software

   B. Monitoring network traffic on the server for signs of compromise

   C. Determining the best time to phase in the primary server into operations

   D. Using a newer operating system with different server software

8. Your team has successfully restored services on the original server and verified that it is free from malware. What activity should be performed as soon as practical?

   A. Preparing the lessons-learned report

   B. Notifying law enforcement to press charges

   C. Notifying industry partners about the incident

   D. Notifying the press about the incident

Answers

1. **D.** Reverse engineering is the process of decomposing malware to understand what it does and how it works.

2. **D.** Additional scanning should be performed during validation, part of the recovery process, to ensure that no additional vulnerabilities exist after remediation.

3. **A.** The goal of containment is to prevent or reduce the spread of this incident while you strive to eradicate it.

4. B. Containment is the set of actions that attempts to deny the threat agent the ability or means to cause further damage.

5. C. Since unauthorized access of computer systems is a criminal act in many areas, it may be useful to take a snapshot of the device in its current state using forensic tools to preserve evidence.

6. B. Generally, the most effective means of disposing of an infected system is a complete reimaging of a system's storage to ensure that any malicious content was removed and to prevent reinfection.

7. D. The goal of the IR process is to get services back to normal operation as quickly and safely as possible. Introducing completely new and untested software may introduce significant challenges to this goal.

8. A. Preparing the lessons-learned report is a vital stage in the process after recovery. It should be performed as soon as possible after the incident to record as much information and complete any documentation that may be useful for the prosecution of the incident and to prevent future incidents from occurring.

Analyze Potential Indicators of Compromise

In this chapter you will learn:

- How to diagnose incidents by examining network symptoms
- How to diagnose incidents by examining host symptoms
- How to diagnose incidents by examining application symptoms

Diagnosis is not the end, but the beginning of practice.

—Martin H. Fischer

The English word "diagnosis" comes from the Greek word *diagignōskein*, which literally means "to know thoroughly." Diagnosis, then, implies the ability to see through the myriad of irrelevant facts, honing in on the relevant ones, and arriving at the true root cause of a problem. Unlike portrayals in Hollywood, in the real world, security incidents don't involve malware in bold-red font, conveniently highlighted for our benefit. Instead, our adversaries go to great lengths to hide behind the massive amount of benign activity in our systems, oftentimes leading us down blind alleys to distract us from their real methods and intentions. The CySA+ exam, like the real world, will offer you plenty of misleading choices, so it's important that you stay focused on the important symptoms and ignore the rest.

Network-Related Indicators

We start our discussion as you will likely start hunting for many of your adversaries—from the outside in. Our network sensors often give us the first indicators that something is amiss. Armed with this information, we can interrogate hosts and the processes running on them. In the discussions that follow, we assume that you have architected your network with a variety of sensors whose outputs we will use to describe possible attack symptoms.

Src IP	Src Port	Dst IP	Dst Port	Protocol	Packets	Bytes/Pkt
10.0.0.3	54902	192.168.0.7	80	TCP	2491	740
10.0.0.6	55097	172.31.21.3	443	TCP	100227	1528
10.0.0.12	993	10.0.0.3	48450	TCP	2210	762
10.0.0.6	443	10.0.0.7	54122	TCP	2271	1040
10.0.0.6	443	10.0.0.3	53112	TCP	1022	810

Figure 17-1   NetFlow report showing suspicious bandwidth use

Bandwidth Utilization

Bandwidth, in computing, is defined as the rate at which data can be transferred through a medium, and it is usually measured in bits per second. Networks are designed to support organizational requirements at peak usage times, but they usually have excess capacity during nonpeak periods. Each network will have its own pattern of utilization with fairly predictable ebbs and flows. Attackers can use these characteristics in two ways: Patient attackers can hide data exfiltration during periods of peak use by using a low-and-slow approach that can make them exceptionally difficult to detect by just looking at network traffic. Most attackers, however, will attempt to download sensitive information quickly and thus generate distinctive signals.

Figure 17-1 shows a suspicious pattern of NetFlow activity. Though one host (10.0.0.6) is clearly consuming more bandwidth than the others, this fact alone can have a multitude of benign explanations. It is apparent from the figure that the host is running a web server, which is serving other hosts in its own subnet. What makes it odd is that the traffic going to the one host in a different subnet is two orders of magnitude greater than anything else in the report. Furthermore, it's puzzling that a web server is connecting on a high port to a remote web server. When looking at bandwidth consumption as an indicator of compromise, you should look not only at the amount of traffic, but also at the endpoints and directionality of the connection.

Beaconing

Another way in which attackers often tip their hands is by using a common approach for maintaining contact with compromised hosts. Most firewalls are configured to be very careful about inbound connection requests but more permissive about outbound ones. The most frequently used malware command and control (C2) schemes have the compromised host periodically send a message or beacon out to a C2 node. *Beaconing* is a periodical outbound connection between a compromised computer and an external controller. This beaconing behavior can be detected by its two common characteristics: periodicity and destination. Though some strains of malware randomize the period of the beacons, the destination address, or both, most have a predictable pattern.

Detecting beacons by simple visual examination is extremely difficult, because the connections are usually brief (maybe a handful of packets in either direction) and easily get lost in the chatter of a typical network node. It is easier to do an endpoint analysis and see how regularly a given host communicates with any other hosts. To do this, you would have to sort your traffic logs first by internal source address, then by destination address, and finally by time. Then the typical beacon will jump out and become apparent.

 NOTE Some legitimate connections will look like beacons on your network. An example from our personal experience is certain high-end software, which periodically checks with a license server to ensure that it is licensed to be used.

Irregular Peer-to-Peer Communication

Most network traffic follows the familiar client/server paradigm in which a (relatively) small number of well-known servers provide services to a larger number of computers that are not typically servers themselves. Obviously, there are exceptions, such as n-tier architectures in which a front-end server communicates with back-end servers. Still, the paradigm explains the nature of most network traffic, at least within our organizational enclaves. It is a rare thing in a well-architected corporate network for two peer workstations to be communicating with each other. This sort of peer-to-peer communication is usually suspicious and can indicate a compromised host.

Sophisticated attackers will oftentimes dig deeper into your network once they compromise their initial entry point. Whether the first host they own is the workstation of a hapless employee who clicked a malicious link or an ill-configured externally facing server, it is rarely the ultimate target for the attacker. *Lateral movement* is the process by which an attacker compromises additional hosts within a network after having established a foothold in one. The most common method of achieving this is by leveraging the trusted tools built into the hosts. All they need is a valid username and password to use tools such as Server Message Block (SMB) and PsExec in Windows or Secure Shell (SSH) in Linux. The required credentials can be obtained in a variety of ways, including cached/stored credentials on a compromised host, password guessing, and pass-the-hash attacks on certain Windows domains. Here is a list of things to look for:

- **Unprivileged accounts connecting to other hosts** Unless a well-known (to you) process is being followed (for example, hosts sharing printers), any regular user connection to a peer host is likely to indicate a compromise and should be investigated.

- **Privileged accounts connecting from regular hosts** It is possible for a system or domain administrator to be working at someone else's computer (for example, fixing a user problem) and needing to connect to another resource using privileged credentials, but this should be rare. These connections should get your attention.

- **Repeated failed remote logins** Many attacks will attempt lateral movement by simply guessing passwords for remote calls. Any incidences of repeated failed login attempts should be promptly investigated, particularly if they are followed by a successful login.

PART IV

Detecting the irregular peer-to-peer communications described here can be extremely difficult because all you would see are legitimate users using trusted tools to connect to other computers within your network. Context matters, however, and the question to ask as an analyst should be this: Does this user account have any legitimate reason to be connecting from this host to this other resource?

Tales from the Trenches: Package Analysis

A colleague once told us about a black-box pen test he was doing on a pretty well-secured organization. After a couple of days of not getting anywhere, he decided to send a package to an employee who (as attested by her social media profile) would be away from the office for a couple of weeks. In the package was a mobile phone with an invoice to the employee for full market value (to ensure it would eventually be returned). The phone was loaded with wireless hacking tools and would connect over the cellular data network back to his workstation. Within a day, the battery was depleted, but our colleague had been able to connect his rogue device to the wireless network undetected and compromised multiple workstations, all from the comfort of his office. And, yes, he did get his phone back in the mail.

Rogue Devices on the Network

One of the best things you can do to build and maintain secure networks is to know what's on them. Hardware and software asset management is the bedrock upon which the rest of your security efforts are built. If you don't know what hosts belong on your network, you won't be able to determine when an unauthorized one connects. Unfortunately, this lack of asset awareness is the case in many organizations, which makes it easy for attackers to join their devices to target networks and compromise them.

An attacker can connect in two main ways: physically through a network plug and wirelessly. Though you would think that it would be pretty easy to detect a shady character sitting in your lobby with a laptop plugged into a wall socket, the real threat here is with employees who connect their own wireless access points to the network to provide their own devices with wireless access where there may have been none before. Rare as this is, it is damaging enough to require a mention it here. The likelier scenario is for an attacker to connect wirelessly.

In either case, you need a way to tell when a new host is connected. The best approach, of course, is to deploy Network Access Control (NAC) to ensure that each device is authenticated, potentially scanned, and then joined to the appropriate network. NAC solutions abound and give you fine-grained controls with which to implement your policies. They also provide you with centralized logs that can be used to detect attempted connections by rogue devices.

If you don't have NAC in your environment, your next best bet is to have all logs from your access points (APs) sent to a central store in which you can look for physical (Media Access Control [MAC]) addresses that you haven't seen before. This process, obviously, is a lot more tedious and less effective. The easiest way for an attacker to get around this surveillance is to change the MAC address to one that is used by a legitimate user. The challenge, of course, is that this could cause problems if that user is also on the network using the same MAC address, but an attacker could simply wait until the user is gone before attempting impersonation.

Scan Sweeps

Some attackers, particularly those more interested in volume than stealth, will use scan sweeps to map out an environment after compromising their first host in it. They may download and run a tool like nmap, or they can use a custom script or even a feature of a hacking toolkit they bought on the Dark Web. Whatever the attacker's approach, the symptoms on the network are mostly the same: one host generating an abnormally large number of connection attempts (but typically no full connections) to a multitude of endpoints. Scan sweeps may use Transport Control Protocol (TCP), User Datagram Protocol (UDP), Internet Control Message Protocol (ICMP), or Address Resolution Protocol (ARP), depending upon how reliable the attacker wants the results to be and how undetectable they want to be, since some scan types are more reliable than others and some are nosier than others.

A good way to detect scan sweeps is by paying attention to ARP messages. ARP is the means by which interfaces determine the address of the next hop toward the ultimate destination of a packet. An ARP request is simply a node broadcasting to every other node in its LAN the question, "Who is responsible for traffic to this IP address?" If the IP address belongs to another host on the same LAN, that host responds by providing its own MAC or physical address. At that point, the source host will send an IP packet encapsulated in a point-to-point Ethernet frame to the interface address from the responder. If, on the other hand, the IP address belongs to a different LAN, the default gateway (that is, IP router) will respond by saying it is responsible for it.

When an attacker attempts a scan sweep of a network, the scanner will generate a large number of ARP queries, as shown on Figure 17-2. In this example, most of the requests will go unanswered, because there are only a handful of hosts on the network segment, though the subnet mask is for 255 addresses. This behavior is almost always indicative of a scan sweep, and, unless it is being done by an authorized security staff member, it should be investigated. The catch, of course, is to ensure that you have a sensor in every subnet that is monitoring ARP messages.

Time	Source	Destination	Protocol	Length	Info
1.88519600	Vmware_4a:58:30	Broadcast	ARP	42	who has 192.168.192.162? Tell 192.168.192.6
1.88528900	Vmware_4a:58:30	Broadcast	ARP	42	who has 192.168.192.163? Tell 192.168.192.6
1.88540000	Vmware_4a:58:30	Broadcast	ARP	42	who has 192.168.192.164? Tell 192.168.192.6
1.88555900	Vmware_4a:58:30	Broadcast	ARP	42	who has 192.168.192.165? Tell 192.168.192.6
1.88566200	Vmware_4a:58:30	Broadcast	ARP	42	who has 192.168.192.166? Tell 192.168.192.6
1.88574400	Vmware_4a:58:30	Broadcast	ARP	42	who has 192.168.192.167? Tell 192.168.192.6
1.88583400	Vmware_4a:58:30	Broadcast	ARP	42	who has 192.168.192.168? Tell 192.168.192.6
1.88591000	Vmware_4a:58:30	Broadcast	ARP	42	who has 192.168.192.169? Tell 192.168.192.6
1.88601800	Vmware_4a:58:30	Broadcast	ARP	42	who has 192.168.192.170? Tell 192.168.192.6
1.88610000	Vmware_4a:58:30	Broadcast	ARP	42	who has 192.168.192.171? Tell 192.168.192.6
1.88618800	Vmware_4a:58:30	Broadcast	ARP	42	who has 192.168.192.172? Tell 192.168.192.6
1.88626800	Vmware_4a:58:30	Broadcast	ARP	42	who has 192.168.192.173? Tell 192.168.192.6
1.88643300	Vmware_4a:58:30	Broadcast	ARP	42	who has 192.168.192.174? Tell 192.168.192.6
1.88654100	Vmware_4a:58:30	Broadcast	ARP	42	who has 192.168.192.175? Tell 192.168.192.6
1.88663100	Vmware_4a:58:30	Broadcast	ARP	42	who has 192.168.192.176? Tell 192.168.192.6
1.88671500	Vmware_4a:58:30	Broadcast	ARP	42	who has 192.168.192.177? Tell 192.168.192.6

Figure 17-2 ARP queries associated with a scan sweep

Common Protocol over a Nonstandard Port

Assigned by the Internet Assigned Numbers Authority (IANA), network ports are communication endpoints that are used to serve specific services. Computers read ports as a 16-bit integer, resulting in a range from 0 to 65535. Although such a range of port options presents many possibilities for communications, only a handful of these ports are commonly used, which are referred to as standard ports. Standard web traffic, for example, uses port 80 for HTTP and 443 for HTTPS, while the Simple Mail Transfer Protocol (SMTP) uses port 25.

There are a few reasons why we may encounter a common protocol over a nonstandard port. Firewalls make many of their decisions based primarily on which ports network traffic is attempting to operate on. Blocking unwanted traffic by targeting the port it usually operates on might be a way to reduce the possibility of malicious or unauthorized traffic. One common example of this practice is the transmission of web traffic over port 8080 instead of the standard port 80. Port 8080 is considered a nonstandard port, because it is used for a purpose other than its default assignment. In some cases, 8080 is indicative of the use of a web proxy. In other cases, developers use port 8080 on personally hosted web servers.

Attackers also know that certain ports are allowed across many networks to ensure that critical services can operate. Domain Name System (DNS), for example, is a foundational protocol that uses port 53 to communicate resolution information via UDP and TCP. As a result, port 53 is almost always allowed on a network. On the TCP side, 443 is used for HTTPS communications and, assuming that it's encrypted, traffic running over 443 will not be inspected at many organizations. In both cases, attackers rely on the assumed lack of visibility on traffic running over these ports to send any received malware C2 information or even perform wholesale data exfiltration. This is an effective way to hide malicious traffic over standard ports.

Host-Related Indicators

After noticing unusual network behaviors such as those we discussed in the previous sections (or after getting an alert from an intrusion detection/protection system or other sensor), your next step is to look at the suspicious host to see if there is a benign explanation for the anomalous behavior. It is important that you follow the evidence and not jump to conclusions, because it is often difficult to get a clear picture of an intrusion simply by examining network traffic or behaviors.

Capacity Consumption

We have already seen how the various indicators of threat activity can consume network resources. In many cases, attacker behavior will also create spikes in capacity consumption on the host, whether it is memory, CPU cycles, disk space, or local bandwidth. Part of your job as an analyst is to think proactively about where and when these spikes would occur based on your own risk assessment or threat model, and then provide the capability to monitor resources so that you can detect the spikes. Many tools can assist

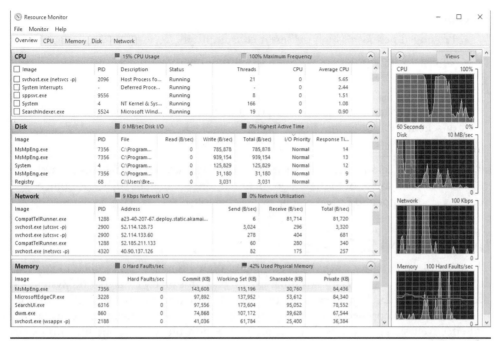

Figure 17-3 Windows 10 Resource Monitor

with establishing your network and system baselines over time. The CySA+ exam will not test you on the proactive aspect of this process, but you will be expected to know how to identify these anomalies in a scenario. You are likely to be presented an image like Figure 17-3 and will be asked questions about the resources being consumed and what they may be indicative of. The figure, by the way, is of a Windows 10 system that is mostly idle and not compromised.

When faced with unexplained capacity consumption, you should refer to the steps we described for analyzing processes, memory, network connections, and file systems. The unusual utilization will be a signal, but your response depends on which specific resource is being used. The following are a few examples of warning signs related to abnormal consumption of various types of system resources.

Memory:

- Consistently low available memory despite low system usage
- Periods of high memory consumption despite no active user interaction or scheduled tasks

Drive Capacity:

- Sudden drop in available free space during a period of increased network activity

Processor:

- Prolonged periods of high processor consumption
- Unusually high processor consumption from unfamiliar tasks

Network:

- Periods of extremely high network throughput despite no active user interaction or scheduled tasks

---

 NOTE Some malware, such as a rootkit, will alter its behavior or the system itself so as not to show signs of its existence in utilities such as Resource Monitor and Task Manager.

Unauthorized Software

The most blatant artifact since the beginning of digital forensics is the illicit binary executable file. Once an adversary saves malware to disk, it is pretty clear that the system has been compromised. There are at least two reasons why threat actors still rely widely on this technique (as opposed to the newer memory-only or fileless malware): convenience and effectiveness. The truth is that it is oftentimes possible to move the file into its target unimpeded, because many defensive systems rely on signature detection approaches that can easily be thwarted through code obfuscation. Even behavioral detection systems (those that look at what the code *does*) are constantly playing catch-up to new evasion techniques developed by the attackers.

Having bypassed the antimalware systems (if any) on the target, this software will continue to do its work until you find it and stop it. This task is made orders of magnitude easier if you have a list of authorized programs that each computer is allowed to run. Depending on your organization's policy over what software can be executed, the mere presence of some kinds of software may be an indicator that a security control was circumvented.

Whitelisting

As we covered in Chapter 12, software *whitelisting* is the process of ensuring that only known-good software is allowed to execute on a system. The much more common alternative is software *blacklisting*, when we prevent known-bad (or suspected-bad) software from running. Whitelisting is very effective at reducing the attack surface for organizations that implement it. However, it is also deeply unpopular with the rank-and-file user, because any new application needs to be approved through the IT and security departments, which delays the acquisition process. Even if you don't (or can't) implement software whitelisting, you absolutely should have an accurate list of the software that is installed in every computer. This software asset inventory is important not only to detect unauthorized (and potentially harmful) software more easily, but also for license auditability and upgrade-planning purposes. Together with a hardware inventory, these two are the most essential steps for ensuring the security of your networks.

Malicious Processes

Someone once said, "Malware can hide, but it has to run." When you're responding to an incident, one of your very first tasks should be to examine the running processes. Every operating system provides a tool to do this, but you must be wary of trusting these tools too much, because the attacker may be using a rootkit that would hide his activities from these tools. Still, most incidents do not involve such sophisticated concealment, so running top or ps in Linux or looking at the Processes tab of the Windows Task Manager (and showing processes from all users) can be very helpful.

 EXAM TIP For many of us, we first look at the processes running on a system before we decide whether to capture its volatile memory. In the exam, it is always preferable to capture memory first and then look at running processes. We reverse the order here for pedagogical reasons.

On a typical system, the list of running processes will likely number a few dozen or so. Many of these will have names like svchost.exe and lsass.exe for Windows or kthreadd and watchdog for Linux. Unless you know what is normal, you will struggle to find suspicious processes. A solution to this challenge is to baseline the hosts in your environment and make note of the processes you normally see in a healthy system. This will enable you to rapidly filter out the (probably) good and focus on what's left. As you do this, keep in mind that attackers will commonly use names that are similar to those of benign processes, particularly if you're quickly scanning a list. Common examples include adding an *s* at the end of svchost.exe, or replacing the first letter of lsass.exe with a numeral 1. Obviously, any such change should automatically be investigated.

Another way in which processes can reveal their nefarious nature is by the resources they utilize, such as network sockets, CPU cycles, and memory. It is exceptionally rare for malware not to have a network socket of some sort at some point. Some of the less-sophisticated ones will even leave these connections up for very long periods of time. So if you have a process with a name you've never seen before and it is connected to an external host, you may want to dig a bit deeper. An easy way to see which sockets belong to which processes is to use the **netstat** command. Unfortunately, each operating system implements this tool in a subtly different way, so you need to use the right parameters, as shown here:

- Windows: **netstat -ano**
- macOS: **netstat -v**
- Linux: **netstat -nap**

Another resource of interest is the processor. If a malicious process is particularly busy (for example, cracking passwords or encrypting data for exfiltration), it will be using a substantial amount of CPU cycles, which will show up on the Windows Task Manager or, if you're using Linux, with the top or ps utilities. In the Linux environment, top is particularly useful because it is interactive, enabling a user to view, sort, and

```
○ ● ●                    🔒 brent — brent@budgie-dev: ~ — ssh budgie — 96×32
top - 13:57:45 up 1 min,  1 user,  load average: 1.21, 0.43, 0.16
Tasks: 213 total,   3 running, 118 sleeping,   0 stopped,   0 zombie
%Cpu(s): 71.7 us,  9.0 sy,  0.0 ni,  0.0 id, 19.3 wa,  0.0 hi,  0.0 si,  0.0 st
KiB Mem :  2017476 total,   177344 free,   296300 used,  1543832 buff/cache
KiB Swap:  2097148 total,  2097148 free,        0 used.  1546188 avail Mem

  PID USER      PR  NI    VIRT    RES    SHR S %CPU %MEM     TIME+ COMMAND
 2004 root      20   0  455936 127864  60280 S  7.3  6.3   0:00.46 unattended-upgr
12106 root      39  19   98884  22364  12376 R  2.0  1.1   0:00.06 apt-check
  619 message+  20   0   50708   5248   3980 S  0.3  0.3   0:00.09 dbus-daemon
 1730 root      20   0  454940 196912 129388 S  0.3  9.8   0:06.24 unattended-upgr
 7189 brent     20   0   47952   4100   3364 R  0.3  0.2   0:00.02 top
    1 root      20   0  159784   9068   6796 S  0.0  0.4   0:00.85 systemd
    2 root      20   0       0      0      0 S  0.0  0.0   0:00.00 kthreadd
    3 root      20   0       0      0      0 I  0.0  0.0   0:00.00 kworker/0:0
    4 root       0 -20       0      0      0 I  0.0  0.0   0:00.00 kworker/0:0H
    5 root      20   0       0      0      0 I  0.0  0.0   0:00.00 kworker/u256:0
    6 root       0 -20       0      0      0 I  0.0  0.0   0:00.00 mm_percpu_wq
    7 root      20   0       0      0      0 S  0.0  0.0   0:00.10 ksoftirqd/0
    8 root      20   0       0      0      0 R  0.0  0.0   0:00.09 rcu_sched
    9 root      20   0       0      0      0 I  0.0  0.0   0:00.00 rcu_bh
   10 root      rt   0       0      0      0 S  0.0  0.0   0:00.00 migration/0
   11 root      rt   0       0      0      0 S  0.0  0.0   0:00.00 watchdog/0
   12 root      20   0       0      0      0 S  0.0  0.0   0:00.00 cpuhp/0
   13 root      20   0       0      0      0 S  0.0  0.0   0:00.00 kdevtmpfs
   14 root       0 -20       0      0      0 I  0.0  0.0   0:00.00 netns
   15 root      20   0       0      0      0 S  0.0  0.0   0:00.00 rcu_tasks_kthre
   16 root      20   0       0      0      0 S  0.0  0.0   0:00.00 kauditd
   17 root      20   0       0      0      0 S  0.0  0.0   0:00.00 khungtaskd
   18 root      20   0       0      0      0 S  0.0  0.0   0:00.00 oom_reaper
   19 root       0 -20       0      0      0 I  0.0  0.0   0:00.00 writeback
   20 root      20   0       0      0      0 S  0.0  0.0   0:00.00 kcompactd0
```

Figure 17-4   Output from the top task manager utility in Ubuntu Linux

act on processes from the same pane without the need for piping or additional utilities. Figure 17-4 shows the output of the utility. By default, the visible columns are process ID, process owner, details about memory and processor consumption, and running time for the processes. Using various keyboard commands, you can sort the columns to identify abnormal processes more easily.

Memory Contents

Everything you can discern about a computer through the techniques discussed in the previous section can also be used on a copy of its volatile memory. The difference, of course, is that a volatile memory analysis tool will not lie, even if the attacker used a rootkit on the original system. The reason why you probably won't go around doing full memory captures of every computer involved in an incident response is that it takes time to capture them and even longer to analyze them.

 NOTE   There are tools used by threat actors that reside only in memory and that have no components stored on the file system. These sophisticated tools all but require incident responders to rely on memory forensics to understand them.

You will need a special tool to dump the contents of memory to disk, and you probably don't want that dump to go to the suspected computer's hard drive. Your best bet is to have a removable hard drive at the ready. Ensure that the device has enough capacity for the largest amount of memory on any system you could be called to investigate, and be sure to wipe its contents before you use it. Finally, install on it one of the many free applications available for memory capture. Among our favorites are AccessData's FTK Imager (which can also acquire file systems) for Windows systems and Hal Pomeranz's Linux Memory Grabber for those operating systems.

Once you have a memory image, you will need to use an analysis suite to understand it, because the layout and contents of memory are large, complex, and variable. They follow predictable patterns, but these are complicated enough to render manual analysis futile. Among the most popular tools for memory forensics is the open source Volatility Framework, which can analyze Windows, Linux, and macOS memory images and runs on any of those three platforms. Having taken an image of the memory from the suspected computer, you will be able to open it in Volatility and perform the same tasks we described in the previous section (albeit in a more trustworthy manner), plus you can conduct a myriad of new analyses that are beyond the scope of this book.

 EXAM TIP You do not need to understand how to perform memory forensics for the exam; you only need to know why memory dumps are valuable to incident response.

Unauthorized Changes

A common technique for an attacker to maintain access to compromised systems is to replace system libraries, such as dynamic-link libraries (DLLs) in Windows systems, with malicious ones. These stand-ins provide all the functionality of the originals, but they also add whatever the attacker needs. Replacing these files requires elevated privileges, but an adversary can accomplish this in a variety of ways. Once the switch is made, it becomes difficult to detect unless you've taken some preparatory actions.

In Windows systems, some built-in features are helpful in detecting unauthorized changes to files or sensitive parts of the system, such as the registry. One of these is automatic logging of access or changes to files in sensitive folders. This feature, called object access auditing, can be applied globally or selectively as a group policy. The explanation of the full range of options is shown in Figure 17-5. Once the feature is in place, Windows will generate an event whenever anyone reads, modifies, creates, or deletes a file in the audited space. For example, modifying an audited file would generate an event with code 5136 (a directory service object was modified) and record the user responsible for the change as well as the time and what the change was. Linux has an equivalent audit system that provides similar features. Obviously, you want to be selective about using this, because you could generate thousands if not millions of alerts if you are too gratuitous about it.

Another way to detect changes to important files is to hash them and store the resulting values in a safe location. This would be useful only for files that are never supposed

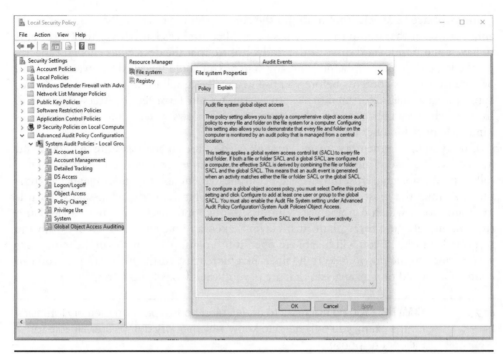

Figure 17-5   Windows 10 local security policy audit options for the file system

to change in any way at all, so you would have to be selective. Still, for most if not all of your libraries and key programs, this approach works well. You would still have to check the hashes manually over time to ensure they haven't changed, but this is a simple process to script and schedule on any system. If you need a more comprehensive solution to file integrity monitoring, commercial organizations such as Tripwire offer solutions in this space.

File System

A *file system* is the set of processes and data structures that an operating system uses to manage data in persistent storage devices such as hard disk drives. These systems have traditionally been (and continue to be) the focal point of incident responses because of the richness of the relevant artifacts that can be found in them. It is extremely difficult for adversaries to compromise a computer and not leave evidence of their actions on the file system.

 NOTE   The word *artifact* is frequently used in forensics, and, though there is no standard definition for it in this context, it generally denotes a digital object of interest to a forensic investigation.

Unauthorized Privileges

Regardless of the type or purpose of the attack, the adversaries will almost certainly attempt to gain elevated privileges. Sometimes, the exploit itself will provide access to a privileged account such as *system* (in Windows) or *root* (in Linux). Some remote execution vulnerabilities, when exploited, place the adversary in a privileged context. More commonly, however, the attacker will have to take some action to get to that status. These actions can be detected and often leave artifacts as evidence.

Privilege escalation is the process by which a user who has limited access to a system elevates that access in order to acquire unauthorized privileges. Note that this could easily apply to an authorized user of the system gaining unauthorized privileges, just as much as it applies to a remote attacker. The means by which this escalation occurs are very system dependent, but they tend to fall into three categories: acquiring privileged credentials, exploiting software flaws, and exploiting misconfigurations. The credentials can be obtained by social engineering or password guessing, but gaining access in this way would be anomalous in that a user would be connecting from or to computers that are not typical for that person. Detecting the exploitation of software flaws requires an awareness of the flaws and monitoring systems that are vulnerable until they can be patched. Obviously, this would not normally be detectable in the event of a zero-day exploit.

Once elevated privileges are detected, your response depends on the situation. The simplest approach is to disable the suspected account globally and place any hosts that have an active session for that user into an isolated VLAN until you can respond. The risk there, however, is that if the account user is legitimate, you may have interfered with a teammate performing an important function for the company, which could have financial impacts. A more nuanced approach would be to monitor all the activities on the account to determine whether they are malicious or benign. This approach can reduce the risk of a false positive, but it risks allowing an attacker to remain active on the system and potentially cause more harm. Absent any other information, you should prioritize the protection of your information assets and contain the suspicious user and systems.

Data Exfiltration

Apart from writing malware to your file systems and modifying files in them for malicious purposes, adversaries will also want to steal data as part of certain attacks. The data that would be valuable to others is usually predictable by the defenders. If you worked in advanced research and development (R&D), your project files would probably be interesting to uninvited guests. Similarly, if you worked in banking, your financial files would be lucrative targets. The point is that we can and should identify the sensitivity of our data before an incident so that we can design controls to mitigate risks to them.

A common approach to exfiltrating data is to consolidate it first in a staging location within the target network. Adversaries don't want to duplicate efforts or exfiltration streams because such duplication would also make them easier to detect. Instead, they will typically coordinate activities within a compromised network. This means that even if multiple agents are searching for sensitive files in different subnets, they will tend to copy those files at the coordination hub at which they are staged, prepared, and relayed to an external repository. Unfortunately, these internal flows will usually be difficult to detect because they may resemble legitimate functions of the organization. (A notable exception is described in the section "Irregular Peer-to-Peer Communication" earlier in this chapter.)

Detecting the flow from the staging base outward can be easier if the amount of data is large or if the adversaries are not taking their time. The exfiltration will attempt to mimic an acceptable transfer such as a web or e-mail connection, which will typically be encrypted. The important aspect of this to remember is that the connection will look legitimate, but its volume and endpoint will not. Even if a user is in the habit of uploading large files to a remote server for legitimate reasons, the pattern will be broken unless the attackers also compromise that habitually used server and use it as a relay. This case would be exceptionally rare unless you were facing a determined nation-state actor. What you should do, then, is to set automated alarms that trigger on large transfers, particularly if they are directed to an unusual destination. NetFlow analysis is helpful in this regard.

For a more robust solution, some commercial entities sell data loss prevention (DLP) solutions, which rely on tamper-resistant labels on files and networks that track them as they are moved within and out of the network. DLP requires data inventories and a data classification system in addition to technical controls. DLP is not explicitly covered in the CySA+ exam, but if data exfiltration is a concern for you or your organization, you should research solutions in this space.

Data Exfiltration: A Real-World Case

One of the tactics used by the threat actor alternatively known as APT28 or Fancy Bear is as effective as it is low tech. This group is known to create a fake Outlook Web Access (OWA) page that looks identical to that of the target organization, has a message indicating the OWA session timed out, and prompts the users to reenter her credentials. The page is hosted on a domain that has a name very similar to the target's (for example, mail.state.qov instead of state.gov). The target is sent a spear-phishing e-mail with a link to the decoy site that seems interesting and appropriate to the target. When the user clicks the link within OWA, it forwards the OWA tab to the fake timed-out page and opens the decoy site on a new tab. When the user closes or switches out of the decoy tab, she sees the (fake) OWA prompt and reenters her credentials. The fake page then forwards them to the (still valid) OWA session. At this point, the threat actor simply creates an IMAP account on some computer and uses the user's credentials to synchronize his folders and messages. This is data exfiltration made easy. Apart from the phony OWA domain name, everything else looks perfectly legitimate and is almost impossible to detect as an attack unless you are aware of this specific tactic.

Registry Change or Anomaly

Registry changes can stand out as obvious indicators because they can lead to significant changes in a system's behavior. The Windows registry is a local database where the system stores all kinds of configuration information. Malware authors want to maintain persistence, or survival of a system restart, whenever possible, and making changes the registry is a common way to achieve lasting presence on a system with minimal effort. Following are a few examples of commonly modified registry keys that attackers use to maintain persistence on Windows machines.

Automatic startup of services or programs:

- HKEY_LOCAL_MACHINE\Software\Microsoft\Windows\CurrentVersion\Run
- HKEY_LOCAL_MACHINE\Software\Microsoft\Windows\CurrentVersion\RunOnce
- HKEY_LOCAL_MACHINE\Software\Microsoft\Windows\CurrentVersion\RunServicesOnce
- HKEY_LOCAL_MACHINE\Software\Microsoft\Windows\CurrentVersion\RunServices
- HKEY_LOCAL_MACHINE\Software\Microsoft\Windows\CurrentVersion\Policies\Explorer\Run

Setting startup folders:

- HKEY_LOCAL_MACHINE\Software\Microsoft\Windows\CurrentVersion\Explorer\Shell Folders
- HKEY_LOCAL_MACHINE\Software
- \Microsoft\Windows\CurrentVersion\Explorer\User Shell Folders

Unauthorized Scheduled Task

A common way for malware to maintain persistence is to employ the user's own operating system's scheduling functionality to run programs or scripts periodically. While this feature is very useful for legitimate power users, it's been used successfully by malware authors for years to maintain persistence and spread viruses. As far as tools, the Windows Task Scheduler is the go-to for those operating the Microsoft OS. The scheduling application is fairly straightforward to operate and provides an overview of the functionality and rundown of the current scheduled tasks for that machine, as shown in Figure 17-6.

For Linux, options such as cron, anacron, and the **at** command have been used for decades to give users fine-grained control over when certain commands would be executed. The cron program can be invoked using the **crontab –e** command to view and edit what commands or shell scripts are queued up for execution. Entries can be made directly to the crontab (cron table) from this point using your preferred text editor. Figure 17-7 shows the overview and basic syntax of entries to the table.

PART IV

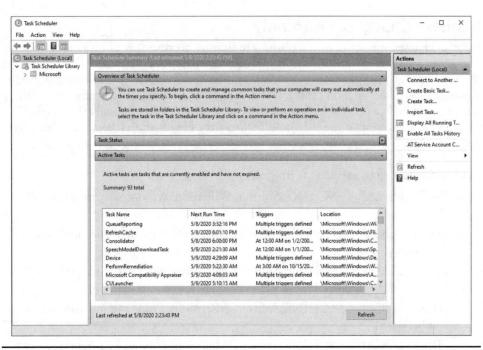

Figure 17-6   Windows 10 Task Scheduler

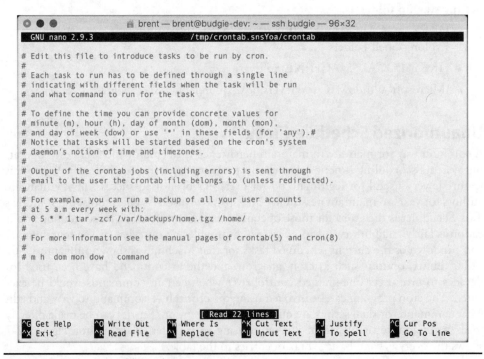

Figure 17-7   The cron table in Ubuntu Linux

result of resource allocation issues induced by malicious software on the host. An examination of the resource manager and log files will help you determine whether or not these symptoms are indicative of malicious activities.

Memory Overflows

Another resource that is often disrupted by exploits or malware is main memory. This is because memory is an extremely complex environment, and malicious activities are prone to disrupt the delicate arrangement of elements in that space. If an attacker is off by even a byte when writing to memory, it could cause memory errors that terminate processes and display some sort of message indicating this condition to the user. This type of symptom is particularly likely if the exploit is based on stack or buffer overflow vulnerabilities. Fortunately, these messages sometimes indicate that the attack failed. Your best bet is to play it safe and take a memory dump so you can analyze the root cause of the problem.

Application Logs

Operating system application logs are a rich source of details about an application's performance. They are a useful data source for detection efforts if they're continuously piped to security information and event management (SIEM) software or a key part of an auditing program in the case of business-critical applications. Application logs are also very important for response efforts, since details of application records can give insight into how an event may have unfolded.

Parsing logs can be quite a challenge and will require some consideration to do it correctly. First is the issue of log volume. Software developers will provide the types of activity they want to record logs from and provide that to the system to be viewed using tools such as the Event Viewer. However, some developers want to record everything that's happening with an application, and for a security team, this could mean more noise than signal. As the logs grow larger, it becomes far more difficult to identify relevant events. The second issue is that logs don't provide actionable output by themselves. Events are provided in standard formats that reflect only what happened, and they never offer intent or an explanation of why it may have occurred. It would be misguided simply to attribute every failed login attempt to malicious activity. An analyst, therefore, must understand the particular role a log plays in telling the story of what may have occurred.

Chapter Review

Like bloodhounds in a hunt, incident responders must follow the strongest scents to track their prey. The analogy is particularly apt, because you too will sometimes lose the scent and have to wander a bit before reacquiring it. Starting from the network level and working your way to the host and then individual applications, you must be prepared for ambiguous indicators, flimsy evidence, and occasional dead ends. The most important consideration in both the real world and the CySA+ exam is to look at the aggregated evidence before reaching any conclusions. As you go through this investigative process, keep in mind Occam's razor: the simplest explanation is usually the correct one.

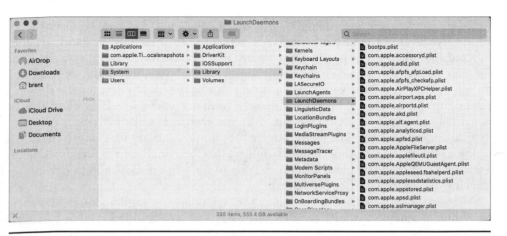

Figure 17-8   Contents of the System/Library/LaunchDaemons/ directory containing system daemons

In the macOS environment, persistence mechanisms use native features such as LaunchAgent, LaunchDaemon, cron, Login Items, and kernel extensions (or kexts). Contents of the System/Library/LaunchDaemons/ directory are shown in Figure 17-8. The entries are stored as a plist, or property list.

Application-Related Indicators

Though most of your work will take place at the network and host levels, you may find it sometimes necessary to examine application symptoms as part of an incident response. By "application," we mean user-level as opposed to system-level features or services. In other words, we are referring to software such as Microsoft Office and not web or e-mail services.

Anomalous Activity

Perhaps the most common symptom of possible infections is unusual behavior in the infected application. Web browsers have long been a focus of attackers not only because of their pervasiveness, but also their complexity. To provide the plethora of features that we have grown accustomed to, browsers are huge, complicated, and often vulnerable applications. Our increasing reliance on plug-ins and the ability to upload as well as download rich content only complicates matters. It is little wonder that these popular applications are some of the most commonly exploited. The first sign of trouble is usually anomalous behavior such as frozen pages, rapidly changing uniform resource locators (URLs) in the address bar, or the need to restart the browser. Because web browsers are common entry points for attacks, these symptoms are likely indicative of the early stages of a compromise.

Other commonly leveraged applications are e-mail clients. Two popular tactics used by adversaries are to send e-mail messages with links to malicious sites and to send infected attachments. The first case was covered in the preceding paragraph, because it is the web browser that would connect to the malicious web resource. In the second case, the application associated with the attachment will be the likely target for the exploit. For example, if the infected file is a Microsoft Word document, it will be Word that is potentially exploited. The e-mail client, as in the link case, will simply be a conduit. Typical anomalous behaviors in the targeted application include unresponsiveness (or taking a particularly long time to load), windows that flash on the screen for a fraction of a second, and pop-up windows that ask the user to confirm a given action (for example, allowing macros in an Office document).

A challenge with diagnosing anomalous behaviors in user applications is that they often mimic benign software flaws. We have all experienced applications that take way too long to load even though there is no ongoing attack. Still, the best approach may be to move the host immediately to an isolated VLAN and start observing it for outbound connection attempts until the incident response team can further assess it.

Introduction of New Accounts

Regardless of the method of infection, the attacker will almost always attempt to elevate the privileges of the exploited account or create a new one altogether. We already addressed the first case, so let's now consider what the second might look like. The new account created by the attackers will ideally be a privileged domain account. This is not always possible in the early stages of an attack, so it is not uncommon to see new local administrator accounts or regular domain accounts being added. The attacker's purpose is twofold: to install and run tools required to establish persistence on the local host, and to provide an alternate and more normal-looking persistence mechanism in a domain account. In either case, you may want to reset the password on the account and immediately log off the user (if a session is ongoing). Next, monitor the account for attempted logins to ascertain the source of the attempts. Unless the account is local, it is not advisable simply to isolate the host, because the attacker could then attempt a connection to almost any other computer.

Unexpected Output

Among the most common application outputs that are indicative of a compromise are pop-up messages of various kinds. Unexpected User Account Control (UAC) pop-ups in Windows, like the one shown in Figure 17-9, are almost certainly malicious if the user is engaged in routine activities and not installing new software. Similarly, when the user is not taking any actions, certificate warnings and navigation confirmation dialogs are inherently suspicious.

Figure 17-9
User Account
Control pop-up
for unsigned
software

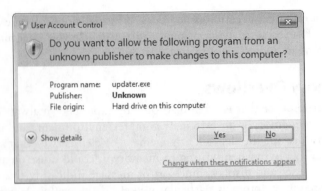

Unexpected Outbound Communication

Perhaps the most telling and common application behavior that indicates a compromise is the unexpected outbound connection. We already spoke about why this is so common in attacks during our discussion of network symptoms. It bears repeating that it is exceptionally rare for a compromise not to involve an outbound connection attempt by the infected host. The challenge in detecting these is that it is normally not possible for a network sensor to tell whether that outbound connection to port 443 was initiated by Internet Explorer or by Notepad. The first case may be benign, but the latter is definitely suspicious. Because most malicious connections will attempt to masquerade as legitimate web or e-mail traffic, you will almost certainly need a host-based sensor or intrusion detection system (IDS) to pick up this kind of behavior. Assuming you have this capability, the best response may be to automatically block any connection attempt from an application that has not been whitelisted for network connections.

Another challenge is that an increasing number of applications are relying on network connectivity, oftentimes over ports 80 or 443, for a variety of purposes. It is also likely that an application that did not previously communicate like this may start doing so as the result of a software update.

 EXAM TIP The fact that an application suddenly starts making unusual outbound connections, absent any other evidence, is not necessarily malicious. During exam simulations, for example, look for indicators of new (authorized) installations or software updates to assess benign behavior.

Service Interruption

Services that start, stop, restart, or crash are always worthy of further investigation. For example, if a user notices that the antimalware icon in the status bar suddenly disappears, this could indicate that an attacker disabled this protection. Similarly, error messages stating that a legitimate application cannot connect to a remote resource may be the

Questions

1. The practice of permitting only known-benign software to run is referred to as what?

   A. Blacklisting

   B. Whitelisting

   C. Blackhatting

   D. Vulnerability scanning

2. Which of the following is *not* considered part of the lateral movement process?

   A. Internal reconnaissance

   B. Privilege escalation

   C. Exfiltration

   D. Pivoting attacks

3. What is a common technique that attackers use to establish persistence in a network?

   A. Buffer overflows

   B. Adding new user accounts

   C. Deleting all administrator accounts

   D. Registry editing

4. Which one of the following storage devices is considered to be the most volatile?

   A. Random-access memory

   B. Read-only memory

   C. Cloud storage

   D. Solid-state drive

5. Which of the following is *not* an area to investigate when looking for indicators of threat activity?

   A. Network speed

   B. Memory usage

   C. CPU cycles

   D. Disk space

6. What is a useful method to curb the use of rogue devices on a network?

   A. SSID

   B. FLAC

   C. WPA

   D. NAC

PART IV

Use the following scenario to answer Questions 7–10:

You receive a call from the head of the R&D division because one of her engineers recently discovered images and promotional information of a product that looks remarkably like one that your company has been working on for months. As she read more about the device, it became clear to the R&D head that this is, in fact, the same product that was supposed to have been kept under wraps in-house. She suspects that the product plans have been stolen. When inspecting the traffic from the R&D workstations, you notice a few patterns in the outbound traffic. The machines all regularly contact a domain registered to a design software company, exchanging a few bytes of information at a time. However, all of the R&D machines communicate regularly to a print server on the same LAN belonging to Logistics, sending several hundred megabytes in regular intervals.

7. What is the most likely explanation for the outbound communications from all the R&D workstations to the design company?

   A. Command and control instructions

   B. Exfiltration of large design files

   C. License verification

   D. Streaming video

8. What device does it make sense to check next to discover the source of the leak?

   A. The DNS server

   B. The printer server belonging to Logistics

   C. The mail server

   D. The local backup of the R&D systems

9. Why is this device an ideal choice as a source of the leak?

   A. This device may not arouse suspicion because of its normal purpose on the network.

   B. This device has regular communications outside of the corporate network.

   C. This device can emulate many systems easily.

   D. This device normally has massive storage resources.

10. What is the term for the periodic communications observed by the R&D workstations?

    A. Fingerprinting

    B. Chatter

    C. Footprinting

    D. Beaconing

Answers

1. **B.** Whitelisting is the process of ensuring that only known-good software can execute on a system. Rather than preventing known-bad software from running, this technique enables only approved software to run in the first place.

2. **C.** Lateral movement is the process by which attackers compromise additional hosts within a network after having established a foothold in one. This is often achieved by leveraging the trust between hosts to conduct internal reconnaissance, privilege escalation, and pivoting attacks. These actions may all be used to facilitate an attacker's end goal, such as data exfiltration, or the removal of sensitive information from the victim network.

3. **B.** A clever way that attackers use for permanence is to add administrative accounts or groups and then work from those new accounts to conduct additional attacks.

4. **A.** Random-access memory (RAM) is the most volatile type of storage listed. RAM requires power to keep its data, and once power is removed, it loses its content very quickly.

5. **A.** Spikes in memory CPU cycles, disk space, or network usage (not necessarily network speed) may be indicative of threat activity. It's important that you understand what the normal levels of usage are so that you can more easily identify abnormal activity.

6. **D.** Network access control (NAC) is a method to ensure that each device is authenticated, scanned, and joined to the correct network. NAC solutions often give you fine-grained controls for policy enforcement.

7. **C.** Some types of software, particularly those for high-end design, will periodically check licensing using the network connection.

8. **B.** A common approach to removing data from the network without being detected is first to consolidate it in a staging location within the target network. As you note the size of the transfers to the print server, it makes sense for you to check to see if it is serving as a staging location and communicating out of the network.

9. **A.** This device is a good choice because an administrator would not normally think to check it. However, because a print server normally has no reason to reach outside of the network, it should alert you to investigate further.

10. **D.** Beaconing is a periodic outbound connection between a compromised computer and an external controller. This beaconing behavior can be detected by its two common characteristics: periodicity and destination. Beaconing is not always malicious, but it warrants further exploration.

Utilize Basic Digital Forensics Techniques

In this chapter you will learn:

- How digital forensics is related to incident response
- Basic techniques for conducting forensic analyses
- Familiarity with a variety of forensic utilities
- How to assemble a forensics toolkit

Condemnation without investigation is the height of ignorance.

—Albert Einstein

Digital forensics is the process of collecting and analyzing data to determine whether and how an incident occurred. The word *forensics* can be defined as an argumentative exercise, so it makes sense that a digital forensic analyst's job is to build compelling, facts-based arguments that explain an incident. The digital forensic analyst answers the questions *what, where, when,* and *how,* but not *who* or *why.* These last two questions are answered by the rest of the investigative process, of which digital forensics is only a part.

The investigation of a security incident need not end up in a courtroom, but it is almost impossible to predict whether a criminal charge is appropriate in the event of a breach. To ensure that we can bring a case to court if necessary, we should treat every digital forensic investigation as if it *will* ultimately be held to the level of scrutiny of a criminal case. We all know, however, that this is not always possible when we're trying to bring critical business processes back online or simply based on our required workload. Still, the closer we stay to the principles of legal admissibility in court, the better off we'll be in the end.

The National Institute of Justice identifies the following three principles that should guide every investigation:

- Actions taken to secure and collect digital evidence should not affect the integrity of that evidence.

- Persons conducting an examination of digital evidence should be trained for that purpose.

- Activity relating to the seizure, examination, storage, or transfer of digital evidence should be documented, preserved, and available for review.

Phases of an Investigation

Forensic investigations, like many other standardized processes, can be conducted in phases. In this case, we normally recognize four: seizure, acquisition, analysis, and reporting. *Seizure* is the process of controlling the crime scene and the state of potential evidentiary items. *Acquisition* is the preservation of evidence in a legally admissible manner. The *analysis* takes place in a controlled environment and without unduly tainting the evidence. Finally, the goal in *reporting* is to produce a report that is complete, accurate, and unbiased.

 EXAM TIP We break down digital forensics into four phases—seizure, acquisition, analysis, and reporting—though many organizations have reduced this to three phases by combining seizure and acquisition. The CySA+ exam will not cover the phases of an investigation but will focus on the techniques and technologies used throughout the process.

Seizure

The goal of seizure is to ensure that neither the perpetrators nor the investigators make any changes to the evidence. An overly simplistic, but illustrative, example of protecting evidence is putting up yellow "Crime Scene" tape and posting guards around the area where a murder took place, so guilty parties can't return to the scene and pick up shell casings with their fingerprints on them. Obviously, the digital crime scene is different from a physical scene, in that the invisible perpetrator may continue to make changes even as the investigators are trying to gather evidence.

Controlling the Crime Scene

Whether the crime scene is physical or digital, it is important that processes be put in place to control who comes in contact with the evidence of the crime to ensure its integrity. The following are just some of the steps that should take place to protect a crime scene:

- Allow only authorized individuals access to the scene.
- Ensure that each person involved in technical tasks is trained and certified for his or her role.
- Document who is present at the crime scene.
- Document who last interacted with the systems.
- If the crime scene does become contaminated, document it. The contamination may not negate the derived evidence, but it will make investigating the crime more challenging.

After you have secured and documented the environment, you can prepare to begin acquiring data. This may involve collecting evidence at the scene or, in some cases, unplugging electronic devices and removing them for transport to a forensics lab. It's often a good idea to photograph the scene and individual elements before you touch or move anything. It is also important that you properly tag, label, and inventory everything you seize to avoid questions later on about evidence tampering or other issues. You'll need access to disassembly and removal tools, such as antistatic bands, pliers, and screwdrivers, with appropriate packaging such as antistatic bags and evidence bags. And keep in mind that weather conditions (for example, extreme temperatures, snow, or rain) may impose additional requirements on your packaging and transportation arrangements.

Chain of Custody

A *chain of custody* is a documented history that shows how evidence was handled, collected, transported, and preserved at every stage of the process. Because digital evidence can be easily modified, a clearly defined chain of custody demonstrates that the evidence has not been tampered with and is trustworthy. It is important to follow very strict and organized procedures when collecting and tagging evidence in every single case. Furthermore, the chain of custody process should follow evidence through its entire lifecycle, beginning with identification and ending with its destruction, permanent archiving, or return to the owner. Figure 18-1 shows a sample form that could be used for this purpose.

Data Acquisition

Forensic acquisition is the process of extracting digital content from seized evidence so that it may be analyzed. This is commonly known as "taking a forensic image of a hard drive" (or any other storage media), but it actually involves more than just that. The main reason you extract the data is to conduct your analysis on a copy of the data evidence and

CHAIN OF CUSTODY

Received from_____By_____
Date_____Time_____A.M./P.M.
Received from_____By_____
Date_____Time_____A.M./P.M.
Received from_____By_____
Date_____Time_____A.M./P.M.
Received from_____By_____
Date_____Time_____A.M./P.M.

WARNING: THIS IS A TAMPER-EVIDENT SECURITY PACKAGE. ONCE SEALED, ANY
ATTEMPT TO OPEN WILL RESULT IN OBVIOUS SIGNS OF TAMPERING.

Figure 18-1   A sample chain of custody form

not on the original; this protects the original content from changes, to ensure that it can be used later as evidence. Throughout the process, preserving the integrity of the original evidence is paramount. To acquire the original digital evidence in a manner that protects and preserves it, the following steps are generally considered best practices:

1. *Prepare the destination media/medium.* Secure any media on which you will store the digital content of your seized evidence. This destination medium may be a removable hard drive or a storage area network (SAN). You must ensure that the destination is free of any content that may taint the evidence. The best way to do this securely is to wipe the media by overwriting it with a fixed pattern of 1's and/or 0's.

2. *Prevent changes to the original.* The simple act of attaching a device to a computer or duplicator will normally cause its contents to change in small but potentially significant ways. To prevent any changes at all, you must use write-protection mechanisms such as hardware write blockers (described in the section "Write Blockers and Drive Adaptors" later in this chapter). Some forensic acquisition software products enable software-based write protection, but it is almost always better to use hardware mechanisms, since a physical barrier or separation could guarantee that no changes could ever be made.

3. *Hash the original evidence.* Before you copy anything, you should take a cryptographic hash of the original evidence. Most products support MD5 and SHA-1 (Secure Hash Algorithm 1) hashes. Though these protocols have been shown to be susceptible to collision attacks and are no longer recommended for general use, we have seen no pushback from the courts on their admissibility with evidence in criminal trials.

4. *Copy the evidence.* You can use a variety of applications to make a forensic copy of digital media, including the venerable dd utility in Linux systems. All these applications perform complete binary copies of the entire source medium. Merely copying the files is insufficient, because you may not acquire relevant data in deleted or unallocated spaces.

5. *Verify the acquisition.* After the copy is complete, compare the cryptographic hash of the copy against the original. If they match, you can perform analyses of the copy and be assured that it is perfectly identical to the original.

6. *Safeguard the original evidence.* Because you now have a perfect copy of the evidence, you must store the original in a safe place and ensure that no one can gain access to it.

Analysis

Analysis is the process of interpreting the extracted data to determine its significance to the case. Though the specific applications and commands you use for analysis may vary depending on the operating or file systems involved, the key issues are the same.

Examples of the types of analysis that may be performed include the following:

- **Timeframe** What happened and when?
- **Data hiding** What has been intentionally concealed?
- **Applications and files** Which applications accessed which files?
- **Ownership and possession** Which user accounts accessed which applications and files?

One of the most important tools to a forensic analyst is the timeframe, or timeline, which establishes a basis for comparing the state of the system at different points in time. For example, you may suspect that a user copied sensitive files to a thumb drive last Friday, but you don't see that drive registered on the system until Monday. Without evidence of the user tampering with the data and time on the system, you can conclude that the exfiltration mechanism was not that particular thumb drive. The timeframe provides a chronologically ordered list of actions taken on the system, which can be categorized as read, write, modify, and delete operations on an item of interest.

Many investigators we know keep track of timelines in a simple spreadsheet with the following columns:

- Data and time
- Time zone
- Source (for example, Windows registry or syslog)
- Item name (for example, registry key name or filename)
- Item location (full path)
- Description

 EXAM TIP You should always regard system timestamps with a healthy dose of skepticism. Threat actors are known to modify the system clock to hide the true sequence of their actions. This practice is known as *timestomping*. Keep an eye out for inconsistencies in timestamps during the CySA+ exam incident simulations, as they may be evidence of tampering.

At every step of the process, you should take copious notes on each specific action you take, down to the command and parameters you use. If you use a forensic analysis suite such as EnCase or Forensic Toolkit (FTK), the tool will record your actions for you. Even so, it is a best practice to keep notes on your own throughout the investigation.

Reporting

If you have been taking notes, you have been writing parts of your report as you conduct the investigation. Once you arrive at sound conclusions based on the available evidence, you can put together narrative statements in a report that present your arguments and conclusions in a readable fashion. As with any form of communication, knowing your

PART IV

audience is crucial. If the report is geared toward executive leaders, for example, the document would be different from one that would be presented in a court of law. If you need help creating a report, all major commercial forensic analysis suites have a feature that will generate a draft report you can customize for your own purposes.

Network

Before you begin to analyze network data for its usefulness in an incident or investigation, it must be collected as completely as possible. Broadly speaking, there are two approaches to capturing packets on a network: header captures and full packet captures. The difference, as the terms imply, is whether you capture only the IP headers or the entire packets, which would include payloads. Although it may be tempting for you to jump on the full-packet bandwagon, you should keep in mind that this approach comes with significant data management, as well as potential legal and privacy issues associated with collecting network traffic. It's important that any collection activity which might involve capturing employee or customer data be brought to the attention of your legal and privacy team. Capturing very large sets of packet data is useful only if you can gain actionable information from them. You may choose to keep the data for reference in case a major incident occurs, but this doesn't do away with the need to be able to handle all this data at collection time.

Many solutions are available for storing and retrieving very large data stores—the point is not that it shouldn't be done, but rather that it should be carefully engineered.

Network Tap

A common option for packet capture is the network tap. Using tap hardware, you may be able to capture traffic between various points on the network for follow-on analysis. Like a phone tap, network taps can be used for diagnostic or monitoring operations. There are two types of network taps: passive and active.

A *passive tap* requires no additional power. A passive tap on copper cable will form a direct connection to wires in the cable and split the signal going across the line; power is still flowing to the destination, but enough is diverted to the tap to be used by the packet sniffer. Similarly, passive optical taps attempt to split the light beam passing though the fiber and divert a portion to a sensor. Although these taps require additional hardware, the original signal is not likely to be impacted greatly should the device fail. There are some disadvantages with this tap method, particularly on Gigabit-speed lines. Gigabit Ethernet connections are much more sensitive to power fluctuations and may experience high error rates, distortion, or failure should a passive tap be installed. For this reason, on Gigabit lines, an *active tap* (or active relay) must be used. Active taps completely terminate the signal in the tap device, sending a copy of the signal to a local interface and moving the original signal to a forwarder. That forwarder then amplifies the original signal, if necessary, and passes it to its original destination. This method works well for Gigabit lines, but at the expensive of adding another electrical device in the chain.

Should the active tap fail, the entire circuit may remain open, alerting the administrator that something is amiss.

CAUTION   Tapping a network using these methods has the potential to change the transmission characteristics of the line.

Hub

An alternate method of collection is to capture the traffic directly from the intermediary device, or hub. Because hubs share traffic coming in and out of all interfaces equally, they rely on the connected hosts to be honest and listen in on only what's addressed to them. On some networks, it may be possible to place a hub at a network chokepoint and collect traffic traversing that location. Hubs are increasingly rare, even in home use, and have been replaced with the more discerning switch.

Switches

In a switched environment, data units called *frames* are forwarded only to destinations they are meant for. As each frame enters the switch, the switch compares the incoming frame's destination Media Access Control (MAC) address with its existing list of addresses and their matching physical ports on the switch. When it finds a match, the switch forwards the data to the appropriate interface and then on to the destination device. Because a device's MAC address is meant to be immutable, collecting from switches requires additional setup steps. Some switches have built-in functionality, called *port mirroring*, that directly supports packet capturing. With port mirroring enabled, the switch sends a copy of all the packets it sees to a monitored port. In some devices, this is referred to as the switched port analyzer (SPAN) port.

Wireshark/TShark

Wireshark and its command-line version, TShark, are network protocol analyzers, or *packet analyzers*. Wireshark is an indispensable tool for network engineers, security analysts, and attackers. Available for macOS, Linux, and Windows, this open source software provides a graphical representation of packet types and advanced filtering. Wireshark can interact directly with some wired and wireless network cards, enabling the user to place the device in promiscuous mode for more complete network capture. For work after the capture is complete, Wireshark provides statistical analysis summary and graphing functionality.

Wireshark is probably the most widely used GUI-based packet analyzer. Despite the many advantages of using Wireshark's graphical front end, you may sometimes find it useful to use TShark if you can't get to a GUI (for example, when connecting over Secure Shell [SSH]) or you want to script a packet capture. Whether you capture the traffic through the GUI or CLI, you can save it and view it on GUI later. You can similarly view captures from other tools (such as tcpdump) provided they were saved

PART IV

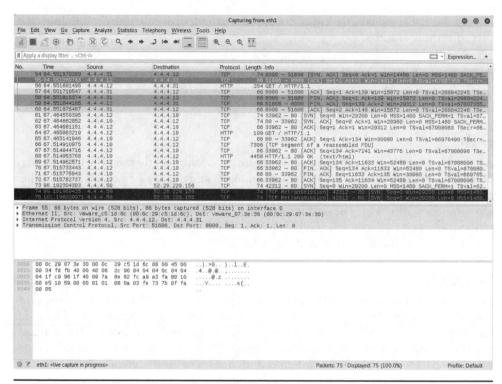

Figure 18-2 Typical Wireshark packet capture

in the packet capture (PCAP) file format. Figure 18-2 shows a typical capture, whereas Figure 18-3 shows how you can drill into a specific packet to get a detailed view that includes the payload.

tcpdump

The tcpdump command-line tool comes standard in many distributions of the Berkeley Software Distribution (BSD), Linux, and macOS, which means you typically don't have to worry about installing it on the platform from which you'd like to capture traffic. As long as you can SSH into a host and run as a privileged user (such as root), you can capture packets on most non-Windows systems. As shown in Figure 18-4, the display is not as easy to read as Wireshark's, but the information captured can be the same. A Windows version, WinDump, is typically not installed by default. Unless you were planning to use a Windows computer for the capture in the scenario, tcpdump would be a good choice, particularly if you couple it with a more robust analysis engine.

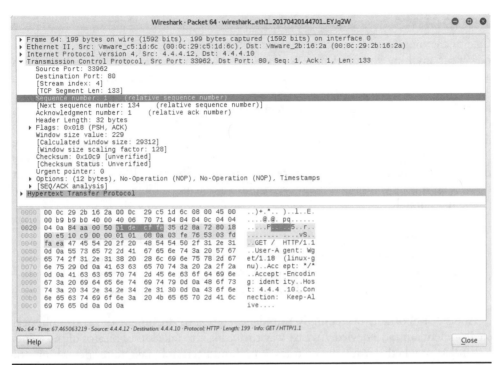

Figure 18-3 Wireshark capture showing packet details

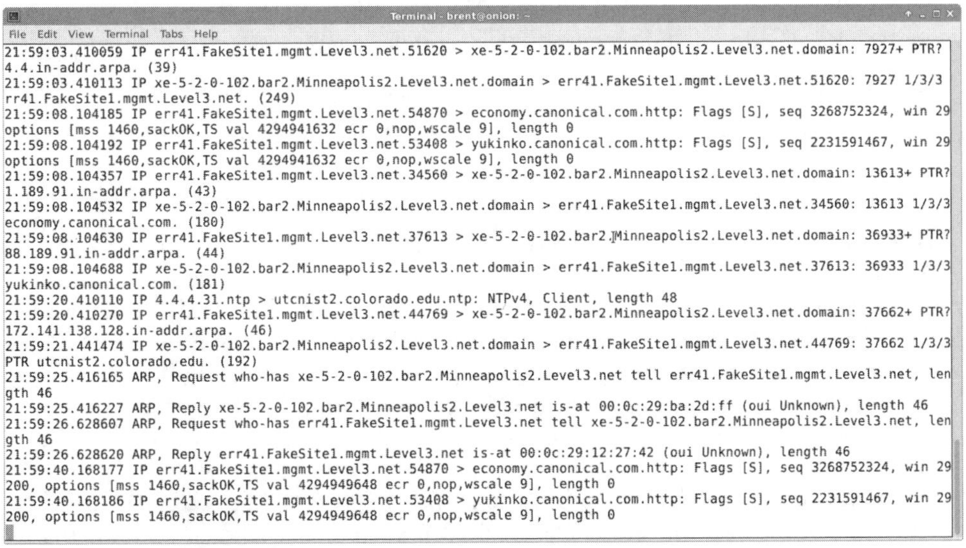

Figure 18-4 Typical tcpdump packet capture

Endpoints

One of the most important steps you can take during a forensic investigation is to *not* power off anything you don't have to. The one universal exception to this rule is if you are pretty sure that there is a running process that is deliberately destroying evidence. There are many reasons for keeping the devices running, but a key one is that memory forensics (that is, digital forensics on the primary storage units of computing devices) has dramatically evolved over the past few years. Although it is possible for a threat actor to install rootkits that hide processes, connections, or files, it is almost impossible to hide tracks in running memory. Furthermore, an increasing number of malware never touches the file system directly and lives entirely in memory. Shutting down a device without first acquiring the contents of memory could make it impossible to piece together the incident accurately.

 NOTE To acquire volatile memory, you will likely have to make some changes to the computer, which typically include connecting an external device and executing a program. As long as you document everything you do, this should not render the evidence inadmissible in court.

Another important, if seemingly mundane, step is to document the entire physical environment around a device. An easy way to do this is to take lots of photos of the scene. Regardless of whether you take pictures, you should certainly take notes describing not only the environment but also each action your team takes to seize the evidence. Specific photos you typically want to take are listed here:

- Computer desktop showing running programs (if the device is unlocked)
- Peripherals connected to the device (for example, thumb drives and external drives)
- Immediate surroundings of the device (for example, physical desktop)
- Proximate surroundings of the device (for example, the room or cubicle)

Tales from the Trenches: A Picture Is Worth...

We were once chatting with a federal law enforcement agent about best practices for photographing crime scenes. He described a case in which he raided a suspect's home and seized a large amount of evidence, including a stack of dozens of CDs. When law enforcement attempted to acquire the contents of the hard drive, they found out it was protected by strong full-disk encryption. They asked the suspect for the passphrase, but he happily informed them that he didn't know it. Incredulous, our friend asked him how that was possible. The suspect responded that he never memorized the passphrase because it was simply the first character in each CD's title. To his horror, our friend realized that he didn't take any photos that showed the stack of CDs in order, and the disks had been shuffled during handling.

Servers

Conducting a forensic analysis of a server requires addressing additional issues compared to workstations. For starters, it may not be possible to take the server offline and remove it to a safe analysis room. Instead, you may have no choice but to conduct an abbreviated analysis onsite. *Live forensics* (or live response) is the process of conducting digital forensics on a device that remains operational throughout the investigation. We already touched on a related issue earlier when we described the importance of capturing the contents of volatile memory before shutting off a device. If you cannot remove the server from a production environment, the next best thing is to capture its memory contents and files of interest (for example, log files).

Another consideration when dealing with servers is that they typically have significantly more storage capacity (both primary and secondary) than workstations. This is guaranteed to make the analysis process take more time, and it may also require special tools. For example, if your server uses a redundant array of inexpensive disks (RAID), you will likely need specialized tools to deal with those disks. Apart from the hardware differences, you will also have to consider the particular architectures of the software running on servers. Microsoft Exchange Server, for example, has a large number of features that help a forensic investigator, but that person will have to know his or her way around Exchange's complex architecture. This point also holds for database management systems (DBMS) and in-house web applications.

OS and Process Analysis

We know that an operating system manages and controls all interactions with a computer. Though there are clearly a variety of operating systems in use today, they all perform the same three basic functions:

- Manage all computer resources such as memory, CPU, and disks
- Provide a user interface
- Provide services for running applications

The first of these is of particular interest to a forensic analyst, because every action that occurs on a computer system is mediated by its OS.

If you are investigating a Microsoft Windows system, two of the most important sources of information are the registry and the event log. The registry is the principal data store where Windows stores most system-wide settings. Though all major analysis suites include viewers for this database, you can also examine it directly on any Windows computer by launching the Registry Editor application. You can find literally hundreds of interesting artifacts in the registry, including the following:

- **Autorun locations** This is where programs tell Windows that they should be launched during the boot process. Malware oftentimes uses this for persistence (for example, HKLM\Software\Microsoft\Windows\CurrentVersion\Run).

- **Most Recently Used lists**   Often referred to as MRUs, this is where you'll find the most recently launched applications, recently used or modified documents, and recently changed registry keys. For example, if you want to see recently used Word documents, you would look in HKEY_CURRENT_USER\Software\ Microsoft\Office\12.0\Common\Open Find.

- **Wireless networks**   Every time a computer connects to a wireless network, this is recorded in the registry, which you can then examine an as investigator in HKLM\ SOFTWARE\Microsoft\Windows NT\CurrentVersion\NetworkList\Profile.

Another useful source of information is the event logs, which you can access by launching the Event Viewer application in any Windows computer. There is actually a collection of logs, the number of which depends on the specific system. All Windows computers, however, will have an application log in which applications report usage, errors, and other information. The OS also maintains in a security log security-related events such as unsuccessful login attempts. Finally, every Windows system has a system log in which the OS records system-wide events.

Although Linux doesn't have the convenience of a centralized registry like Windows, it has its own rich set of sources of artifacts for a forensic investigator. For starters, a lot of relevant data can be found in plaintext files, which (unlike Windows) makes it easy to search for strings. Linux also typically includes a number of useful utilities such as dd, sha1sum, and ps, which can help you acquire evidence, hash it, and get a list of running processes (and resources associated with them), respectively. You can do all this in Windows, but you'll need to install additional tools first.

The Linux file system starts in the root directory, which is denoted by a slash. As an analyst, you need to be familiar with certain directories. We highlight a few of these, but you should build up your own list from this start:

- **/etc**   This primary system configuration directory contains a subdirectory for most installed applications.

- **/var/log**   All well-behaved Linux applications will keep their log files in plaintext files in this directory, making it a gold mine for analysts.

- **/home/$USER**   Here, $USER is a variable name that you should replace with the name of a given user. All user data and configuration data are kept here.

Log Viewers

Every major OS provides the means to view the contents of its log files. The reason you may need a dedicated log viewer is that the built-in tools are meant for cursory examination and not for detailed analysis, particularly when the logs number in the thousands. Like most other features described so far, this functionality is oftentimes found in the forensic analysis suites. If you need a dedicated log viewer, there is no shortage of options, including many free ones.

A scenario in which a standalone log viewer would make sense is when you are trying to aggregate the various logs from multiple computers to develop a holistic timeline of events. You would want a tool that enables you to bring in multiple files (or live systems) and filter their contents in a variety of ways. Some tools that help you do this include Splunk, SolarWinds Event Log Consolidator/Manager, and Ipswitch's WhatsUp.

Mobile Device Forensics

These days, it is uncommon for criminal investigations not to include mobile device forensics. Though this is somewhat less true in the corporate world, you should still be aware of the unique challenges that mobile devices present. Chief among these is that the device will continue to communicate with the network unless you power it off (which we already said may not be a good idea). This means that a perpetrator can remotely wipe the device or otherwise tamper with it. A solution to this problem is to place the device into a Faraday container that prevents it from communicating over radio waves. Faraday bags have special properties that absorb radiofrequency (RF) energy and redistribute it, preventing communication between devices in the container and those outside. Obviously, you will also need a larger Faraday facility in which to analyze the device after you seize it.

Although you can do some amount of forensic analysis on a live Windows or Linux system, mobile devices require dedicated forensic tools. The exceptions to this rule are jailbroken iPhones or iPads and rooted Android devices, because both of these expose an OS that is very similar to Linux and includes some of the same tools and locations. To make things a bit more interesting, many phones require special cables, although the migration toward USB-C in recent years is simplifying this as more devices adopt this interface.

Among the challenges involved in mobile forensics is simply getting access to the data. The mobile OS is not designed to support acquisition, which means that the forensic analyst must first get the device to load an alternate OS. This usually requires a custom bootloader, which is an almost-essential feature of any mobile forensics toolkit. Another peculiarity of mobile devices is that much of their data is stored in miniature DBMSs such as SQLite. These systems require special tools to view their data properly. Their advantage, however, is that the systems almost never delete data when the user asks them to. Instead, they mark the rows in the database table as deleted and keep their entire contents intact until new data overwrites them. Even then, the underlying file system may allow recovery of this deleted information. As with the bootloader, any common analysis suite will include the means to analyze this data.

Virtualization and the Cloud

All forensic efforts require that investigators follow specific procedures to collect evidence in a careful, verifiable, and repeatable manner. By following these procedures, if the evidence needs to be admissible in a court of law, you can be confident that you did the work to protect its integrity. Performing forensics on virtual environments has some significant benefits because the entire OS, memory contents, and in some cases networking and other infrastructure are all stored as files on a disk. When you combine this with the fact that virtualization technologies enable you to take a snapshot of the state of the OS, you can see how this could speed up a forensics analyst's workflow as it related to acquiring the data.

VMware's vSphere Hypervisor, for example, uses virtual devices such as network cards, memory, and certain peripherals. When creating a virtual machine, the hypervisor will create several important files. Among the most useful are the machine's configuration in a VMX file, virtual hard drive in a VMDK file, BIOS state in a NVRAM file, and main memory in a VMEM file. While the vSphere Hypervisor may be useful if the security team has control of the hypervisor and supporting hardware, the challenge increases in a cloud environment. In preparing for forensics analysis, it's critical that the identification of cloud computing occur as early as possible, because this may significantly affect the resources required to acquire that data, depending on its location.

Even with the rapid adoption of new distributed technologies, traditional forensics concepts are still generally applicable to cloud storage. As in traditional environments, cloud storage contains everything from the network and system configurations to files and user information. What gets particularly tricky is identifying and tracking data associated with virtualized devices and functionality. Because this is not normally exposed to the end user, and is often volatile and ephemeral, it will take more effort to track down and verify. Another aspect of cloud computing that complications the forensics model is the actual location of the data in question. The physical location of data in the cloud environment may pose a challenge, particularly if the investigation extends beyond the security team and involves legal and law enforcement efforts.

Given the existing strong auditing policies that are in place at many cloud service providers, you may have opportunities to take advantage of those offerings to support your forensics efforts. You may recall from our discussion of cloud technologies in Chapter 6 that many providers use the shared responsibility model when providing services to customers. Depending on your organization's agreement with the provider, both in terms of which model is used as well as outlined by any service level agreements (SLAs), your security team may be entitled to a great deal of access to the underlying hardware, or you may have none at all. Additionally, it's worth noting that in highly virtualized environments, it may be trivial to recover full snapshots of a system before, during, and after an attack. This can be incredibly useful in piecing together what occurred. It may also be the case that the same challenges that exist for auditors will exist for you, in that artifacts may sometimes not be traceable or available because of the distributed nature and rapid turnover of compute and storage resources.

Procedures

Among the core principles of forensics is maintaining the integrity of data regardless of whether it will be presented in a courtroom or kept in the security team's archives. Security analysts, therefore, must take extra steps to document the process as completely as possible. Checklists and standard operating procedures ensure that the entire team is prepared to conduct at least a baseline level of forensics if needed.

Building Your Forensic Kit

There is no one-size-fits-all answer for what you should include in your forensic kit. It really depends on your environment and workflow processes. Still, there are some general tool types that almost everyone should have available if their work includes forensic analyses.

The jump bag is a prepackaged set of tools that is always ready to go. This is your first line of help when you are asked to drop everything you're doing and respond to an incident that may involve a forensic examination. Because you want to ensure that the bag is always ready, you'll probably want to develop a packing list that you can use to inventory the bag after each use to ensure that it is ready for the next run. You'll probably want to include each of the following items in the jump bag.

Live Response Tools

Some live response tools enable responders to collect live volatile data quickly from a system using a USB stick, optical discs, or external drive. This is a useful solution for data that may be lost forever if the system is powered down.

Write Blockers and Drive Adapters

Hardware write blockers come in many flavors and price points, but they all do essentially the same thing: they prevent modifications to a storage device while you acquire its contents. Your most important consideration is the type of interfaces they support. You should consult your asset inventory to see how many different types of storage interfaces are in use in your environment. Some tools support SCSI and ATA, but not SATA, and others may not support USB devices. As long as you have an adapter and cable for each type of storage device interface in your organization, you should be in good shape.

Cables

A good part of your jump bag will be devoted to cables of various types. A good rule of thumb is that if you've ever needed a particular cable for one investigation, you should probably keep it in your jump bag forever. Here are some ideas for cables to include:

- Ethernet cables (crossover, straight-through, one-way)
- Serial cables (various flavors of USB and RS-232)
- Power cables
- Common proprietary cables (Lightning, Thunderbolt)

Wiped Removable Media

You may not have a few hours to wipe a hard drive before you must respond to an incident, so it pays to keep a few packed and ready. The type of interface doesn't much matter (as long as it is supported by your write blocker), but the capacity does. In general, look into your asset inventory and find the largest workstation or external drive in your organization, and pack at least twice that amount of storage in your bag. Servers tend to have significantly larger drives than workstations, so if that is a concern, you may have to invest in a portable RAID solution such as Forensic Data Monster by Forensic Computers. Solutions like these are portable and designed to facilitate the acquisition of evidence.

A common approach in organizations that deal with fairly frequent investigations is to set up a network-attached storage (NAS) solution specifically for forensic images. As long as you have a fast network connection, you'll be able to image any workstation or server with ease. An added advantage is that the NAS can serve as an archival mechanism for past investigations that may still be pending in court. In these cases, it is important to abide by your organization's data-retention policies.

Camera

A camera is an often-overlooked but critical item in your jump bag. It is important to photograph the crime or incident site, and pretty much any digital camera with a flash will do. A useful addition to your camera is a small ruler that you can include in shots whenever you need to capture a sense of distance or scale. Ideally, the ruler should have a matte surface to minimize glare.

Crime Scene Tape

This may sound like overkill, but having some means of notifying others in the area that they should not enter is critical to the seizure process. Crime scene (or other restricted area–labeled) tape does the job nicely and inexpensively.

Tamper-Proof Seals

When the amount of evidence you collect, or the distance you have to transport it, requires the assistance of others (for example, drivers), you probably want to seal the evidence containers with a tamper-resistant seal. In a pinch, you can use tape and sign your name across it. However, if you can afford them, dedicated lockable containers will be best.

Documentation and Forms

Digital forms and other documents may be required by your organization during an investigation. It is a good idea to print hard copies and keep them in your jump bag, because you never know whether you'll be able to access your corporate data store in the middle of an incident response. Following are some items most of us would keep in our bags.

Chain of Custody Form   Earlier in the chapter, Figure 18-1 showed an example chain of custody form, but you should tailor this to your own organization's requirements if you don't already have your own form. The important aspect is to ensure that there are enough copies to match with each seized piece of evidence. Ideally, your evidence transport containers have a waterproof pouch on the outside into which you can slide a form for the container (individual items in it may still need their own forms).

Incident Response Plan   It is not unusual for an incident response to start off as one thing and turn into something else. Particularly when it comes to issues that may have legal implications (for example, forensic investigations), it is a good idea to have a copy of the incident response plan in your jump bag. This way, even if you are disconnected from your network, you will know what you are expected or required to do in any situation you encounter.

Incident Log   Every good investigator takes notes. When you're performing a complex investigation, as most digital forensics investigations are, it is important that you document every action you take and every hypothesis you are considering. The most important reason for this level of thoroughness is that your conclusions are only as valid as your processes are repeatable. In other words, any qualified individual with access to the same evidence you have should be able to follow your notes and get the same results that you did. Keeping a notebook and pen in your jump bag ensures that you are always ready to write down exactly what you do.

Call/Escalation List   If the conditions on the ground are not what you thought they'd be when you started your investigation, you may have to call someone to notify him or her of an important development or request authorization to perform some action. Though the call/escalation list should really be part of your incident response plan, it bears singling it out as an important item to carry in your jump bag.

Cryptography Tools

It is often the case that you must ensure the confidentiality of an investigation and its evidence. To accomplish this, you can turn to a variety of cryptography tools that are available for multiple platforms. Perhaps the simplest approach to encrypting files is to use the compression utilities available in most operating systems, but you must ensure that they are password protected. The advantage is that these applications are ubiquitous, and the files are mostly usable across platforms.

If you need something a little more robust, you can try any number of available encryption tools. One of the most popular and recommended open source solutions is VeraCrypt, which is based on the now defunct TrueCrypt. This tool is free and available for Windows, macOS, and Linux systems. VeraCrypt supports multiple cryptosystems, including AES, Twofish, and Serpent. It also supports the creation of hidden, encrypted volumes within other volumes.

Acquisition Utilities

The acquisition phase of a forensic investigation is perhaps the most critical point in terms of ensuring the admissibility of evidence, analysis, and conclusions in court. This is where you want to slow down, use a checklist, and ensure that you make no mistakes at all, because doing so could possibly invalidate all the work that follows.

Forensic Duplicators

Forensic duplicators are systems that copy data from a source to a destination, while ensuring that not even a single bit gets altered in the process. What sets them apart from other copying utilities is that they do not rely on file system operations, which means they can recover file system artifacts such as the Master File Table (MFT) in Windows systems and the inode table in Linux. This means that a hard drive running the macOS, Windows, or Linux can be copied in the same way using the same utility. Imaging tools usually allow for the entire contents of the drive to be duplicated to a single file in a remote destination. Unlike regular file copies, forensic duplicates also include the file system's slack and free space, where the remnants of deleted files may reside.

dd Utility

Using the dd utility is just about the easiest way to make a bit-for-bit copy of a hard drive. You can find the program in nearly every Linux distribution as well as in macOS. Its primary purpose is to copy or convert files, and accordingly there are several options for block sizes and image conversion during the imaging process that may assist in follow-on analysis. Because almost everything in the extended file system (ext) used in Linux is a "file" (even network connections and peripheral devices), dd can duplicate data across files, devices, partitions, and volumes. The following command will do a bit-for-bit copy of hard drive hda to a file called case123.img using a block size of 4096 bytes, and it will fill the rest of a block with null symbols if it encounters an error:

```
dd if=/dev/hda of=case123.img bs=4k conv=noerror,sync
```

FTK Imager

FTK Imager is a free data preview and imaging tool developed by AccessData. Unlike the dd utility, this imager is a full-featured product that enables you to perform a forensically sound acquisition, verify it by generating MD5 and/or SHA-1 hashes, and even preview the files and folders in a read-only fashion. FTK Imager will also read registry keys from Windows and lets you preview them and their values. It also supports compression, encryption, and multiple output formats, including EnCase evidence file format (E01) and the raw format generated by dd (001).

Password Crackers

It is increasingly common to find encrypted files or drives in everything from mobile devices to back-end servers. If a suspect is unable or unwilling to provide the password, or if there is no suspect to interrogate in the first place, you may have to resort to specialized software that is designed to guess passwords and decrypt the protected resources. A popular commercial solution in this space is Passware Kit Forensic. It can operate on its own or be integrated with EnCase. Passware Kit Forensic can decrypt more than 280 different types of protected files, including BitLocker, FileVault, iCloud, and Dropbox. Additionally, because password cracking can take a very long time, this tool can take advantage of graphics processing units (GPUs) and multiple networked computers to accelerate the process. The two most popular password-cracking tools among security professionals are John the Ripper and Hashcat. Though their feature sets are very similar, there are subtle differences with which you may want to become acquainted.

John the Ripper

John the Ripper is an open source password-cracking tool, initially developed for Unix, that now has variations for many other operating systems. Figure 18-5 shows options for usage with the command-line tool. John runs attacks with wordlists, which reference a precompiled list of possible passwords, or by brute force, which tries many possible combinations in the character space. Additionally, John supports autodetection of password hash types, the protective measure used by operation systems to prevent unauthorized viewing of the password file. The commercial version expands on the already impressive selection of hashes supported.

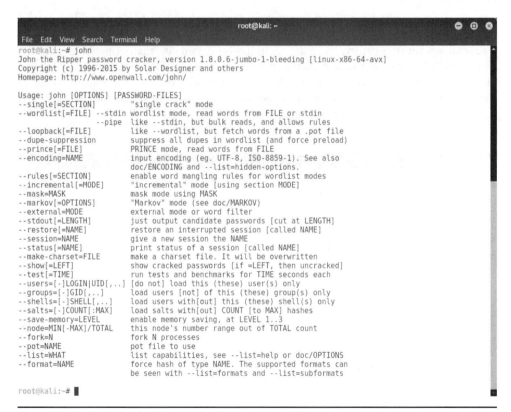

```
                                    root@kali: ~
File  Edit  View  Search  Terminal  Help
root@kali:~# john
John the Ripper password cracker, version 1.8.0.6-jumbo-1-bleeding [linux-x86-64-avx]
Copyright (c) 1996-2015 by Solar Designer and others
Homepage: http://www.openwall.com/john/

Usage: john [OPTIONS] [PASSWORD-FILES]
--single[=SECTION]          "single crack" mode
--wordlist[=FILE] --stdin   wordlist mode, read words from FILE or stdin
                  --pipe    like --stdin, but bulk reads, and allows rules
--loopback[=FILE]           like --wordlist, but fetch words from a .pot file
--dupe-suppression          suppress all dupes in wordlist (and force preload)
--prince[=FILE]             PRINCE mode, read words from FILE
--encoding=NAME             input encoding (eg. UTF-8, ISO-8859-1). See also
                            doc/ENCODING and --list=hidden-options.
--rules[=SECTION]           enable word mangling rules for wordlist modes
--incremental[=MODE]        "incremental" mode [using section MODE]
--mask=MASK                 mask mode using MASK
--markov[=OPTIONS]          "Markov" mode (see doc/MARKOV)
--external=MODE             external mode or word filter
--stdout[=LENGTH]           just output candidate passwords [cut at LENGTH]
--restore[=NAME]            restore an interrupted session [called NAME]
--session=NAME              give a new session the NAME
--status[=NAME]             print status of a session [called NAME]
--make-charset=FILE         make a charset file. It will be overwritten
--show[=LEFT]               show cracked passwords [if =LEFT, then uncracked]
--test[=TIME]               run tests and benchmarks for TIME seconds each
--users=[-]LOGIN|UID[,..]   [do not] load this (these) user(s) only
--groups=[-]GID[,..]        load users [not] of this (these) group(s) only
--shells=[-]SHELL[,..]      load users with[out] this (these) shell(s) only
--salts=[-]COUNT[:MAX]      load salts with[out] COUNT [to MAX] hashes
--save-memory=LEVEL         enable memory saving, at LEVEL 1..3
--node=MIN[-MAX]/TOTAL      this node's number range out of TOTAL count
--fork=N                    fork N processes
--pot=NAME                  pot file to use
--list=WHAT                 list capabilities, see --list=help or doc/OPTIONS
--format=NAME               force hash of type NAME. The supported formats can
                            be seen with --list=formats and --list=subformats

root@kali:~#
```

Figure 18-5   The John the Ripper utility in the command line

PART IV

Hashcat and oclHashcat

As we covered in Chapter 4, Hashcat and its GPU-optimized variant, oclHashcat, are powerful password-cracking utilities that support a vast number of attack modes, such as brute-force, dictionary, and rule-based. Combined with their ability to target various hash types, databases, and full-disk encryption schemes, these software products have become the go-to for many forensics analysts.

 NOTE Password-cracking software has been used successfully for many years, but the trend of increasingly affordable hardware has ushered in the age of hardware-accelerated password cracking. Using "rigs" composed of several GPUs, a user can brute-force passwords orders of magnitude faster than traditional CPU-only methods.

Hashing Utilities

The most popular hashing algorithms for forensic analysis are MD5 and SHA-1, and they are supported by all the popular tools we've discussed in this chapter. If you need a standalone hashing utility, these are included by default by many operating systems. The macOS has the md5 tool available from the command line. Linux typically has the md5 tool as well as sha1sum. Finally, Microsoft provides the File Checksum Integrity Verifier (FCIV) command-line tool as a free but unsupported download. FCIV is able to compute both MD5 and SHA-1 hashes, as shown in Figure 18-6.

```
PS C:\Users\Brent\Desktop> .\fciv.exe
//
// File Checksum Integrity Verifier version 2.05.
//

Usage:  fciv.exe [Commands] <Options>

Commands: ( Default -add )

        -add    <file | dir> : Compute hash and send to output (default screen).
                dir options:
                -r      : recursive.
                -type   : ex: -type *.exe.
                -exc file: list of directories that should not be computed.
                -wp     : Without full path name. ( Default store full path)
                -bp     : specify base path to remove from full path name

        -list           : List entries in the database.
        -v              : Verify hashes.
                        : Option: -bp basepath.

        -? -h -help     : Extended Help.
Options:
        -md5 | -sha1 | -both    : Specify hashtype, default md5.
        -xml db                 : Specify database format and name.

To display the MD5 hash of a file, type fciv.exe filename
PS C:\Users\Brent\Desktop> .\fciv.exe .\ReadMe.txt
//
// File Checksum Integrity Verifier version 2.05.
//
79ac8d043dc8739f661c45cc33fc07ac .\readme.txt
```

Figure 18-6   The Windows File Checksum Integrity Verifier options and usage

Forensic Suites

Reconstructing what happened after the fact is inherently a difficult task, but we have many tools at our disposal to assist with the entire process, from documentation to reporting. *Forensic suites* include a range of tools to uncover data thought to be lost, or data that may be lost easily, such as in the case of volatile memory. Because documentation is an important part of forensics, particularly in criminal investigations, some suites automatically document the evidence analysis progression and technical tasks that have been performed by the analyst.

EnCase

The EnCase suite of tools is very popular with law enforcement and government agencies for forensics missions because of its easy-to-use GUI and chain of custody features. The EnCase suite includes tools for forensic acquisition, analysis, and report generation. Its evidence file format (E01) is among the most common types of forensic imaging formats, in part because of its high portability. The imaged volume's data, metadata, and hashes are all included in a single file.

FTK

AccessData Forensic Toolkit, or FTK, is a popular choice for investigators needing to create forensic images of hard drives. FTK is a favorite for forensics analysis because of its built-in logging features, which make the process of documentation easier for investigators looking to preserve details of the analysis itself. One of the more popular tools included in the FTK suite is the FTK Imager, a data preview and volume imaging tool.

Cellebrite

Cellebrite is a company that developed data transfer solutions for mobile carriers and has since moved into the mobile forensics market. Its flagship product, the Universal Forensic Extraction Device (UFED), is a handheld hardware device primarily marketed to law enforcement and military communities. With the UFED, a user can extract encrypted, deleted, or hidden data from select mobile phones. Cellebrite also provides evidence preservation using techniques such as write blocking during the data extraction procedure.

File Carving

File carving is a technique used to fully recover partially recovered files or those discovered to be damaged. Because carving techniques don't depend on the file system in use, file carving is a common method for data recovery when all else fails. The basic concept of carving is that specified file types are searched for and extracted from raw binary data by looking at file structure and content without any matching file system metadata.

A popular multiplatform carving utility, PhotoRec, is among the fastest and most reliable free tools available. While originally designed to recover media files from damaged digital camera memory, the program is capable of extracting files, including system files and documents, from hard disks, optical discs, and external media. Figure 18-7 shows the progress screen from a PhotoRec recovery attempt. Notice that the utility lists the types and quantity of the recovered files. It's able to determine the file type by reading

```
● ● ●                    brent — brent@budgie-dev: ~ — ssh budgie — 96×32
PhotoRec 7.0, Data Recovery Utility, April 2015
Christophe GRENIER <grenier@cgsecurity.org>
http://www.cgsecurity.org

Disk /dev/sda - 128 GB / 120 GiB (RO) - VMware Virtual disk
     Partition                  Start          End    Size in sectors
 1 * Linux                     0  32 33  15664 222 46  251654144


Pass 1 - Reading sector    2272856/251654144, 1025 files found
Elapsed time 0h00m16s - Estimated time to completion 0h29m15
txt: 953 recovered
elf: 37 recovered
xz: 11 recovered
gz: 8 recovered
tx?: 5 recovered
icc: 4 recovered
a: 2 recovered
zip: 2 recovered
dat: 1 recovered
png: 1 recovered
others: 1 recovered
 Stop
```

Figure 18-7   PhotoRec command-line utility actively recovering files from a volume

the media block by block, looking for patterns associated with certain document types. JPEG files, for example, can be identified by looking for blocks with any of the following byte sequences:

- 0xff, 0xd8, 0xff, 0xe0
- 0xff, 0xd8, 0xff, 0xe1
- 0xff, 0xd8, 0xff, 0xfe

Chapter Review

Digital forensic investigations require a very high degree of discipline and fixed adherence to established processes. A haphazard approach to these activities can mean the difference between successfully resolving an incident and watching a threat actor get away with criminal behavior. The challenge is in striking the right balance between quick responses to incidents that don't require this level of effort and identifying those that do require the effort early enough to adjust the team's approach. Because you may not know which events can escalate to forensic investigations, you should always be ready to perform investigations in a forensically sound way, even if you must conduct the investigation with little or no notice.

The CySA+ exam will require you to be familiar with the techniques associated with seizure, acquisition, analysis, and reporting in digital forensics. For example, you may see questions that present a scenario in which some part of the process has already been completed, and you are asked to decide what should be the next thing to do. This may require familiarity with the way in which you would use some of the most common tools, such as the Linux dd utility. Though you will probably not see questions that require you to issue commands with arguments, you may have to interpret the output of such tools and perform some sort of simple analysis of what may have happened.

Questions

1. In the event of a serious incident, which task is *not* a critical step to take in controlling the crime scene?

   A. Record any interactions with digital systems.

   B. Verify roles and training for individuals participating in the investigation.

   C. Remove power from currently running systems.

   D. Carefully document who enters and leaves the scene.

2. What is the practice of controlling how evidence is handled to ensure its integrity during an investigation called?

   A. Chain of control

   B. Chain of concern

   C. Chain of command

   D. Chain of custody

3. As part of the forensic analysis process, what critical activity often includes a graphical representation of process and operating system events?

   A. Registry editing

   B. Timeline analysis

   C. Network mapping

   D. Write blocking

4. The practice of modifying details about a file's creation, access, and modification times is referred to as what?

   A. Timestomping

   B. Timestamping

   C. Timelining

   D. Timeshifting

Use the following command-line input to answer Questions 5–7:

```
dd if=/dev/sda of=/dev/sdc bs=2048 conv=noerror,sync status=progress
```

5. How many bits of data are read and written at a time?

    A. 2048

    B. 16384

    C. 256

    D. 512

6. What is the destination of the dd operation?

    A. noerror

    B. /dev/sda

    C. sync

    D. /dev/sdc

7. What is the purpose of the command?

    A. To copy the primary partition to an image file

    B. To restore the contents of a hard drive from an image file

    C. To copy the entire contents of the hard drive to an image file

    D. To delete the entire contents of /dev/sda

Use the following scenario to answer Questions 8–10:

You are called to the scene of a high-profile incident and asked to perform forensic acquisition of digital evidence. The primary objective is a Linux server that runs several services for a small company. The former administrator is suspected of running illicit services using company resources and is refusing to provide passwords for access to the system. Additionally, several company-owned mobile phones appear to be functioning and are sitting on the desk beside the servers.

8. What utility will enable you to make a bit-for-bit copy of the hard drive contents?

    A. MFT

    B. dd

    C. MD5

    D. GPU

9. What type of specialized software might you use to recover the credentials required to get system access?

    A. Forensic duplicator

    B. dd

    C. Password cracker

    D. MD5

10. You want to take the mobile phones back to your lab for further investigation. Which two tools could you use to maintain device integrity as you transport them?

   A. Faraday bag and a tamper-evident seal

   B. Write blocker and crime scene tape

   C. Thumb drive and crime scene tape

   D. Forensic toolkit and tamper-evident seal

Answers

   1. **C.** Removing power should not be done unless it's to preserve life or limb, or for other exigent circumstances. In many cases, it's possible to recover evidence residing in running memory.

   2. **D.** A chain of custody is a history that shows how evidence was collected, transported, and preserved at every stage of the investigation process.

   3. **B.** Timeline, or timeframe, analysis is the practice of arranging extracted data from a Unix file system, the Windows registry, or a mobile device in chronological order to better understand the circumstances of a suspected incident.

   4. **A.** Timestomping is a technique that attackers use to modify details about a file's creation, access, and modification times.

   5. **B.** The **bs** argument indicates the number of bytes transferring during the process. Because there are 8 bits in a byte, you multiply 2048 by 8 to get 16384 bits.

   6. **D.** The **of** argument indicates /dev/sdc as the the *output* file, or destination, of the process.

   7. **C.** This command will duplicate the contents of the entire hard drive, indicated by the argument **/dev/sda**. You should be careful to double-check the spelling of both input and output files to avoid overwriting the incorrect media.

   8. **B.** dd is a common utility included in most Linux-based systems that enables you to make bit-for-bit copies of hard drive contents. It can duplicate data across files, devices, partitions, and volumes.

   9. **C.** Password crackers are specialized software designed to guess passwords and decrypt the protected resources. The software can be very resource intensive since cracking usually requires a lot of processing power or storage capacity.

   10. **A.** A Faraday container will prevent the devices from communicating over radio waves by absorbing and redistributing their RF energy. You should secure the bag with a tamper-evident seal to help you identify whether its contents have been interfered with during transport.

PART IV

PART V

Compliance and Assessment

The Importance of Data Privacy and Protection

In this chapter you will learn:

- The difference between privacy and security
- About common data types and laws governing their protection
- How nontechnical controls protect data
- How technical controls protect data

> *When it comes to privacy and accountability, people always demand the former for themselves and the latter for everyone else.*
>
> —David Brin

As security practitioners, we should all have a good understanding of privacy and our obligations with regard to it. After all, we frequently have privileged access to information that could easily be used to infringe on the privacy of others in our organizations, even if done unintentionally. Overstepping that line could very well result in the erosion of the trust in us that enables us to do our jobs protecting our teammates. Worse, it could result in significant financial fines or even jail time. In this chapter, we introduce you to some of the key privacy concepts you need to understand not only for the CySA+ certification, but for the rest of your cybersecurity career.

Privacy vs. Security

Firstly, we should differentiate these two interrelated, but distinct, terms. *Privacy* indicates the amount of control to which an individual is entitled over how others view, share, or use the individual's personal information. Privacy is a difficult thing to achieve, because it is difficult for most of us to know who has what piece of information about us where, and what they're doing with it. You may agree to share your name, address, phone number, and credit card information with a vendor, but it's difficult to know who, exactly, is able to view or use this information. Reputable vendors will have privacy policies that detail what they will (and won't) do with your data, but there is always a possibility that a data breach lands your personal information on the Dark Web.

Security is the protection of information assets against unauthorized access, modification, or destruction. Going back to the vendor example, the vendor could (and should) protect your privacy by using appropriate security controls, such as vulnerability management, access controls, and encryption of data at rest and in transit. It is generally easy to have security without privacy but almost impossible to have privacy without security.

It is important to note that security can provide privacy, but it can also limit it. Encrypting all our customers' information certainly provides a measure of privacy for them. On the other hand, if our security policy specifies that all data processed or stored on a corporate information system belongs to the company, and then we implement tools to monitor the activities of our employees, we are definitely reducing their privacy in the workplace.

Types of Data

It's worth highlighting that our discussion in this chapter centers on the compliance and assessment implications of data privacy and protection. In other words, we are mostly concerned with data types for which existing laws or regulations impose organizational requirements. Here are four types of data that fit this constraint:

- **Personal data**   Any information relating to a natural person (as opposed to a corporation), including name, identification number, address, and phone number
- **Personal health data**   Individually identifiable information relating to the past, present, or future health status of an individual
- **Financial data**   Individually identifiable information relating to the transactions, assets, and liabilities of an individual
- **Copyrighted data**   Data protected under copyright law

Legal Requirements for Data

No single statute or law specifies an organization's legal requirements with regard to data security and privacy for protected data. Instead, multiple local, national, and international laws, when taken together, dictate what we should do with the data that is entrusted to our care. If this sounds confusing or difficult to address, there is hope. Most laws speak of "reasonable" measures that ought to be taken but don't define what these should be. Legal developments over the last several years point toward a legal standard for "reasonableness" that hinges on assessing risk and then implementing controls that reduce that risk to levels that are acceptable to the organization. It is about process, more so than about specific controls. Still, there are certain laws with which you should be aware.

In the following sections, we provide notable examples of laws that govern each of the four types of data: personal, health, financial, and copyrighted. It is important to remember that this list is not meant to be all-inclusive. It is simply intended as a start point for understanding legal and regulatory requirements on these types of data.

General Data Protection Regulation

The *General Data Protection Regulation* (GDPR) is perhaps the most important privacy law affecting organizations around the world today. It affects any organization holding personal data on a European Union (EU) citizen and requires, among other things, that these organizations not use the data without the subject's consent, that they delete the data when a subject requests them to do so, and that they report a data breach within 72 hours of becoming aware of it. A key provision of the GDPR is that data be retained only as long as it is necessary for its intended purpose, and no longer.

Health Insurance Portability and Accountability Act

In the United States, the *Health Insurance Portability and Accountability Act* (HIPAA) is a federal law that covers the storage, use, and transmission of personal medical information and healthcare data. It outlines how security should be managed for any facility that creates, accesses, shares, or destroys medical information. HIPAA mandates steep federal penalties for noncompliance. If medical information is used in a way that violates the privacy standards dictated by HIPAA, even by mistake, monetary penalties of at least $100 per violation are enforced, up to $1.5 million per year, per standard. If protected health information is obtained or disclosed knowingly, the fines can be as much as $50,000 per violation, plus one year in prison. If the information is obtained or disclosed under false pretenses, the fine can be up to $250,000, with ten years in prison if there is intent to sell or use the information for commercial advantage, personal gain, or malicious harm. This is serious business.

Payment Card Industry Data Security Standard

The *Payment Card Industry Data Security Standard* (PCI DSS) is a private-sector industry initiative, not a law. Noncompliance with or violations of the PCI DSS may result in financial penalties or possible revocation of merchant status within the credit card industry, but not jail time. The PCI DSS applies to any entity that processes, transmits, stores, or accepts credit card data. Varying levels of compliance and penalties exist and depend on the size of the customer and the volume of transactions. However, credit cards are used by tens of millions of people and are accepted almost anywhere, which means just about every business in the world is affected by the PCI DSS.

US Copyright Law

US copyright law is discussed in Chapters 1–8 and 10–12 of Title 17 of the United States Code. But it is all builds on the Copyright Act of 1976, which provides the basic framework for the current law. Copyright law protects the right of the creator of an original work to control the public distribution, reproduction, display, and adaptation of that original work. The laws cover many categories of work: pictorial, graphic, musical, dramatic, literary, pantomime, motion picture, sculptural, sound recording, and architectural. A person's creation is provided copyright protection for the duration of his or her life, plus 70 years. If a copyrighted work was created jointly by multiple authors, the 70 years start counting after the death of the last surviving creator.

PART V

Nontechnical Controls

Security controls are measures that counter specific risks by reducing their probability of occurring. For this reason, they are also called countermeasures. There are two general types of controls that apply to data: technical and nontechnical. The difference between the two hinges on whether or not the control depends on hardware or software to function properly. A nontechnical control is not implemented through technical means such as hardware or software. Instead, these controls are normally implemented through policies and procedures. They are sometimes called administrative or soft controls. Let's see some examples of these.

Data Ownership

Who "owns" the data? *Data ownership* refers both to having possession of the data and to having responsibility over it. For example, you are the owner of your own personal data. When you provide some of it to a vendor, such as Equifax, that entity has a legal responsibility to safeguard it for you. That company also becomes an owner of the data you shared with it.

For purposes of the CySA+ exam, the data owner (information owner) is usually a member of management who is in charge of a specific business unit, and who is ultimately responsible for the protection and use of a specific subset of information. The data owner has due care responsibilities and thus will be held responsible for any negligent act that results in the corruption or disclosure of the data. The data owner decides upon the classification of the data she is responsible for and alters that classification if the business need arises. This person is also responsible for ensuring that the necessary security controls are in place, defining security requirements per classification and backup requirements, approving any disclosure activities, ensuring that proper access rights are being used, and defining user access criteria.

 EXAM TIP If you're asked about the data owner in a company, look for the response that mentions a manager or executive in charge of whatever part of the business uses the data.

The formal assignment of data ownership is an important nontechnical control because it establishes responsibility with an individual. This is an important first step we must take before we can classify the data, which is the next control we'll discuss.

Data Classification

Classification indicates that something belongs to a certain class. We could say, for example, that your personnel file belongs to the class named "private," and that your company's marketing brochure for the latest appliance belongs to the class "public." Right away, you would have a sense that your personnel file has more value to your company than the brochure. The rationale behind assigning values to different assets and data is that this enables a company to gauge the amount of resources that should go toward protecting each data class, because not all assets and data have the same value to a company.

Information can be classified by sensitivity, criticality, or both. Either way, the classification aims to quantify how much loss an organization would likely suffer if the information was lost. The *sensitivity* of information is commensurate with the losses to an organization if that information was revealed to unauthorized individuals. This kind of compromise made headlines in 2017, for example, with the Equifax data breach. The compromise of personal data on 147 million people ended up costing the company approximately $1.4 billion.

The *criticality* of information, on the other hand, is an indicator of how the loss of the information would impact the fundamental business processes of the organization. In other words, critical information is essential for the organization to continue operations. For example, in September 2019, Danish hearing-aid manufacturer Demant suffered a devastating ransomware attack that forced it to shut down its entire IT infrastructure. The incident ended up costing the company in excess of $80 million, with a good portion of that amount stemming from lost sales and the inability to service its end users.

Data Confidentiality

Once we've classified our data, even if it's simply "confidential" and "everything else," we can make informed decisions about what we share with whom. We'll talk more about how we restrict sharing data using technical controls later in this chapter (in the section "Technical Controls"), but we must not forget that protected information can be communicated verbally. For example, when we tell someone something "in confidence," we expect the other person to share it only with authorized individuals, or with nobody. But how can we ensure that others keep their end of that deal? If we have an existing contract with that party, it may (and really should) include a confidentiality clause that limits what each party can share with whom. Apart from that, the most common means of enforcing confidentiality is the nondisclosure agreement.

Nondisclosure Agreement

A *nondisclosure agreement* (NDA) is a legally binding document that restricts the manner in which two (or more) parties share information about each other with any other entity. Suppose you want to work with a vendor who could provide you a service or product that you need for your organization. You want to figure out how much this is going to cost you, but in order to give you an accurate cost, the vendor needs confidential information about your organization. The best way forward would be to enter into an NDA with the vendor so you can share your sensitive information, knowing the vendor could not divulge it to anyone else. If they did, they'd be held liable in civil court.

Another angle to an NDA extends beyond willful disclosure. The party receiving the sensitive information from another is required to apply the same security controls to it that they would use with similar information of their own. In other words, if you reveal confidential information to someone or some entity, that person or entity must protect it just as they would their own confidential information. If they suffer a breach and it turns out in the ensuing investigation that they didn't protect your information in this manner, they would be liable for any losses you suffer. This is why you should always establish an NDA before sharing sensitive information.

PART V

Data Sovereignty

Suppose you are a citizen of Country A, a nation with very stringent data privacy laws. One day, you sign up for a free webmail account with a company based in Country B. Could the government of Country B now compel your webmail provider to hand over all your messages? Well, that depends on which country is sovereign over your data. *Data sovereignty* is the notion that the country in which data is collected has supreme legal authority over it. Clearly, if Country A decided to establish its data sovereignty, Country B could well be within its rights to disregard such claims. However, if companies based in Country B want to do business in Country A, they must abide by its laws or face penalties. This is the principle behind the GDPR, the EU law that claims sovereignty over data on citizens of the block of countries for which it applies.

Data sovereignty is a particularly tricky concept when it comes to cloud services. One of the benefits of storing data in the cloud is that your data and services are free to move around, sort of like real clouds in the sky. If the computing resources in one country are saturated or otherwise unavailable, your service provider can seamlessly move you to a different set of physical resources, perhaps in a different country. Now, what if that country doesn't respect your country's claims of data sovereignty? The answer is complicated. Some countries that claim data sovereignty prefer to avoid the whole mess by requiring cloud service providers to use only in-country assets to provide services to their citizens. This is one of the key drivers for major cloud providers to invest in data centers in countries that otherwise might seem like odd choices for this.

Data Minimization

Besides data sovereignty, another key principle in the GDPR has to do with how much data you can collect and use. *Data minimization* is the principle that you can acquire and retain only the minimum amount of data required to satisfy the specific purpose for which the owner has authorized use of that data. For example, suppose that we run an online shoe store. If you want to buy a pair of shoes, we could reasonably be expected to ask you all sorts of personal questions, such as your name, address, payment information, and shoe size. Now suppose we want to keep our options open in case we ever want to start selling clothing, so we gather and retain your body dimensions. We would then be violating the principle of data minimization. If this were a provision of a privacy law such as the GDPR (which it is), we could be fined for doing this.

Data minimization may seem like an undue limitation on our pursuit of business opportunities, but it starts to make a lot of sense when you consider the impacts of a data breach. If I've been gathering all sorts of personal information about you and our systems are breached, then your exposure could be huge. If we had stuck to the minimal set of data that we needed to do our job of selling you shoes, your exposure would be smaller.

Data Purpose Limitation

Purpose limitation, another key principle of the GDPR, states that data may be used only for the purpose for which it was collected and not for any other incompatible purpose. Going back to the example of an online shoe store, suppose that we start analyzing your shoe purchasing behaviors to infer other traits about you. For example, you buy running shoes more frequently than the average shopper, so we assume that you are an avid runner. Because runners tend to be healthy individuals, we share this information with a partner in the health insurance business. They pursue you as a prospect, land your business, and we make a healthy referral fee. In this example, we would've violated the principle of purpose limitation because you authorized us to use your personal information to sell you shoes, and we used it to refer you to a healthcare insurer.

Data Retention

Data retention is the deliberate preservation and protection of digital data in order to satisfy business or legal requirements. It has nothing to do with haphazardly not deleting files. Instead, this intentional effort is aimed at ensuring that your business processes run smoothly and that you are able to satisfy any regulatory or legal requests for information.

If you are legally responsible for maintaining the security of a set of data, it might make sense that you'd want to get rid of it as soon as possible to reduce your personal risk. Or you might prefer to hold on to data as long as possible in hopes that it will be useful at some future point in time. There is no universal agreement on how long an organization should retain data. Legal and regulatory requirements (where they exist) vary among countries and business sectors. What is universal is the need to ensure that your organization has and follows a documented data retention policy. Doing otherwise is flirting with disaster, particularly when dealing with pending or ongoing litigation. It is not enough, of course, simply to have a policy in place; you must ensure that it is being followed, and you must document this through regular audits.

 NOTE   When outsourcing data storage, you must specify in the contract language how long the storage provider will retain your data after you stop doing business with them and what process they will use to eradicate your data from their systems.

Data Retention Standards

At their core, data retention standards answer three fundamental questions:

- What data do we keep?
- How long do we keep this data?
- Where do we keep this data?

Most security professionals understand the first two questions—the "what" and the "how long" are easy. After all, many of us are accustomed to keeping tax records for three years in case we get audited. The last question, however, surprises more than a few of us. The twist is that the question is not so much about the location per se, but rather the manner in which the data is kept at that location. To be useful to us, retained data must be easy to locate and retrieve.

Think about it this way: Suppose your organization had a business transaction with Acme Corporation, in which you learned that Acme was involved in the sale of a particular service to a client in another country. Two years later, you receive a third-party subpoena asking for any information you may have regarding that sale. You know you retained all your data for three years, but you have no idea where the relevant data may be located. Was it an e-mail, a recording of a phone conversation, the minutes from a meeting, or something else? Where would you go looking for it?

In order for retained data to be useful, it must be accessible in a timely manner. It really does us no good to have data that takes an inordinate (and perhaps prohibitive) amount of effort to query. To ensure this accessibility, you'll find it helpful to specify, at a minimum, the following requirements as part of your data retention standards:

- **Taxonomy** A *taxonomy* is a scheme for classifying data. This classification can be made using a variety of categories, including functional (such as human resources, product development), chronological (such as 2018, 2019, and so on), organizational (such as executives, union employees), or any combination of these or other categories.

- **Normalization** Retained data will come in a variety of formats, including word processing documents, database records, flat files, images, PDF files, video, and so on. Simply storing the data in its original format will not suffice in any but the most trivial cases. Instead, you need to develop tagging schemas that will make the data searchable. As part of this step, it is useful to eliminate duplicates.

- **Indexing** Retained data must be searchable if you must be able to pull out specific items of interest in a reasonable timeframe. The most common approach to making data searchable is to build indexes for it. Many archiving systems implement this feature, but others do not. Any indexing approach must support the likely future queries on the archived data.

Ideally, archiving occurs in a centralized, regimented, and homogenous manner. We all know, however, that this is seldom the case. We may have to compromise to arrive at solutions that meet our minimum requirements within our resource constraints. Still, as we plan and implement our retention standards, we must remain focused on how we will efficiently access archived data many months or years.

Technical Controls

A technical control is a security control or countermeasure implemented through the use of an IT asset. This asset is usually, but not always, some sort of software that is configured in a particular way so as to mitigate a specific set of risks. This linkage between controls and the risks they are meant to mitigate is important, because we need to understand the context in which specific controls were implemented. Let us now discuss some examples of technical controls intended for data protection and privacy.

Access Controls

One of the best ways to protect data is to control who has access to it. Many types of technical access controls enable a user to access a system and the resources within that system. A technical access control may be a username and password combination, a Kerberos implementation, biometrics, public key infrastructure (PKI), RADIUS, TACACS+, or authentication using a smart card through a reader connected to a system. These technologies verify the user is who he says he is by using different types of authentication methods. Once a user is properly authenticated, he can be authorized and allowed access to protected data.

Encryption

Encryption, which we discussed in some detail in Chapter 8, can be used to control logical access to information by protecting it as it passes throughout a network and resides on computers. Encryption ensures that the information is received by the correct entity and that it is not modified during transmission. This is an example of a technical control that can preserve the confidentiality and integrity of protected data and enforce specific paths for communication to take place.

Sharing Data While Preserving Privacy

Sometimes, things are not black and white when it comes to who gets to see private data. Suppose you have a customer service representative whose job involves answering questions about customers' orders. That person should have some way of ensuring that she and the customer are talking about the same order or which payment card that was used for it. She needs access to some of the private data, but does she need access to all of it? There are a number of approaches to providing limited access to just enough data for someone to do their job while simultaneously protecting most of the private information. These approaches include data masking, deidentification, and tokenization.

Data Masking

Data masking is the process of covering or replacing parts of sensitive data with data that is not sensitive. It is a simple and effective approach to preventing someone from viewing confidential information. A very common example of this is how password fields show dots or asterisks for each character you type in them. The idea is to prevent anyone around you from seeing what you're typing on the screen. Another common use of data masking is for payment card information. Our customer service representative could be allowed to view only the last four digits of the credit card used to place an order, with the other twelve digits replaced (or masked) with x's. This would preserve the privacy of the customer's sensitive information while allowing her to do her job.

Deidentification

Sometimes, you need access to a lot of private information in a way that does not reveal to whom that information belongs. For example, you may be running a clinical study on the effectiveness of a vaccine. You'll need to know a lot of private information about each subject of the test, including medical history, age, and personal habits involving alcohol and tobacco. You wouldn't, however, need to know their identities.

Deidentification is the process of making it impossible (or at least very hard) to determine the individual to whom a specific data record belongs. It is one of the main approaches to data privacy protection, particularly in healthcare applications, and is sometimes called data anonymization. A common way to deidentify a record is to remove its identifier fields (such as name and Social Security number). The problem with this approach, however, is that there is no way to know if two records belong to the same individual, which means you couldn't track changes over time. Another approach is to assign a pseudorandom identifier to each individual so that all related records share that identifier. This is called tokenization, and it has its pros and cons, as we discuss in the next section.

A challenge with deidentification is that if sufficient data is available about some individuals, that data can be correlated to data from other sources to deanonymize them. This is particularly true with the abundance of personal data being shared in today's highly connected society.

Tokenization

Tokenization is the replacement of sensitive data with a nonsensitive equivalent value that has no value to an adversary. This replacement value, or token, enables the party doing the tokenization to map it back to the original (sensitive) value. Figure 19-1 shows how this works in a near-field communication (NFC) payment context. Note that tokenization requires a token service that is trusted by all parties involved.

At a glance, tokenization may look like data masking, but this is not the case. Data masking is mostly used for sanitizing data that is being displayed. It does not enable someone who has only the masked version of the data to map it back to the original version. Tokenization is all about mapping the two versions to each other so that untrusted parties can act on sensitive data without having direct access to that data.

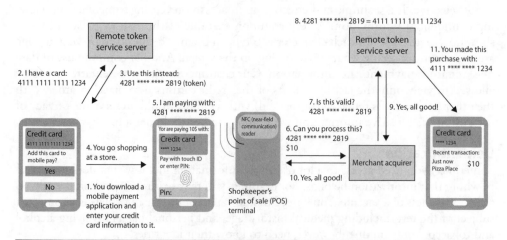

Figure 19-1 How mobile payment tokenization works (source: https://commons.wikimedia.org/wiki/File:How_mobile_payment_tokenization_works.png)

Digital Rights Management

Digital rights management (DRM) refers to a set of technologies that is applied to controlling access to copyrighted data. The technologies themselves don't need to be developed exclusively for this purpose. It is the use of a technology that makes it DRM, not its design. In fact, many of the DRM technologies in use today are standard cryptographic ones. For example, when you buy a Software as a Service (SaaS) license for, say, Office 365, Microsoft uses standard user authentication and authorization technologies to ensure that you install and run only the allowed number of copies of the software. Without these checks during the installation (and periodically thereafter), most of the features will stop working after a period of time. A potential problem with this approach is that the end user's device may not have Internet connectivity.

An approach to DRM that does not require Internet connectivity is the use of product keys. When you install your application, the key you enter is checked against a proprietary algorithm, and if it matches, the installation is activated. It may be tempting to equate this approach to symmetric key encryption, but in reality, the algorithms employed are not always up to cryptographic standards. Since the user has access to both the key and the executable code of the algorithm, the latter can be reverse-engineered with a bit of effort. This could enable a malicious user to develop a product-key generator with which to bypass DRM effectively. A common way around this threat is to require a one-time online activation of the key.

DRM technologies are also used to protect documents. Adobe, Amazon, and Apple, for example, all have their own approaches to limiting the number of copies of an electronic book (e-book) that you can download and read.

Watermarking

One approach to DRM is the use of digital watermarking. *Watermarking* is the practice of embedding specific data into a file for the purpose of identifying its characteristics (such as its origin or owner). Suppose, for example, that you buy a sensitive and very expensive report from us, and we want to ensure you don't post it on the Internet for anyone to download. We could embed in it a hidden mark (a set of data) that identifies the owner and provides other useful information such as the date of purchase. If we come across the report, we know where to look to find the watermark, and we can check it to ensure that the correct owner is using it properly. If you share the file with other parties, we can find all the instances of that shared file and then come after you through legal channels for compensation.

The example just covered is one in which the watermark is invisible. So how can we hide it in a file? One approach is *steganography*, which is the practice of hiding data inside other data. Suppose you have a small image that measures 100 pixels by 100 pixels. Each pixel uses 8 bits for each of the colors red, blue, and green (RBG). Since 8 bits enable you to represent values from 0 to 255, you would represent a bright red pixel as the value 255,0,0 (red = 255, blue = 0, green = 0). Our eyes cannot notice the difference between the version of red that results from 255,0,0 and the one that results from 254,1,1. In other words, we can safely use the last (or least significant) bit each of the three colors of each pixel to our hearts' content. Three bits times 1000 pixels in the image in our example

PART V

comes up to 3000 bits, or 375 bytes. This is more than enough space to write the hidden message "Copyright 2020 by McGraw Hill, licensed to John Doe on December 1, 2020" with plenty of room to spare.

 EXAM TIP When answering questions about steganography, keep in mind that it can be used for watermarking but not every watermark uses steganography.

This example is helpful for illustration purposes but is not practical for watermarking. One problem with it is that the watermark can be damaged or destroyed if the image is modified by cropping it, enhancing the colors, or even compressing it. The robustness of a watermark is a measure of its resistance to attacks, whether those are deliberate or not. A good watermark would still be detectable even if someone is actively trying to get rid of it. Another problem with our example is the fact that we want to watermark more than just images. The approach we described would not work for portable data format (PDF) files, for example.

 NOTE Watermarks will not stop someone from illegally copying and distributing files; they just help the owner track, identify, and prosecute the perpetrator.

Geographic Access Requirements

Another approach to DRM is *geoblocking*, which is the practice of limiting the locations in which the content is available. For instance, movie studios are very deliberate about the way in which they release their films in order to optimize their profits. If a new release is available around the world at the same time, the studio's bottom line may very well suffer. To prevent this, they often restrict the countries in which the movie is viewable at any point in time. Geoblocking can also be necessary when you license the content from someone else and the terms of the licensing agreement restrict where you can share it. Another reason to apply geoblocking could be to ensure that you comply with local laws governing what content may not be shown in a given country.

Geoblocking can be implemented either on a local device or on a remote media server. A local implementation requires special software or firmware to verify that the content is consumable in that location. You see this in e-book readers, media players, and satellite receivers. It is also common to have the server make the check before allowing the downloading of restricted media. This could also take advantage of location services on mobile devices but is most frequently done by checking the geographic location corresponding to the IP address of the client.

Geoblocking can be bypassed in a number of ways but is most frequently done by using virtual private network (VPN) providers that enable you to choose the exit point for your connection. As long as you choose an exit point in an allowed region or country, you may be able to fool the server into allowing you to proceed. Another approach is to route your traffic through a proxy server in the appropriate region. Content owners counter these attempts by disallowing VPN or proxy traffic.

Data Loss Prevention

Data loss prevention (DLP) comprises the actions that organizations take to prevent unauthorized external parties from gaining access to sensitive data. That definition has some key terms you should know. First, the data has to be considered *sensitive*, and we spent a good chunk of the beginning of this chapter discussing its meaning. We can't keep every single datum safely locked away inside our systems, so we focus our attention, efforts, and funds on the truly important data. Second, DLP is concerned with *external parties*. If somebody in the accounting department, for example, gains access to internal research and development data, that is a problem, but technically it is not considered a data leak. Finally, the external party gaining access to our sensitive data must be *unauthorized* to do so. If former business partners have some of our sensitive data that they were authorized to get at the time they were employed, that is not considered a data leak. Although this emphasis on semantics may seem excessive, it is necessary to approach this tremendous threat to our organizations properly.

There is no one-size-fits-all approach to DLP, but there are tried-and-true principles that can be helpful. One important principle is the integration of DLP with our risk-management processes. This enables us to balance out the totality of risks we face and favor controls that mitigate those risks in multiple areas simultaneously. Not only is this helpful in making the most of our resources, but it also keeps us from making decisions in one silo with little or no regard to their impacts on other silos. In the sections that follow, we will look at key elements of any approach to DLP.

Data Inventories

A good first step is to find and characterize all the data in your organization before you even look at DLP solutions. If you've implemented the data classification nontechnical control we covered earlier in this chapter, you may already know what data is most important to your organization. Once you figure this out, you can start looking for that data across your servers, workstations, mobile devices, cloud computing platforms, and anywhere else it may live. Once you get a handle on your high-value data and where it resides, you can gradually expand the scope of your search to include less valuable, but still sensitive, data. As you keep expanding the scope of your search, you will reach a point of diminishing returns in which the data you are inventorying is not worth the time you spend looking for it.

Data Flows

Data that stays put is usually of little use to anyone. Most data will move according to specific business processes through specific network pathways. Understanding data flows at this intersection between business and IT is critical to implementing DLP. Many organizations put their DLP sensors at the perimeter of their networks, thinking that is where the leakages would occur. But if these sensors are placed in that location only, a large number of leaks may not be detected or stopped. Additionally, as we will discuss shortly when we cover network DLP, perimeter sensors can often be bypassed by sophisticated attackers.

Implementation, Testing, and Tuning

Assuming we've done our administrative homework and have a good understanding of our true DLP requirements, we can evaluate products. Once we select a DLP solution, the next interrelated tasks are integration, testing, and tuning. Obviously, we want to ensure that bringing the new toolset online won't disrupt any of our existing systems or processes, but testing needs to cover a lot more than that. The most critical elements when testing any DLP solution are to verify that it allows authorized data processing and to ensure that it prevents unauthorized data processing.

Finally, we must remember that everything changes. The solution that is exquisitely implemented, finely tuned, and effective immediately is probably going to be ineffective in the near future if we don't continuously maintain and improve it. Apart from the efficacy of the tool itself, our organizations change as people, products, and services come and go. The ensuing cultural and environmental changes will also change the effectiveness of our DLP solutions. And, obviously, if we fail to realize that users are installing rogue access points, using thumb drives without restriction, or clicking malicious links, then it is just a matter of time before our expensive DLP solution will be circumvented.

Network DLP

Network DLP (NDLP) applies data protection policies to data in motion. NDLP products are normally implemented as appliances that are deployed at the perimeter of an organization's networks. They can also be deployed at the boundaries of internal subnetworks and could be deployed as modules within a modular security appliance. Figure 19-2 shows how an NDLP solution might be deployed with a single appliance at the edge of the network and communicating with a DLP policy server.

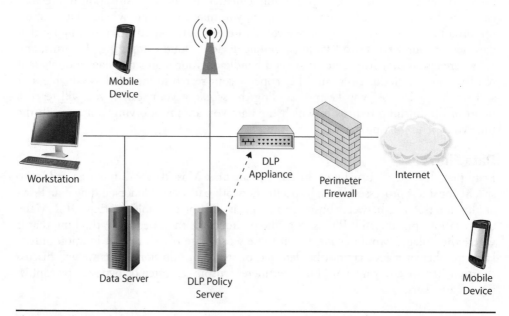

Figure 19-2 Network DLP

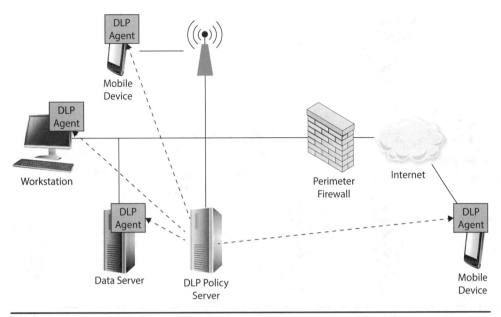

Figure 19-3 Endpoint DLP

Endpoint DLP

Endpoint DLP (EDLP) applies protection policies to data at rest and data in use. EDLP is implemented in software running on each protected endpoint. This software, usually called a DLP agent, communicates with the DLP policy server to update policies and report events. Figure 19-3 illustrates an EDLP implementation.

EDLP provides a degree of protection that is normally not possible with NDLP. The reason is that the data is observable at the point of creation. When a user enters personally identifiable information (PII) on the device during an interview with a client, for example, the EDLP agent detects the new sensitive data and immediately applies the pertinent protection policies to it. Even if the data is encrypted on the device when it is at rest, it will have to be decrypted whenever it is in use, which allows for EDLP inspection and monitoring. Finally, if the user attempts to copy the data to a non-networked device such as a thumb drive, or if it is improperly deleted, EDLP will pick up on these possible policy violations. None of these examples would be possible using NDLP.

Hybrid DLP

Another approach to DLP is to deploy both NDLP and EDLP across the enterprise. Obviously, this approach is the costliest and most complex. For organizations that can afford it, however, it offers the best coverage. Figure 19-4 shows how a hybrid NDLP/EDLP deployment might look.

PART V

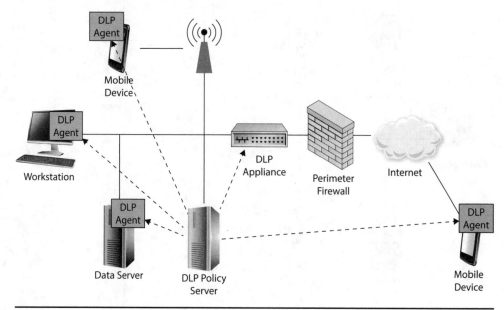

Figure 19-4 Hybrid DLP

Chapter Review

Protecting sensitive data, particularly when it involves the privacy of individuals, is critical to any organization. This protection will probably require the coordination of different controls, both technical and nontechnical. Rather than trying to protect all information equally, our organizations need classification standards that help us identify, handle, and protect data according to its sensitivity and criticality. A key element of our approach must be the protection of privacy of personal data. For various legal, regulatory, and operational reasons, we want to limit what data we acquire, who has access to it (and for what purpose), and how long we retain it.

Questions

1. Which of the following is the most important criterion in determining the classification of data?

   A. The level of damage that could be caused if the data were disclosed

   B. The likelihood that the data will be accidentally or maliciously disclosed

   C. Regulatory requirements in jurisdictions within which the organization is not operating

   D. The cost of implementing controls for the data

2. Information classification is most closely related to which of the following?

 A. The source of the information

 B. The information's destination

 C. The information's value

 D. The information's age

3. The data owner is most often described by all of the following *except* which one?

 A. Manager in charge of a business unit

 B. Ultimately responsible for the protection of the data

 C. Financially liable for the loss of the data

 D. Ultimately responsible for the use of the data

4. Which of the following is *not* addressed by data retention standards?

 A. What data to keep

 B. For whom data is kept

 C. How long data is kept

 D. Where data is kept

5. What approach should you consider if you need to allow a business partner to reference specific credit card transactions without gaining access to the credit card numbers involved?

 A. Tokenization

 B. Data minimization

 C. Nondisclosure agreement (NDA)

 D. Deidentification

6. You want to license a series of action films from a third party and stream them to your customers, but you know that some scenes violate acceptable standards in some countries. What technical control would best enable you to avoid violating those countries' standards?

 A. Data sovereignty

 B. Watermarking

 C. Geoblocking

 D. Data minimization

7. When considering the implementation of a data loss prevention solution, which of the following would be least helpful?

 A. Inventorying your data

 B. Establishing data classification standards

 C. Performing data flow analyses

 D. Implementing data retention standards

PART V

Answers

1. **A.** There are many criteria for classifying information, but it is most important to focus on the value of the data or the potential loss from its disclosure. The likelihood of disclosure, irrelevant jurisdictions, and cost considerations should not be central to the classification process.

2. **C.** Information classification is very strongly related to the information's value and/or risk. For instance, trade secrets that are the key to a business's success are highly valuable, which will lead to that data having a higher classification level. Similarly, information that could severely damage a company's reputation presents a high level of risk and is similarly classified at a higher level.

3. **C.** The data owner is not usually financially liable for the loss of data. The data owner is typically the manager in charge of a specific business unit and is ultimately responsible for the protection and use of a specific subset of information. In most situations, this person is not financially liable for the loss of his or her data.

4. **B.** The data retention policy should address what data to keep, where to keep it, how to store it, and for how long to keep it. The policy is not concerned with "for whom" the data is kept.

5. **A.** Tokenization is the replacement of sensitive data with a nonsensitive equivalent value that enables you to map it back to the original (sensitive) value.

6. **C.** A geoblocking solution placed on your streaming servers would enable you to limit the locations in which the content is available.

7. **D.** Although data retention standards are critically important nontechnical controls for any organization, they are not particularly helpful in implementing a data loss prevention solution. The other three activities listed, on the other hand, are all necessary for this.

Security Concepts in Support of Organizational Risk Mitigation

In this chapter you will learn:

- The importance of a business impact analysis
- How to perform risk assessments to select effective controls
- How to evaluate the effectiveness of security staff and controls
- Important sources of supply chain risk

Risk comes from not knowing what you're doing.

—Warren Buffett

Risk is a constant companion in life. We can take measures to reduce it or transfer it. We might even accept it as inevitable and all but ignore it, but in the end, it is always there. As a cybersecurity analyst, part of your job is to manage risk consciously for your organization. This means that you must understand the business of your organization and determine how it relies on the information systems you are defending before you can assess the risks you face. Armed with an understanding of the risks within the context of your specific organization, you can then select and implement controls to mitigate them. These controls will, of course, have to be assessed periodically to ensure that they remain effective and, just as importantly, your team will have to be periodically tested on their ability to identify and respond to the risks that do manifest themselves in your organization.

Business Impact Analysis

A *business impact analysis* (BIA) is a functional analysis in which a team collects data through interviews and documentary sources; documents business functions, activities, and transactions; develops a hierarchy of business functions; and finally applies a classification scheme to indicate each individual function's criticality level. In creating a BIA, you consider the following issues:

- Maximum tolerable downtime and disruption for activities
- Operational disruption and productivity

- Financial considerations
- Regulatory responsibilities
- Reputation

As a cybersecurity analyst, you may not fully understand all business processes, the steps that must take place, or the resources and supplies these processes require. So you need to gather this information from the people who do know—department managers and specific employees throughout the organization. The first step is identifying the people who will be part of the BIA data-gathering sessions. You then have to figure out how you will collect the data from the selected employees, through surveys, interviews, or workshops. Next, you need to collect the information by actually conducting surveys, interviews, and/or workshops. Later on, during analysis, you'll use data points obtained as part of the information gathering. It is important that you ask about how different tasks—whether processes, transactions, or services, along with any relevant dependencies—are accomplished within the organization. Process flow diagrams are great tools to help visualize this.

Upon completion of the data collection phase, you need to classify the business functions according to their criticality. For instance, your organization may be able to tolerate a few days of downtime in personnel management processes, but if it loses the ability to process customer orders for even a few hours, it could be dead in the water. For each function, you conduct a risk assessment in which you identify the required assets as well as their respective vulnerabilities and threats. After conducting a BIA, you will know what technology assets are critical to the business as well as how long they could be offline without crippling your organization.

BIA Steps
The more detailed and granular steps of a BIA are outlined here:

1. Select individuals to interview for data gathering.
2. Create data-gathering techniques (surveys, questionnaires, and qualitative and quantitative approaches).
3. Identify the company's critical business functions.
4. Identify the resources these functions depend upon.
5. Calculate how long these functions can survive without these resources.
6. Identify vulnerabilities and threats to these resources.
7. Calculate the risk for each different business function.
8. Document findings and report them to management.

Risk Assessment

Through a *risk assessment*, you can identify vulnerabilities and threats and assess their possible impacts to determine what security controls to implement where. The goal is to ensure that security is cost-effective, relevant, timely, and responsive to threats. It is easy to apply too much security, not enough security, or the wrong security controls and to spend too much money in the process without attaining the necessary objectives. Risk assessments help companies prioritize their risks and show management the amount of resources that should be applied to protecting against risks in a sensible manner.

Before we get started, let's talk about exactly what risk is. Back in Chapter 8, we discussed that *assets* are defined as anything of worth to our organizations, including people, partners, equipment, facilities, reputation, and information. Getting back to BIA, we implicitly treat our critical business functions as assets and, for each, we need a risk assessment. The point to remember is that before we can assess risk, we have to ask ourselves, *what, exactly, is at risk?* Once we have identified what may be at risk, we can look for vulnerabilities in that asset.

We also need to know the source of the threat. Recall that we discussed threat modeling in Chapter 2. Are we concerned with natural events such as hurricanes? Employee errors? Cyber criminals? Collectively, these threats may deliberately or accidentally exploit vulnerabilities in our assets.

A risk assessment has at least four main goals:

- Identify vulnerabilities.
- Determine the probability that a threat will exploit a vulnerability.
- Determine the potential business impact of each threat, should it materialize.
- Provide an economic balance between the impact of the threat and the cost of the countermeasure.

A risk assessment helps integrate the security program objectives with the company's business objectives and requirements. The more the business and security objectives are in alignment, the more successful both will be. The assessment also helps the company draft a proper budget for a security program and its constituent security components. Once a company knows how much it could lose from the impact of threats exploiting vulnerabilities, it can make intelligent decisions about how much money it should spend protecting against those threats.

A risk assessment program must be supported and directed by senior management if it is to be successful. Management must define the purpose and scope of the effort, appoint a team to carry out the assessment, and allocate the necessary time and funds to conduct it. It is essential for senior management to review the outcome of the risk assessment and to act on its findings. After all, what good is it to go through all the trouble of a risk assessment and *not* react to its findings? Unfortunately, this does happen all too often.

Asking the Right Questions

When looking at risk, it's good to keep several questions in mind. These questions can be used to help ensure that the risk assessment team and senior management know what is important. Team members must ask the following:

- What event could occur (threat event)?
- What could be the potential impact (magnitude)?
- How likely is it to happen (probability)?
- What level of confidence do we have in the answers to the first three questions (certainty)?

A lot of this information is gathered through internal surveys, interviews, or workshops. Viewing threats with these questions in mind helps the team focus on the tasks at hand and assists in making the decisions more accurate and relevant.

Risk Identification Process

Let's circle back to the first three goals of a risk assessment: identifying vulnerabilities, determining the likelihood that a threat agent will exploit them, and determining the business impact of such exploitations. Risk identification is all about figuring out what could happen, how likely it is to actually happen, and how bad that would be for the organization. So how do we come up with the "what could happen" part of that statement? Most organizations leverage one or more of the following sources:

- **Cyber threat intelligence**   If you have a threat intelligence program (or even read the news to see how other organizations are being attacked), then you'll have no shortage of potential risks to consider.

- **Vulnerability assessment**   This is a staple for most organizations. Run a vulnerability scan, and for each finding, answer this question: What could an adversary do with that?

- **Cybersecurity operations**   There is a lot to be learned in terms of risk from just watching events on your own systems. Failed attacks can provide useful information about adversaries' capabilities and objectives.

- **Brainstorming sessions**   Put together a small group of people who can think like adversaries and ask them, "How would you attack our organization?" This works best if you keep the session length short (to put pressure on them) and don't discard any ideas (to keep them from self-censoring).

The risk identification process is an ongoing effort, not a one-time thing. A good approach is to establish a *threat working group* (TWG) that meets periodically (such as monthly) to discuss new risks that have been identified since the last meeting.

This approach enables the organization to have a steady cadence for risk identification that leverages the various information sources at its disposal. This TWG meeting can also help keep the spotlight on severe risks until they are properly mitigated.

A critical tool for tracking risks is the *risk register*. This is simply a formal list of risks that an organization has identified. Each risk in the register typically includes the following fields:

- Unique identifier
- Short name
- Description
- Owner
- Probability
- Magnitude
- Risk value (or rating)
- Disposition

A bunch of other fields may also be included, but these are the fields we've seen most often being used. It is important to assign an owner to each risk, so that someone is held accountable for tracking it and ensuring that it is addressed. Of course, there are many ways of dealing with a particular risk. If its rating is pretty low, you can accept the risk, which means you do nothing (except maybe monitoring and responding) and hope it is never realized. This may sound silly, but keep in mind that you may have identified some pretty small risks along the way. On the other end of the spectrum, a risk's disposition may require significant investments of staff hours and money. Most risks fall somewhere in between and require some minor configuration changes or patches that someone needs to ensure happen in a timely manner.

Whatever approach you use to identify risks, you should end up with a pretty lengthy risk register. Don't feel overwhelmed; no organization is able to address every possible risk it faces. The goal is to identify as many risks as you can, and then figure out which ones you care most about. That's where calculation comes in.

Risk Calculation

There are two approaches to calculating risks: quantitative and qualitative. A *quantitative* approach is the most rigorous and is used to assign monetary and numeric values to all elements of the risk assessment process. Each element within the analysis (such as threat frequency, impact damage) is quantified and entered into equations to determine total risk. For example, suppose you identify a hundred small organizations that are similar to yours, and on average, eight of them experience ransomware incidents every year. You can determine from industry studies that the average cost per incident is $713,000. So you have an 8 percent chance of suffering a ransomware attack. Using the formula of probability times impact, you could determine that your quantitative risk due to ransomware is $57,040 (0.08 × $713,000) per year.

Probability	Magnitude				
	Negligible	Minor	Moderate	Major	Severe
Almost Certain	Moderate	High	High	Extreme	Extreme
Likely	Moderate	Moderate	High	High	Extreme
Possible	Low	Moderate	Moderate	High	Extreme
Unlikely	Low	Moderate	Moderate	Moderate	High
Rare	Low	Low	Moderate	Moderate	High

Figure 20-1   Qualitative risk matrix

A *qualitative* risk analysis uses a "softer" approach to the data elements of a risk analysis. It does not quantify that data, which means that it does not assign precise numeric values to the data so that it can be used in equations. Instead, it uses categories that describe the qualities of risk elements. For instance, the probability of the ransomware risk can be categorized as unlikely, but its impact is severe, which could yield a high risk for that incident. Figure 20-1 shows a matrix that is typically used to calculate risk qualitatively.

It is worth noting that some organizations use a quasi-quantitative approach by assigning values to the probability and magnitude classifications. For example, a Negligible magnitude impact could be assigned a value of 1, a Minor impact is 2, and so on, until reaching a value of 5 for a Severe impact. Likewise, a Rare probability could be given a value of 1, while an Almost Certain one would get a 5. This enables the assignment of numeric values to risks so that a Likely (4) risk with a Moderate (3) magnitude would be valued at 12, while a Possible (3) one with a Severe (5) impact would get a 15. Despite the fact that this approach uses numbers and a bit of arithmetic, it is still qualitative and should not to be confused with a much more thorough quantitative analysis.

Probability

So far, we've established that a risk is defined by the likelihood of an incident and the magnitude of its impact on an organization. Let's turn our attention to the first factor: the probability of a given incident. There are many ways to estimate this, but, again, they follow either a quantitative or a qualitative approach. (And you really shouldn't mix those two.) You already saw one way of applying a quantitative approach to determining probability in our earlier example about a ransomware attack. In that example, we counted the frequency of such attacks on organizations similar to ours and came up with eight successful attacks per year. This let us know that, on average, organizations like yours have an 8 percent chance of suffering a ransomware incident.

It may be tempting to use this approach, but it is not without some challenges. For starters, you may not have good data on a particular risk. Ransomware is not always reported, which means you would be underestimating its frequency if you rely on public information. It also doesn't take into account (at least in the way we've used it in the example) the fact that some organizations may have better defenses against it. What if all eight organizations that got hit last year were those without good controls in place? Then the 8 percent probability would have no bearing on you, because you would either be way likelier to get hit (if your controls are poor) or way less likely (if you have good defenses). Finally, using last year's statistics fails to take into account the ever-changing threat landscape. There are ways to deal with all these issues, but they require a level of maturity in risk management that most organizations lack. This is why most of us end up using a qualitative approach.

When dealing with risk probability qualitatively, most organizations rely on expert opinions and, quite honestly, intuition. There are three common approaches to estimating the probability of a risk. The first is simply to have an open discussion (typically in the TWG) and agree on the likelihood. This is probably the fastest way to estimate the probability but is prone to groupthink, which could skew the results. An alternative approach, the Delphi technique, involves sharing anonymous opinions repeatedly until consensus is reached. This avoids groupthink while allowing individuals to be persuaded to change their opinions along the way. It does, however, require more time than the other two approaches. Finally, a common middle ground between open discussion and the Delphi technique is to ask everyone to submit his or her own probability estimate secretly. All the estimates are then averaged out to assign a final probability.

The Delphi Technique

The Delphi technique is a group decision method used to ensure that each member gives an honest opinion of what he or she thinks the result of a particular risk will be. This avoids a group of individuals feeling pressured to go along with others' thought processes and enables them to participate in an independent and anonymous way. Each member of the group provides his or her opinion of a certain threat and turns it in to the team that is performing the analysis. The results are compiled and distributed to the group members, who then write down their comments anonymously and return them to the analysis group. The comments are compiled and redistributed for more comments until a consensus is formed. This method is used to obtain an agreement on probabilities of occurrence and magnitude of consequences without individuals having to agree verbally.

Rating	Value	Interpretation
Severe	5	Organizational survival is at risk; damages exceed $1M
Major	4	Prolonged (4+ hrs.) disruption to key business functions; damages do not exceed $1M
Moderate	3	Brief (<4 hrs.) disruptions to key business functions; damages do not exceed $100K
Minor	2	Some disruption of nonessential functions; damages do not exceed $10K
Negligible	1	Individual, nonessential disruptions; damages do not exceed $1K

Table 20-1 Sample Impact Interpretation Map

Magnitude

Estimating the magnitude of the impact of a risk using a quantitative approach is a bit easier than doing so to determine the probability of the risk, because the organization already knows a lot of the factors that would be needed. For example, if a core business function is online sales, then a denial-of-service attack of a certain duration against the website would interrupt a predictable volume of sales. The dollar value of those lost (or, at least, delayed) sales is fairly easy to calculate. Things get a bit more complex when dealing with less-tangible assets such as reputation or intellectual property, but, again, there are ways for mature risk management programs to estimate these impacts. Still, most organizations follow a qualitative approach, so let's focus on that.

The same three approaches we described earlier for qualitatively estimating probability (open discussion, Delphi technique, and secret voting) can be used for estimating risk magnitude. The trick is to have a shared and fairly precise definition of what each magnitude category means. These definitions are sometimes displayed in an *impact interpretation map*. Table 20-1 provides an example of one.

The criteria for interpreting the magnitude of a risk's impact will vary among organizations. These could include impacts on the health and welfare of people, staff hours to remediate, or the number of customers affected. The point is that you should establish the criteria early as part of your risk identification process and then update it periodically as needed.

 EXAM TIP You should be able to categorize impacts and likelihoods as being high, medium, or low for the exam. Using those categories, you should then be able to evaluate the fitness of specific security controls.

Putting It All Together

Let's look at a simple example of a qualitative risk calculation using the secret-voting approach. The risk analysis team presents a scenario explaining the threat of a hacker accessing confidential information held on the five file servers within the company. The risk analysis team then distributes the scenario in a written format to a team of five people (the cybersecurity analyst, database administrator, application programmer, system

Risk: Data Breach	Probability of Risk Taking Place	Magnitude of Loss to the Company
Cybersecurity analyst	2	4
Database administrator	4	4
Application programmer	3	3
System operator	4	3
Operational manager	4	4
Results	3.4	3.6

Table 20-2 Example of a Qualitative Risk Calculation

operator, and operational manager), who are also given a sheet to rank the risk's probability and magnitude. Table 20-2 shows the results.

Avoiding Group Biases

It is ideal to perform risk evaluations as a group, but this could also introduce the problem of group biases. These biases, sometimes called groupthink, can cause the group members to arrive at similar results when evaluating issues such as likelihood, impact, and effectiveness. The following are tips for ensuring everyone presents their own opinions untarnished by those of others:

- Keep the group small (no more than 12).
- Have participants quietly write down opinions before sharing them.
- Encourage respectful disagreements.
- Include plenty of time for discussion, but also put a time limit.

Communication of Risk Factors

It is worth repeating that a risk assessment is intended to integrate the security program objectives with the company's business objectives and requirements. The team involved in this process includes a variety of technical, security, and business stakeholders (among others), so it is important to be able to present the risk factors to a variety of audiences. If you follow the risk identification and risk calculation approaches we just covered, you'll be most of the way there. The risk matrix will provide a high-level view of risks, how likely they are to occur, and how they may affect the business. If a stakeholder wants to drill down into the probabilities, you should have either meeting minutes (from an open discussion) or the scores and notes provided by individual members of the team (from the Delphi technique or secret voting and tabulation). The other risk factor is the magnitude, the meaning of which is documented in the impact interpretation map. How the factors are combined is dictated by the qualitative risk matrix. Together, these products document and communicate the risk factors in ways that should be understandable to technical and nontechnical staff alike.

Risk Prioritization

It may seem obvious at this point how we'd prioritize the risks we've identified. After all, each risk has a risk value that is documented in the risk register. Wouldn't that be enough to prioritize risks and assign resources to controlling them? Maybe, but we have to keep in mind that risk can be dealt with in four basic ways: transfer it, avoid it, reduce it, or accept it. If a company decides the total risk is too high to gamble with, it can purchase insurance, which would *transfer the risk* to the insurance company. If a company decides to terminate the activity that is introducing the risk, this is known as *risk avoidance*. For example, if a company allows employees to use instant messaging (IM), there are many risks surrounding this technology. The company could decide not to allow any IM activity by users because there is not a strong enough business need for its continued use. Discontinuing this service is an example of risk avoidance. Another approach is *risk mitigation*, where the risk is reduced to a level considered acceptable enough to continue conducting business. The implementation of firewalls, training, and intrusion detection/ prevention systems or other control types represent types of risk mitigation efforts. The last approach is to *accept the risk*, which means the company understands the level of risk it is faced with, as well as the potential cost of damage, and decides to just live with it and not implement the countermeasure. Many companies will accept risk when the cost/ benefit ratio indicates that the cost of the countermeasure outweighs the potential loss of value if the risk becomes a reality.

Keep in mind that risk can never be completely eliminated. This means that, even after we implement the best controls, there is still some residual risk left. Risk acceptance, therefore, can happen after each of the other treatment options is employed.

 EXAM TIP Watch out for answers that imply (or state) that a risk is eliminated, since this can never happen. We can mitigate risk but never eliminate it altogether.

Security Controls

Security controls are put into place to mitigate the risk an organization faces, and they can come in three main flavors: administrative, technical, and physical. *Administrative controls* are commonly referred to as soft controls, because they are more management oriented. Examples of administrative controls are policies, procedures, personnel security, and training. *Technical controls* (also called logical controls) are software or hardware components, as in firewalls, IDSs, encryption, and identification and authentication mechanisms. And *physical controls* are items put into place to protect facility, personnel, and resources. Examples of physical controls are security guards, locks, fencing, and lighting.

Once we have our risks prioritized and separated according to how we want to handle them (transfer, avoid, mitigate, or accept), we finally get to talk about what specific controls we should have (or should put) in place for the risks we want to mitigate. There are three important concepts to consider here: The first is that the cure cannot be worse than the disease. If we think a particular risk could have a magnitude of, say, just under $10,000, but the control to mitigate it will cost us more than that, then it is not a cost-effective control and it makes no sense from business perspective to use it. The other important concept to keep in mind is that the risk will never completely go away, regardless of how effective our controls may be. We can reduce the probability and/or magnitude of a risk, but never to zero. Finally, the third consideration is that, even if the controls are cost-effective and mitigate the risks to the degree that you want them to, over time, they may become less effective as the risk landscape changes. For that reason, it is essential to review the controls' effectiveness periodically.

Technical Control Review

A *technical control review* is a deliberate assessment of the effectiveness of technical controls and how they are implemented and managed. You may have decided during your risk evaluation that a firewall may be the best control against a particular risk, but several months later, how do you know it is working as you intended it to? Apart from ensuring that it is still the best choice against a given risk, the review considers issues like these:

- Is the control version up-to-date?
- Is it configured properly to handle the risk?
- Do the right people (and no others) have access to manage the control?
- Are licenses and/or support contracts current?

Even in organizations that practice strict configuration management, it is common to find hardware or software that were forgotten or that still have account access for individuals who are no longer affiliated with the organization. Additionally, the effectiveness of a technical control can be degraded or even annulled if the threat actor changes procedures. These are just some of the reasons why it makes sense to review technical controls periodically.

Administrative Control Review

Like technical controls, our policies can become outdated. Furthermore, it is possible that people are just not following them or that they are attempting to do so in the wrong way. An *administrative control review* is a deliberate assessment of operational control choices and how they are implemented and managed. The review first validates that the controls are still the best choices against a given risk, and then considers issues like these:

- Is the control consistent with all applicable laws, regulations, policies, and directives?
- Are all affected members of the organization aware of the control?
- Is the control part of newcomer or periodic refresher trainer for the affected personnel?
- Are all affected staff members abiding by the control?

Administrative controls, like technical ones, can become ineffective with time. Taking the time to review their effectiveness and completeness is an important part of any security program.

Engineering Tradeoffs

Security controls have costs that are measured not only in dollars or staff hours. Perhaps one of the trickiest aspects of selecting and implementing technical security controls is that of balancing their impact on the performance of our systems. For example, to protect our endpoints, we may deploy a robust endpoint detection and response (EDR) solution that is constantly scanning memory and disks for patterns of malicious activity. With all its features activated (so as to provide maximum security), the EDR solution may cause noticeable system slowdown for the users, particularly those using older devices. An engineering tradeoff is a deliberate balancing of system security and performance aimed at ensuring that, while neither solution option is optimal, both are acceptable to the organization. We see this also with rule-based IDSs that have very large rule bases. We want to process as many rules as we can get away with, without bogging down the network.

Documented Compensating Controls

As we discussed earlier, sometimes leaders will knowingly choose to take actions that leave vulnerabilities in their information systems. This usually happens either because the fix is too costly (perhaps, for example, a patch would break a critical business process), or because there is no feasible way to fix the vulnerability directly (for example, with an older X-ray machine at a hospital). *Compensating security controls* are not directly applied to a vulnerable system, but they compensate for the lack of a direct control. For example, if you have a vulnerable system that is no longer supported by its vendor, you may put it in its own VLAN and create ACLs that allow it to communicate with only one other host, which has been hardened against attacks. You may also want to deploy additional sensors to monitor traffic on that VLAN and activity on the hardened host. The process by which these decisions are made and the compensation controls developed should be documented in a separate procedure or included in another, related procedure.

Systems Assessment

We all know the old Russian proverb, "Trust, but verify." It is not uncommon for organizations to put significant amounts of effort into selecting and implementing controls only to discover (sometimes years later) that their security posture is not what they thought. Every implementation should be followed with verification to ensure that it was done properly. Just as importantly, there should be an ongoing periodic effort to ensure that the safeguards are still being done right and that they are still effective in the face of ever-changing threats.

An *assessment* is any process that gathers information and makes determinations based on it. This rather general term encompasses audits and a host of other evaluations such as vulnerability scans and penetration tests. More important than your remembering this definition is your understanding the importance of continuous assessments to ensure that the security of your systems remains adequate to mitigate the risks in your environment. Among the more popular assessments are these:

- Vulnerability assessment
- Penetration test
- Red team assessment
- Risk assessment
- Tabletop exercises

Every organization should have a formal assessment program that specifies how, when, where, why, and with whom the different aspects of its security will be evaluated. This is a key component that drives organizations toward continuous improvement and optimization. This program is also an insurance policy against the threat of obsolescence caused by an ever-changing environment.

Supply Chain Risk Assessment

Many organizations fail to consider their supply chain when assessing their risks, despite the fact that it often presents a convenient and easier backdoor to an attacker. So what is a supply chain anyway? A *supply chain* is a sequence of suppliers involved in delivering some good or service. If your company manufactures laptops, your supply chain will include the vendor that supplies your video cards. It will also include whoever makes the integrated circuits that go on those cards, as well as the supplier of the raw chemicals that are involved in that process. The supply chain also includes suppliers of services, such as the company that maintains the heating, ventilation, and air conditioning (HVAC) systems needed to keep your assembly line employees comfortable.

The various organizations that make up your supply chain will have a different outlook on security than you do. For one thing, their threats are probably different from yours. Why would a criminal looking to steal credit card information target an HVAC service provider? This is exactly what happened in 2013 when chain discount store Target had more than 40 million credit cards compromised. Target had done a reasonable job at securing its perimeter, but not its internal networks. The attacker, unable (or maybe just unwilling) to penetrate Target's outer shell head-on, decided to exploit the vulnerable network of one of Target's HVAC service providers and steal its credentials. Armed with this information, the thieves were able to gain access to the point-of-sale terminals and, from there, to the credit card information.

The basic processes you'll need to assess risk in your supply chain are the same ones you use in the rest of your risk management program. The differences are mainly in what you look at (that is, the scope of your assessments) and what you can do about it (legally and contractually). One of the first things you'll need to do is to create a supply chain map for your organization. This is essentially a network diagram of who supplies what to whom down to your ultimate customers. Figure 20-2 depicts a simplified systems

PART V

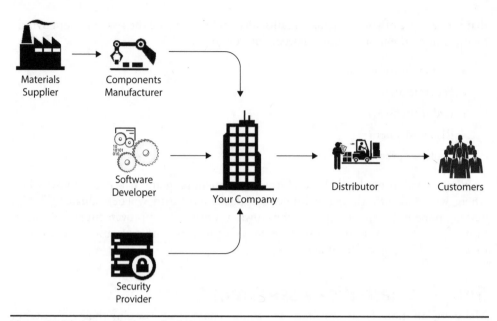

Figure 20-2 A simplified supply chain

integrator company (named Your Company). A hardware components manufacturer supplies its hardware, which, in turn, is supplied by a materials supplier. Your Company receives software from a developer and receives managed security from an external service provider. The hardware and software components are integrated and configured into Your Company's product, which is then shipped to its distributor and then on to its customers. In this example, Your Company has four suppliers on which to base its supply chain risk assessment. It is also considered a supplier to its distributor.

Vendor Due Diligence

Most organizations rely on a multitude of vendors that enable them to perform their core business functions. Companies rely on hardware and software suppliers to provide their IT systems, hosting companies to maintain websites and e-mail servers, service providers for various telecommunication connections, disaster recovery companies for colocation capabilities, cloud computing providers for infrastructure or application services, developers for software creation, and security companies to carry out vulnerability management. It is important to realize that although you can outsource goods and services, you cannot outsource risk.

Due diligence is the exercise of care that a reasonable person is expected to take in a particular situation, and failing to exercise it could leave an organization legally liable for someone else's losses. So if one of your vendors is not duly diligent and this results in your losses, you may be able to recover damages after the fact. But wouldn't it be better to ensure that the vendor is duly diligent before a crisis ensues?

This is a key part of performing supply chain risk assessments: to determine your risk that results from what your vendors and suppliers are or are not doing to protect themselves. Let's look at some things an organization may look at to determine whether its vendors are practicing due diligence and, if not, what the level of risk might be:

- Review references and communicate with former and existing customers.
- Review Better Business Bureau reports.
- Ensure that contracts/agreements include requirements for adequate security controls.
- Ensure that service level agreements are in place if appropriate.
- Review the vendor's security program before signing an agreement, and periodically thereafter.
- Review internal and external audit reports and third-party reviews.
- Conduct onsite inspection and interviews after signing the agreement.
- Ensure that the vendor has a business continuity plan (BCP) in place.
- Implement a nondisclosure agreement (NDA).

Vendor relationships are prevalent within organizations today, but they are commonly forgotten about when it comes to assessing risks. It is very important to check, before *and* after getting into a contract or agreement with any vendor, to ensure that they are exercising due diligence when it comes to cybersecurity.

Hardware Source Authenticity

Another frequently overlooked source of risk is acquired hardware, particularly when it could be (or may actually be known to be) counterfeit. In 2012, there were multiple reports in the media of counterfeit networking products finding their way into critical networks in both industry and government. By one account, some of these fakes were even found in sensitive military networks. *Source authenticity*, or the assurance that a product was sourced from an authentic manufacturer, is important for all of us, but particularly so if we handle sensitive information. Two particular problems with fake products affect a cybersecurity analyst: malicious features and lower quality.

It is not hard to imagine organizations or governments that would want to insert their own fake or modified version of a popular router into a variety of networks. Apart from a source of intelligence or data theft, it could also provide them with remote "kill" switches that could be leveraged for blackmail or in case of hostilities. The problem, of course, is that detecting these hidden features in hardware is often well beyond the means of most organizations. Ensuring that your devices are legitimate and came directly from the vendor can greatly decrease this risk.

Another problem with counterfeit hardware is that, even if there is no malicious design, it is probably not built to the same standard as the genuine hardware. It makes no sense for a counterfeiter to invest the same amount of resources into quality assurance and quality control as the genuine manufacturer. Doing so would increase their footprint (and chance of detection) as well as drive their costs up and profit margins down.

For most of us, the greatest risk in using counterfeits is that they will fail at a higher rate and in more unexpected ways than the originals. And when they do fail, you won't be able to get support from the legitimate manufacturer.

Counterfeit Products: What Can You Do?

Counterfeits are a very real and growing problem. Fortunately, there are some very basic steps you can take to significantly reduce your exposure to this threat. Here are our top three principles for dealing with counterfeit products:

- *You get what you pay for.* If the price of a device seems too good to be true, it probably is. Much of the appeal of counterfeits is that they can lure customers looking for bargains.

- *Buy from authorized retailers.* Most major manufacturers of networking and security equipment will have a network of retailers authorized to sell their products. Purchasing from these retailers will both decrease your chances of buying a counterfeit and improve your odds of remediation.

- *Check the serial number.* Most manufacturers will have a mechanism for you to verify that a serial number maps to a legitimate product. If the serial number is copied, they will alert you to the duplicate as well.

Training and Exercises

General George Patton is famously quoted as having said, "You fight like you train," but this idea, in various forms, has spread to a multitude of groups beyond the Army. It speaks to the fact that each of us has two mental systems: the first is a fast and reflexive one, and the second is slow and analytical. Periodic, realistic training develops and maintains the "muscle memory" of the first system, ensuring that reflexive actions are good ones. In the thick of a fight, bombarded by environmental information in the form of sights, sounds, smells, and pain, soldiers don't have the luxury of processing it all and must almost instantly make the right calls. So do we, when we are responding to security incidents on our networks.

Admittedly, the decision times in combat and incident response are orders of magnitude apart, but you cannot afford to learn or rediscover the standard operating procedures when you are faced with a real incident. We have worked with organizations in which seconds can literally mean the loss of millions of dollars. The goal of your programs for training and exercises should then be to ensure that all team members have the muscle memory to handle the predictable issues quickly and, in so doing, create the time to be deliberate and analytical about the others.

The general purpose of a training event is to develop or maintain a specific set of skills, knowledge, or attributes that enable individuals or groups to do their jobs effectively or better. An *exercise* is an event in which individuals or groups apply relevant skills, knowledge, or attributes in a particular scenario. Although it could seem that training is a prerequisite for exercises (and, indeed, many organizations take this approach), it is also possible for exercises to be training events in their own right.

All training events and exercises should start with a set of goals or outcomes, as well as a way to assess whether or not those were achieved. This makes sense on at least two levels: at an operational level, it tells you whether you were successful in your endeavor or need to do it again (perhaps in a different way), and at a managerial level, it tells decision-makers whether or not the investment of resources is worth the results. Training and exercises tend to be resource-intensive and should be applied with prudence.

Types of Exercises

Though cybersecurity exercises can have a large number of potential goals, they tend to be focused on testing tactics, techniques, and procedures (TTPs) for dealing with incidents and/or assessing the effectiveness of defensive teams in dealing with incidents. Either way, a key to success is to choose scenarios that facilitate the assessment process. The two major types of cybersecurity exercises are tabletop and live-fire.

Tabletop Exercises

Tabletop exercises (TTXs) may or may not happen at a tabletop, but they do not involve a technical control infrastructure. TTXs can happen at the executive level (for example, CEO, CIO, or CFO), at the team level (for example, security operations center or SOC), or anywhere in between. The idea is usually to test out procedures and ensure that they actually do what they're intended to and that everyone knows their role in responding to an event. TTXs require relatively few resources apart from deliberate planning by quali-fied individuals and the undisturbed time and attention of the participants.

After determining the goals of the exercise and vetting them with the senior leadership of the organization, the planning team develops a scenario that touches on the important aspects of the response plan. The idea is normally not to cover every contingency but to ensure that the team is able to respond to the likeliest and/or most dangerous scenarios. As they develop the exercise, the planning team will consider branches and sequels at every point in the scenario. A *branch* is a point at which the participants may choose one of multiple approaches to the response. If the branches are not carefully managed and controlled, the TTX could wander into uncharted and unproductive directions. Conversely, a *sequel* is a follow-on to a given action in the response. For instance, as part of the response, the strategic communications team may issue statements to the news media. A sequel to that could involve a media outlet challenging the statement, which in turn would require a response by the team. Like branches, sequels must be carefully used in order to keep the exercise on course. Senior leadership support and good scenario development are critical ingredients to attract and engage the right participants. Like any contest, a TTX is only as good as the folks who show up to play.

Live-Fire Exercises

In a live-fire exercise (LFX), the participants are defending real or simulated informa-tion systems against real (though friendly) attackers. There are many challenges in orga-nizing one of these events, but the major ones are developing an infrastructure that is representative of the real systems, getting a good red (adversary) team, and getting the right blue (defending) team members in the room for the duration of the exercise. Any one of these, by itself, is a costly proposition. However, you need all three for a successful event.

On the surface, getting a good cyber range does not seem like a major challenge. After all, many or our systems are virtualized to begin with, so cloning several boxes should be easy. The main problem is that you cannot use production boxes for a cyber exercise because you would compromise the confidentiality, integrity, and perhaps availability of real-world information and systems. Manually creating a replica of even one of your subnets takes time and resources, but it is doable given the right level of support. Still, you won't have any pattern-of-life (POL) traffic on the network. POL is what makes networks realistic. It's the usual chatter of users visiting myriads of websites, exchanging e-mail messages, and interacting with data stores. Absent POL traffic, every packet on the network can be assumed to come from the red team.

A possible solution would be to have a separate team of individuals who simply provide this by simulating real-world work for the duration of the event. Unless you have a bunch of interns with nothing better to do, this gets cost-prohibitive really fast. A reasonable compromise is to have a limited number of individuals logged into many accounts, thus multiplying the effect. Another approach is to invest in a traffic generator that automatically injects packets. Your mileage will vary, but these solutions are not very realistic and will be revealed as fake by even a cursory examination. A promising area of research is in the creation of autonomous agents that interact with the various nodes on the network and simulate real users. Through the use of artificial intelligence, the state of the art is improving, but we are not there just yet.

Red Team

A *red team* is a group that acts as adversaries during an exercise. The red team need not be "hands on keyboard" because red-teaming extends to TTXs as well as LFXs. These individuals need to be very skilled at whatever area they are trying to disrupt. If they are part of a TTX and trying to commit fraud, they need to know fraud and anti-fraud activities at least as well as the exercise participants. If the defenders (or blue team members) as a group are more skilled than the red team, the exercise will not be effective or well received.

This requirement for a highly skilled red team is problematic for a number of reasons. First of all, skills and pay tend to go hand-in-hand, which means these individuals will be expensive. Because their skills are so sought after, they may not even be available for the event. Additionally, some organizations may not be willing or able to bring in external personnel to exploit flaws in their systems, even if they have signed a nondisclosure agreement (NDA). These challenges sometimes cause organizations to use their own staff for the red team. If your organization has people whose full-time job is to red team or pen test, then this is probably fine. However, few organizations have such individuals on their staff, which means that using internal assets may be less expensive but will probably reduce the value of the event. In the end, you get what you pay for.

Blue Team

The *blue team* is the group of participants who are the focus of an exercise. They perform the same tasks during a notional event as they would perform in their real jobs if the scenario was real. Though others will probably also benefit from the exercise, it is the blue team that is tested and/or trained the most. The team's composition depends on the scope of the event. However, because responding to events and incidents typically requires coordinated actions by multiple groups within an organization, it is important to ensure that each of these groups is represented in the blue team.

The biggest challenge in assembling the blue team is that they will not be available to perform their daily duties for the duration of the exercise as well as for any pre- or post-event activities. For some organizations, the cost of doing this is too high, and they end up sending the people they can afford to be without, rather than those who really should be participating. If this happens, the exercise may be of great training value for these participants, but it may not enable the organization as a whole to assess its level of readiness. Senior or executive leadership involvement and support will be critical to keep this from happening.

White Team

The *white team* consists of anyone who will plan, document, assess, or moderate the exercise. Although it is tempting to think of the members of the white team as the referees, they do a lot more than that. These individuals come up with the scenario, working in concert with business unit leads and other key advisors. They structure the schedule so that the goals of the exercise are accomplished and every participant is gainfully employed. During the conduct of the event, the white team documents the actions of the participants and interferes as necessary to ensure they don't stray from the flow of the exercise. They may also delay some participants' actions to maintain synchronization. Finally, the white team is normally in charge of conducting an after-action review by documenting and sharing their observations (and, potentially, assessments) with key personnel.

Chapter Review

This chapter was all about proactive steps you can take to ensure the security of your corporate environment. The implication of this discussion is that risk is not something you address once and then walk away from. It is something that needs to be revisited periodically and deliberately. You may have done a very through risk assessment and implemented appropriate controls, but six months later many of those may be moot. You wouldn't know this to be the case unless you periodically (and formally) review your controls for continued effectiveness. It is also wise to conduct periodic systems assessments to ensure that the more analytical exercise of managing risks actually translates to practical results on the real systems. Finally, the human component of your information systems must also be considered. Training is absolutely essential both to maintain skills and to update awareness to current issues of concern. However, simply putting the right information into the heads of your colleagues is not necessarily enough. It is best to test their performance under conditions that are as realistic as possible in either table-top or live-fire exercises. This is where you will best be able to tell whether the people are as prepared as the devices and software to combat the ever-changing threats to your organization.

Questions

1. Which of the following is one of the key reasons why you would perform a business impact analysis?

   A. Determine the company's market penetration.

   B. Balance the impact of security controls on the business functions.

   C. Identify effective technical controls for the critical business functions.

   D. Identify the company's critical business functions.

2. Which document is typically used to list all known risks to an organization and show their probabilities and magnitudes?

   A. Delphi ledger

   B. Qualitative risk matrix

   C. Impact interpretation map

   D. Risk register

3. Which of the following is true about a qualitative risk assessment?

   A. It is of higher quality than other types of assessment.

   B. It is more rigorous than a quantitative one.

   C. It is not as rigorous as a quantitative one.

   D. It is not commonly used.

4. In an exercise, which type of team is the focus of the exercise, performing their duties as they would normally in day-to-day operations?

   A. Gray team

   B. White team

   C. Red team

   D. Blue team

Use the following scenario to answer Questions 5–8:

Your company does not have a risk management program in place, but your boss is concerned about the risk of a data breach. Nobody has the necessary experience, but she knows that organizational risk mitigation was covered in your CySA+ certification exam, so she asks you to lead a risk assessment and ensure that appropriate security controls are in place.

5. You decide to begin with a business impact analysis (BIA). Which of following would not be part of this analysis?

   A. The company's critical business functions

   B. Calculating risks

   C. Identifying vulnerabilities

   D. Reviewing administrative controls

6. Which approach would you follow in performing a risk assessment?

   A. Delphi

   B. Quantitative

   C. Qualitative

   D. Hybrid

7. You determine that the risk of a data breach is severe but also discover that you already have security appliances in place that are supposed to mitigate this risk. What could help you determine whether the risk is truly mitigated?

   A. Engineering tradeoff analysis

   B. Technical control review

   C. Administrative control review

   D. Tabletop exercise

8. Satisfied that the security appliances are working as expected, you decide to assess whether your staff, policies, and procedures are also up to the task of handling a data breach incident. Which of the following would be *least* effective in this effort?

   A. Administrative control review

   B. Tabletop exercises

   C. Live-fire exercises

   D. Red team

Answers

1. **D.** A business impact analysis (BIA) is a functional analysis that develops a hierarchy of business functions and determines each individual function's criticality level. It should eventually inform the selection of both technical and administrative controls, but that is not part of the BIA.

2. **D.** A risk register is a formal list of risks that an organization has identified. Qualitative risk matrix is normally a table that assigns a risk category to each possible combination of probabilities and magnitudes.

3. **C.** A qualitative risk analysis uses a "softer" approach that, unlike the more rigorous quantitative analysis, does not attempt to assign precise values to the variables.

4. **D.** The blue team is the group of participants who are the focus of an exercise and will be tested the most while performing the same tasks in a notional event as they would perform in their real jobs.

5. **D.** Security controls are not normally part of a business impact analysis. Instead, you will focus on understanding the critical business functions, their required resources, what vulnerabilities may impact either, and what the overall risk is for each critical function.

6. **C.** Since the organization is very immature with regard to risk management, it is probably best to follow a qualitative approach, which is not as rigorous as other approaches and doesn't require extensive expertise or data.

7. **B.** A technical control review is a deliberate assessment of the effectiveness of technical controls and how they are implemented and managed. This is what you would do to verify your level or risk mitigation with regard to a technical control such as a security appliance.

8. **C.** While a live-fire exercise could potentially assess the effectiveness of administrative controls, it would require an investment of significant resources. Administrative control reviews and tabletop exercises are both focused on assessing staff, policies, and procedures. Red teams can support either live-fire or tabletop exercises, so if you have one, they would definitely be effective in this effort.

The Importance of Frameworks, Policies, Procedures, and Controls

In this chapter you will learn:

- Common information security management frameworks
- Common policies and procedures
- Considerations in choosing controls
- How to verify and validate compliance

Innovation and best practices can be sown throughout an organization—but only when they fall on fertile ground.

—Marcus Buckingham

It is never a good idea to reinvent the wheel. It wastes time and you could end up with a worse wheel. This is particularly true in cybersecurity, where many great minds have spent years curating best practices and organizing them in useful ways. One of the byproducts of their efforts is the development of frameworks that provide structure to collections of these best practices. In this chapter, we'll introduce some cybersecurity frameworks with which you should be familiar, and then we'll delve into the policies, procedures, and controls that bring those frameworks to life in your organization. We close this last chapter with a brief discussion of the means by which you can verify that all this work is actually protecting your organization.

Security Frameworks

As you probably well know, security is not a one-size-fits-all proposition. Though there are certainly best practices, how we implement them ultimately depends on a variety of factors in our environments. Still, we need to have a way to conceptually frame the various issues we must consider as we protect our information systems. This is where security frameworks come in. The *Oxford English Dictionary* defines *framework* as a

Framework Type	Focus	Examples
Risk-based	Holistic	NIST RMF ISO 27005
Prescriptive	Security program	ISO 27001 NIST Cybersecurity Framework (CSF)
	Security controls	NIST 800-53 CIS Controls

Table 21-1 Types of Cybersecurity Frameworks

basic structure underlying a system, concept, or text. By extension, a cybersecurity framework can be defined as a structure on which we build cybersecurity systems. There is no shortage of them in cybersecurity, but, fortunately, you don't have to remember them all.

Generally speaking, there are two types of cybersecurity frameworks: those that are risk-based and those that aren't. This second category is sometimes referred to as *prescriptive* security frameworks, because they consist of sets of requirements that must be implemented without an explicit relationship to risk management. Prescriptive frameworks, in turn, can be divided into those that are focused on overarching security programs and those that are focused on specific controls. Table 21-1 captures these distinctions and highlights some of the frameworks with which you should be familiar, not only for the CySA+ exam but in your daily job. We delve into these in the sections that follow.

 EXAM TIP The CySA+ exam differentiates between risk-based and prescriptive frameworks, even though the latter can (and typically) address risk as well. The distinction lies in whether the focus of the framework is on risk.

NIST

One of the most influential organizations in terms of cybersecurity standards and frameworks is the National Institute for Standards and Technology (NIST). It is the part of the US Department of Commerce, which is charged with promoting innovation and industrial competitiveness. As part of this mission, NIST develops and publishes standards and guidelines aimed at improving practices, including cybersecurity across a variety of sectors. Though it is certainly worth your time to familiarize yourself with the many NIST publications, you should be aware of three in particular as a CySA+ candidate: the Risk Management Framework (RMF) described in NIST Special Publication (SP) 800-37, "Risk Management Framework for Information Systems and Organizations: A System Life Cycle Approach for Security and Privacy"; the whitepaper "Framework for Improving Critical Infrastructure Cybersecurity" (the Cybersecurity Framework, or CSF); and NIST SP 800-53, "Recommended Security Controls for Federal Information Systems."

Risk Management Framework

The NST RMF is defined in multiple publications. However, the foundational document for understanding the RMF is SP 800-39, "Managing Information Security Risk." This publication defines three tiers to risk management, which are depicted in Figure 21-1:

- **Organization tier**   This tier is concerned with risk to the business as a whole, which means it frames the rest of the conversation and sets important parameters such as the risk tolerance level. It focuses primarily on governance and program-related risk management.

- **Mission/Business Processes tier**   This tier deals with the risk to the major functions of the organization, such as defining the criticality of the information flows between the organization and its partners or customers. It focuses more on cross-organizational process risk common to multiple business units.

- **Information Systems tier**   This tier addresses risk from an information systems perspective. This is where you will likely spend most of your time, though you need to be aware of the higher two tiers.

The tiers described should highlight the first advantage of using a security framework: it enables you to use multiple layers of abstraction to describe issues at different levels. The lowest tier is the tactical world in which cybersecurity analysts spend most of their time. Those working in this tier provide information to the business tier that, in turn, provides inputs to the more strategic organization tier. Directives and constraints flow down, taking into account the inputs that went up the hierarchy.

PART V

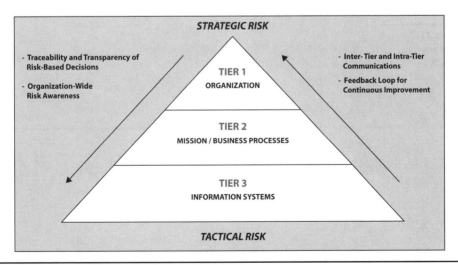

Figure 21-1   Tiers of the organization-wide Risk Management Framework

Another advantage of using a security framework is that it enables you to organize activities into related groupings. As shown in Figure 21-2, the RMF has four components that fully describe it at any one (or all) of the tiers described. The arrows denote communication flows between the functional groups, which are described here:

- **Frame risk**   Risk framing defines the context within which all other risk activities take place. What are our assumptions and constraints? What are the organizational priorities? What is the risk tolerance of senior management?

- **Assess risk**   Before we can take any action to mitigate risk, we have to assess it. This is perhaps the most critical aspect of the process, and one that we will discuss at length. If your risk assessment is spot-on, then the rest of the process becomes pretty straightforward.

- **Respond to risk**   By now, we've done our homework. We know what we should, must, and cannot do (from the framing component), and we know what we're up against in terms of threats, vulnerabilities, and attacks (from the assess component). Responding to the risk becomes a matter of matching our limited resources with our prioritized set of controls. Not only are we mitigating significant risk, but, more importantly, we can tell our bosses what risk we can't do anything about because we're out of resources.

- **Monitor risk**   No matter how diligent we've been so far, we probably missed something. But even if we didn't miss anything, the environment will likely change (perhaps a new threat source emerged or a new system brought new vulnerabilities). To stay one step ahead of the bad guys, we need to monitor the effectiveness of our controls continuously against the risks for which we designed them.

Figure 21-2
Components
of the RMF

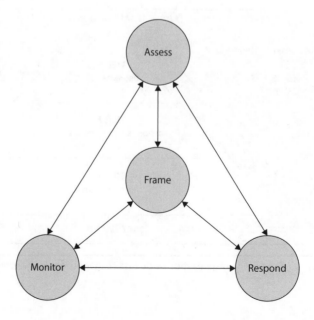

So what can you do with this? For starters, you can line up tasks with the appropriate framework components. For example, assessing risk at each of the tiers will be different. At the tactical level, you may consider technical vulnerabilities in your platforms. Assessing business-level risk, on the other hand, may involve dealing with threats that could prevent online orders from being processed. Finally, at the organizational level, the risks may be to brand image. The framework, then, gives you a way to look holistically at all the issues from the multiple different perspectives and in a structured way.

As we mentioned earlier, not every framework is explicitly concerned with risks. Let us now turn our attention to one of the most popular, prescriptive security program frameworks, which was also developed by the NIST.

Framework for Improving Critical Infrastructure Cybersecurity

On February 12, 2013, US President Barack Obama signed Executive Order 13636, calling for the development of a voluntary cybersecurity framework for organizations that are part of the critical infrastructure. The goal of this construct was for it to be flexible, repeatable, and cost-effective so that it could be prioritized for better alignment with business processes and goals. A year to the day later, NIST published the "Framework for Improving Critical Infrastructure Cybersecurity," which was the result of a collaborative process with members of the government, industry, and academia. As of this writing, it is estimated that at least 52 percent of worldwide organizations have adopted some portion of or all of the framework. The framework is divided into three main components:

- **Framework Core** Consists of the various activities, outcomes, and references common to all organizations. These are broken down into five functions, 22 categories, and 98 subcategories.

- **Implementation Tiers** Categorize the degree of rigor and sophistication of cybersecurity practices, which can be Partial (tier 1), Risk Informed (tier 2), Repeatable (tier 3), or Adaptive (tier 4). The goal is not to force an organization to move to a higher tier, but rather to inform its decisions so that it can do so if it makes business sense.

- **Framework Profile** Describes the state of an organization with regard to the CSF categories and subcategories. It enables decision-makers to compare the "as-is" situation to one or more "to-be" possibilities, so that they can align cybersecurity and business priorities and processes in ways that make sense to that particular organization. An organization's profile is tailorable based on the requirements of the industry segment within which it operates and the organization's needs.

 NOTE Avoid confusing the CSF's implementation tiers with the RMF risk tiers. The word "tier" means two entirely different things in the context of these frameworks. CSF tiers speak to the level of depth or rigor that an organization has put into implementing the CSF core practices. The RMF tiers describe different levels or context of risk within an organization.

The core practices organize cybersecurity activities into five higher level functions with which you should be familiar. Everything we do can be aligned with one of these:

- **Identify**   Understand your organization's business context, resources, and risks.
- **Protect**   Develop appropriate controls to mitigate risk in ways that make sense.
- **Detect**   Discover in a timely manner anything that threatens your security.
- **Respond**   Quickly contain the effects of anything that threatens your security.
- **Recover**   Return to a secure state that enables business activities after an incident.

 EXAM TIP   For the exam, you should remember the five functions of the NIST Cybersecurity Framework and the fact that it is voluntary and not a one-size-fits-all solution to cybersecurity.

SP 800-53

One of the standards that NIST has been responsible for developing is SP 800-53, "Security and Privacy Controls for Federal Information Systems and Organizations," currently in its fourth revision. It outlines controls that agencies need to put into place to be compliant with the Federal Information Processing Standards (FIPS). It is worth noting that, although this publication is aimed at federal government organizations, many others have voluntarily adopted it to help them better secure their systems.

Basically, SP 800-53 provides specific guidance on how to select security controls. It prescribes a four-step process for applying controls:

1. Select the appropriate security control baselines.
2. Tailor the baselines.
3. Document the security control selection process.
4. Apply the controls.

The first step starts off with categorizing your information systems based on criticality and sensitivity of the information to be processed, stored, or transmitted by those systems. If this sounds familiar, that's because we discussed these concepts in Chapter 19 when we covered data classification. SP 800-53 applies sensitivity and criticality to each security objective (confidentiality, integrity, and availability) to determine a system's criticality. For example, suppose you have a customer relationship management (CRM) system. If its confidentiality were to be compromised, this would cause significant harm to your company, particularly if the information fell into the hands of your competitors. The system's integrity and availability, on the other hand, would probably not be as critical to your business, so they would be classified as relatively low. The format for describing the security category (SC) of this CRM would be as follows:

$$SC_{CRM} = \{(\textbf{confidentiality}, \textit{high}),(\textbf{integrity}, \textit{low}),(\textbf{availability}, \textit{low})\}$$

ID	Family	ID	Family
AC	Access Control	MP	Media Protection
AT	Awareness and Training	PE	Physical and Environmental Protection
AU	Audit and Accountability	PL	Planning
CA	Security Assessment and Authorization	PS	Personnel Security
CM	Configuration Management	RA	Risk Assessment
CP	Contingency Planning	SA	System and Services Acquisition
IA	Identification and Authentication	SC	System and Communications Protection
IR	Incident Response	SI	System and Information Integrity
MA	Maintenance	PM	Program Management

Table 21-2 Security Control Identifiers and Family Names

SP 800-53 uses three SCs: low, moderate, and high impact. A low-impact system is defined as an information system in which all three of the security objectives are low. A moderate-impact system is one in which at least one of the security objectives is moderate and no security objective is greater than moderate. Finally, a high-impact system is an information system in which at least one security objective is high. In our example, the SC of the CRM system would be high, because at least one objective (confidentiality) is rated high.

This exercise in categorizing your information systems is important because it enables you to prioritize your work. It also determines which of the 965 controls listed in SP 800-53 you need to apply to it. These controls are broken down into 18 families, as shown on Table 21-2.

We can go through the entire catalog of controls and see which of them apply to our hypothetical CRM. In the interest of brevity, we will only look at the first two controls (AC-1 and AC-2) in the first family (Access Control, or AC). You can see in Table 21-3 how these controls apply to the different SCs. Since the CRM is SC high, both controls are required for it. You can also see that AC-2 has five control enhancements listed. (There are actually more, but we're trying to keep this brief.)

Let's dive into the first control and see how we would use it. Appendix F of SP 800-53 is the "Security Control Catalog" that describes in detail what each control is. If we go to the description of the baseline AC-1 (Access Control Policy and Procedures) control, we'll see that it requires that the organization do the following:

a. Develops, documents, and disseminates to [Assignment: organization-defined personnel or roles]:

  1. An access control policy that addresses purpose, scope, roles, responsibilities, management commitment, coordination among organizational entities, and compliance; and

  2. Procedures to facilitate the implementation of the access control policy and associated access controls; and

Cntl. No.	Control Name *Control Enhancement Name*	Control Baselines		
		Low	**Mod**	**High**
AC-1	Access Control Policy and Procedures	X	X	X
AC-2	Account Management	X	X	X
AC-2(1)	*Account Management | Automated System Account Management*		X	X
AC-2(2)	*Account Management | Removal of Temporary/ Emergency Accounts*		X	X
AC-2(3)	*Account Management | Disable Inactive Accounts*		X	X
AC-2(4)	*Account Management | Automated Audit Actions*		X	X
AC-2(5)	*Account Management | Inactivity Logout*			X

Table 21-3   Sample Mapping of Security Controls to the Three Security Categories in SP 800-53

b. Reviews and updates the current:

   1. Access control policy [Assignment: organization-defined frequency]; and

   2. Access control procedures [Assignment: organization-defined frequency].

You noticed that there are assignments in square brackets in three of these requirements. These are parameters that enable an organization to tailor the baseline controls to its own unique conditions and needs. In this case, we get to specify who receives the policies and procedures pertaining to access control and how frequently these are reviewed.

The decisions we make in tailoring the controls must have sound rationales and should not be taken lightly. In the case of AC-1, how did we arrive at the list of staff/roles who would receive the policy, and on what factors did we base our decision to review the policy and procedures at a particular frequency? Sometimes, why we make decisions is almost as important as what decisions are made. Documenting the control selection process is important to support our assessment of the controls' effectiveness, as well as for auditing purposes.

 NOTE   Revision 5 of SP 800-53 is in final review as of the writing of this book and introduces some significant improvements to the standard.

ISO/IEC 27000 Series

The International Organization for Standardization (ISO) and the International Electrotechnical Commission (IEC) 27000 series serves as industry best practices for the management of security controls in a holistic manner within organizations around the world. The list of standards that makes up this series grows each year. Each standard has a specific focus (such as metrics, governance, auditing, and so on). The currently published standards (with more than a few omitted) include the following:

- **ISO/IEC 27000**   Overview and vocabulary
- **ISO/IEC 27001**   ISMS requirements
- **ISO/IEC 27002**   Code of practice for information security controls
- **ISO/IEC 27003**   ISMS implementation
- **ISO/IEC 27004**   ISMS measurement
- **ISO/IEC 27005**   Risk management
- **ISO/IEC 27006**   Certification body requirements
- **ISO/IEC 27007**   ISMS auditing
- **ISO/IEC 27008**   Guidance for auditors
- **ISO/IEC 27011**   Telecommunications organizations
- **ISO/IEC 27014**   Information security governance
- **ISO/IEC 27015**   Financial sector
- **ISO/IEC 27031**   Business continuity
- **ISO/IEC 27032**   Cybersecurity
- **ISO/IEC 27033**   Network security
- **ISO/IEC 27034**   Application security
- **ISO/IEC 27035**   Incident management
- **ISO/IEC 27037**   Digital evidence collection and preservation
- **ISO/IEC 27799**   Health organizations

ISO 27001

One of these ISO/IEC standards that is particularly relevant to our discussion is 27001. ISO/IEC 27001 applies to any organization, regardless of size, that wants to formalize its security activities through the creation of an information security management system (ISMS). This system documents the following, at a minimum:

- Who the stakeholders are and what are their expectations?
- What are the information security objectives?
- What are the risks to information systems?
- Which controls will be used to handle those risks?
- How are these controls to be implemented?
- How will the controls' effectiveness be continuously monitored?
- What is the process to continuously improve the ISMS?

PART V

Similarly to how NIST SP 800-53 has 965 controls organized into 18 families, ISO/IEC 27001 has 114 controls organized into 14 domains. While the two standards are compatible with each other (SP 800-53 even has an appendix that maps its controls to these 14 domains), the NIST standard is focused on controls, and the ISO/IEC standard deals with the overarching program as well. The 14 domains in ISO/IEC 27001 are listed here:

- Information security policies
- Organization of information security
- Human resource security
- Asset management
- Access control
- Cryptography
- Physical and environmental security
- Operations security
- Communications security
- System acquisition, development, and maintenance
- Supplier relationships
- Information security incident management
- Information security aspects of business continuity management
- Compliance

Compare this to the families of controls shown in Table 21-2, and you should be able to see that ISO/IEC 27001 is concerned with much more than just the security controls. That is why, although it includes risk considerations as well as security controls, this framework is considered a prescriptive program–level framework.

ISO 27005

ISO/IEC 27005 is a risk-based framework, akin to the NIST RMF we discussed already. It provides guidelines for information security risk management in an organization but does not provide any specific method for information security risk management. In that regard, it is best used in conjunction with ISO/IEC 27001.

The risk management process defined by this standard is illustrated in Figure 21-3. It all hinges on establishing the context in which the risk exists. This is similar to the business impact analysis (BIA) we discussed in Chapter 20, but it adds new elements, such as evaluation criteria for risks as well as the organizational risk appetite. The risk assessment box in the middle of the figure should look familiar, since we also discussed this process (albeit with slightly different terms) in the last chapter. The risk treatment and risk acceptance steps are very similar to the "Respond to risk" functional grouping in the RMF, and the risk monitoring and review step is equivalent to the RMF's "Monitor risk" group.

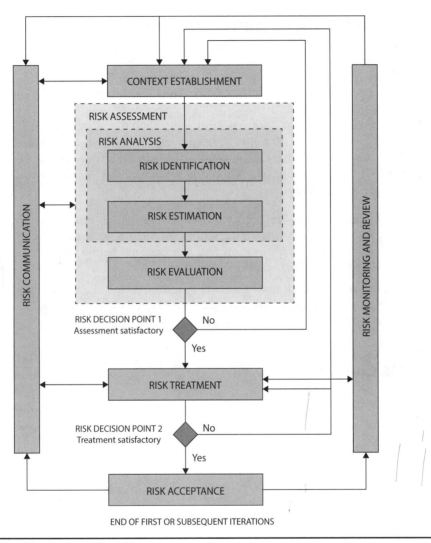

Figure 21-3 The ISO/IEC 27005 Risk Management Process

Finally, risk communication is an essential component of any risk management methodology, since we cannot enlist the help of senior executives, partners, or other stakeholders if we cannot effectively convey our message to a variety of audiences.

As you can see, this framework doesn't really introduce anything new to the risk conversation we've been having over the last two chapters; it just rearranges things a bit. Of course, despite these high-level similarities, the two risk-based frameworks we've discussed differ in a lot of the details. If you ever need to support your organization's efforts to implement either of these, you probably want to get into these details.

Center for Internet Security Controls

The Center for Internet Security (CIS) is a nonprofit organization that, among other things, maintains a list of 20 critical security controls designed to mitigate the threat of the majority of common cyber attacks. It is another example (together with NIST SP 800-53) of a prescriptive controls framework. The CIS controls, currently in version 7.1, are shown in Figure 21-4.

Despite their use of the word "controls," you should really think of these like the 18 families of controls in SP 800-53: as groupings. Under these 20, there are a total of 171 subcontrols that have similar granularity as those established by the NIST. For example, if we look into control 3 (Continuous Vulnerability Management), we can see the following seven subcontrols:

- 3.1 Utilize an up-to-date Security Content Automation Protocol (SCAP) compliant vulnerability scanning tool to automatically scan all systems on the network on a weekly or more frequent basis to identify all potential vulnerabilities on the organization's systems.

- 3.2 Perform authenticated vulnerability scanning with agents running locally on each system or with remote scanners that are configured with elevated rights on the system being tested.

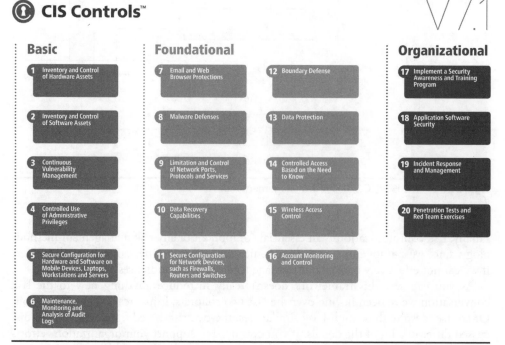

Figure 21-4 The CIS controls

- 3.3 Use a dedicated account for authenticated vulnerability scans, which should not be used for any other administrative activities and should be tied to specific machines at specific IP addresses.

- 3.4 Deploy automated software update tools in order to ensure that the operating systems are running the most recent security updates provided by the software vendor.

- 3.5 Deploy automated software update tools in order to ensure that third-party software on all systems is running the most recent security updates provided by the software vendor.

- 3.6 Regularly compare the results from consecutive vulnerability scans to verify that vulnerabilities have been remediated in a timely manner.

- 3.7 Utilize a risk-rating process to prioritize the remediation of discovered vulnerabilities.

The CIS recognizes that not every organization will have the resources (or face the risks) necessary to implement all controls. For this reason, they are grouped into three categories, listed next. While every organization should strive for full implementation, this approach provides a way to address the most urgent requirements first and then grow over time.

- **Basic**　These key controls should be implemented by every organization to achieve minimum essential security.

- **Foundational**　These embody technical best practices to improve an organization's security.

- **Organizational**　These controls focus on people and processes to maintain and improve cybersecurity.

Policies and Procedures

The implementation of a security framework in an organization usually starts with a set of policies and procedures. For this effort to be successful, it must start at the top level and be useful and functional at every single level within the organization. Senior management needs to define the scope of security and identify and decide what must be protected and to what extent. Management must understand the regulations, laws, and liability issues it is responsible for complying with regarding security and ensure that the company as a whole fulfills its obligations. Senior management also must determine what is expected from employees and what the consequences of noncompliance will be. These decisions should be made by the individuals who will be held ultimately responsible if something goes wrong. But it is a common practice to bring in the expertise of the security officers to collaborate in ensuring that sufficient policies and controls are being implemented to achieve the goals being set and determined by senior management.

A security program contains all the pieces necessary to provide overall protection to an organization and lays out a long-term security strategy. A security program's documentation should include security policies and procedures. The more detailed the rules are, the easier it is to know when one has been violated. However, overly detailed documentation and rules can prove to be more burdensome than helpful. The business type, its culture, and its goals must be evaluated to make sure the proper language is used when writing security documentation.

In the following sections, we discuss some of the foundational policies and procedures that any organization should have. The list is not all-inclusive, however, and simply mirrors the ones with which you must be familiar for the CySA+ exam.

Ethics and Codes of Conduct

Ethics are the moral principles that define what is right and wrong for a particular group of people, such as a corporation. Corporate ethics establish the "tone at the top," which means that the executives need to ensure not only that their employees are acting ethically, but also that they themselves are following their own rules. The main goal is to ensure that the motto "Succeed by any means necessary" is not the spoken or unspoken culture of a work environment. Certain structures can be put in place to create a breeding ground for unethical behavior. If the CEO's salary increases are based on stock prices, for example, then she may find ways to inflate stock prices artificially, which can directly hurt the investors and shareholders of the company. If managers can be promoted only based on the amount of sales they bring in, these numbers may be fudged and may not represent reality. If an employee can get a bonus only if a low budget is maintained, he might be willing to take shortcuts that could hurt company customer service or product development. Although ethics seem like things that float around in the ether and make us feel good when we talk about them, they have to be actually implemented in the real corporate world through proper business processes and management styles.

A *code of conduct* defines how a company's employees should act on a day-to-day basis in accordance with their organization's ethics. Most organizations have mission and vision statements; the code of conduct describes the manner in which employees are to conduct themselves in pursuit of the goals outlined in those statements. The code enables people to make the right decision when faced with an ethical dilemma. It also communicates to the outside world what the organization is all about. In some cases, legislation and other regulations actually require that organizations have codes of conduct formally published, in part because it is recognized that this reduces the risk of misbehaviors.

Acceptable Use Policy

The acceptable use policy (AUP) specifies what the organization considers an acceptable use of the information systems that are made available to the employees. Using a workplace computer to view pornography, send hate e-mail, and hack other computers, for example are almost always forbidden. On the other hand, many organizations allow their employees limited personal use of their networks, such as checking personal e-mail and surfing the Web during breaks. The AUP is a useful first line of defense, because it documents when each user was made aware of what is and is not acceptable use of computers (and other resources) at work. This makes it more difficult for a user to claim ignorance if he or she subsequently violates the AUP.

Password Policy

The password policy is perhaps the most visible of security policies, because every user will have to deal with its effects on a daily basis. A good password policy should motivate users to manage their passwords securely, describe to them how this should be accomplished, and prescribe the consequences of failing to comply. The three main elements in most password policies relate to generation, duration, and use.

When creating passwords, users should be informed of the requirements of an acceptable one. Commonly, these standards include some or all of the following:

- Minimum length (for example, eight characters or greater)
- Requirement for specific types of characters (such as uppercase, lowercase, numbers, and special characters)
- Prohibition against reuse (for example, cannot be any of the last four passwords)
- Minimum age (to prevent flipping in order to reuse an old password)
- Maximum age (for example, 90 days)
- Prohibition against certain words (such as user's name or company name)

The policy should also cover the use of different passwords. For example, users should not use the same password for multiple systems, so that a compromise of one does not automatically lead to the compromise of all user accounts. Admittedly, this is a difficult provision to enforce, particularly when it comes to the reuse of passwords for personal and organizational use. Still, it may be worth including for educational purposes as well as to potentially mitigate some liability for the organization.

Data Ownership

Data ownership policies are typically combined with data classification policies because it is difficult to separate the two issues. The main reason is that the data is classified by the person who "owns" it. Data ownership policies establish the roles and responsibilities of data owners within the organization. The data owner (information owner) is usually a member of management who is in charge of a specific business unit and who is ultimately responsible for the protection and use of a specific subset of information. The data owner has due care responsibilities and thus will be held responsible for any negligent act that results in the corruption or disclosure of the data. Data owners decide the classification of the data for which they are responsible and alter classifications if business needs arise.

PART V

These individuals are also responsible for ensuring that the necessary security controls are in place, defining security requirements per classification and backup requirements, approving any disclosure activities, ensuring that proper access rights are being used, and defining user-access criteria. The data owners approve access requests or may choose to delegate this function to business unit managers. Also, data owners will deal with security violations pertaining to the data they are responsible for protecting.

A key issue to address in a data ownership policy is who owns the personal data that an employee brings into an organizational information system. For example, if employees are allowed to check e-mail or social media sites from work, their personal data will traverse and be stored, albeit temporarily, on corporate information systems. Does it now belong to the company? Are there expectations of privacy? What about personal e-mail received by employees at their work accounts? These issues should be formally addressed in a data ownership policy.

Data Retention

No universal agreement states how long you should retain data that you own. Legal and regulatory requirements (where they exist) vary among countries and sectors. What is universal is the need to ensure that your organization has and follows a documented data retention policy. Doing otherwise is flirting with disaster, particularly when dealing with pending or ongoing litigation. It is not enough, of course, simply to have a policy; you must ensure that it is being followed and you must document this through regular audits.

A very straightforward and perhaps tempting approach would be to look at the lengthiest legal or regulatory retention requirement that affects your organization and then apply that timeframe to all your data retention. The problem with this, however, is that it will probably make your retained data set orders of magnitude larger than it needs to be. Not only does this impose additional storage costs, but it also makes it more difficult to comply with electronic discovery (e-discovery) orders. When you receive an e-discovery order from a court, you are typically required to produce a specific amount of data (usually pretty large) within a given timeframe (usually very short). Obviously, the more data you retain, the more difficult and expensive this process will be.

A better approach is to find the specific data sets that have mandated retention requirements and handle those accordingly. Everything else has a retention period that minimally satisfies the business requirements. You will find that different business units within medium and large organizations will probably have different retention requirements. For instance, you may want to keep data from your research and development (R&D) division for a much longer period than you keep data from the customer service division. R&D projects that are not particularly helpful today may be so at a later date, but audio recordings of customer service calls probably don't have to hang around for a few years.

Work Product Retention

Work products are materials collected or prepared by an organization in anticipation of litigation. Just like conversations with legal counsel are protected under attorney–client privilege, work products enjoy a similar (but lesser) form of protection from discovery by the other party in the United States. For example, suppose your organization suffers a breach that *likely* exposes it to a lawsuit and you, as a cybersecurity analyst, are directed by your boss (ostensibly under some advice from legal counsel) to collect log files that could be used to defend the company in court. A plaintiff would not be entitled to those work products even though you were the one putting them together and you are not an attorney, nor did you directly communicate with one.

The issue becomes, how long can and should you retain those work products? You may (and should) have a data retention policy, but work products should be treated differently. There is no single best answer. The best thing to do is to consult with your legal counsel and establish what the right work product retention policy should be.

Account Management

A preferred technique of attackers is to become "normal" privileged users of the systems they're compromising as soon as possible. They can accomplish this in at least three ways: compromise an existing privileged account, create a new privileged account, and elevate the privileges of a regular user account. The first approach can be mitigated through the use of strong authentication (for example, strong passwords and two-factor authentication) and by having administrators use privileged accounts only for specific tasks and only from jump boxes. The second and third approaches can be mitigated by paying close attention to the creation, modification, or misuse of user accounts. These controls all fall within the scope of an account management policy.

When new employees arrive, they should follow a well-defined process that is aimed at ensuring not only that they understand their duties and responsibilities, but also that they are assigned the required company assets and that these are properly configured, protected, and accounted for. Among these assets is a user account that grants them access to the information systems and authorization to create, read, modify, execute, or delete resources (for example, files) within it. The policy should dictate the default expiration date of accounts, the password policy (unless it is a separate document), and the information to which a user should have access. This last part becomes difficult, because the information needs of the users will typically vary over time.

Adding, removing, or modifying user permissions should be a carefully controlled and documented process. When is the new permission effective? Why is it needed? Who authorized it? Organizations that are mature in their security processes will have a change-control process in place to address user privileges. While many auditors will focus on who has administrative privileges in the organization, there are many custom sets of permissions that approach the level of an admin account. It is important, then, to have and test the processes by which elevated privileges are issued.

Another important practice in account management is the suspension of accounts that are no longer needed. Every large organization eventually stumbles across one or more accounts that belong to users who are no longer part of the organization. In some extreme cases, these users left the organization several months ago and had privileged accounts. The unfettered presence of these accounts on a network gives adversaries a powerful means to become a seemingly legitimate user, which makes our job of detecting and repulsing them that much more difficult.

PART V

The Problem with Running as Root

It is undoubtedly easier to do all our work from one user account, especially if that account has all the privileges we could ever need. The catch, as you may well know, is that when the account is compromised, the malicious processes will run with whatever privileges the account has. If you run as root (or admin) all the time, you can be certain that when your attackers compromise your box, they will instantly have the privileges to do whatever they need or want to do.

A better approach is to do as much of our daily work as we can using a restricted account, and elevate to a privileged account only when we must. Consider the following:

- Windows operating systems enable you to right-click any program and select Run As to elevate your privileges. From the command prompt, you can just use the command **runas /user:<*AccountName*>** to accomplish the same goal.

- In Linux operating systems, you can simply type **sudo <*SomeCommand*>** at the command line to run a program as the super (or root) user. With a GUI program, you need to start it from the command line using the command **gksudo** (or **kdesudo** for Kubuntu). Linux has no way to run a program with elevated privileges directly from the GUI; you must start from the command line.

- In macOS, you use **sudo** from the Terminal app just like you would do from a Linux terminal. However, if you want to run a GUI app with elevated privileges, you need to use **sudo open –a <*AppName*>** since there is no **gksudo** or **kdesudo** command.

Continuous Monitoring

In SP 800-137, NIST defines information security *continuous monitoring* as "maintaining ongoing awareness of information security, vulnerabilities, and threats to support organizational risk management decisions." A continuous-monitoring procedure, therefore, would describe the process by which an organization collects and analyzes information in order to maintain awareness of threats, vulnerabilities, compliance, and the effectiveness of security controls. Obviously, this is a complex and very broadly scoped effort that requires coordination across multiple business units and tight coupling with the organization's risk management processes.

When continuous monitoring reveals actionable intelligence (for example, a new threat or vulnerability), an established process should be in place to deal with this situation. The *remediation plan* describes the steps that an organization takes whenever its security posture worsens. This plan will likely have references to multiple procedures, some of which we discuss in this chapter. For example, if the issue is a newly discovered vulnerability in an application, the remediation plan would point the security team to the patching procedure. If, on the other hand, the change is due to an awareness that a security control is not as effective as it was thought to have been, the team would have to consider whether the control-testing procedure was effective or should be updated.

Control Types

Controls are put into place to reduce the organization's risk, and they come in three main classes: managerial, technical (or logical), and operational (or administrative). Managerial controls are those that enable the overarching administration of the security of an organization. Examples of managerial controls are planning, risk assessment, security assessments, and systems acquisition processes. Technical controls are software or hardware components, as in firewalls, IDSs, encryption processes, and identification and authentication mechanisms. Finally, operational controls are those that are implemented by people (rather than by systems, as are technical controls). Examples of operational controls are incident response, personnel security, and security awareness.

 EXAM TIP Many controls have features of all three classes (managerial, technical, and operational), so if you are asked to classify one, look for the dominant feature(s) and use that to decide your answer.

These three control classes are used in NIST documents, but there is another class of control that we already discussed in Chapter 20. Physical security controls are tangible means of protecting facilities, devices, or people. They may not directly secure the data on a disk but they keep the disk from falling into the wrong hands. Examples of physical controls are fences, locks, cameras, and security guards.

Controls need to be put into place to provide *defense-in-depth*, which is the coordinated use of multiple security controls in a layered approach. A multilayered defense system minimizes the probability of successful penetration and compromise because an attacker would have to get through several different types of protection mechanisms before she gained access to the critical assets.

Technical controls that are commonly used to provide this type of layered approach include the following:

- Firewalls
- Intrusion detection systems
- Intrusion prevention systems
- Antimalware
- Access control
- Encryption

The types of controls that are actually implemented must map to the threats the company faces, and the number of layers that are included must map to the sensitivity of the asset. The rule of thumb is the more sensitive the asset, the more layers of protection must be in place.

So, the different *classes* of controls that can be used are managerial, technical, operational, and physical. But what *functions* do these controls actually perform? We need to understand the different functionality that each control type can provide us in our quest to secure our environments. By having a better understanding of the different control functions, you will be able to make more informed decisions about what controls will be best used in specific situations. The five different control functions are as follows:

- **Preventative controls**   Intended to avoid an incident from occurring
- **Detective controls**   Help identify an incident's activities and potentially an intruder
- **Corrective controls**   Repair or restore components or systems after an incident has occurred
- **Deterrent controls**   Intended to discourage a potential attacker
- **Compensating controls**   Controls that provide an alternative measure of control

---

 EXAM TIP   The exam splits up the *corrective* function into two parts: *responsive* and *corrective* controls. Responsive controls are used to contain an incident and limit its impact. Corrective controls (in the exam) are intended to restore the system back to full functionality after responding to an incident. Most organizations do not make that distinction in the real world.

When you're implementing security for an environment, it is most productive to use a preventive model and then use detective and corrective mechanisms to help support this model. Basically, you want to stop any trouble before it starts, but you must be able to react quickly and combat trouble if it does find you. It is not feasible to prevent everything; therefore, what you cannot prevent, you should be able to detect quickly. That's why preventive and detective controls should always be implemented together and should complement each other. To take this concept further, remember this: what you can't prevent, you should be able to detect, and if you detect something, it means you weren't able to prevent it, so therefore you should take corrective action to make sure it is indeed prevented the next time around.

One control functionality that some people struggle with is a compensating control, which we introduced in Chapter 20. Let's look at some examples of compensating controls to best explain their function. If your company needed to implement strong physical security, you might suggest to management that they employ security guards. But after calculating all the costs of security guards, your company might decide to use a compensating (alternative) control that provides similar protection but is more affordable—as in a fence. In another example, let's say you are a security administrator and you are in charge of maintaining the company's firewalls. Management tells you that a certain protocol that you know is vulnerable to exploitation has to be allowed through the firewall for business reasons. The network needs to be protected by a compensating (alternative) control pertaining to this protocol, such as setting up a proxy server for that specific traffic type to ensure that it is properly inspected and controlled. So a compensating control is just an alternative control that provides similar protection as the original control but has to be used because it is more affordable or allows specifically required business functionality.

Audits and Assessments

Frameworks, policies, procedures, and controls are only useful if they are properly implemented. Otherwise, they can lead to a false sense of security, which can do more harm than good. How do we ensure they are properly implemented? We already discussed assessments in Chapter 20 in the context of risk management. Recall that an assessment is simply a process that gathers information and makes determinations based on it. We could, for example, conduct an assessment of our cybersecurity program to see if the controls we've implemented satisfy the requirements of SP 800-53. We could also hire a third party to do it, presumably in a more objective manner. Generally speaking, a cybersecurity assessment is any voluntary process that examines our programs (or a portion thereof) and determines whether we are meeting some sort of standard. In certain cases, however, this is not a voluntary decision.

An *audit* is a systematic inspection by an independent third party to determine whether the organization is in compliance with some set of external requirements. Think of it as a formal and rigorous assessment performed by an auditing firm that typically costs a bunch of money. Companies pay that money because either they want their stakeholders to know they are in compliance with some standard (e.g., "We are ISO 27001 certified") or they are forced to do so by an industry or government regulation. Let's explore each case in a bit more detail.

Standards Compliance

There are lots of reasons why companies would voluntarily put forth the effort to comply with external standards even if they are not required to do so. For starters, it can significantly reduce their risk and help establish due diligence. It may also be a precondition for sales to certain customers. For example, if you are a cloud service provider and want to sell your services to the US federal government, you must demonstrate that you are in compliance with the Federal Risk and Authorization Management Program (FedRAMP). Another reason is simply to differentiate yourself in a crowded market and show your potential customers that you care about protecting their information.

The most common standards compliance certifications, at least in the United States, are the ISO 27000 family (especially 27001) and FedRAMP. Both require significant investment both in preparing for an audit, and in supporting the auditors. For this reason, most organizations hire a third party to conduct one or more assessments before the auditors are brought in. This approach usually provides the best chance of getting successfully certified on the first audit.

NOTE While CySA+ exam objectives differentiate regulatory and compliance, whenever you hear the term "compliance" at work, it will almost always mean regulatory compliance.

Regulatory Compliance

Some organizations are subject to governmental statutes and regulations that may impose threshold requirements on securing information systems. Typically, being noncompliant with applicable regulations can lead to fines, penalties, and even criminal charges. While describing all aspects of regulatory compliance is beyond the scope of this book (and the CySA+ exam), you should be familiar with the more common laws and regulations, highlighted here:

- **Sarbanes-Oxley Act (SOX)** This law, enacted in 2002 after the Enron and WorldCom financial crises, is intended to protect investors and the public against fraudulent and misleading activities by publicly traded companies. Its effect on information security controls is mostly in the area of integrity protections. SOX-regulated organizations have a higher bar set when it comes to ensuring that digital records are not improperly altered.

- **Payment Card Industry Data Security Standard (PCI DSS)** This industry standard applies to any organization that handles credit or debit card data. We already discussed it in Chapter 5, but it bears repeating that its main impact on security controls is focused on vulnerability scanning.

- **The Gramm-Leach-Bliley Act (GLBA)** This 1999 law applies to financial institutions and is intended to protect consumers' personal financial information. Notably, it includes what is known as the Safeguards Rule, which requires financial institutions to maintain safeguards to protect the confidentiality and integrity of personal consumer information.

- **Federal Information Security Management Act (FISMA)** Enacted in 2002, FISMA applies to information systems belonging or operated by federal agencies or contractors working on their behalf. Among its key provisions are requirements on the minimum frequency of risk assessments, security awareness training, incident response, and continuity of operations.

- **Health Insurance Portability and Accountability Act (HIPAA)** This law mostly deals with improving privacy in the healthcare system, but it has important elements that impact information security policies and procedures. Significantly, it includes the Security and Privacy Rules, which place specific requirements on protecting the confidentiality, integrity, availability, and privacy of patient data.

EXAM TIP You do not need to memorize all these regulations, but you do need to be aware of the general nature of regulatory requirements and their impact on the formulation of organizational policies and procedures as well as the selection of controls.

Chapter Review

This chapter is packed with important information that will enable you to understand security frameworks, policies, procedures, and controls. Although you will probably see a handful of specific questions on these topics in the CySA+ exam, you will definitely see their influence not only on most exam questions, but in your work in a real-world organization. Though many of us prefer to spend our time fighting our cyber foes or improving our technical defenses, it is just as important to develop the formal documents that will ensure that the entire organization is pulling in the right direction. Without the appropriate policies and procedures in place, our efforts at securing our systems may very well ultimately be doomed to fail.

Questions

1. Which of the following is *not* a category for access controls and their implementation?

   A. Managerial

   B. Operational

   C. Virtual

   D. Logical

2. Which publication outlines various security controls for government agencies and information systems?

   A. NIST SP 800-53

   B. NIST SP 800-37

   C. ISO/IEC 27001

   D. ISO/IEC 27005

3. Which of the following publications describes a voluntary cybersecurity structure for organizations that are part of the critical infrastructure?

   A. "Framework for Improving Critical Infrastructure Cybersecurity"

   B. ISO/IEC 27005

   C. ISO/IEC 27023

   D. NIST SP 800-37

4. ISO/IEC 27001 describes which of the following?

   A. The Risk Management Framework (RMF)

   B. Information security management system (ISMS)

   C. Work product retention (WPR) standards

   D. International Electrotechnical Commission (IEC) standards

PART V

5. Which component of the NIST Cybersecurity Framework describes the degree of sophistication of cybersecurity practices?

   A. Framework Core

   B. Implementation Tiers

   C. NIST SP 800-53 control categories

   D. ISO/IEC 27001

6. Which are the key functions of the Framework Core of the NIST Cybersecurity Framework?

   A. Identify, Protect, Detect, Respond, Recover

   B. Identify, Process, Detect, Respond, Recover

   C. Identify, Process, Detect, Relay, Recover

   D. Identify, Protect, Detect, Relay, Recover

7. Which of the following would be least influential in developing a data retention policy?

   A. Legal and regulatory requirements

   B. Acceptable use

   C. Business needs

   D. E-discovery

8. Which of the following is *not* a good example of a preventative control?

   A. Guard dogs

   B. Policies and procedures

   C. Audit logs

   D. Biometrics

Answers

1. **C.** Virtual controls are not one of the categories of access controls, which are the mechanisms put in place to protect the confidentiality, integrity, and availability of systems. Access controls are categorized as managerial, operational, and technical (aka logical).

2. **A.** NIST SP 800-53, "Security and Privacy Controls for Federal Information Systems and Organizations," aims to establish a unified information security framework for the federal government and related organizations.

3. **A.** The NIST "Framework for Improving Critical Infrastructure Cybersecurity" focuses on aligning cybersecurity activities with business processes and including cybersecurity risks as part of the organization's risk management processes. The Cybersecurity Framework (CSF) consists of three parts: the Framework Core, the Framework Profile, and the Framework Implementation Tiers.

4. B. ISO/IEC 27001 provides best practice recommendations on information security management systems (ISMS).

5. B. CSF Implementation Tiers categorize the degree of rigor and sophistication of cybersecurity practices, which can be Partial (tier 1), Risk Informed (tier 2), Repeatable (tier 3), or Adaptive (tier 4).

6. A. The Framework Core consists of five functions that can provide a high-level view of an organization's management of cybersecurity risk: Identify, Protect, Detect, Respond, Recover.

7. B. Although acceptable use violations may influence a data retention policy to some extent, the other three options *must* be considered when developing it.

8. C. Audit logs are a good example of a detective control, in that they help us detect when a security incident has occurred. However, unless a threat actor is somehow aware of and discouraged by our logging policy, it is extremely unlikely that this would be a good preventive control.

PART VI

Appendixes and Glossary

Objective Map

Exam CS0-002

	Official Exam Objective	Ch #	All-in-One Coverage
1.0	**Threat and Vulnerability Management**		
1.1	**Explain the importance of threat data and intelligence**		
	Intelligence sources	1	Foundations of Intelligence
	• Open-source intelligence	1	Open Source Intelligence
	• Proprietary/closed-source intelligence	1	Proprietary/Closed Source Intelligence
	• Timeliness	1	Timeliness
	• Relevancy	1	Relevancy
	• Accuracy	1	Accuracy
	Confidence levels	1	Confidence Levels
	Indicator management	1	Indicator Management
	• Structured Threat Information eXpression (STIX)	1	Structured Threat Information Expression
	• Trusted Automated eXchange of Indicator Information (TAXII)	1	Trusted Automated Exchange of Indicator Information
	• OpenIoC	1	OpenIOC
	Threat classification	1	Threat Classification
	• Known threat vs. unknown threat	1	Known Threats vs. Unknown Threats
	• Zero-day	1	Zero Day
	• Advanced persistent threat	1	Advanced Persistent Threat
	Threat actors	1	Threat Actors
	• Nation-state	1	Nation-State Threat Actors
	• Hacktivist	1	Hacktivists
	• Organized crime	1	Organized Crime
	• Insider threat	1	Insider Threat Actors

Official Exam Objective	Ch #	All-in-One Coverage
• Intentional	1	Intentional
• Unintentional	1	Unintentional
Intelligence cycle	1	Intelligence Cycle
• Requirements	1	Requirements
• Collection	1	Collection
• Analysis	1	Analysis
• Dissemination	1	Dissemination
• Feedback	1	Feedback
Commodity malware	1	Commodity Malware
Information sharing and analysis communities	1	Information Sharing and Analysis Communities
• Healthcare	1	Information Sharing and Analysis Communities
• Financial	1	Information Sharing and Analysis Communities
• Aviation	1	Information Sharing and Analysis Communities
• Government	1	Information Sharing and Analysis Communities
• Critical infrastructure	1	Information Sharing and Analysis Communities
1.2 Given a scenario, utilize threat intelligence to support organizational security		
Attack frameworks	2	Attack Frameworks
• MITRE ATT&CK	2	MITRE ATT&CK
• The Diamond Model of Intrusion Analysis	2	The Diamond Model of Intrusion Analysis
• Kill chain	2	Kill Chain
Threat research	2	Threat Research
• Reputational	2	Reputational
• Behavioral	2	Behavioral
• Indicator of compromise (IOC)	2	Indicator of Compromise
• Common vulnerability scoring system (CVSS)	2	Common Vulnerability Scoring System
Threat modeling methodologies	2	Threat Modeling Methodologies
• Adversary capability	2	Adversary Capability
• Total attack surface	2	Total Attack Surface
• Attack vector	2	Attack Vector
• Impact	2	Impact
• Likelihood	2	Likelihood

Official Exam Objective	Ch #	All-in-One Coverage
Threat intelligence sharing with supported functions	2	Threat Intelligence Sharing with Supported Functions
• Incident response	2	Incident Response
• Vulnerability management	2	Vulnerability Management
• Risk management	2	Risk Management
• Security engineering	2	Security Engineering
• Detection and monitoring	2	Detection and Monitoring
1.3 Given a scenario, perform vulnerability management activities		
Vulnerability identification	3	Vulnerability Identification
• Asset criticality	3	Critical Assets
• Active vs. passive scanning	3	Active vs. Passive Scanning
• Mapping/enumeration	3	Mapping/Enumeration
Validation	3	Validation
• True positive	3	True Positives
• False positive	3	False Positives
• True negative	3	True Negatives
• False negative	3	False Negatives
Remediation/mitigation	3	Remediation
• Configuration baseline	3	Endpoints
• Patching	3	Patching
• Hardening	3	Hardening
• Compensating controls	3	Compensating Controls
• Risk acceptance	3	Risk Acceptance
• Verification of mitigation	3	Verification of Mitigation
Scanning parameters and criteria	3	Scanning Parameters and Criteria
• Risks associated with scanning activities	3	Risks Associated with Scanning Activities
• Vulnerability feed	3	Vulnerability Feed
• Scope	3	Scope
• Credentialed vs. non-credentialed	3	Credentialed vs. Noncredentialed
• Server-based vs. agent-based	3	Server Based vs. Agent Based
• Internal vs. external	3	Internal vs. External
• Special considerations	3	Special Considerations
• Types of data	3	Types of Data
• Technical constraints	3	Technical Constraints
• Workflow	3	Workflow

Official Exam Objective	Ch #	All-in-One Coverage
Wireless assessment tools	4	Wireless Assessment Tools
• Aircrack-ng	4	Aircrack-ng
• Reaver	4	Reaver
• oclHashcat	4	oclHashcat
Cloud infrastructure assessment tools	4	Cloud Infrastructure Assessment Tools
• ScoutSuite	4	Scout Suite
• Prowler	4	Prowler
• Pacu	4	Pacu

1.5 Explain the threats and vulnerabilities associated with specialized technology

Mobile	5	Mobile Devices
Internet of Things (IoT)	5	Internet of Things
Embedded	5	Embedded Systems
Real-time operating system (RTOS)	5	Real-Time Operating Systems
System-on-Chip (SoC)	5	System on a Chip
Field programmable gate array (FPGA)	5	Field Programmable Gate Array
Physical access control	5	Physical Access Control
Building automation systems	5	Industrial Control Systems
Vehicles and drones	5	Connected Vehicles Drones
• CAN bus	5	CAN Bus
Workflow and process automation systems	5	Process Automation Systems
Industrial control system	5	Industrial Control Systems
Supervisory control and data acquisition (SCADA)	5	SCADA Devices
• Modbus	5	Modbus

1.6 Explain the threats and vulnerabilities associated with operating in the cloud

Cloud service models	6	Cloud Service Models
• Software as a Service (SaaS)	6	Software as a Service
• Platform as a Service (PaaS)	6	Platform as a Service
• Infrastructure as a Service (IaaS)	6	Infrastructure as a Service
Cloud deployment models	6	Cloud Deployment Models
• Public	6	Public
• Private	6	Private
• Community	6	Community
• Hybrid	6	Hybrid

PART VI

Official Exam Objective	Ch #	All-in-One Coverage
• Race condition	7	Race Condition
• Broken authentication	7	Authentication Attacks
• Sensitive data exposure	7	Sensitive Data Exposure
• Insecure components	7	Insecure Components
• Insufficient logging and monitoring	7	Insufficient Logging and Monitoring
• Weak or default configurations	7	Weak or Default Configurations
• Use of insecure functions	7	Use of Insecure Functions
• strcpy	7	strcpy

2.0 Software and Systems Security

2.1 Given a scenario, apply security solutions for infrastructure management

Cloud vs. on-premises	8	Cloud vs. On-Premises Solutions
Asset management	8	Asset Management
• Asset tagging	8	Asset Tagging
Segmentation	8	Network Segmentation
• Physical	8	Physical Segmentation
• Virtual	8	Virtual Local Area Networks
• Jumpbox	8	Jump Boxes
• System isolation	8	System Isolation
• Air gap	8	System Isolation
Network architecture	8	Network Architecture
• Physical	8	Physical Network
• Software-defined	8	Software-Defined Network
• Virtual private cloud (VPC)	8	Virtual Private Cloud Network
• Virtual private network (VPN)	8	Virtual Private Network
• Serverless	8	Serverless Network
Change management	8	Change Management
Virtualization	8	Virtualization
• Virtual desktop infrastructure (VDI)	8	Virtual Desktop Infrastructure
Containerization	8	Containerization
Identity and access management	8	Identity and Access Management
• Privilege management	8	Privilege Management
• Multifactor authentication (MFA)	8	Multifactor Authentication
• Single sign-on (SSO)	8	Single Sign-On
• Federation	8	Identity Federation
• Role-based	8	Role-Based Access Control
• Attribute-based	8	Attribute-Based Access Control

Official Exam Objective	Ch #	All-in-One Coverage
Service-oriented architecture	9	Service-Oriented Architecture
• Security Assertions Markup Language (SAML)	9	Security Assertions Markup Language
• Simple Object Access Protocol (SOAP)	9	Simple Object Access Protocol
• Representational State Transfer (REST)	9	Representational State Transfer
• Microservices	9	Microservices
2.3	**Explain hardware assurance best practices**		
Hardware root of trust	10	Hardware Root of Trust
• Trusted platform module (TPM)	10	Trusted Platform Module
• Hardware security module (HSM)	10	Hardware Security Module
eFuse	10	eFuse
Unified Extensible Firmware Interface (UEFI)	10	Unified Extensible Firmware Interface
Trusted foundry	10	Trusted Foundry
Secure processing	10	Secure Processing
• Trusted execution	10	Trusted Execution Environment
• Secure enclave	10	Trusted Execution Environment
• Processor security extensions	10	Processor Security Extensions
• Atomic execution	10	Atomic Execution
Anti-tamper	10	Anti-Tamper Techniques
Self-encrypting drive	10	Self-Encrypting Drive
Trusted firmware updates	10	Trusted Firmware Updates
Measured boot and attestation	10	Measured Boot and Attestation
Bus encryption	10	Bus Encryption

3.0 Security Operations and Monitoring

3.1	Given a scenario, analyze data as part of security monitoring activities		
Heuristics	11	Heuristics
Trend analysis	11	Trend Analysis
Endpoint	11	Endpoint Security
• Malware	11	Malware
• Reverse engineering	11	Decomposition
• Memory	11	Fileless Malware
• System and application behavior	11	Behavioral Analysis
• Known-good behavior	11	Anomaly Analysis

Official Exam Objective	Ch #	All-in-One Coverage
• Anomalous behavior	11	Anomaly Analysis
• Exploit techniques	7	Buffer Overflow Attacks Privilege Escalation Authentication Attacks Rootkits
• File system	11	Fileless Malware
• User and entity behavior analytics (UEBA)	11	User and Entity Behavior Analytics
Network	11	Network
• Uniform Resource Locator (URL) and domain name system (DNS) analysis	11	Embedded Links Domain Name System Analysis
• Domain generation algorithm	11	Domain Generation Algorithms
• Flow analysis	11	Flow Analysis
• Packet and protocol analysis	11	Packet Analysis
• Malware	11	Malware
Log review	11	Log Review
• Event logs	11	Event Logs
• Syslog	11	Syslog
• Firewall logs	11	Firewall Logs
• Web application firewall (WAF)	11	Web Application Firewall Logs
• Proxy	11	Proxy Logs
• Intrusion detection system (IDS)/ Intrusion prevention system (IPS)	11	Intrusion Detection/Prevention Systems
Impact analysis	11	Impact Analysis
• Organization impact vs. localized impact	11	Impact Analysis
• Immediate vs. total	11	Impact Analysis
Security information and event management (SIEM) review	11	Security Information and Event Management Review
• Rule writing	11	Security Information and Event Management Review
• Known-bad Internet protocol (IP)	12	Blacklisting
• Dashboard	11	Security Information and Event Management Review
Query writing	11	Query Writing
• String search	11	Query Writing
• Script	11	Query Writing
• Piping	11	Query Writing

Official Exam Objective	Ch #	All-in-One Coverage
E-mail analysis	11	E-mail Analysis
• Malicious payload	11	Malicious Payload
• Domain Keys Identified Mail (DKIM)	11	DomainKeys Identified Mail
• Domain-based Message Authentication, Reporting, and Conformance (DMARC)	11	Domain-Based Message Authentication, Reporting, and Conformance
• Sender Policy Framework (SPF)	11	Sender Policy Framework
• Phishing	11	Phishing
• Forwarding	11	Forwarding
• Digital signature	11	Digital Signatures and Encryption
• E-mail signature block	11	Digital Signatures and Encryption
• Embedded links	11	Embedded Links
• Impersonation	11	Impersonation
• Header	11	Header

3.2   Given a scenario, implement configuration changes to existing controls to improve security

Permissions	12	Permissions
Whitelisting	12	Whitelisting
Blacklisting	12	Blacklisting
Firewall	12	Firewalls
Intrusion prevention system (IPS) rules	12	Intrusion Prevention System Rules
Data loss prevention (DLP)	12	Data Loss Prevention
Endpoint detection and response (EDR)	12	Endpoint Detection and Response
Network access control (NAC)	12	Network Access Control
Sinkholing	12	Sinkholing
Malware signatures	12	Malware Signatures
• Development/rule writing	12	Malware Signatures
Sandboxing	12	Sandboxing
Port security	12	Port Security

3.3   Explain the importance of proactive threat hunting

Establishing a hypothesis	13	Establishing a Hypothesis
Profiling threat actors and activities	13	Profiling Threat Actors and Activities
Threat hunting tactics	13	Threat-Hunting Tactics
• Executable process analysis	11	Decomposition

PART VI

Official Exam Objective	Ch #	All-in-One Coverage
Response coordination with relevant entities	15	Response Coordination with Relevant Entities
• Legal	15	Legal Counsel
• Human resources	15	Human Resources
• Public relations	15	Public Relations
• Internal and external	15	Internal Staff Contractors and External Parties
• Law enforcement	15	Law Enforcement
• Senior leadership	15	Senior Leadership
• Regulatory bodies	15	External Communications
Factors contributing to data criticality	15	Factors Contributing to Data Criticality
• Personally identifiable information (PII)	15	Personally Identifiable Information
• Personal health information (PHI)	15	Personal Health Information
• Sensitive personal information (SPI)	15	Personally Identifiable Information
• High value asset	15	High-Value Assets
• Financial information	15	Payment Card Information
• Intellectual property	15	Intellectual Property
• Corporate information	15	Corporate Confidential Information

4.2 Given a scenario, apply the appropriate incident response procedure

Preparation	16	Preparation
• Training	16	Training
• Testing	16	Testing
• Documentation of procedures	16	Documentation
Detection and analysis	16	Detection and Analysis
• Characteristics contributing to severity level classification	16	Characteristics of Severity Level Classification
• Downtime	16	Downtime
• Recovery time	16	Recovery Time
• Data integrity	16	Data Integrity
• Economic	16	Economic Impacts
• System process criticality	16	System Process Criticality
• Reverse engineering	16	Reverse Engineering
• Data correlation	16	Data Correlation
Containment	16	Containment
• Segmentation	16	Segmentation
• Isolation	16	Isolation

PART VI

Official Exam Objective	Ch #	All-in-One Coverage
Eradication and recovery	16	Eradication and Recovery
• Vulnerability mitigation	16	Vulnerability Mitigation
• Sanitization	16	Sanitization
• Reconstruction/reimaging	16	Reconstruction
• Secure disposal	16	Secure Disposal
• Patching	16	Patching
• Restoration of permissions	16	Restoration of Permissions
• Reconstitution of resources	16	Reconstruction Restoration of Permissions Restoration of Services and Verification of Logging
• Restoration of capabilities and services	16	Restoration of Services and Verification of Logging
• Verification of logging/communication to security monitoring	16	Restoration of Services and Verification of Logging
Post-incident activities	16	Post-Incident Activities
• Evidence retention	16	Eradication and Recovery
• Lessons learned report	16	Lessons-Learned Report
• Change control process	16	Change Control Process
• Incident response plan update	16	Updates to Response Plan
• Incident summary report	16	Summary Report
• Indicator of compromise (IOC) generation	16	Indicator of Compromise Generation
• Monitoring	16	Monitoring
4.3 Given an incident, analyze potential indicators of compromise		
Network-related	17	Network-Related Indicators
• Bandwidth consumption	17	Bandwidth Utilization
• Beaconing	17	Beaconing
• Irregular peer-to-peer communication	17	Irregular Peer-to-Peer Communication
• Rogue device on the network	17	Rogue Devices on the Network
• Scan/sweep	17	Scan Sweeps
• Unusual traffic spike	17	Bandwidth Utilization
• Common protocol over non-standard port	17	Common Protocol over a Nonstandard Port
Host-related	17	Host-Related Indicators
• Processor consumption	17	Capacity Consumption
• Memory consumption	17	Capacity Consumption

Official Exam Objective	Ch #	All-in-One Coverage
• Drive capacity consumption	17	Capacity Consumption
• Unauthorized software	17	Unauthorized Software
• Malicious process	17	Malicious Processes
• Unauthorized change	17	Unauthorized Changes
• Unauthorized privilege	17	Unauthorized Privileges
• Data exfiltration	17	Data Exfiltration
• Abnormal OS process behavior	17	Malicious Processes
• File system change or anomaly	17	File System
• Registry change or anomaly	17	Registry Change or Anomaly
• Unauthorized scheduled task	17	Unauthorized Scheduled Task
Application-related	17	Application-Related Indicators
• Anomalous activity	17	Anomalous Activity
• Introduction of new accounts	17	Introduction of New Accounts
• Unexpected output	17	Unexpected Output
• Unexpected outbound communication	17	Unexpected Outbound Communication
• Service interruption	17	Service Interruption
• Application log	17	Application Logs
4.4 Given a scenario, utilize basic digital forensics techniques		
Network	18	Network
• Wireshark	18	Wireshark/TShark
• tcpdump	18	tcpdump
Endpoint	18	Endpoints
• Disk	18	Servers OS and Process Analysis
• Memory	18	Endpoints
Mobile	18	Mobile Device Forensics
Cloud	18	Virtualization and the Cloud
Virtualization	18	Virtualization and the Cloud
Legal hold	18	Tamper-Proof Seals
Procedures	18	Procedures
Hashing	18	Hashing Utilities
• Changes to binaries	18	Data Acquisition
Carving	18	File Carving
Data acquisition	18	Acquisition Utilities

Official Exam Objective	Ch #	All-in-One Coverage
Systems assessment	20	Systems Assessment
Documented compensating controls	20	Documented Compensating Controls
Training and exercises	20	Training and Exercises
• Red team	20	Red Team
• Blue team	20	Blue Team
• White team	20	White Team
• Tabletop exercise	20	Tabletop Exercises
Supply chain assessment	20	Supply Chain Risk Assessment
• Vendor due diligence	20	Vendor Due Diligence
• Hardware source authenticity	20	Hardware Source Authenticity
5.3		**Explain the importance of frameworks, policies, procedures, and controls**
Frameworks	21	Security Frameworks
• Risk-based	21	Security Frameworks
• Prescriptive	21	Security Frameworks
Policies and procedures	21	Policies and Procedures
• Code of conduct/ethics	21	Ethics and Codes of Conduct
• Acceptable use policy (AUP)	21	Acceptable Use Policy
• Password policy	21	Password Policy
• Data ownership	21	Data Ownership
• Data retention	21	Data Retention
• Account management	21	Account Management
• Continuous monitoring	21	Continuous Monitoring
• Work product retention	21	Work Product Retention
Category	21	Control Types
• Managerial	21	Control Types
• Operational	21	Control Types
• Technical	21	Control Types
Control type	21	Control Types
• Preventative	21	Control Types
• Detective	21	Control Types
• Corrective	21	Control Types
• Deterrent	21	Control Types
• Compensating	21	Control Types
• Physical	21	Control Types

PART VI

Official Exam Objective	Ch #	All-in-One Coverage
Audits and assessments	21	Audits and Assessments
• Regulatory	21	Regulatory Compliance
• Compliance	21	Standards Compliance

About the Online Content

This book comes complete with TotalTester Online customizable practice exam software with more than 200 practice exam questions, including ten simulated performance-based questions.

System Requirements

The current and previous major versions of the following desktop browsers are recommended and supported: Chrome, Microsoft Edge, Firefox, and Safari. These browsers update frequently, and sometimes an update may cause compatibility issues with the TotalTester Online or other content hosted on the Training Hub. If you run into a problem using one of these browsers, please try using another until the problem is resolved.

Your Total Seminars Training Hub Account

To get access to the online content you will need to create an account on the Total Seminars Training Hub. Registration is free, and you will be able to track all your online content using your account. You may also opt in if you wish to receive marketing information from McGraw Hill or Total Seminars, but this is not required for you to gain access to the online content.

Privacy Notice

McGraw Hill values your privacy. Please be sure to read the Privacy Notice available during registration to see how the information you have provided will be used. You may view our Corporate Customer Privacy Policy by visiting the McGraw Hill Privacy Center. Visit the **mheducation.com** site and click **Privacy** at the bottom of the page.

Single User License Terms and Conditions

Online access to the digital content included with this book is governed by the McGraw Hill License Agreement outlined next. By using this digital content you agree to the terms of that license.

Access   To register and activate your Total Seminars Training Hub account, simply follow these easy steps.

1. Go to this URL: **hub.totalsem.com/mheclaim**

2. To register and create a new Training Hub account, enter your e-mail address, name, and password on the **Register** tab. No further personal information (such as credit card number) is required to create an account.

   If you already have a Total Seminars Training Hub account, enter your e-mail address and password on the **Log in** tab.

3. Enter your Product Key: `q6m0-w5zq-5wt9`

4. Click to accept the user license terms.

5. For new users, click the **Register and Claim** button to create your account. For existing users, click the **Log in and Claim** button.

   You will be taken to the Training Hub and have access to the content for this book.

Duration of License   Access to your online content through the Total Seminars Training Hub will expire one year from the date the publisher declares the book out of print.

Your purchase of this McGraw Hill product, including its access code, through a retail store is subject to the refund policy of that store.

The Content is a copyrighted work of McGraw Hill, and McGraw Hill reserves all rights in and to the Content. The Work is © 2021 by McGraw Hill.

Restrictions on Transfer   The user is receiving only a limited right to use the Content for the user's own internal and personal use, dependent on purchase and continued ownership of this book. The user may not reproduce, forward, modify, create derivative works based upon, transmit, distribute, disseminate, sell, publish, or sublicense the Content or in any way commingle the Content with other third-party content without McGraw Hill's consent.

Limited Warranty   The McGraw Hill Content is provided on an "as is" basis. Neither McGraw Hill nor its licensors make any guarantees or warranties of any kind, either express or implied, including, but not limited to, implied warranties of merchantability or fitness for a particular purpose or use as to any McGraw Hill Content or the information therein or any warranties as to the accuracy, completeness, correctness, or results to be obtained from, accessing or using the McGraw Hill Content, or any material referenced in such Content or any information entered into licensee's product by users or other persons and/or any material available on or that can be accessed through the licensee's product (including via any hyperlink or otherwise) or as to non-infringement of third-party rights. Any warranties of any kind, whether express or implied, are disclaimed. Any material or data obtained through use of the McGraw Hill Content is at your own discretion and risk and user understands that it will be solely responsible for any resulting damage to its computer system or loss of data.

Neither McGraw Hill nor its licensors shall be liable to any subscriber or to any user or anyone else for any inaccuracy, delay, interruption in service, error or omission, regardless of cause, or for any damage resulting therefrom.

In no event will McGraw Hill or its licensors be liable for any indirect, special or consequential damages, including but not limited to, lost time, lost money, lost profits or good will, whether in contract, tort, strict liability or otherwise, and whether or not such damages are foreseen or unforeseen with respect to any use of the McGraw Hill Content.

TotalTester Online

TotalTester Online provides you with a simulation of the CompTIA CySA+ CS0-002 exam. Exams can be taken in Practice Mode or Exam Mode. Practice Mode provides an assistance window with hints, references to the book, explanations of the correct and incorrect answers, and the option to check your answer as you take the test. Exam Mode provides a simulation of the actual exam. The number of questions, the types of questions, and the time allowed are intended to be an accurate representation of the exam environment. The option to customize your quiz allows you to create custom exams from selected domains or chapters, and you can further customize the number of questions and time allowed.

To take a test, follow the instructions provided in the previous section to register and activate your Total Seminars Training Hub account. When you register you will be taken to the Total Seminars Training Hub. From the Training Hub Home page, select **CompTIA CySA+ All-in-One (CS0-002) TotalTester** from the Study drop-down menu at the top of the page, or from the list of Your Topics on the Home page. You can then select the option to customize your quiz and begin testing yourself in Practice Mode or Exam Mode. All exams provide an overall grade and a grade broken down by domain.

Performance-Based Questions

In addition to multiple-choice questions, the CompTIA CySA+ (CS0-002) exam includes performance-based questions (PBQs), which, according to CompTIA, are designed to test your ability to solve problems in a simulated environment. More information about PBQs is provided on CompTIA's website.

You can access the PBQs included with this book by navigating to the Resources tab and selecting the quiz icon. You can also access them by selecting **CompTIA CySA+ All-in-One (CS0-002) Resources** from the Study drop-down menu at the top of the page or from the list of Your Topics on the Home page. The menu on the right side of the screen outlines all of the available resources. After you have selected the PBQs, an interactive quiz will launch in your browser.

Technical Support

For questions regarding the TotalTester or operation of the Training Hub, visit **www.totalsem.com** or e-mail **support@totalsem.com**.

For questions regarding book content, visit **www.mheducation.com/customerservice**.

access control list (ACL) A list of rules that control the manner in which a resource may be accessed.

active defense Adaptive measures aimed at increasing the amount of effort attackers need to exert to be successful, while reducing the effort for the defenders.

advanced persistent threat (APT) The name given to any number of stealthy and continuous computer-hacking efforts, often coordinated and executed by an organization or government with significant resources over a longer period of time.

anomaly analysis Any technique focused on measuring the deviation of some observation from some baseline and determining whether that deviation is statistically significant.

assessment A process that gathers information and makes determinations based on it.

asymmetric cryptography A cryptosystem that uses two different but complementary keys for encryption and decryption.

atomic execution An approach to controlling the manner in which certain sections of a program run so that they cannot be interrupted between the start and end of the section.

audit A systematic inspection by an independent third party to determine whether the organization is in compliance with some set of external requirements.

beaconing A periodical outbound connection between a compromised computer and an external controller.

blue team The group of participants who are the focus of a training event or exercise; they are usually involved with the defense of the organization's infrastructure.

cloud access security broker (CASB) A software system that sits between each user and each cloud service, monitoring all activity, enforcing policies, and alerting when something seems to be wrong.

cloud computing The use of shared, remote computing devices for the purpose of providing improved efficiencies, performance, reliability, scalability, and security.

common vulnerability scoring system (CVSS) A well-known standard for quantifying severity ratings for vulnerabilities.

compensating control A security control that satisfies the requirements of some other control when implementing the latter is not possible or desirable.

containerization   A virtualization technology that abstracts the operating system for the applications running above it, allowing for low overhead in running many applications and improved speed in deploying instances.

containment   Actions that attempt to deny the threat agent the ability or means to cause further damage.

Control Objectives for Information and Related Technologies (COBIT)   A framework and set of control objectives developed by ISACA and the IT Governance Institute that define goals for the controls that should be used to manage IT properly and to ensure that IT maps to business needs.

Controller Area Network (CAN) bus   A serial bus that enables embedded devices to communicate directly with each other, often in an industrial or vehicular environment.

credential stuffing   A type of brute-force attack in which credentials obtained from a data breach of one service are used to authenticate to another system in an attempt to gain access.

cross-site scripting (XSS)   A vulnerability in a web application that provides an opportunity for malicious users to execute arbitrary client-side scripts.

dereferencing   A common flaw that occurs when software attempts to access a value stored in memory that does not exist, which sometimes enables attackers to bypass security measures or learn more about how the program works by reading the exception information.

DevSecOps   A combination of the terms development, security, and operations that denotes the practice of incorporating development, security IT, and quality assurance (QA) staff into software development projects to align their incentives and enable frequent, efficient, and reliable releases of software products.

digital certificate   A file that contains information about the certificate owner, the certificate authority (CA) who issued it, the public key, its validity timeframe, and the CA's signature of the certificate itself, typically following the X.509 standard defined by the Internet Engineering Task Force (IETF) in its RFC 5280.

digital signature   A short sequence of data that proves that a larger data sequence (say, an e-mail message or a file) was created by a given person and was not modified by anyone else after being signed.

domain generation algorithm (DGA)   A threat actor technique used to generate domain names rapidly using seemingly random, but predictable processes. This enables malware to connect eventually with its command and control infrastructure without providing defenders the opportunity to identify and block the domains.

eFuse   A single bit of nonvolatile memory that, once set to 1, can never be reverted to 0.

embedded system   Systems that are characterized by lightweight software running specialized tasks on low-power microprocessors.

event   Any occurrence that can be observed, verified, and documented.

eXtensible Markup Language (XML)   A markup language that defines a set of rules for encoding documents in a format that is both human-readable and machine-readable.

false positive   A report that states that a given condition is present when in fact it is not.

field-programmable gate array (FPGA)   A programmable chip that enables programmers to reconfigure the hardware itself to accommodate new software functionality.

firewall   A device that permits the flow of authorized data through it while preventing unauthorized data flows.

firmware   Software that is stored in read-only, nonvolatile memory in a device and is executed when the device is powered on.

forensic acquisition   The process of extracting the digital contents from seized evidence so that they may be analyzed.

fuzzing   A technique used to discover flaws and vulnerabilities in software by sending large amounts of malformed, unexpected, or random data to the target program in order to trigger failures.

hardening   The process of securing information systems by reducing their vulnerabilities and functionality.

hardware security module   A removable expansion card or external device that can generate, store, and manage cryptographic keys, used to improve encryption/decryption performance by offloading these functions to a specialized module.

hashing function   A one-way function that takes a variable-length sequence of data such as a file and produces a fixed-length result called a "hash value"; sometimes referred to as a digital fingerprint.

heuristic   A "rule of thumb" or any other experience-based, imperfect approach to problem solving.

heuristic analysis   The application of heuristics to find threats in practical, if imperfect, ways.

honeynet   A network of devices that is created for the sole purpose of luring an attacker into trying to compromise it.

host-based intrusion detection system (HIDS)   An IDS that is focused on the behavior of a specific host and packets on its network interfaces.

incident   One or more related events that compromise the organization's security posture.

incident response   The process of negating the effects of an incident on an information system.

indicator of compromise (IOC)   An artifact that indicates the possibility of an attack or compromise.

industrial control system (ICS)   A cyber-physical system that enables specialized software to control the physical behaviors of some system.

Information Technology Infrastructure Library (ITIL)   A customizable framework that provides the goals of internal IT services, the general activities necessary to achieve these goals, and the input and output values for each process required to meet these determined goals.

Infrastructure as a Service (IaaS)   A cloud computing model in which a service provider offers direct access to a cloud-based infrastructure on which customers can build and configure their own devices.

input validation   An approach to protect systems from abnormal user input by testing the data provided against appropriate values.

International Organization for Standardization (ISO)   An independent, nongovernmental international organization that is the world's largest developer and publisher of international standards.

Internet of Things (IoT)   The broad term for Internet-connected, nontraditional computing devices such as televisions and fridges.

intrusion detection system (IDS)   A system that identifies violations of security policies and generates alerts.

intrusion prevention system (IPS)   A form of IDS that is able to stop any detected violations.

isolation   A state in which a part of an information system, such as a compromised host, is prevented from communicating with the rest of the system.

jump box   A computer that serves as a jumping-off point for external users to access protected parts of a network.

mandatory access control (MAC)   A policy in which access controls are always enforced on all objects and subjects.

man-in-the-middle (MITM) attack   An attack in which an adversary intercepts communications between two endpoints to obtain illicit access to message contents and potentially alter them.

MITRE ATT&CK   The MITRE Corporation's Adversarial Tactics, Techniques, and Common Knowledge (ATT&CK) framework is a model that enables organizations to document and exchange attacker tactics, techniques, and procedures (TTPs).

multifactor authentication (MFA)   Authentication techniques that require multiple pieces of information to authenticate a user.

National Institute for Standards and Technology (NIST)   An organization within the U.S. Department of Commerce that is charged with promoting innovation and industrial competitiveness.

network segmentation   The practice of separating various parts of the network into subordinate zones to thwart adversaries' efforts, improve traffic management, and prevent spillover of sensitive data.

network-based intrusion detection system (NIDS)   An IDS that is focused on the packets traversing a network.

nmap   A popular open source tool that provides the ability to map network hosts and the ports on which they are listening.

open source intelligence (OSINT)   The collection and analysis of publicly available information appearing in print or electronic form.

Open Web Application Security Project (OWASP)   An organization that promotes web security and provides development guidelines, testing procedures, and code review steps.

OpenIOC   A framework to organize indicators of compromise (IOC) in a machine-readable format for easy sharing and automated follow-up.

operational control   Security mechanisms implemented primarily through people and procedures.

packet analyzer   A tool that captures network traffic, performs some form of analysis on it, and reports the results; also known as a network or packet sniffer.

password spraying   A type of brute-force technique in which an attacker tries a single password against a system, and then iterates though multiple systems on a network using the same password.

patch management   The process by which fixes to software vulnerabilities are identified, tested, applied, validated, and documented.

patching   The application of a fix to a software defect.

Payment Card Industry Data Security Standard (PCI DSS)   A global standard for protecting stored, processed, or transmitted payment card information.

penetration test   The process of simulating attacks on a network and its systems at the request of the owner or senior management for the purpose of measuring an organization's level of resistance to those attacks and to uncover any exploitable weaknesses within the environment.

personal health information (PHI)   Information that relates to an individual's past, present, or future physical or mental health condition.

personally identifiable information (PII)   Information, such as Social Security number or biometric profile, that can be used to distinguish an individual's identity.

phishing   The use of fraudulent e-mail messages to induce recipients to provide sensitive information or take actions that could compromise their information systems; a form of social engineering.

physical control   A safeguard that deters, delays, prevents, detects, or responds to threats against physical property.

Platform as a Service (PaaS)   A cloud computing model in which a service provider offers cloud-based platforms on which customers can either use preinstalled applications or install and run their own.

Public Key Infrastructure (PKI)   A framework of programs, procedures, communication protocols, and public key cryptography that enables a diverse group of individuals to communicate securely.

real-time operating system (RTOS)   An operating system designed to provide low-latency responses on input, usually used in vehicle electronics, manufacturing hardware, and aircraft electronics.

red team   A group that acts as adversaries during a security assessment or exercise.

regression testing   The formal process by which code that has been modified is tested to ensure that no features and security characteristics were compromised by the modifications.

regulatory environment   An environment in which the way an organization exists or operates is controlled by laws, rules, or regulations put in place by a formal body.

remediation   The application of security controls to a known vulnerability to reduce its risk to an acceptable level.

Remote Authentication Dial-In User Service (RADIUS)   An authentication, authorization, and accounting (AAA) remote access protocol.

representational state transfer (REST)   A software architectural style that defines a set of constraints to be used for creating web services.

reverse engineering   The process of deconstructing something in order to discover its features and constituents.

risk   The possibility of damage to or loss of any information system asset, as well as the ramifications should this occur.

risk acceptance   The decision that the potential loss from a risk is not severe enough to warrant spending resources to avoid it.

risk appetite   The amount of risk that senior executives are willing to assume.

rootkit   A typically malicious software application that interferes with the normal reporting of an operating system, often by hiding specific resources such as files, processes, and network connections.

sandbox   A type of control that isolates processes from the operating system to prevent security violations.

sanitization   The process by which access to data on a given medium is made infeasible for a given level of effort.

Security Assertion Markup Language (SAML)   An open standard for exchanging authentication and authorization data between parties, specifically, between an identity provider and a service provider.

Security Content Automation Protocol (SCAP)   A protocol developed by NIST for the assessment and reporting of vulnerabilities in the information systems of an organization.

security information and event management (SIEM)   A software product that collects, aggregates, analyzes, reports, and stores security information.

security policy   An overall general statement produced by senior management (or a selected policy board or committee) that dictates what role security plays within the organization or that dictates mandatory requirements for a given aspect of security.

service-oriented architecture (SOA)   A set of interconnected but self-contained software components that communicate with each other and with their clients through standardized protocols called application program interfaces.

session hijacking   A class of attacks by which an attacker takes advantage of valid session information, often by stealing and replaying it.

Simple Object Access Protocol (SOAP)   An SOA messaging protocol that uses XML over HTTP to enable clients to invoke processes on a remote host in a platform-agnostic way.

single sign-on (SSO)   An authentication mechanism that enables a user to log in once with a single set of credentials and gain access to multiple related but separate systems.

social engineering   The manipulation of people with the intent of deceiving or persuading them to take actions that they otherwise wouldn't take, and that typically involve a violation of a security policy or procedures.

Software as a Service (SaaS)   A software distribution model in which a service provider hosts applications for customers and makes them available to customers via the Internet.

software-defined networking (SDN)   A network architecture in which software applications are responsible for deciding how best to route data (the control layer) and then for actually moving those packets around (the data layer).

spear phishing   Phishing attempts directed at a specific individual or group.

PART VI

static code analysis   A technique that is meant to help identify software defects or security policy violations and is carried out by examining the code without executing the program.

stress test   A test that places extreme demands that are well beyond the planning thresholds of the software in an effort to determine how robust it is.

Structured Threat Information eXpression (STIX)   A standardized language for conveying data about cybersecurity threats in a way that can be easily understood by humans and security technologies.

supervisory control and data acquisition (SCADA) system   A system for remotely monitoring and controlling physical systems such as power and manufacturing plants over large geographic regions.

symmetric cryptography   A cryptosystem that uses the same shared secret key for both encryption and decryption.

syslog   A popular protocol used to communicate event messages.

system on a chip (SoC)   The integration of software and hardware onto a single integrated circuit and a processor, similar to microcontrollers but usually involving more complicated circuitry.

technical control   A software or hardware tool used to restrict access to objects; also known as a logical control.

Terminal Access Controller Access Control System (TACACS)   An authentication, authorization, and accounting (AAA) remote access protocol.

trend analysis   The study of patterns over time in order to determine how, when, and why they change.

Trusted Automated eXchange of Indicator Information (TAXII)   An application protocol that defines how cyber threat intelligence, specifically that formatted in accordance with the STIX standard, may be shared among participating partners.

trusted foundry   An organization capable of developing prototype- or production-grade microelectronics in a manner that ensures the integrity of its products.

Trusted Platform Module (TPM)   A system on a chip installed on the motherboard of modern computers that is dedicated to carrying out security functions involving the storage of cryptographic keys and digital certificates, symmetric and asymmetric encryption, and hashing.

Unified Extensible Firmware Interface (UEFI)   A software interface standard that describes the way in which firmware executes its tasks.

virtual desktop infrastructure (VDI)   A virtualization technology that separates the physical devices that the users are touching from the systems hosting the desktops, applications, and data, typically resulting in a thin client environment.

virtual private network (VPN)   A system that connects two or more devices that are physically part of separate networks and enables them to exchange data as if they were connected to the same local area network.

vulnerability   A flaw in an information system that can enable an adversary to compromise the security of that system.

whaling   Spear phishing aimed at high-profile targets such as executives.

white team   The group of people who plan, document, assess, or moderate a training exercise.

write blocker   A device that prevents modifications to a storage device while its contents are being acquired.

zero-day   A vulnerability or exploit that is unknown to the broader community of software developers and security professionals.

INDEX

A

acceptable use policy (AUP), 514–515
acceptance testing, 224
access control lists (ACLs), 7, 70–71, 175, 397
access controls, 468–469
access points, 122
account management, 517
  running as root, 518
accounting data, 383
accuracy of intelligence data, 13
active defense, 210
active tap, 438
Address Resolution Protocol (ARP), 413
Advanced Encryption Standard (AES), 246
advanced persistent threats (APTs), 22, 382
adversary capability, 48
air gaps, 194
Aircrack-ng, 110, *111*
alert fatigue, 33–34
Amazon Lambda, 151–152
Amazon Simple Storage Service (S3), 157
Amazon Web Services, 149
American Fuzzy Lop (AFL), 103, *104*
American Registry for Internet Numbers
  (ARIN), 7
Amin, Rohan, 40
anomalous activity, 425–426
anomaly analysis, 263
anti-tamper techniques, 251
Apache Lucene, 289
API, 220–221
  automating API calls, 356–358
  insecure, 154–156
  integration, 354–358
  restful APIs, 354–356
API Security Project, 154
application logs, 428
application programming interface. *See* API
approved scanning vendors (ASVs), 59

APT28, 422
Arachni, 91
artifacts, 420
assemblers, 102
assembly language, 102
asset inventory, 61–62, 195
asset management, 195–196
  improper, 156
asset reporting format (ARF), 69
asset tagging, 195–196
assumption of breach, 329
asymmetric cryptography, 208, 209
asynchronous attacks, 249
atomic execution, 249
ATT&CK framework, 35–38, 331–332
attack patterns, 14–15
attack types, 163
  asynchronous attacks, 249
  authentication attacks, 169–173
  buffer overflow attacks, 166–169
  credential stuffing, 170
  cross-site scripting (XSS) attacks,
    165–166
  cyclic redundancy check (CRC)
    attacks, 169
  directory traversal attacks, 166, *167*
  downgrade attacks, 177
  eXtensible Markup Language (XML)
    attack, 164–165
  heap-based attacks, 168
  impersonation, 170–171
  injection attacks, 164–166
  integer attacks, 168–169
  jailbreaking, 169
  man-in-the-middle, 171–172
  Pass-the-Hash (PtH) attacks, 334
  password spraying, 170
  privilege escalation, 169
  remote code execution (RCE), 128, 164
  rooting, 169